Drug
Level
Monitoring

DRUG LEVEL MONITORING

Analytical Techniques, Metabolism, and Pharmacokinetics

Wolfgang Sadée
Geertruida C. M. Beelen
University of California
San Francisco, California

A Wiley-Interscience Publication

JOHN WILEY & SONS
New York • Chichester • Brisbane • Toronto

Library of Congress Cataloging in Publication Data

Sadée, Wolfgang, 1942–
  Drug Level Monitoring

  "A Wiley–Interscience publication."

  Includes index.
  1. Drugs—Analysis. 2. Drug metabolism.
3. Pharmacokinetics. I. Beelen, Geertruida C. M.,
1946— joint author. II. Title. [DNLM:
1. Drug therapy. 2. Monitoring, Physiologic drug—
Analysis. QV25 W859d]

RS189.S15   615.7   79-22652
ISBN 0-471-04881-X

lw
11-12-80

PREFACE

Drug level monitoring represents an important aspect of pharmacologic research. Moreover, the value of drug serum concentrations in optimizing individual drug dosages has led to therapeutic drug level monitoring as a standard clinical practice for many important drugs. This book demonstrates how the principles of drug analysis and drug disposition are combined in drug level monitoring. It serves as a guide to the analytical techniques applicable to drug assays in biological samples, and as a reference source for metabolic and pharmacokinetic data, and is addressed to the analytical and clinical chemist, to the health-care professional specialized in drug therapy—for example the clinical pharmacologist and clinical pharmacist—and to researchers in all areas of pharmacology.

The first four chapters form a general summary of the principles of drug metabolism, pharmacokinetics, clinical pharmacokinetics, therapeutic drug level monitoring, and analytical techniques for biological samples. The specific part, Chapter 5, reviews in detail the analysis and metabolic disposition of approximately 100 selected drugs in a series of monographs. We emphasize the drugs that are currently analyzed in clinical laboratories for therapeutic purposes. Furthermore, representative drugs have been chosen from a large variety of classes—in particular, drugs of abuse, anticancer drugs, antibiotics, cardiovascular drugs, and centrally active drugs. The pertinent literature published before May 1979 is included in the monographs and updated to November 1979 in an addendum.

WOLFGANG SADÉE
GEERTRUIDA C. M. BEELEN

San Francisco, California
January 1980

v

ACKNOWLEDGMENTS

We are indebted to the Clinical Chemistry Section at the University of California at San Francisco and to the staff of the Clinical Pharmacokinetics Laboratory, in particular to J. L. Powers, B. R. Stafford, and R. A. Marques, who assisted in the development of this laboratory. The CPL provides services in therapeutic drug level monitoring at UCSF and represents the basis for this book. Several sections and drug monographs were written by J. L. Powers, B. R. Stafford, T. Guentert, and T. L. Ding. Finally, graduate students enrolled in the elective course "Bioanalytical Theory and Techniques" taught by W. Sadée at UCSF in 1978, have contributed to many monographs, and they are mentioned appropriately.

CONTENTS

Drug
Level
Monitoring

INTRODUCTION

The already classical studies on the nature and significance of drug metabolism and disposition by Axelrod, Brodie, Dost, Krüger-Thiemer, and others, conducted only a few decades ago, stimulated an enormous activity in this area. We now recognize that pharmacokinetic parameters determine to a large extent the pharmacological responses of individual patients. Moreover, differences in drug disposition account for a major share of interindividual differences in drug response. Drug doses required to achieve the same response in different patients may vary by more than one order of magnitude. It is for these reasons that drug level monitoring has assumed its present importance in drug research and therapy.

The definition of drug level monitoring also outlines the scope of this book; it is the quantitative determination of drugs and their metabolites in biological specimens and the interpretation of such data using principles of pharmacokinetics and pharmacodynamics. Drug level monitoring thus incorporates the following fields:

Drug metabolism
Pharmacokinetics
Pharmacodynamics
Pathophysiology
Drug analysis
Clinical chemistry
Clinical toxicology
Clinical pharmacokinetics and therapeutic drug level monitoring

Detailed investigations on the disposition of a new drug are now required before it can be applied to human clinical trials. Pharmacologic research often utilizes drug level monitoring to study the mechanism of drug action, including the contribution of metabolites to the drug effect. Furthermore, bioavailability studies depend on drug level monitoring; new drug formulations have to be tested for their bioequivalence in terms of established standard preparations. Drug level monitoring in the area of drug abuse and overdose is again based on a different set of criteria and objectives. However, the most rapid growth at present occurs in the application of drug level monitoring as a guide to optimize individual drug dosage regimens. We will refer to this area as "clinical pharmacokinetics and therapeutic drug level monitoring." Most major hospitals have established clinical chemistry sections devoted to the analysis of therapeutic drug concentrations ("Drug Level Laboratory," "Clinical Pharmacokinetics Laboratory").

We have approached the complex area of drug level monitoring by selecting approximately 100 important, representative drugs. Selection criteria were as follows:

1. Drugs that are currently measured in clinical pharmacokinetics laboratories.
2. Drugs that are representative of a class of chemical or pharmacological agents (e.g., ε-aminocaproic acid, chloroquin, clofibrate, ethynylestradiol, isosorbide dinitrate, succinylcholine, warfarin).
3. Drugs that belong to the following major classes: antimicrobials, anticancer drugs, antiepileptics, cardiovascular drugs, psychotropic drugs, analgesics, and drugs of abuse.

The metabolism, pharmacokinetics, and, in more detail, the analytical assays of biological specimens are reviewed in separate monographs for these 100 drugs. In addition, many more drugs listed in the register are close chemical analogs of the selected drugs, and the literature cited often contains detailed information on these analogs as well. The monographs evaluate the literature that is currently available and assist health-care professionals, pharmacologists, and analytical chemists to utilize drug level monitoring to its full potential.

Brief general chapters are designed to introduce the reader to the terminology and scope of pharmacokinetics, drug metabolism, and analytical techniques, while tables enumerate the information contained in the drug monographs for ready cross-referencing. Detailed discussions of the various general areas of interest can be found in the textbooks cited.

1

DRUG METABOLISM

Drugs are eliminated from the body either in the unchanged form, usually via renal excretion, or as metabolites. While it is generally assumed that drug metabolites are inactive and more readily excreted by the kidneys than is the parent drug, there are many examples of active or even toxic metabolites. Table I contains the therapeutically relevant metabolites of the listed drugs. Many of these drugs give rise to active or toxic metabolites that need to be considered in drug level monitoring. Major inactive metabolites are also included in Table I, since they may interfere with the drug level assay or can be utilized to indirectly determine the fate of the active species in the body. For example, the urinary excretion of the glucuronide of 4-OH-phenytoin, the major inactive product of phenytoin, can serve to differentiate between rapid metabolism and noncompliance in patients who do not respond to therapy (see phenytoin monograph).

Table I also includes the major mode of elimination of the active drug from the body, that is, renal or metabolic, which is an important parameter in clinical pharmacokinetics. Metabolic elimination means that the active species, either parent drug or active metabolite, is predominantly cleared by metabolic conversion to inactive products, regardless of whether or not these inactive metabolites are then excreted into the urine. For instance, diazepam is sequentially metabolized to the pharmacologically active N-desmethyldiazepam and oxazepam, followed by glucuronidation to the inactive oxazepam-3-O-glucuronide as the major urinary product. Thus the mode of elimination of active diazepam is by metabolism (see diazepam monograph). The predominance of the metabolic route as the major mode of elimination is striking. Many of the rather inert, lipophilic drugs would possess exceedingly long half-lives in the body were it not for the surprising capacity of mammalian species to metabolize almost any ingested chemical substance.

3

Table I. Summary of Pharmacokinetic and Metabolic Data on 102 Selected Drugs
(For further details see individual drug monographs.)

Drug	Active metabolites[a]	Toxic metabolites	Inactive metabolites[a]	Plasma elimination half-life[b]	Plasma levels		Major mode of elimination[d]
					Therapeutic[c]	Toxic	
Acetaminophen		Oxidized intermediates	S-conjugates (overdose)	2 hr, longer after toxic doses	1–10 µg/ml	>10 µg/ml for days	M
Acetazolamide				2 (α) and 13 (β) hr	~10 µg/ml		R
ε-Aminocaproic acid				~1 hr	100–400 µg/ml		R
Aminopyrine	4-Aminoanti-pyrine	4-Formyl-aminopyrine (?), dimethyl-nitrosamine	Methylru-bazoic acid	2.7 hr	5 µg/ml (peak level)		M
Amphetamine			Phenyl-acetone	7 hr, acidic urine; 20 hr, basic urine	10 ng/ml (peak level)		M + R
Anticonvulsants[e]							
Mephobarbital	Phenobarbital			24–45 hr	Therapeutic levels of phenobarbital		M + R
Phenobarbital				84–108 hr	10–30 µg/ml		M + R
Primidone	Phenobarbital, phenylethyl-malondiamide			12 ± 6 hr	Therapeutic levels of phenobarbital		M + R
Ethotoin					15–50 µg/ml		M
Mephenytoin	Phenylethyl-hydantoin				15–40 µg/ml (sum of mephenytoin and phenylethyl-hydantoin		M

4

Drug	Metabolite	Half-life	Therapeutic level	Toxic level	
Phenytoin	4-OH-phenytoin glucuronide	24 ± 12 hr	10-20 μg/ml		M
Paramethadione	Ethylmethyl-oxazolidinedione				
Trimethadione	Dimethadione	8 hr (dimethadione)	>100 μg/ml (as dimethadione)		M
Ethosuximide	Oxidative metabolites	24-72 hr	40-100 μg/ml		
Methsuximide	N-Desmethyl-methsuximide	2-4 hr	10-40 μg/ml (as N-desmethyl-methsuximide)		M
Phensuximide		4 hr	5-15 μg/ml		
Atropine		2(α) and 13-38 (β) hr	Low ng/ml range		
Barbiturates			1-5 μg/ml	10 μg/ml (hypnotics)	
Amobarbital	N-Desmethyl-hexobarbital	14-42 hr			M
Hexobarbital		3-7 hr			M
Pentobarbital sodium	Hydroxylated metabolites	23-30 hr			M
Thiopental sodium		>8 hr			M
BCNU	Alkylating and carbamoylating species	<15 min			M
Carbamazepine	10,11-Epoxide	18-65 hr (single dose) 10-20 hr (maintenance)	5-10 μg/ml		M
Chloramphenicol		1.5-3.5 hr	20-40 μg/ml (peak levels)		M
Chlordiazepoxide	N-Desmethyl-chlordiazepoxide, demoxepam	20-24 hr (14-95 hr for demoxepam)	1-3 μg/ml	>5.5 μg/ml	M

6

Table I. (Continued)

Drug	Active[a] metabolites	Toxic metabolites	Inactive[a] metabolites	Plasma elimination[b] half-life	Plasma levels Therapeutic[c]	Plasma levels Toxic	Major mode of elimination[d]
Chloroquine	N-Desethyl chloroquine			3 (α) and 18 (β) days	None defined		M
Chlorpromazine	Monodesmethyl-chlorpromazine, 7-hydroxy-chlorpromazine		Many metabolites	6 hr (α), β-phase possibly much longer	50–300 ng/ml		M
Clofibrate	Clofibrinic acid		Clofibrinic acid glucuronide	12 hr (clofibrinic acid)	80–200 μg/ml (clofibrinic acid)		M
Clonazepam		7-Amino-clonazepam (?)	7-Aminoclon-azepam	13–60 hr	5–50 μg/ml		M
Clonidine	4-Hydroxy-clonidine			5–23 hr	~1–2 ng/ml		M
Cocaine			Benzoylecgonine and methylecgo-nine	30–70 min	200 ng/ml (peak levels)		M
Cyclophosphamide	Aldophosphamide, phosphoramide mustard			5.6–8.4 hr	10–150 ng/ml		M
Dexamethasone			Many metabolites	2.5–6.5 hr	Low ng/ml range		M
Diatrizoate				4 hr	>>100 μg/ml (diagnostic)		R
Diazepam	N-Desmethyl-diazepam		Oxazepam glucuronide	26–53 hr	0.1–1.0 μg/ml		M
Diazoxide				24–36 hr	15–50 μg/ml (in hypoglycemia)		M + R
Digoxin				1.6 days	0.5–2 ng/ml	>2 ng/ml	R

Drug	Metabolites	Half-life	Concentration	Toxic level	M/R
Diphenhydramine	(Diphenylmethoxy) acetic acid				M
Doxorubicin	Doxorubicinol; Aglycone metabolites	0.3–1.5 hr (α), 14–30 hr (β)	Low ng/ml range in the β-phase		M
Ethambutol	Aldehydes and carboxylic acids	4–5 hr	3–5 µg/ml		R
Ethynylestradiol	Estradiol; Glucuronides	6.5 hr	60–500 pg/ml		M
5-Fluorouracil	FdUMP; FUTP; α-Fluoro-β-alanine (toxic?)	10 min	0.1–1 µg/ml (slow i.v. infusions)		M
Gentamicin		2–4 hr	4–12 µg/ml	>12 µg/ml	R
Haloperidol	β'(p-Fluorobenzoyl)-propionic acid	12–39 hr	6–245 ng/ml		M
Hydralazine	Methyltriazolophthalazine	1–2 hr	ng/ml range		M
Hydrochlorothiazide	Pyruvate and β-ketoglutarate hydrazones	3–4 hr (α), 7–10 hr (β)	0.5 µg/ml (peak levels)		R
Hydrocortisone		60 min	2–26 µg/100 ml (physiolog. conc.)		M
Imipramine	Desipramine; Many metabolites	4–20 hr	15–500 ng/ml		M
Indomethazin	N-Deschlorobenzoyl and O-desmethyl metabolites	2.6–11.2 hr	0.5–3 µg/ml		M
Isoniazid	Acetylhydrazine; N-Acetylisoniazid	45–80 min (fast acetyl.), 140–200 min (slow acetyl.)	~10 µg/ml		M + R
Isosorbide dinitrate	Isosorbide mononitrates (active?)	30–50 min	2–9 ng/ml (peak levels)		M
Lidocaine	Monoethylglycine xylidide; Monoethylglycine-xylidide and glycinexylidide (convulsants)	17 min (α), 100 min (β)	1.5–7 µg/ml	>7 µg/ml	M

Table I. (Continued)

Drug	Active metabolites	Toxic metabolites	Inactive[a] metabolites	Plasma elimination[b] half-life	Plasma levels Therapeutic[c]	Toxic	Major mode of elimination[d]
Lithium				15–20 hr	0.6–1.2 mEq/l	>2 mEq/l	R
Lysergide			2-Oxy-LSD	3.5 hr	1–5 ng/ml (peak levels)		M
Melphalan	Monohydroxy-melphalan			30 min (dogs)	1 µg/ml (peak level)		M
Meperidine	Normeperidine	Normeperidine (convulsant)		3.2–3.7 hr	0.6 µg/ml	>5 µg/ml	M
6-Mercaptopurine	Nucleotides of 6-mercaptopurine, 6-methyl mercaptopurine, and 6-thioguanine		Thiouric acid				M
Methadone			N-Demethylated, cyclized metabolites	29 ± 5 hr	100–400 ng/ml		M
Methaqualone			Hydroxylated metabolites	26 hr (α), 37 hr (β), 72 hr (γ)	2–3 µg/ml (peak level)	5–30 µg/ml	M
Methotrexate		7-OH-methotrexate (?)		4–24 hr		$>4.5 \times 10^{-6}\,M$ for 48 hr	R
Methyldopa	α-Methyldopamine, α-methylnorepinephrine				6 µg/ml (after 1 hr; 250 mg i.v.)		M
Metronidazole		2-Hydroxymethyl-metronidazole, reductive metabolites		6–14 hr	5 µg/ml (antimicrobial), 100–200 µg/ml (radio-sensitizer)		M
Morphine	Normorphine, etc.		Morphine-3-glucuronide	2 hr (α)	70 ng/ml (peak level) after 10 mg (s.c.)		M

Drug	Metabolite / measurement	Half-life	Level	Level (overdose)	M/R
Nicotine	Cotinine (to assess smoking behavior)	<1 hr (α)	15–38 ng/ml mid-morning levels in smokers		M
Nitrofurantoin	Cotinine (?) Reductive intermediate metabolites	0.3–1.0 hr	~1.8 µg/ml		M + R
Oxazepam	3-Glucuronide	5.9–25 hr	1–2 µg/ml		M
Penicillins	Penicilloic acids		Peak level after 500 mg oral dose		
Amoxicillin		1–1.3 hr	7–8 µg/ml		R
Ampicillin		1.3 hr	2–6 µg/ml		R
Carbenicillin		1 hr	—		R
Cloxacillin		0.5 hr	7–14 µg/ml		R
Dicloxacillin		0.8 hr	15–18 µg/ml		R
Methicillin		0.43 hr			R
Nafcillin		0.55–1 hr			Bile
Oxacillin		0.4–0.7 hr	5–6 µg/ml		M + R
Penicillin G		0.7 hr	1.5–2.7 µg/ml		R
Penicillin V			3–5 µg/ml		M + R
Phencyclidine	Oxidative metabolites	11 hr		6–240 ng/ml after nonfatal overdoses	M
Phenylbutazone	Oxyphenbutazone	72 hr	100 µg/ml		M
Procainamide	N-Acetylprocainamide	2.5–5 hr	4–10 µg/ml		M + R
Procarbazine	Azo and azoxy metabolites	<10 min			M
Propoxyphene	N-Norpropoxyphene	3.5 hr	50–200 ng/ml	>5 µg/ml	M
Propanolol	4-OH-propranolol, propranolol glycol, isopropylamine	2–3 hr	50–100 ng/ml		M
Quinidine	Oxidative and hydrolytic metabolites	6–7 hr	2–3 µg/ml (mean)		M
Reserpine	3-OH-quinidine	4.5 hr (α) 46 hr (β)	Low ng/ml range		M

Table I. (Continued)

Drug	Active metabolites	Toxic metabolites	Inactive[a] metabolites	Plasma elimination[b] half-life	Plasma levels Therapeutic[c]	Plasma levels Toxic	Major mode of elimination[d]
Salicylic acid			Glycine and glucuronic acid conjugates	2–4 hr (low dose) 15–30 hr (high dose)	15–30 mg/100 ml (anti-inflammatory)	>30 mg/100 ml	M
Spironolactone	Canrenone, sulfur retaining metabolites			13.5–24 hr (canrenone)	415 ± 154 ng/ml (canrenone and other active metabolites)		M
Succinylcholine			Succinate monocholine, choline, succinate	5 min	μg/ml range		M
Sulfonamides			N⁴-Acetylated metabolites		Between 30 and 150 μg/ml		M + R
Sulfacetamide				12–14 hr			
Sulfanilamide				8.8–11 hr			
Sulfapyridine				9.4–12 hr			
Sulfathiazole				3.5–4 hr	100–150 μg/ml		
Sulfisomidine				6–7 hr			
Sulfisooxazole				6–10 hr	90–150 μg/ml		
Tetracyclines					Range 0.55–5 μg/ml		
Doxycycline				12–15 hr			R + extrarenal
Minocycline							R
Oxytetracycline				8–10 hr			R
Tetracycline				8–9 hr			R
Tetrahydrocannabinol			Hydroxylated metabolites (activity?)	30–40 hr (α), 50–60 hr (β)	1–50 ng/ml (after marijuana smoking)		M

Theophylline	3-Methylxanthine	5–6 hr (2.9–20.7 range)	5–20 µg/ml	25 µg/ml	M
Tolbutamide	1-Methyl- and 1,3-dimethyluric acid				
	Carboxytol-butamide	4–6 hr	50–100 µg/ml		M
Valproic acid	Many metabolites	6–16 hr	20–100 µg/ml		M
Warfarin	Warfarin alcohols	33 hrs (S), 45 hrs (R)	1–10 µg/ml		M

[a] Inactive metabolites that are potentially useful in drug level monitoring (usually major metabolites).
[b] Mean or range, as available. Most values are based on results obtained in adult patients with normal renal, hepatic, and cardiovascular functions.
[c] For interpretation of "therapeutic plasma levels" see drug monograph and references cited.
[d] Major mode of elimination of the active drug (see text). M: metabolic, R: renal.
[e] Carbamazepine, clonazepam, diazepam, and valproic acid are separately listed.
[f] Mephobarbital, phenobarbital, and primidone are listed under "Anticonvulsants."

11

The discovery of special enzymes in the liver, which are responsible for the chemical conversion of foreign substances to polar excretable metabolites, marked a significant advance in our knowledge of drug action. Brodie and co-workers have shown that the activity of these drug metabolizing enzymes toward certain barbiturates directly determines the duration of pharmacological response by eliminating the barbiturate either rapidly (e.g., hexobarbital) or slowly (e.g., phenobarbital). The drug metabolizing activity is mainly located in the smooth-surfaced endoplasmic reticulum (SER) of hepatic parenchymal cells. Upon tissue homogenization the SER breaks up into microsomes which can be isolated by centrifugation. The microsomal enzymes contain a mixed function oxygenase system (cytochrome P-450 monooxygenases) and the conjugating enzyme, glucuronyl transferase.

The mixed function oxygenases are rather nonspecific and are capable of oxidizing a large variety of substrates, particularly aromatic and aliphatic hydrocarbons, N-, O-, and S-alkyl functions, and so on. The oxidative reactions often occur at thermodynamically unfavorable sites of the molecule; they require energy in the form of NADPH, rather than producing energy, as do oxidations in glycolysis or fatty acid metabolism. The conjugating enzymes include glucuronyl transferases and sulfate transferring enzymes. The functional groups generated by the mixed function oxidases, for example, OH, can be readily conjugated with glucuronic acid and sulfate by the neighboring conjugating enzymes. These products are rather polar and rapidly excreted into the urine. Until very recently it was assumed that practically all of the polar drug conjugates were inactive; however, evidence is now available that oxidative conjugates of aromatic amide carcinogens or cytotoxins (acetamidofluorene, acetaminophen) actually represent the more proximate toxic intermediates.

The drug metabolizing capacity of an individual patient is determined by two major factors; these are genetic determinants of the enzyme activity and environmental factors which may induce or inhibit enzyme activity. The genetic aspects of drug metabolism are part of the rapidly developing field of pharmacogenetics. Using fraternal and identical twins, as well as other genetic methods, it has been shown that genetic factors by far outweigh environmental factors for many of the lipophilic drugs. However, enzyme induction of the microsomal drug metabolizing activity may also play a major role in determining the half-life of a drug in the body. Enzyme inducibility itself is genetically determined and may vary considerably from one patient to another.

It has not been possible as yet to design a clinically applicable test for individual drug metabolizing activity, although some methods already available may have predictive value (see aminopyrine monograph). However, the existence of several cytochrome P-450 monooxygenases with different distributions, inducibilities, and substrate specificities limits the clinical utility of such tests.

Some of the selected drugs (Table I) are metabolized by specialized enzymes normally responsible for the conversion of endogenous substrates. These drugs include theophylline, the antimetabolites (5-fluorouracil, 6-mercaptopurine), and biogenic amine analogs (amphetamine, α-methyldopa). A few drugs are metabolized by the INH *N*-acetyltransferase, that is, several sulfonamides, INH, hydralazine, and procainamide. Still others are hydrolyzed by esterases, in particular, atypical serum cholinesterase (i.e., succinylcholine, cocaine, acetylsalicylic acid). All of these enzyme systems are differently regulated; their contributions to the drug disposition and response have to be considered in drug level monitoring (see individual drug monographs).

We can conclude that, at present, drug level monitoring of the individual patient provides the only reliable means to assess drug exposure at a given dosage, unless pharmacological effects can be readily determined. Interindividual differences in drug elimination half-life and steady-state plasma levels can span more than one order of magnitude (see imipramine monograph), so that equally divergent dosage levels are required to obtain the desired effects. Without drug level monitoring, rapid metabolizers are not likely to receive the full benefit of drug therapy, while slow metabolizers are prone to toxic side effects.

Bibliography

Beyer, K.-H., *Biotransformation der Arzneimittel*, Wiss. Verlagsges. m.b.H., Stuttgart, 1975.

Gorrod, J. W., *Biological Oxidation of Nitrogen*, Elsevier/North-Holland Biomedical Press, Amsterdam, New York, Oxford, 1978.

LaDu, B. N., H. G. Mandel, and E. L. Way, *Fundamentals of Drug Metabolism and Disposition*, The Williams and Wilkins Co., Baltimore, 1971.

Testa, B. and D. Jenner, *Drug Metabolism: Chemical and Biochemical Aspects*, Marcel Dekker, New York, 1976.

Zimmerman, H. J., *Hepatotoxicity: The Adverse Effects of Drugs and Other Chemicals on the Liver*, Appleton-Century-Crofts, New York, 1978.

2

PHARMACOKINETICS

The term "pharmacokinetics" was first introduced by F. H. Dost (*Der Blutspiegel*, 1953) and propagated by E. Krüger-Thiemer in several research articles.

The following definition of pharmacokinetics is taken from the first issue of the *Journal of Pharmacokinetics and Biopharmaceutics* (Vol. 1, No. 1, 1973): "*Pharmacokinetics* is the study of the kinetics of absorption, distribution, metabolism, and excretion of drugs and their corresponding pharmacologic, therapeutic, or toxic response in animals and man."

The time course of drug concentrations in the body is often regulated by first-order kinetic processes that can be readily described by mathematical models. This in turn allows one to utilize mathematical models to accurately predict plasma level-time curves, once the individual pharmacokinetic parameters are known. Furthermore, the significance of active metabolites can be determined by pharmacokinetic methods, knowing the formation and elimination rates of such metabolites.

There are two simple mathematical models that can be used to describe the plasma level-time curves for a majority of the drugs listed in Table I. These are the one-compartment and the two-compartment open body models:

One-Compartment Open Body Model

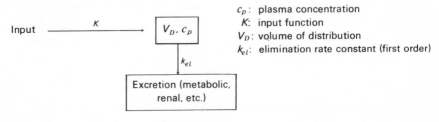

c_p: plasma concentration
K: input function
V_D: volume of distribution
k_{el}: elimination rate constant (first order)

Two-Compartment Open Body Model

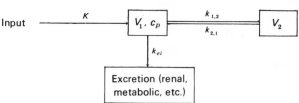

V_1, V_2: volume of distribution of the central and peripheral compartments

$k_{1,2}$, $k_{2,1}$: tissue distribution rate constants

The two-compartment model takes into account the distribution of drugs into slowly equilibrating tissues as the potentially rate limiting process. Distribution is particularly important for the lipophilic drugs, which sequester into poorly perfused adipose tissues and may remain there for a long time (see thiopental, methaqualone, and chloroquine monographs).

We assume, when using first-order rate constants in pharmacokinetic models, that the drug levels are well below the concentrations that saturate the drug metabolizing enzymes. This is indeed correct for a very large number of the drugs listed in Table I; their plasma concentration-time curves are linear beyond the tissue distribution phase if plotted on a log plasma concentration-time scale (linear kinetics). However, the assumption of linear kinetics is not always correct. Most notably, phenytoin and salicylic acid have therapeutic plasma concentrations close to the K_m values of the enzymes responsible for the elimination of these two drugs. Consequently, their log plasma concentration-time curves are curvilinear, and increasing doses result in disproportionally higher steady-state plasma levels. We refer to this phenomenon as nonlinear kinetics.

Upon peroral administration a drug has to pass through the liver before reaching the systemic circulation. Thus the drug metabolizing enzymes in the liver are exposed to rather high drug levels. On the assumption that all of the drug is absorbed from the gastrointestinal tract, a portion of the absorbed drug will be extracted from the portal vein blood during the first liver passage after absorption. The "first-pass effect" may be variable, depending on the rate of drug delivery to the liver. If drug concentrations in the liver exceed the K_m value, a larger fraction of the drug will survive first passage through the liver. The first-pass effect is a kinetic process that is rather difficult to accurately predict by pharmacokinetic models, although it may be therapeutically important (see propranolol monograph as an example).

Table I includes some pharmacokinetic parameters, that is, the plasma elimination half-lives and therapeutic-toxic plasma levels in human beings. The half-lives are usually obtained from subjects with normal hepatic and renal functions and may be altered under pathophysiological conditions. The large difference in the half-life of salicylate at low analgesic doses, compared to high antipyretic doses, exemplifies the therapeutic importance of nonlinear kinetics for this drug.

Today, pharmacokinetic methods are routinely applied in pharmacological research, bioavailability studies, and clinical Phase I–III trials of new drugs. A growing area of interest exists in the clinical application of pharmacokinetics as an aid in optimizing individual drug therapy (clinical pharmacokinetics and therapeutic drug level monitoring), which will be discussed in the next chapter.

Bibliography

Curry, S. H., *Drug Disposition and Pharmacokinetics*, 2nd edition, Blackwell Sci. Publ., Oxford, London, Edinburgh, Melbourne, 1977.

Dost, F. H., *Grundlagen der Pharmakokinetik*, 2nd edition, Georg Thieme Verlag, Stuttgart, 1968.

Gibaldi, M., *Introduction to Biopharmaceutics*, Lea and Febiger, Philadelphia, 1971.

Gibaldi, M. and D. Perrier, *Pharmacokinetics*, Marcel Dekker, New York, 1975.

Notari, R. E., *Biopharmaceutics and Pharmacokinetics: An Introduction*, Marcel Dekker, New York, 1975.

Plakiogiannis, F. M. and A. J. Cutie, *Basic Concepts in Biopharmaceutics*, Brooklyn Medical Press, Brooklyn, N.Y., 1977.

Ritschel, W. A., *Handbook of Basic Pharmaceutics*, 1st edition, Drug Intelligence Publ., Hamilton, Ill., 1976.

Swarbrick, J., *Current Concepts in the Pharmaceutical Sciences: Biopharmaceutics*, Lea and Febiger, Philadelphia, 1970.

Teorell, T., R. L. Dedrick, and P. G. Condliffe, *Pharmacology and Pharmacokinetics*, Plenum Press, New York, London, 1974.

Wagner, J. G., *Biopharmaceutics and Relevant Pharmacokinetics*, 1st edition, Drug Intelligence Publ., Hamilton, Ill., 1971.

3

CLINICAL PHARMACOKINETICS AND THERAPEUTIC DRUG LEVEL MONITORING

Following the in-depth investigation of metabolism, pharmacokinetics, and pharmacodynamics and their relationships to the therapeutic response in patients, clinical application of these newly established disciplines became imminent. The following definition is taken from the introduction to *Clinical Pharmacokinetics: a Symposium*, edited by Levy (1974): "Clinical pharmacokinetics is a health sciences discipline which deals with the application of pharmacokinetics to the safe and effective therapeutic management of the individual patient."

Large intersubject variations in drug response have already been addressed in preceding chapters. Clinically, there are many factors that may cause variability in drug disposition and drug plasma levels resulting from standard dosage regimens:

1. Individual capacity to metabolize or excrete the drug.
2. Drug absorption from various pharmaceutical dosage forms.
3. Disease state (in particular, of the liver and kidney).
4. Age, body weight, and other pertinent patient data.
5. Drug-drug interactions.
6. Patient compliance.

All these factors demand dosage adjustments for the individual patient and contribute to the key observation with many drugs that the correlation between plasma level and response is much better than that between the total dose and response among a given patient population. Pharmacokinetic-pharmacodynamic studies in patients yield information on the therapeutic plasma concentration range of a drug that should be maintained during a course of therapy in order to achieve maximum benefits. This leads to the concept of the "target level" of a particular drug in plasma, which is paramount to the application of clinical pharmacokinetics in drug therapy ("quantitative therapeutics"). There are two avenues by which "target levels" can be utilized to optimize the individual drug dosage regimen; first, calculation of the dose based on predictive pharmacokinetic models for the individual patient, and, second, inclusion of drug level monitoring to correct the predictive model for unaccountable intersubject variation.

Predictive pharmacokinetic models are particularly useful for drugs that are mainly eliminated by renal excretion, since good correlations between clinical kidney function tests and drug clearance have been established. Thus dosage adjustments, based on individual kidney functions, are an integral part of modern drug therapy with most antibiotics, digoxin, and lithium. Typically, a therapeutic decision process follows the pattern depicted below:

Diagnosis.
|
Selection of drug.
|
Individual predictive pharmacokinetic model based on clinical patient data.
|
┌→Calculation of the dosage schedule designed to establish a given taget plasma level.
| |
| Drug administration.
| |
| Observation of clinical effects.
| |
└─Correct individual pharmacokinetic model.

Computer packages are now routinely available that simplify the application of clinical pharmacokinetics. However, a series of shortcomings is inherent in this clinical process. First of all, our ability to predict individual pharmacokinetic parameters is still essentially nonexistent in the case of most drugs that are eliminated by metabolism. The regulation of individual drug metabolizing capacity is too complex to allow the clinical applications of predictive models. Moreover, a major share of interindividual variability cannot be accounted for on the basis of renal function alone, even for drugs that are mainly eliminated by the kidneys (see Digoxin monograph). Consequently, the use of predictive pharmacokinetic models reduces, but does not eliminate, the incidence of toxic side effects of digoxin therapy.

The second approach, which includes actual drug level determinations, should ideally be based on the following scheme:

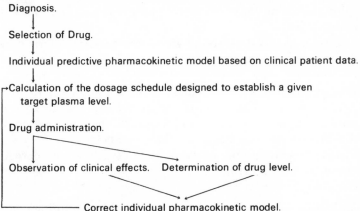

Diagnosis.
↓
Selection of Drug.
↓
Individual predictive pharmacokinetic model based on clinical patient data.
↓
Calculation of the dosage schedule designed to establish a given target plasma level.
↓
Drug administration.

Observation of clinical effects.   Determination of drug level.

Correct individual pharmacokinetic model.

Determination of drug levels as an integral part of this feedback system may be called "therapeutic drug level monitoring." If performed properly, it allows the rapid and safe attainment of target plasma levels and, in conjunction with observations of the clinical drug effects, should provide the safest approach to drug therapy.

Nevertheless, scepticism prevails among many clinicians who contend that the cost of therapeutic drug level monitoring outweighs the potential benefits to the patient. Several limiting factors, which indeed may seriously affect the interpretation of drug level data, are the basis for this scepticism. Included are the following:

1. There is no clearly defined therapeutic concentration range or toxic concentration range. Optimum effects for an individual patient may at times be achieved outside the established therapeutic range. Moreover, it should be understood that drug level data are only additional aids, to be used together with clinical observations, to optimize individual therapy. If, however, the clinical response cannot be readily monitored, use of drug target levels may be the next best approximation to optimum therapy.

2. The formation of active metabolites, frequently with unknown efficacy in human beings, complicates the interpretation of drug level data (e.g., lidocaine, procainamide, and quinidine, to be discussed below).

3. Analytical assay error may be substantial. Moreover, different assay techniques are utilized in different hospitals, resulting in unpredictable differences in assay precision and specificity.
4. Interpretation of drug level data depends on the following information: dosage regimen, time of last dose, and time of blood sampling. Information on any of these variables may be inaccurate or unavailable.
5. The time of blood sampling giving the greatest information value depends on the dose schedule and the half-life of the drug. Since the half-life may differ greatly between patients, blood samples may be obtained at time points that do not allow kinetic interpretation of the data. This can occur more frequently in patients with unusual drug clearance values, where the information is most urgently needed (see lidocaine, discussed below).
6. Plasma level data may be improperly used to adjust dosage levels, or may be altogether ignored.

It seems, therefore, that therapeutic drug level monitoring can be successful only if the organization of sample and data collection, the interpretation of drug levels, and the implementation of appropriate therapeutic decisions are all supervised by suitably trained personnel, that is, the specialist physician, clinical pharmacologist, clinical chemist, and clinical pharmacist.

Furthermore, selection of drugs for routine therapeutic drug level monitoring must obey rather stringent criteria. The following are suitable:

1. Drugs that have a narrow therapeutic index.
2. Drugs from which serious side effects may be imminent when reaching the toxic range.
3. Drugs whose pharmacological effects are not readily measurable.
4. Drugs that are used under special circumstances (e.g., drug treatment of infants, noncompliance problems, unusual drug reactions).

Table II lists the drugs for which routine drug level assays were available at University of California–San Francisco (UCSF) during 1978. This table also contains information on the frequency of drug level requests, the current assay method, and the total body clearance (Cl) under normal and physiological conditions. The clearance concept is particularly valuable in clinical pharmacokinetics, since Cl is inversely proportional to the steady-state drug serum level. Assuming that there is complete or a constant degree of drug absorption, and that the drug clearance is dose independent, one has the following simple relationship:

Dosage Regimen = Clearance × Steady-State Serum Level (mean)

$$\frac{D}{t} = Cl \times C_{pss}; \qquad \text{fraction of the drug absorbed, } F = 1$$

$$\left[\frac{\text{mg}}{\text{hr}}\right] = \left[\frac{1}{\text{hr}}\right] \times \left[\frac{\text{mg}}{1}\right]$$

The total body clearance is related to the half-life, elimination rate (k_{el}), and volume of distribution in the following way:

$$Cl = k_{el} \times V_D = \frac{0.693}{t_{1/2}} \times V_D$$

Knowing the dosage regimen and the steady-state serum level, one can calculate the total body clearance. If the total body clearance of a drug is known in a particular patient, one can readily calculate the dosage regimen required to achieve a target serum level of the drug. This simple relation is model independent, meaning that it can be applied to drugs which obey first-order kinetics according to a one-compartment, two-compartment, or even multicompartment open model. However, drugs with nonlinear kinetics have to be treated differently, for example, by using Michaelis-Menten kinetics (see phenytoin monograph).

The most frequent assay requests were for the anticonvulsants digoxin, theophylline, methotrexate, and gentamicin. Many of the anticonvulsants have narrow therapeutic indices, and clinical response is not readily measurable (suppression of seizure activity). Therapeutic concentrations of phenytoin close to the K_m value of the eliminating hepatic enzymes make the clinical management of this major anticonvulsant even more difficult; therefore determination of the phenytoin serum level is among the most frequently requested tests.

Digoxin serum levels are clinically valuable because of the narrow therapeutic range and the prompt onset of sometimes severe toxicity when the toxic range is reached. The primary objectives of therapeutic drug level monitoring of digoxin are the prevention of drug toxicity and of undermedication, which may not be noticeable early enough by monitoring drug effects.

Theophylline also has a narrow therapeutic index with severe side effects. Furthermore, intersubject variability is quite large, requiring sizable dosage adjustments for individual patients. Theophylline level monitoring appears to be rather important in pediatrics as well.

Table II. Drugs for Which Routine Plasma Level Determination Were Available at UCSF During 1978

Some additional pharmacokinetic parameters in adults with normal renal, hepatic, and cardiovascular functions are also given.

Drug[a]	Number of assays requested in 1978	Assay method	Metabolites clinically significant	Percentage of drug excreted unchanged in urine	V_D (l/70 kg)	Cl (l/hr·70 kg)	Usual therapeutic serum levels
Acetaminophen	22	HPLC	S-conjugates (overdose)	<5			N.A.[b]
Carbamazepine	226	HPLC	10,11-Epoxide	<5			3–12 µg/ml
Digoxin	1668	RIA		60–80	760	12	0.5–2 ng/ml
Ethosuximide	112	HPLC		10–20			40–100 µg/ml
Gentamicin[c]	636	RIA, HPLC		>96			4–12 µg/ml
Isoniazid	58	COL	N-Acetyliso-niazid				N.A.
Lidocaine	31	GC	N-Desethyl-lidocaine	<5	110	46	1.5–7 µg/ml
Lithium	452	AA		>95	52	1.9	0.6–1.2 mEq/l
Methotrexate	262	PBA, HPLC	7-Hydroxy-methotrexate	90			N.A.
Phenobarbital	569	HPLC		20–50	50	0.28	10–30 µg/ml
Phenytoin	1411	HPLC	4-OH-phenytoin glucuronide	<5	50	1.2–3	10–20 µg/ml

Primidone	48	HPLC	Phenobarbital, phenylethyl-malondiamide	20–50			5–10 µg/ml (10–30 µg/ml phenobarbital)
Procainamide	81	HPLC	N-Acetylpro-cainamide	40–55	140	34	4–10 µg/ml
Propranolol	40	HPLC	4-OH-propranolol	<5	180	50	50–100 ng/ml
Quinidine	233	HPLC	3-OH-quinidine; further metabolites (?)	<20	170	20	2–3 µg/ml
Salicylate	250	COL		3–20	9–30	0.6–2.3	15–30 mg %
Sulfonamides	9	COL					
Theophylline	1326	HPLC		<15	32	4.3	5–20 µg/ml

[a] Additional available assays (Number of assays in 1978): anticonvulsants, general (669), barbiturates (38), diazepam (8), digitoxin (2, discontinued), mephenytoin (10), mephobarbital (13), methsuximide (0).

[b] N.A. = not applicable.

[c] Tobramycin (0); HPLC (substituted for gentamicin, 1979).

The early detection of potentially severe toxicity caused by high-dose methotrexate-leucovorin rescue treatment is the major concern of methotrexate serum level monitoring. If methotrexate is retained in the body with an unusually long half-life, severe toxicity can be prevented by increasing the leucovorin rescue treatment, which bypasses the toxic mechanism of action of methotrexate. Interpretation of methotrexate serum levels may vary with the clinical protocols using a variety of high-dose methotrexate-leucovorin rescue schedules.

Gentamicin level monitoring is generally accepted for its clinical utility, because of the high renal toxicity and often irreversible ototoxicity of this drug. The rather low number of gentamicin assay requests (Table II) may reflect the impending substitution of gentamicin by tobramycin, which is potentially less toxic. A tobramycin serum level assay will be made available through the Clinical Pharmacokinetics Laboratory at UCSF.

The antiarrhythmic drugs, lidocaine, procainamide, propranolol, and quinidine present special problems for therapeutic drug level monitoring. Lidocaine may be used as an antiarrhythmic agent by intravenous infusion in life threatening situations. Steady-state serum levels should be achieved within 3 to 6 hr if the half-life falls within the normal range. However, in patients with much longer half-lives than the normal value of approximately 1.5 hr, it may take considerably longer to reach steady state. Lidocaine serum levels obtained within the first 6 hr may therefore be rather poor indicators of the final steady-state levels unless sampled and assayed with great care. Moreover, plasma level measurements need to be performed within a short time (2 to 5 hr) in order to serve as a guide for adjustments of the infusion rate; such rapid data flow may not be possible throughout the day in most hospitals. Finally, the antiarrhythmic and convulsant metabolite of lidocaine, monoethylglycinexylidide, may accumulate in some patients to serum levels higher than those of the parent drug. Several plasma level determinations may be needed before the contribution of this metabolite can be estimated and considered in eventual dosage adjustments. The sum of these adverse factors render therapeutic drug level monitoring of lidocaine impractical, unless a special interest of the clinical staff exists. Because of the low number of lidocaine assay requests (Table II), this drug has now been deleted from the list of available assays at UCSF.

Procainamide is metabolized to the active *N*-acetylprocainamide, which is also determined by the HPLC method. However, the therapeutic and toxic concentration ranges of *N*-acetylprocainamide are not yet well defined. The *N*-acetyl metabolite may be present at higher concentrations than the parent drug, particularly in rapid INH-acetylators; we have observed levels up to and above 50 μg/ml without clinically apparent drug toxicity, indicating that N-actylprocainamide may be less toxic than procainamide. Other reports suggest that N-acetylprocainamide may be equally active, as an antiarrhythmic, as the parent drug. Currently, serum assay results are reported for procainamide alone, omitting the N-acetyl metabolite, unless unusually high levels are observed. It becomes apparent that some potentially useful information is lost by this procedure, and further research is needed to accurately determine the metabolite contribution to the drug effects. Similar considerations apply to propranolol and its active 4-hydroxyme-tabolite; however, 4-hydroxypropranolol appears to accumulate to a lesser extent.

Our current knowledge on the therapeutic contribution of quinidine metabolites is equally fragmented. It appears that 3-OH-quinidine may contribute to the quinidine effects, but the significance of other metabolites remains unknown. With the availability of analytical methods of increasing specificity for quinidine in the presence of its metabolites, the therapeutic concentration range has been steadily decreasing. At present, it is prudent to select the most specific assay for quinidine, until the pharmacological significance of quinidine metabolites is well understood. Then a suitable assay method can be selected which delivers plasma level data with the greatest information value.

Many antibiotics have rather high therapeutic indices, for example, the penicillins. Nevertheless, drug level monitoring of these agents may be appropriate under special conditions. For example, it is difficult to predict pharmacokinetic parameters in newborn infants. Most penicillins are cleared mainly by the kidneys and exhibit dramatically increased half-lives in patients with renal failure. In such cases, drug level monitoring may again be indicated.

Isoniazid is used in the Clinical Pharmacokinetics Laboratory at UCSF to determine the INH-acetylator phenotype of an individual patient. The population of the United States is equally divided into a group that acetylates isoniazid rapidly and a group that acetylates it slowly. The enzyme activity is genetically determined (see Chapter 1) and may be a useful parameter to design dosage regimens of drugs that are mainly eliminated by the INH–N-acetyltransferase (see isoniazid monograph).

Conclusions

Clinical pharmacokinetics and therapeutic drug level monitoring are useful and necessary tools for optimizing individual drug therapy.

Further research is needed, however, to fully exploit the potential benefits from these disciplines. Moreover, the process of therapeutic drug level monitoring must be supervised by suitably trained health-care professionals if benefits to the patient are to be obtained. This book is designed to provide the necessary background information and specific data on selected drugs for these health-care professionals.

Bibliography

Anderson, R. J., J. G. Gambertoglio, and R. W. Schrier, *Clinical Use of Drugs in Renal Failure*, Charles C Thomas, Springfield, Ill., 1976.

Benet, L. Z., *The Effect of Disease States on Drug Pharmacokinetics*, Amer. Pharmac. Assoc., Acad. Pharmac. Sci., Washington, D.C., 1976.

Davies, D. S. and B. N. C. Pritchard, *Biological Effects of Drugs in Relation to Their Plasma Concentrations*, University Park Press, Baltimore, London, Tokyo, 1973.

Gibaldi, M., *Biopharmaceutics and Clinical Pharmacokinetics*, 2nd edition, Lea and Febiger, Philadelphia, 1977.

Goldstein, A., L. Aronow, and S. M. Kalman, *Principles of Drug Action: The Basis of Pharmacology*, 2nd edition, John Wiley and Sons, New York, London, Sydney, Toronto, 1974.

Goodman, L. S. and A. Gilman, *The Pharmacological Basis of Therapeutics*, 5th edition, The Macmillan Co., New York, 1975.

Levine, R. R., *Pharmacology, Drug Actions and Reactions*, Little, Brown, and Co., Boston, 1973.

Levy, G., *Clinical Pharmacokinetics: A Symposium*, American Pharmac. Assoc., Acad. Pharmac. Sci., Washington, 1974.

Melmon, K. L. and H. F. Morelli, *Clinical Pharmacology: Basic Principles in Therapeutics*, 2nd edition, The Macmillan Co., New York, 1978.

Smith, S. E. and M. D. Rawlins, *Variability in Human Drug Response*, Butterworth and Co., London, 1973.

Wagner, J. G., *Fundamentals of Clinical Pharmacokinetics*, Drug Intelligence Publ., Hamilton, Ill., 1975.

4

ANALYTICAL TECHNIQUES

4.1. CURRENT SCOPE AND FUTURE TRENDS

Drug level monitoring is based on analytical techniques suitable for the quantitative detection of drugs and their metabolites in biological samples. Qualitative methods, although often similar, are not discussed here. We have reviewed the assays currently available for 102 representative drugs in the extensive literature on this subject; the applicability of these assays to specific drugs is evaluated in the monograph section (Chapter 5) of this book. A survey of the analytical techniques for drug analysis in biological specimens is given in Table III, which includes the following methods:

High performance liquid chromatography (HPLC)
Gas chromatography (GC)
Mass spectrometry (MS)
Thin-layer chromatography (TLC)
Ultraviolet spectrophotometry (UV)
Fluorescence (Fluor)
Colorimetry (Col)
Polarography (Pol)
Radioimmunoassay (RIA)
Enzyme immunoassay (EIA)
Other immunoassays (IA)
Protein binding assays (PBA) (excluding immunoassays)
Enzymatic assays (EA)
Microbiological assays (Micro)
Various assays

Table III. Survey of Analytical Techniques

Drug	Analytical technique														
	HPLC	GC	MS	TLC	UV	Fluor	Col	Pol	RIA	EIA	IA	PBA	EA	Micro	Various
Acetaminophen	+	+	+		+		+								
Acetazolamide	+	+						+							
ε-Aminocaproic acid	+	+													
Aminopyrine	+	+	+	+	+		+								
Amphetamine	+	+	+	+		+	+		+	+			+		
Anticonvulsants	+	+	+	+	+		+			+					
Atropine							+		+						+
BCNU		+	+				+								
Barbiturates	+	+	+	+	+				+	+					
Carbamazepine	+	+	+	+	+					+					
Chloramphenicol	+	+	+	+		+	+	+					+	+	
Chlordiazepoxide	+	+		+		+	+	+	+						
Chloroquine	+	+		+		+	+		+						
Chlorpromazine	+	+	+	+			+	+	+						
Clofibrate	+	+		+	+				+						
Clonazepam	+	+	+	+	+			+	+						
Clonidine	+	+	+						+						
Cocaine	+	+	+	+					+	+					
Cyclophosphamide		+	+				+								+
Dexamethasone	+								+						
Diatrizoate sodium					+		+								
Diazepam	+		+	+	+				+						
Diazoxide	+			+	+										
Digoxin	+	+	+	+	+		+		+	+		+	+		
Diphenhydramine	+	+	+	+					+				+		
Doxorubicin	+	+	+			+			+					+	
Ethambutol	+	+					+								

Drug	1	2	3	4	5	6	7	8	9	10	11	12	13	14	15	16
Ethosuximide											+					
Ethynylestradiol				+		+					+					
5-Fluorouracil	+			+		+					+					
Gentamicin			+			+	+				+					
Haloperidol			+			+	+				+	+				
Hydralazine		+				+	+				+	+				
Hydrochlorothiazide			+			+	+				+	+				
Hydrocortisone			+			+	+				+	+				
Imipramine			+			+	+				+	+				
Indomethacin					+		+				+					
Isoniazid				+							+					
Isosorbide dinitrate		+				+					+					
Lidocaine					+						+					
Lithium	+															
Lysergide						+					+					
Melphalan			+	+	+	+	+				+					
Meperidine			+	+	+	+	+				+					
6-Mercaptopurine			+	+	+	+	+				+					
Methadone			+		+	+	+				+					
Methaqualone							+				+					
Methotrexate			+		+	+	+				+					
Methyldopa			+		+	+	+				+					
Metronidazole				+		+					+					
Morphine			+	+	+	+	+				+					
Nicotine			+				+				+					
Nitrofurantoin						+	+				+					
Oxazepam	+					+	+				+					
Penicillin			+			+					+					
Phencyclidine							+				+					
Phenylbutazone			+		+	+	+				+					
Phenytoin	+				+	+	+				+					
Primidone		+			+	+	+				+					
Procainamide						+	+				+					

29

Table III. (Continued)

Drug	Analytical technique														
	HPLC	GC	MS	TLC	UV	Fluor	Col	Pol	RIA	EIA	IA	PBA	EA	Micro	Various
Procarbazine	+														
Propoxyphene		+	+			+									
Propranolol	+	+	+	+		+			+						
Quinidine	+	+		+	+	+									
Reserpine				+					+						
Salicylic acid	+	+		+	+	+	+								
Spironolactone	+	+				+	+								
Succinylcholine		+		+											
Sulfonamides	+	+				+	+								
Tetracyclines	+					+									+
Tetrahydrocannabinol	+	+	+	+					+	+					
Theophylline	+	+	+		+					+	+				
Tolbutamide	+	+	+				+								
Valproic acid		+													
Warfarin	+	+			+	+								+	+
Total number of assays	52	58	40	24	22	26	26	10	29	17	5	4	6	8	5
Percent of listed drugs	70	77	53	32	29	35	35	13	39	23	7	5	8	11	7

Column chromatography, paper chromatography, and similar procedures are regarded as preparatory methods for the suitable purification of drugs from biological samples and are therefore not listed as independent analytical procedures, whereas TLC, usually coupled with densitometry, does represent a complete quantitative assay technique.

We can directly obtain an estimate of the relative applicability of the various methods for the 75 drugs or drug classes listed in Table III: (1) GC (77%); (2) HPLC (70%); (3) MS (53%); (4) RIA (39%); (5, 6) Fluor and Col (35% each); (7) TLC (32%); (8) UV (29%); (9) EIA (23%). The remaining methods are applicable to only 13% of the drugs or less. The prevalence of GC attests to its versatility. The even more widely applicable HPLC technique ranks only second because it was introduced to drug level monitoring several years later than GC; however, HPLC can be expected to supersede GC as the most widely applicable method in the near future. Also, the ranking order obtained from Table III does not include the suitability of the various methods. For example, simple UV assays often lack the specificity required for clinical applications.

The major techniques used in clinical laboratories fall into three general classes:

1. Chromatographic methods, combined with a variety of detection modes (HPLC, GC, TLC).
2. Spectroscopic analyses (UV, Fluor, Col).
3. Competitive protein binding assays (RIA, EIA, IA, PBA).

These three classes are applicable to 89, 67, and 45%, respectively, of the listed drugs. Mass spectrometry, in combination with GC or by direct analysis, accounts for a surprising 52% of the drugs. Many MS assays have been developed because of the potentially high sensitivity and specificity of this method, although the instrumentation is considerably more elaborate than that of other techniques. Since the development of highly sensitive and specific HPLC, GC, and RIA drug assays, however, MS as a quantitative tool has become somewhat obsolete, unless employed in special research or quality control situations. The various techniques will be discussed in more detail in subsequent sections.

The current trend is undoubtedly directed toward even more sensitive, specific, and reproducible techniques for the quantitation of drugs and their metabolites in biological specimens. There is a need for further knowledge of the pharmacokinetics, metabolism, and pharmacodynamics of many drugs requiring the application of sophisticated methods. However, we may soon be able to select simple assay systems, particularly in therapeutic drug level monitoring, once sufficient knowledge has accumulated on the pharmacodynamics of many drugs under a variety of pathophysiological conditions. Quinidine could become an example where judicious choice of a simple fluorescence assay is appropriate, although such assays are nonspecific for quinidine in the presence of its metabolites. First, we need to know the pharmacologic contributions of the various quinidine metabolites, assayed by sophisticated HPLC techniques. Expecting a possible return to simpler methodology in the future, we have paid some attention to the spectroscopic methods in spite of their obsolescence in many cases. In fact, spectroscopic assays are applicable to 67% of the drugs listed. Indeed, current developments of simple but effective methods for the rapid purification of drugs from biological samples (minicolumns, etc.) may bring spectroscopic methods back into clinical and research laboratories.

Bibliography

Christian, G. D., *Analytical Chemistry*, 2nd edition, John Wiley and Sons, New York, 1977.

Florey, K., *Analytical Profiles of Drug Substances*, Vols. 1–7, Academic Press, New York, San Francisco, London, 1972–1978.

Kalman, S. M. and D. R. Clark, *Drug Assay: The Strategy for Therapeutic Monitoring*, Masson Publ., New York, 1979.

Karger, B., L. R. Snyder, and C. Horvath, *An Introduction to Separation Science*, Wiley-Interscience, New York, 1973.

Knevel, A. M. and F. E. Digangi, *Jenkins' Quantitative Analytical Chemistry*, 7th edition, McGraw-Hill Book Co., New York, 1977.

Pecsok, R. L., L. D. Shields, T. Cairns, and I. G. McWilliam, *Modern Methods of Chemical Analysis*, 2nd edition, John Wiley and Sons, New York, 1976.

Raffauf, R. F. and V. D. Warner, *Introduction to Drug Analysis*, F. A. Davis Co., Philadelphia, 1978.

Siest, G. and D. S. Young, *Drug Interference and Drug Measurement in Clinical Chemistry*, Karger Verlag, Basel, 1976.

Skoog, D. A. and D. M. West, *Fundamentals of Analytical Chemistry*, 2nd edition, John Wiley and Sons, New York, 1969.

Special Issues: Toxicology and Drug Assay, *Clin. Chem.*, **20**(2), 111–316 (1974); Monitoring Drugs in Biological Fluids, *Clin. Chem.*, **22**(6), 711–956 (1976).

Sunshine, I., *Methodology for Analytical Toxicology*, CRC Press, West Palm Beach, Fla., 1975.

Tsuji, K and W. Morozowich, *GLC and HPLC Determination of Therapeutic Agents* (Part I); *Therapeutic Drug Monitoring* (Part II), Marcel Dekker, New York, 1978–1979.

4.2. PREPARATION OF BIOLOGICAL SPECIMEN FOR ANALYSIS

4.2.1. Composition of Blood, Plasma, Serum, Saliva, and Urine. The most readily accessible body fluids are blood, saliva, and urine. While all of these fluids are utilized for drug assays, plasma or serum measurements may yield a better correlation between drug concentrations and effects. The analysis of drugs in whole blood should not be encouraged, since the erythrocyte/plasma concentration ratio is dependent on a number of variables that may impede the pharmacologic interpretation of the results. Blood samples should therefore be centrifuged to obtain either plasma or serum, if the coagulation reaction has proceeded. There appears to be very little drug binding to the protein that forms the blood clot, that is, fibrin, and assay results using plasma and those using serum are usually identical. Plasma specimens can be obtained by preventing coagulation with various agents, including heparin and the Ca^{2+} binding agents, EDTA, citrate, and fluoride. Fluoride also serves as a preservative and inhibits serum cholinesterase, which may cause *in vitro* degradation of several drugs (see cocaine and acetylsalicylic acid monographs). Choice of the anticoagulant may affect assay results; similarly, blood collection tubes (e.g., plastic Vacutainers) may release substances, such as plasticizers, that can also interfere with the assay. Most clinical laboratories now use serum for drug analysis. The composition of plasma and serum varies considerably with pathophysiological condition; therefore drug assays that are recommended for clinical use have to be validated in samples obtained from patients with a variety of disease states. Most notably, hyperlipidemia and hyperbilirubinemia are likely to interfere with many drug level assays.

The use of saliva for drug level monitoring is based on its ready availability and the notion that saliva concentrations may reflect the concentration of free drug in the serum, presumed to represent the active portion of the total serum concentration. However, the various salivary glands (e.g., parotid and submandibular) secrete fluids of different and varying composition; thus drug saliva concentrations may vary relative to free serum levels, and only a few drugs show constant saliva/serum concentrations ratios.

All of the drug assays cited in Chapter 5 are designed to measure total concentration rather than the free fraction of a drug in serum; yet many drugs are highly bound to serum proteins. Albumin accounts for most of the drug binding, while globulins may significantly contribute in some instances. When defining the range of therapeutic serum concentrations of individual drugs, we consider only total drug concentrations; however the interindividual variability of drug protein binding may be large under normal conditions and even greater under pathophysiological ones (hyperbilirubinemia, hypoalbuminemia). The correlation between total drug level and response deteriorates under conditions of variable drug protein binding; phenytoin represents an example of clinically important binding variability, and drug level monitoring is best performed on the free drug fraction in the serum. Because of the additional analytical work-up required (e.g., ultracentrifugation or ultrafiltration) most clinical laboratories do not perform such assays.

Urine drug analysis has normally a different set of objectives, which include mass balance studies, determination of urinary clearance, bioavailability studies, and monitoring individual compliance with the prescribed dosage regimen (see phenytoin monograph). Because of large fluctuations in urinary drug concentrations under different diuresis conditions, we are usually interested in the amount of drug excreted over a given time period, which requires measurement of the drug concentration and the total urine volume excreted during the period in question.

Some of the major components in normal serum and saliva are listed in Table IV. Plasma contains between 2 and 3 g fibrinogen/l. Serum enzymes include alkaline and acid phosphatases, dehydrogenases, and transaminases. Major urinary components consist of nitrogenous substances (7 to 16 g/24 hr), of which urea accounts for the major portion ($\sim$60%). Upon storage of urine samples, the enzyme urease converts urea to ammonia, causing a pH shift to higher values which may decompose alkali-labile drugs. Protein excretion is normally in the range of 47 to 76 mg/24 hr.

Bibliography

Diem, K. and C. Lentner, Eds., *Documenta Geigy: Scientific Tables*, 7th edition, Geigy S.A., Basel, 1970.

4.2.2. Work-up Procedures Prior to Analysis. Suitable preparation of biological specimens is decisive in the successful application of an analytical technique. The preparation should be as simple as possible, yet allow the specific assay of a drug in the presence of numerous biological components. The extent of sample work-up is therefore largely determined by the selectivity of the analytical technique. Potentially interfering endogenous substrates need to be removed before analysis. A second objective of the preparation of biological specimens is to protect the analytical apparatus from contamination by lipids, proteins, and undissolved particles. Biological sample preparation has

Table IV. Some Normal Components of the Serum and Saliva of Adults

Components	Serum	Saliva
Water (g/l)	930–955	
Dry substance (g/l)	80	
Total proteins (g/l)	65–80	1.4–6.4
Albumin (% of total protein)	50–65	
Globulins (% of total proteins)	35–50	
Total cations (mEq/l)	149–159	
Sodium (mEq/l)	132–151	5.2–36
Bicarbonate (mEq/l)	21.3–28.5	2–13
Chloride (mEq/l)	99–111	15–32
Total nitrogen (g/l)	12–14	0.2–1
Nonprotein nitrogen (mg/l)	139–307	
Urea (mg/l)	230–426	140–750
Creatinine (mg/l)	6.6–18.2	5–20
Free amino acids (as α-amino-N) (mg/l)	28–50	
Alanine (mg/l)	22–45	
Lysine (mg/l)	13–31	
Proline (mg/l)	13–51	
Catecholamines	ng/ml range	
Histamine	ng/ml range	
Serotonine	ng/ml range	
Bilirubin (total) (mg/l)	2.6–14	
Uric acid (mg/l)	18–76	5–29
Total lipids (g/l)	3.5–8.5	
Fatty acids (g/l)		
Total	1–5	
Free	0.1–0.35	
Cholesterol (g/l)		
Total	1–3	0.025–0.5
Free	0.3–1	
Phosphatides (g/l)	1.5–3.5	
Triglycerides	0.5–2.2	
Bile acids (mg/l)	<10	
Glucose (mg/l)	750–1170	2–300
Glucuronic acid (mg/l)	20–44	
Heparin (mg/l)	1–2.4	
Glucoproteins (g/l)	2.7	
Succinate (mg/l)	5	
Citrate (mg/l)	17–31	<20
Pyruvate (mg/l)	2.6–10.2	
Acetone (mg/l)	2.3–3.5	
Vitamins	Varying	
Ascorbic acid (mg/l)	2–14	0.6–3.8

to vary according to the technical demands of the various analytical instruments. Since the advent of highly selective analytical methods that combine chromatographic separation and detection in one unit (e.g., GC, HPLC), the second objective has gained in importance.

For each drug discussed in a separate monograph in Chapter 5, one selected assay is described in full detail. The work-up procedures used in these selected assays should provide us with an estimate of their relative utility (Table V).

The high inherent specificity and sensitivity of radioimmunoassays usually allows the direct detection of drugs in very small samples (<100 μl serum). High-performance liquid chromatographic methods require at least protein precipitation in order to protect the column against the deposit of proteins on the stationary phase. Alternatively, a small precolumn, onto which plasma samples can be injected without any preparation, can be used. Such precolumns filter out the protein fraction and have to be regularly exchanged. Protein precipitation methods are rapid and include mixing the sample (serum, urine) with water-miscible organic solvents or acids ($HClO_4$, trichloroacetic acid). Of the organic solvents, acetonitrile yields a protein precipitate that can be readily centrifuged into a small pellet. Choice of the protein precipitating reagent may affect assay results by potential coprecipitation of varying amounts of the drug, a possibility that has to be tested in each case. Use of protein precipitation alone without further work-up is rapidly rising because of the increased application of HPLC. However, sensitivity is usually limited to the micrograms per milliliter range

Table V. **Analytical Work-up Procedures Used in the Selected Assays That Are Described in Detail for Each Drug in Chapter 5**

Work-up procedure	Number of assays	Most frequently used analytical technique
None	9	RIA
Protein precipitation	13	HPLC
Organic solvent extraction		
Single extraction	34	HPLC + GC
Multiple extractions	17	GC
Column chromatography	2	HPLC, PBA[a]

[a] Competitive protein binding assay.

when using UV detection, since only small aliquots of the biological sample can be taken for analysis without overloading the column or causing assay interferences.

Organic solvent extractions are still by far the most useful methods for the preparation of biological samples for subsequent analysis. Many drugs are lipophilic, while a majority of the small molecular weight constituents of serum and urine are polar. Thus a single solvent partitioning step removes many impurities, allowing the analysis of larger aliquots of the biological sample and, thereby, greater sensitivity. Removal of polar, nonvolatile substances is also needed to protect GC columns from unwanted contaminations. Three major variables should be considered in the design of suitable organic solvent extraction procedures: the organic solvent, the pH of the aqueous phase, and the volumes of the organic and aqueous phases. A higher pH is often desirable, since many endogenous substances are acidic and are not extractable at alkaline pH. Consideration of the pH is therefore important even when assaying for neutral drugs. Lipophilic bases are quite uncommon in body fluids, and it is relatively easy to analyze many of the lipophilic basic drugs. Figure 1 shows an HPLC-UV record obtained from a serum extract. The serum sample was extracted with ether at pH 2, 7, and 13. The lack of UV absorbing substrates following extraction at the alkaline pH is striking. However, one solvent partitioning step alone is not capable of separating bases from acids and neutral compounds. If this is required, multiple extraction steps have to be employed, for example, back-extractions from the organic medium into water at the proper pH. Several extractions are often needed for the GC analysis of very low serum concentrations of lipophilic basic drugs (e.g., clonidine, diazepam, haloperidol, chlorpromamine, nicotine).

Special modes of organic solvent extractions include ion-pair extractions (succinylcholine, methotrexate) and extractive alkylations (oxazepam, 6-mercaptopurine). The second approach involves the simultaneous extraction and chemical derivatization of the drug. This technique is important for a drug that might otherwise be difficult to extract because of its amphoteric polar nature.

Sample preparation by chromatographic techniques is infrequently employed for the selected assays, since chromatographic separation often represents an integral part of the analytical technique (GC, HPLC). Particularly difficult problems may still require column chromatography, even for highly developed routine assays (gentamicin, ethylnylestradiol).

A more detailed, systematic discussion of the preparation of analytical samples for analysis is not appropriate in this book. Unfortunately, few general textbooks are available on this subject which

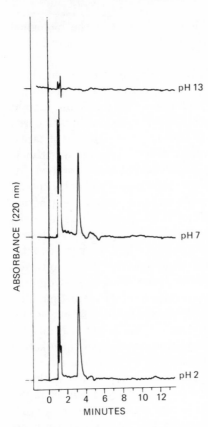

ABSORBANCE (220 nm)

pH 13

pH 7

pH 2

0 2 4 6 8 10 12
MINUTES

Fig. 1. HPLC records of serum extracts obtained with ether at pH 2, pH 7, and pH 13. The column is a reverse phase μ-Bondapak C_{18}, and the eluents were monitored at 220 nm. The chromatographic conditions were those used for the determination of procainamide [Shukur et al., *Clin. Chem.*, **23**, 636–638 (1977)]. The single peak at 3.5 min represents an acidic endogenous component with the same retention time as procainamide. No interferences are present after the ether extraction at pH 13.

consider the recent developments of analytical methodology. Easy access to many examples in Chapter 5, however, should be of value in the selection of suitable methods for analogous drug assays.

In the future, we are likely to witness significant advances in sample preparation techniques that can be used together with simple spectroscopic methods. Our experience in HPLC drug assays could lead to the development of short, hand-operated columns with sufficient separation efficacy to allow the use of the rather nonspecific UV, fluorescence, and colorimetry methods.

4.2.3. Chemical Derivatizations. The purpose of chemical derivatization of drugs prior to analysis are to (1) achieve suitable separation characteristics, (2) provide a means of sensitive detection, (3) stabilize the drug, and (4) separate enantiomers or solve special problems. The most common derivatization reagents are listed in Table VI, together with the functional group to be derivatized, the major purpose of derivatization, and drug examples selected from Chapter 5. The majority of the reactions serve to induce physicochemical characteristics suitable for GC analysis. This is achieved by derivatizing polar functions (i.e., COOH, NH, OH) with lipophilic masking groups in order to increase the vapor pressure, which, in turn, reduces the required GC temperatures. The major reactions consist of alkylations, acylations, and silylations. The fluorinated reagents not only increase the vapor pressure but also convey electronegative labels to the drug molecule, which can be detected by GC with electron capture with high sensitivity. Such reagents also affect the electron impact-induced fragmentations during MS analysis and usually produce an abundant ion peak for this sensitive assay.

Silylation reactions are among the most versatile derivatization methods and certainly have contributed to the broad applicability of GC analysis, which would otherwise be limited to compounds with relatively high vapor pressures. Trimethylsilyl derivatives are readily formed with a large variety of functional groups (i.e., OH, NH, COOH). Selectivity of the silylation reaction between primary and secondary, and hindered and unhindered, OH and NH functions can be achieved by selecting the proper reaction conditions and silylating reagent. Trimethylsilyl reagents differ in their leaving group X in $(CH_3)_3Si—X$, which determines the chemical reactivity of the reagent. Common reagents are trimethylchlorosilane (TMCS), bis(trimethylsilyl)acetamide (BSA), bis(trimethylsilyl)trifluoroaacetamide (BSTFA), and trimethylsilyl imidazole (IMTS). Substitution of the trimethylsilyl group with *t*-butyldimethylsilyl yields hydrolysis resistant, less volatile derivatives which may have a superior electron impact fragmentation pattern for GC-MS analysis.

Many drugs are racemic mixtures, and separation of the enantiomers may be important, since they often display different pharmacological activity and metabolism. One method for the separate analysis of enantiomers consists of derivatization with an asymmetric reagent in order to produce diastereomers that are separable by regular chromatographic procedures (Table VI). Gas chromatography is the predominant separation mode because of its high efficiency; however, HPLC can also be utilized to separate diastereomeric mixtures, for instance, in the case of propranolol (B. Silber, unpublished).

There are numerous additional derivatization reactions, only a few of which are of importance to drugs assays in biological samples (Table VI). The versatility of HPLC obviates the need of derivatization for the purpose of enhanced chromatographic behavior in most cases. Rather,

Table VI. Selected Derivatization Reactions for Analytical Purposes (Excluding Colorimetric and Fluorescence Reactions)

Reagent	Functional group	Analytical purpose	Example
Alkylations			
CH_3I/base	OH, NH	GC	Anticonvulsants, barbiturates, 6-mercaptopurine, oxazepam
CH_2N_2	COOH, OH, NH (acidic)	GC, MS	Clofibrate, diazoxide, tolbutamide, warfarin
Alcohol/acid	COOH, OH	GC, MS	ϵ-Aminocaproic acid, α-methyldopa (metabolites)
$(CH_3)_2SO_4$	COOH, OH, NH	GC	Tolbutamide
$C_6H_5N(CH_3)_3^+$	COOH, OH, NH	GC	Acetazolamide, anticonvulsants, barbiturates, quinidine, clonidine
CH_3CH_2I and other alkyl iodides	OH, NH	GC	Acetaminophen, barbiturates, theophylline
Aldehydes + borohydride	NH, NH_2	GC	Lidocaine
$CH_3CH_2N_2$	COOH	GC	Indomethacine
$C_6F_5-CH_2Br$	COOH, NH	GC-EC	Cocaine (metabolite), carbamazepine, indomethacine, warfarin
Acylations			
$(CH_3CO)_2O$	NH, NH_2, OH	GC	Lidocaine (metabolites)
$(CF_3CO)_2O$	NH, NH_2, OH	GC-EC, GC-MS	Amphetamine, morphine, cyclophosphamide, propranolol, ethambutol, hydralazine
$(CF_3CF_2CO)_2O$	OH, NH	GC-EC	Desipramine, α-methyldopa

Reagent	Functional groups	Method	Compounds
$(CF_3CF_2CF_2CO)_2O$	OH, NH	GC-EC	Digoxigenin, morphine, meperidine (metabolite), oxyphenbutazone
C_6F_5—COCl	NH	GC-EC	Clonidine, primidone, Δ^9-THC
Asymmetric Reagents			
(S)-N-Trifluoro-acetylprolyl chloride	OH, NH	GC	Propranolol
(S)-N-penta-fluoro-benzoyl-prolyl chloride	OH, NH_2, NH	GC-EC, GC-MS	Amphetamine, α-methyldopa
(S)-α-Methoxy-α-(tri-fluoromethyl)phenylacetyl chloride	NH_2, NH	GC-MS	Amphetamine
Various			
Ketones and aldehydes	NH_2, $NHNH_2$	GC	Amphetamine, isoniazid
Amines, hydrazines, hydroxylamines	C=O	GC	Doxorubicin (aglycone), hydrocortisone
Dansyl chloride	NH, NH_2, OH	HPLC-fluorescence	Gentamicin, ε-aminocaproic acid
o-Phthalaldehyde	NH_2	HPLC-fluorescence	Gentamicin
Silylations	COOH, OH, NH	GC, GC-MS	Acetaminophen, carbamazepine, chloramphenicol, ethynylestradiol, 5-fluorouracil (metabolites), methaqualone, metronidazole, morphine, salicylic acid

derivatizations are commonly used to increase the sensitivity toward a specific HPLC detector, for example, UV and fluorescence. The potential for using spectroscopic labels in conjunction with HPLC has not been fully realized in drug analysis in biological samples, and Table VI contains only a few examples utilizing fluorescent reagents.

Special precautions have to be taken to avoid a loss of specificity through the derivatization reaction. While methylation is a preferred mode of derivatization of many drugs, metabolic methylation and demethylation reactions are frequently observed and may contribute to assay interferences. For example, theophylline is N-demethylated to 1- and 3-methylxanthine; moreover, caffeine is a common constituent of biological samples and is also metabolized by oxidative N-demethylations. Therefore derivatization with methylating reagents yields caffein from all methylated xanthines and has to be avoided for the analysis of theophylline. Instead, ethylation, propylation, and butylation yield unique derivatives for each methylated xanthine and are suitable for theophylline analysis. Similar problems are encountered with the analysis of mephobarbital and other N-methylated barbiturates, cocaine, and indomethacine, to name just a few.

Acetaminophen serves to demonstrate the variety of derivatization reactions that can be successfully utilized to monitor its plasma levels in the presence of the precursor phenacetin (Fig. 2).

Bibliography

Blau, K. and G. King, *Handbook of Derivatives for Chromatography*, Heyden, London, Philadelphia, Rheine, 1978.

Lawrence, J. F. and R. W. Frei, *Chemical Derivatization in Liquid Chromatography*, Elsevier Sci. Publ. Co., Amsterdam, 1976.

4.3. HIGH PERFORMANCE LIQUID CHROMATOGRAPHY

The recent explosion of HPLC drug assays in biological samples is based on the versatility, efficacy, precision, and speed of this technique. Rapid technological advancements make it possible to employ HPLC as a routine method for quantitative drug analysis for many purposes, including therapeutic drug level monitoring in clinical laboratories.

The introduction of small particle sizes for the column packing material with pressured column systems to shorten analysis time (high pressure liquid chromatography) has contributed the major share to this advancement. Furthermore, new column packing materials, improved pressurized packing procedures, highly sensitive detectors,

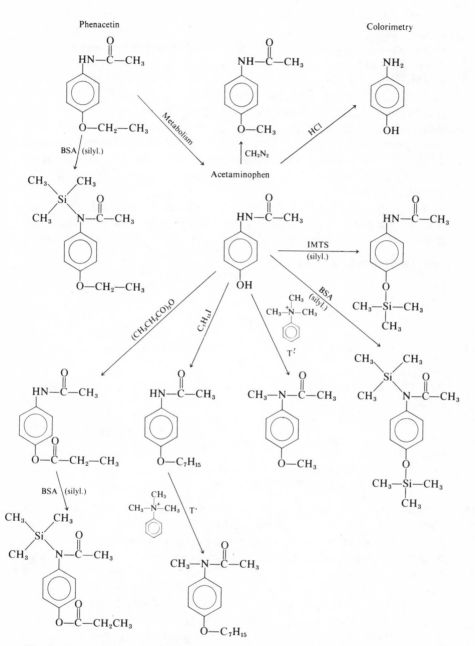

Fig. 2. Derivatization reactions of acetaminophen and phenacetin used for quantitative analysis (see acetaminophen monograph).

43

and reliable, quantitative injection systems have all contributed to the current utility of HPLC in drug level monitoring. Despite the rather recent development of this technique, HPLC has already become superior to gas chromatography—not yet in number, but rather in utility, of assays. Table VII lists the drugs included in Chapter 5 for which HPLC assays in biological samples have already been published. Serum sample preparation may not be required at all; otherwise, simple protein precipitation or single organic solvent extraction is usually sufficient preparation for final HPLC analysis. It is also apparent from Table VII that chemical derivatizations are not needed for the majority of drug level assays, and that a single detector (UV) is almost universally applicable.

The four basic types of chromatography are liquid-solid, partitioning, ion-exchange, and exclusion chromatography. The following diagram can serve as a guide to the selection of the proper HPLC mode.

Selection Criteria for Chromatography

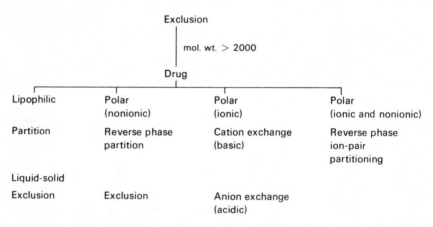

Liquid-solid chromatography, often referred to as adsorption chromatography, represents the earliest form of chromatography and mainly uses silica gel particles as the stationary phase. Silica gel HPLC is quite similar to silica gel thin-layer chromatography as far as separation efficiency is concerned. Polar compounds are retained longer than lipophilic material on a silica gel column. Many of the relatively lipophilic drugs can be readily analyzed by silica gel HPLC with UV detection (Table VII). However, an organic solvent extraction of the drug has to precede HPLC analysis, since biological samples contain large quantities of polar UV absorbing substances that keep eluting after the drug from the silica column unless separated before analysis.

Partition chromatography is now mainly performed on columns containing a stationary phase chemically bonded to inert support material (bonded phase chromatography). It is called "normal phase" if the stationary phase is more polar than the mobile phase. An example is the alkylnitrile bonded phase used for the analysis of doxorubicin and sulfonamides. In "reverse phase" partition chromatography, the stationary bonded phase is more lipophilic than the mobile phase. Table VII demonstrates that reverse phase HPLC (RP) is the most widely used mode of chromatography today in drug level monitoring. There are four reasons for this dominance of reverse phase HPLC:

1. Separation based on lipophilicity is advantageous, since many drugs are more lipophilic than potentially interfering endogenous substrates. This criterion, however, applies equally to normal phase HPLC.
2. Since polar compounds are less well retained on the column, the polar endogenous substrates elute prior to the lipophilic drugs. Usually, no further UV absorbing material elutes after the drug, thus allowing rapid analysis of multiple samples. Furthermore, organic solvent extraction may not be necessary, as is the case with normal phase HPLC. Figure 3 illustrates this elution pattern with a quinidine HPLC-UV record obtained from serum samples following protein precipitation.
3. Several chemically different reverse phase column packings are applicable to a surprisingly large fraction of the selected drugs. Stationary bonded phases include alkyl side chains (C_2, C_8, C_{18}, C_{22}) and alkylphenyl side chains (Fig. 4).
4. In conjunction with ion-pair chromatography, a variety of polar ionic and nonionic drugs can be analyzed by reverse phase HPLC (gentamicin, isoniazid, 6-mercaptopurine, Table VII).

Reverse phase HPLC is capable of directly measuring therapeutic serum concentrations of the following drugs without organic solvent extraction: ϵ-amino caproic acid, anticonvulsants, chloramphenicol, hydrochlorothiazide, indomethacin, melphalan, metronidazole, nitrofurantoin, penicillins, quinidine, sulfonamides, theophylline, and warfarin.

Ion-exchange chromatography depends upon the exchange of ions between the mobile phase and the ionic sites of the packing, for example, sulfonic acids and quaternary ammonium groups. Both cationic and anionic drugs can be analyzed (e.g., acetaminophen, 5-fluorouracil, methotrexate, procainamide, Table VII). The actual mode

Table VII. Summary of HPLC Assays

Drug	Column[a]	Detection	Comments
Acetaminophen	RP, CE, silica polyamide	UV	
Acetazolamide	RP	UV	
ε-Aminocaproic acid	RP, CE	Fluor, Col	Dansylation, ninhydrin
Anticonvulants	RP, silica	UV	195 nm UV
Barbiturates	RP	UV	
Carbamazepine	RP, silica	UV	
Chloramphenicol	RP	UV	
Chlordiazepoxide	RP	UV	
Chlorpromazine	RP	UV	
Clonazepam	Silica	UV	
Clonidine	RP	UV	
Dexamethasone	Silica	UV	
Diazepam	RP	UV	
Diphenhydramine	CN	Fluor	Postcolumn ion-pair extraction by autoanalyzer
Doxorubicin	RP, silica, CN	UV, Fluor	
5-Fluorouracil	RP, AE	UV	
Gentamicin	RP	Fluor	o-Phthalaldehyde, dansyl chloride, ion-pair chromatography
Hydralazine	RP	UV	
Hydrochlororthiazide	RP	UV	
Hydrocortisone	RP	UV	
Imipramine	Silica	UV	
Indomethacin	RP	UV	
Isoniazid	RP, silica	UV	Ion-pair chromatography
Lidocaine	RP	UV	205 nm UV
Melphalan	RP	UV	
6-Mercaptopurine	Various columns	UV, Fluor	Permanganate oxidation (fluorescence), ion-pair chromatography, dithioerythritol stabilizer in eluent
Methotrexate	RP, AE	UV, Fluor	Permanganate oxidation (fluorescence), ion-pair chromatography
Methyldopa	RP, CE	UV, electrochem.	
Metronidazole	RP	UV	
Morphine	RP, silica	UV, Fluor	$K_3Fe(CN)_6$ oxidation (fluorescence)
Nicotine	Silica	UV	
Nitrofurantoin	RP	UV	
Oxazepam	RP	UV	
Penicillins	RP	UV	
Phenylbutazone	Silica	UV	

Table VII. (Continued)

Drug	Column[a]	Detection	Comments
Phenytoin	RP	UV	Acid hydrolysis of 4-OH-phenytoin glucuronide
Procainamide	RP, silica CE	UV	
Procarbazine	RP	UV	
Propranolol	RP	UV, Fluor	
Quinidine	RP, silica, CE	UV, Fluor	
Salicylic acid	RP	UV	
Spironolactone	RP	UV	
Sulfonamides	RP, AE, CN	UV	
Tetracyclines	RP, AE	UV, electrochem.	
Theophylline	RP, silica, CE	UV	
Tolbutamide	RP	UV	

[a] RP: reverse phase; AE: anion exchange; CE: cation exchange; CN: alkylnitrile bonded phase.

of separation may not depend exclusively on the ionic strength of the drug, with lipophilicity also contributing to the chromatographic behavior.

Exclusion chromatography is based on the molecular size of the solute. It is potentially useful for the separation not only of macromolecules but also of small (<1000) molecular weight substances. Gel permeation and gel filtration are synonyms for exclusion chromatography. Steroids are commonly separated on Sephadex LH-20 at ambient pressure, using organic eluents (see ethynylestradiol monograph). The HPLC assay of acetaminophen can be performed on a polyamide column as an example of exclusion chromatography.

Of the many existing and potentially useful HPLC detection systems, only three have actually been employed for the analysis of the selected drugs in biological fluids. These are the flow cell detectors based on UV absorbance, fluorescence, and electrochemical reactions. The UV detector is by far the most widely used system because of its relatively high sensitivity and general applicability. Typically, HPLC-UV analysis is presently sensitive to about 10 pmol of a substance (ϵ = 10,000) injected onto the column. Most columns possess sufficient capacity to tolerate injections of large amounts of endogenous substrates, allowing the analysis of large aliquots of a biological specimen without causing column overloading. Moreover, reverse phase HPLC-UV drug assays are useful even in the UV absorbance range below 200 nm, where a single amide bond provides sufficient absorp-

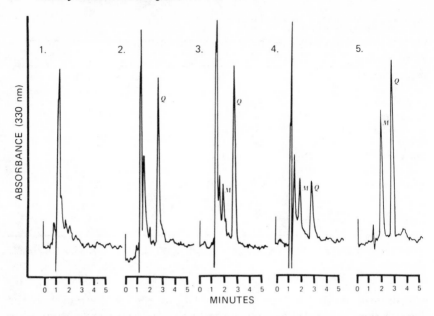

Fig. 3. HPLC records of serum samples, using a reverse phase column and UV detection at 330 nm according to the quinidine assay of Powers and Sadée [*Clin. Chem.*, **24,** 299–302 (1978)]. *Q:* quinidine; *M:* quinidine metabolites, 3'-hydroxyquinidine and 2-quinidinone. Samples 1 to 4 were directly analyzed after serum protein precipitation, while sample 5 was assayed after ether extraction at pH 13. Note the absence of UV absorbing serum components after elution of polar material. (1) Control serum. (2) Quinidine (5 µg/ml) added to control serum. (3) Patient's serum with low relative metabolite content. (4) Patient's serum with high relative metabolite content. (5) Same serum sample as shown in (4), after alkaline extraction resulting in a significant increase of the signal/noise ratio.

tivity. The separation efficacy is great enough to permit the sensitive and specific detection in serum of drugs such as the anticonvulsants (195 nm) and lidocaine (205 nm) at very low wavelengths (Table VII).

Fluorescence detection increases the specificity and sensitivity of the assay method. Whereas a 50 µl serum sample is needed for the analysis of quinidine by UV detection, less than 5 µl of serum is required for the same assay based on fluorescence detection (Fig. 5). However, few drugs display sufficiently high native fluorescence under the HPLC conditions (doxorubicin, propranolol, quinidine), and most HPLC-fluorescence procedures include chemical manipulations in order to induce high fluorescence yields. After oxidation reactions, 6-mercaptopurine and its metabolites, methotrexate, and morphine can be assayed by HPLC-fluorescence in serum. Primary amines (ε-aminocaproic acid, gentamicin) can be reacted with a variety of reagents, including dansyl chloride and o-phthalaldehyde. An elegant, though technically more complicated HPLC-fluorescence method is the analysis of gentamicin, which employs postcolumn on-line derivatization of the drug with the nonfluorescent o-phthalaldehyde to form a fluorescent product. This reaction can also be performed off-line, precolumn, since it is quantitative; however, postcolumn derivatization separates the underivatized sample, providing greater separation efficiency and potentially fewer assay interferences.

The sensitivity of electrochemical flow cell detectors can match that of the fluorescence detector (<1 pmol injected), yet few examples are available of the assay of drugs in biological samples (e.g., methyldopa, tetracyclines). The utility of this detector system for the analysis of endogenous catecholamines is well established, and future developments will certainly include the analysis of several drugs with suitable redox potentials.

In summary, HPLC is one of the most useful techniques in drug level monitoring. Technical advances in the near future are likely to include the development of columns with a substantially larger number of theoretical plates for the analysis of complex mixtures, and the enhancement of existing detection systems by electronic and statistical means and computerization.

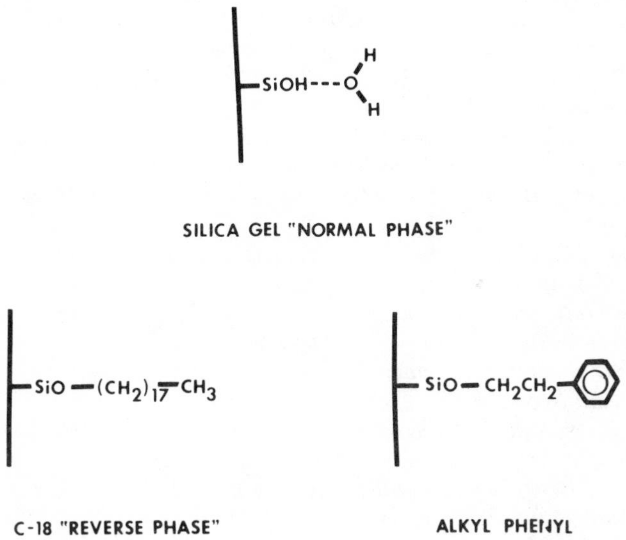

SILICA GEL "NORMAL PHASE"

C-18 "REVERSE PHASE"   ALKYL PHENYL

Fig. 4. Chemical structures of three HPLC column packing materials that are widely used in drug analysis.

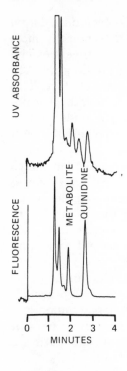

Fig. 5. Comparison of UV and fluorescence detection in a reverse phase HPLC assay of serum quinidine [HPLC-UV: Powers and Sadée, *Clin. Chem., 24*, 299–302 (1978); see Fig. 3]. Serum was analyzed directly after protein pecipitation; 50 μl serum was used for the HPLC-UV record, while only 5 μl of the same serum was analyzed by HPLC-fluorescence (Schoeffel fluorescence detector). Quantitative results with external standards were equivalent for the two methods when assaying 20 patient sera for quinidine content (r = .991; slope of the regression line = 1.063; y intercept = 0.097 μg/ml; HPLC-UV: x, HPLC-fluorescence: y). The metabolite peak can represent either 3-hydroxyquinidine or 2'-quinidinone. Since 2'-quinidinone fluoresces only weakly under the assay conditions, the metabolite peak consists primarily of 3-hydroxyquinidinone.

Bibliography

Brown, P. R., *High Pressure Liquid Chromatography: Biochemical and Biomedical Applications*, Academic Press, New York, 1973.

Dixon, P. F., C. H. Gray, C. K. Lim, and M. S. Stall, *High Pressure Liquid Chromatography in Clinical Chemistry*, Academic Press, New York, 1976.

Done, D. N., J. H. Knox, and J. Loheac, *Applications of High-Speed Liquid Chromatography*, John Wiley and Sons, London, 1974.

Engelhardt, H., *Hochdruck-Flüssigkeits Chromatographie*, Springer-Verlag, Berlin, 1975 (in German).

Giddings, J. C., E. Grushka, R. A. Keller, and J. A. Cates, *Advances in Chromatography*, Vols. 12 and 13, Marcel Dekker, New York, 1975.

Giddings, J. C., E. Grushka, J. Cazes, and P. R. Brown, *Advances in Chromatography*, Vol. 16, Marcel Dekker, New York, 1978.

Grushka, E., *Bonded Stationary Phases in Chromatography*, Ann Arbor Sci. Publ., Ann Arbor, Mich., 1974.

Hamilton, J. R. and P. A. Sewell, *Introduction to High Performance Liquid Chromatography*, Chapman and Hall, London, 1978.

Hawk, G. L., *Biological/Biomedical Applications of Liquid Chromatography*, Chromatographic Science Series, Vol. 10, Marcel Dekker, New York, 1979.

Johnson, E. L., *Liquid Chromatography: Bibliography*, Varian Assoc., Palo Alto, Calif., 1977.

Johnson, E. L. and R. Stevenson, *Basic Liquid Chromatography*, Varian Assoc., Palo Alto, Calif., 1978.

Kissinger, P. T., Recent Developments in the Clinical Assessment of Metabolites of Aromatics by High Performance Reverse Phase Chromatography with Amperometric Detection, *Clin. Chem.*, **23**, 1449–1455 (1977).

Rajcsanyi, P. J., *High Speed Liquid Chromatography*, Elsevier Sci. Publ. Co., Amsterdam, 1976.

Simpson, C. F., *Practical High Performance Liquid Chromatography*, Heydon and Son, London, 1976.

Snyder, L. R. and J. J. Kirkland, *Introduction to Modern Liquid Chromatography*, Wiley-Interscience, New York, 1974.

Weals, B. B. and I. Jane, Analysis of Drugs and Their Metabolites by High Performance Liquid Chromatography, *Analyst*, **102**, 625–644 (1977).

4.4. GAS CHROMATOGRAPHY—BY J. L. POWERS

The utility of GC analysis in drug level monitoring is attested to by the large number of drugs that can be analyzed by this technique (~80%, Table VIII). There are two basic modes of gas chromatography, gas-liquid partitioning (GLC) and gas-solid adsorption (GSC). Adsorption gas chromatography (GSC) employs an adsorptive solid column packing material and is mainly applied to gases and highly volatile compounds, for example, volatile anesthetics and ethanol. The GSC assay of volatile compounds is usually performed by head-space analysis of biological samples without organic solvent extraction. However, the predominant GC mode is by partitioning (GLC); since Chapter 5 contains only assay examples utilizing gas-liquid chromatography, this technique is referred to as GC throughout this book (i.e., GC = GLC).

The liquid stationary GC phase consists of silicone plastomers or long-chain hydrocarbons with low vapor pressure. These materials are actually solid at room temperature and melt at ~100°C; GC analysis is usually performed at >100°C in order to increase the vapor pressure of the organic compounds. Separation of components of a mixture occurs by partitioning between the liquid stationary phase and the mobile gas phase (nitrogen, helium). Retention in the column is a function of the physicochemical interactions of the compound with the liquid phase, the temperature, and the flow rate of the gas phase.

Partitioning GC is performed either on columns packed with an inert, microparticulate support material that is coated with the stationary liquid phase, or on capillary long tubes coated on the inside wall with the liquid phase. Capillary GC columns are highly efficient, since turbulence of the carrier gas is minimized. With a column length of

Table VIII. Gas Chromatography

Drug	Derivatization	Detection
Acetaminophen	Alkylation, acylation, silylation	FID
Acetazolamide	Flash methylation	EC
ϵ-Aminocaproic acid	Trifluoroacetylation and *n*-butylation	FID
Aminopyrine	None	FID
Amphetamine	None; acetone imine	FID
Anticonvulsants	None; alkylations	FID, N-FID
Barbiturates	None; alkylations	FID, N-FID
Carbamazepine	Alkylation, silylation	FID, N-FID, EC
Chloramphenicol	Silylation	FID, EC
Chlordiazepoxide	None	EC
Chloroquine	None	FID, EC
Chlorpromazine	None	EC
Clofibrate	Methylation	FID
Clonazepam	Methylation, hydrolysis	EC
Clonidine	Flash methylation, pentafluoro-benzylation	N-FID, EC
Cocaine	Various (metabolites)	N-FID, EC
Cyclophosphamide	Trifluoroacetylation	FID, EC
Diazepam	None	FID, N-FID, EC
Digoxin	Hydrolysis and heptafluoro-butyrylation	EC
Diphenhydramine	None	FID
Doxorubicin	Hydrolysis, silylation, trifluoroacetyl-ation, methoxime formation	FID, EC
Ethosuximide	None; alkylations, pentafluorobenzoylation	FID EC
5-Fluorouracil	Silylation, methylation	FID, N-FID, EC
Haloperidol	None	EC
Hydralazine	Trifluoroacetylation, formation of tetrazolophthalazine with HNO₂	EC
Hydrochlorothiazide	Alkylation	EC
Imipramine	Trifluoroacetylation (desipramine)	N-FID
Indomethacin	Ethylation, pentafluorobenzylation	EC
Isoniazid	*p*-Chlorobenzaldehyde hydrazone formation, silylation	FID, N-FID
Isosorbide dinitrate	None	EC
Lidocaine	Acylation (metabolites)	FID, N-FID
Meperidine	Heptafluorobutyrylation, other acylations (*N*-normeperidine)	FID, N-FID
6-Mercaptopurine	Methylation	FID
Methadone	None	FID
Methaqualone	Silylation (metabolites)	FID
Methyldopa	(S)-*N*-Pentafluorobenzoylprolylation, various reactions	FID
Metronidazole	Silylation	EC
Morphine	Silylation, acylation	FID, EC

52

Table VIII. (Continued)

Drug	Derivatization	Detection
Nicotine	None; hydrogenation, acylation	FID, N-FID, EC
Nitrofurantoin	Azomethine bond hydrolysis	EC
Oxazepam	Methylation, hydrolysis, thermolysis	EC
Phencyclidine	None	FID
Phenylbutazone	Oxidation to azobenzene, heptafluoro-butyrylation (oxyphenbutazone)	FID
Phenytoin	None; methylation	FID
Primidone	None; methylation, pentafluoro-benzoylation	FID, EC
Procainamide	None	FID
Propoxyphene	Various reactions, including chemical rearrangements	FID
Propranolol	Acylation (asymmetric reagent), silylation	FID
Quinidine	Flash methylation	FID, N-FID
Salicylic acid	Silylation	FID
Spironolactone	None (canrenone)	FID, EC
Succinylcholine	Thermolysis, hydrolysis, and alkylation of succinate	FID
Sulfonamides	Methylation	EC
Tetrahydrocannabinol	Pentafluorobenzoylations, heptafluorobutyrylation, diethylphosphate formation	EC
		Flame photometric detection
Theophylline	None; alkylation	FID, N-FID
Tolbutamide	Methylation, thermolysis	FID
Valproic acid	None	FID
Warfarin	Methylation, pentafluorobenzoylation	FID, EC

more than 100 m, capillary GC offers the greatest separation potential at present. Packed columns, on the other hand, are quite versatile and are sufficient for most applications in drug level monitoring.

The major limitations of GC applications to organic material are the requirements for a sufficiently high vapor pressure and for thermostability at the GC temperatures. The column temperatures range between 100 and 350°C, depending on the votality of the compounds. Upper temperature limits are set by the thermostability of the organic compound and the liquid phase and by the vapor pressure of the liquid phase (bleeding). Polar functions (OH, NH, COOH, etc.) reduce the vapor pressure and may prevent effective GC analysis. Therefore many drugs have to be derivatized prior to GC in order to lower their vapor pressure (Table VIII; see also Section 4.2.3). Thermolytic reactions during GC separation are quite common (carbamazepine, oxazepam, sulfonamides, propoxyphene, succinylcholine, phenobarbital, chlordiazepoxide) and should be carefully checked for quantitative GC analysis. Biological samples contain large amounts of polar nonvolatile materials such as peptides, sugars, and amino acids, which have to be removed prior to GC in order to prevent column degradation. Organic solvent extractions selectively remove nonpolar substances from the biological sample and are among the most common sample purification procedures for GC assays.

The GC detection techniques are versatile and highly sensitive. The following is a list of common detectors:

Thermoconductivity detector.
Flame ionization detector (FID).
Nitrogen (phosphorus)-sensitive flame ionization detector (N-FID).
Electron capture (EC) detector.
Flame photometric detector (see tectrahydrocannabinol monograph).
Mass spectrometer (see Section 4.5).

The detectors most frequently used in drug level monitoring are the FID, N-FID, and EC detector (Table VIII). While the sensitivity of the FID is high, assay sensitivity is usually limited to injected drug amounts in the nanogram range, since the FID is nonselective and measures any organic compound capable of flame ionization. Thus FID sensitivity is background noise limited, normally providing an usable assay range above 100 ng drug/ml serum. However, some lipophilic basic volatile drugs can be effectively purified and separated on the column and are therefore measurable at concentrations of $\lesssim$10 ng/ml serum (meperidine, nicotine). The N-FID, while inherently no more sensitive than the regular FID, yields greater assay sensitivity for nitrogenous drugs in biological samples because of greater selectivity (anticonvulsants, cocaine, 5-fluorouracil, nicotine). Electron capture detection is not only more sensitive than the FID but also more selective than the N-FID, in this case for electron affinity. Only a few drugs possess sufficient electron affinity to permit sensitive EC detection (acetazolamide, chlorpromazine—but not promazine, clonazepam, isosorbide dinitrate, nitrofurantoin, etc.). However, the polyfluorinated derivatizing reagents convey excellent sensitivity toward EC detection as well as suitable GC characteristics of the derivative; these reagents are frequently employed for drugs with low therapeutic or active plasma concentrations ($\leq$10 ng/ml) (clonidine, cocaine, digoxin, tetrahydrocannabinol). The lower limit for detection of a drug in serum may be as little as 100 to 500 pg/ml (clonazepam, clonidine, morphine).

Because of the many steps involved in GC sample preparation, an internal standard is needed to correct for extraction losses and injection inaccuracies. An internal standard should be a close chemical congener subject to the same derivatization reaction. It also should be added at a concentration similar to that of the analyte and well separated not only from the drug but also from other interfering peaks. For instance, a drug with an alkyl ($\geq C_2$) side chain may serve as the internal standard for a drug containing a methyl group (cocaine, lidocaine, methaqualone). Selection of a suitable internal standard contributes to the overall assay performance and is therefore discussed in many of the drug assay monographs in Chapter 5.

Since the success of GC assays is also dependent on instrumental parameters, sample injection, and a number of analytical "tricks," the following paragraphs give a brief technical review of GC analysis.

Instrumentation. The basic GC system consists of the following components: carrier gas, injector, column oven, column, detector, electrometer, and recorder. Furthermore, automated and computerized GC analysis requires an autosampler, data processor, and controller with automatic printout.

Carrier Gas. The carrier gas must be inert to minimize sample degradation at the high GC temperatures. For most applications nitrogen, helium, or argon is a suitable carrier gas. Nitrogen gas cannot be used for the nitrogen-selective detector. Trace amounts of water or air which may be impurities in the carrier gas are detrimental to GC analysis and column life; therefore a water and oxygen scavenger should be connected to the gas regulator. These gas purifiers also filter any trace corrosives and pump oil that may be in the gas tanks. Most gas chromatographs have a low-pressure controller built into the machine, and a suitable two-stage regulator is required to deliver gases from pressurized tanks at a manageable pressure of about 40 psig. Gas flow must be monitored by a rotameter or by the simple "soap bubble" method, using a 10 ml volumetrical pipet.

Proper operation of the gas chromatograph demands a leak-free system. Small leaks (e.g., at the column fittings), which can be detected with soapy water or high boiling leak detector fluids, drastically reduce column performance.

Injector. The injection port consists of a metal tube in an independent heater assembly and a septum, which seals the exterior end of the assembly. The GC column tubing fits inside the injector tube with a small gap between the injector wall and the column tube so that carrier gas can flow into the column. The column is sealed into the injector by a swagelok fitting on the inside wall of the oven, which must be gastight. Liquid samples (<10 μl) are injected through a septum secured by an assembly at the open end of the injector. The septum usually consists of a soft rubber or silicone pellet which allows a syringe needle to penetrate without causing a gas leak when the syringe is withdrawn. It is important to select a low bleed, high temperature septum. High septum bleed may cause ghost peaks, detector contamination, and a yellow burning flame. Septa can be conditioned inside the column oven for several days before use in order to evaporate the softeners, which may interfere with chromatography.

Proper injection technique is important for good precision and accuracy. The sample, which is usually extracted into a volatile organic solvent such as carbon disulfide or dichloromethane, should be injected as a plug, that is, it should not linger in the injection module. Broad or tailing peaks and shortened column life will result from improper injection. The operator should develop a style and repeat that style for each injection. A recommended method for the typical Hamilton 701N syringe, which ensures injection reproducibility, is as follows:

1. Hold the syringe by the barrel, being careful not to touch the plunger or the needle.
2. Draw the sample into the syringe, and pump the plunger a few times to force out any air bubble.
3. Guide the syringe with the barrel, and press through the septum. Quickly inject the sample, touching only the syringe plunger button.
4. Hold the syringe in the injector for a count of 2, and then quickly withdraw the syringe.

Oven. The column oven contains a heating element, thermostat, temperature readout, and sufficient insulation to maintain a regulated temperature within the chamber. The oven must be large enough to fit the various columns, which are most commonly coiled or U-shaped. Retention times of the compounds are dependent on column temperature: the higher the temperature, the shorter the retention time, and vice versa. Temperature gradients can be used to perform difficult separations or speed analysis time for mixtures of compounds with greatly different vapor pressures.

Columns. The most widely used columns consist of open glass tubes, 3 to 9 ft long, containing a solid support material coated with the liquid phase. Although many of the versatile silicone liquid phases, for example, SE-30 and OV-17, are available in prepacked columns, it is advisable to learn the packing procedure in any GC laboratory.

Newly packed columns need to be conditioned at elevated temperatures to bleed off any volatile contaminants that may be present in the packing or absorbed onto the glassware. The procedure is as follows. Connect the inlet side of the column to the injector, and let the carrier gas flow through the column. To prevent detector contamination, leave the column unattached to the detector. Condition the column at 50°C above the expected operating temperature range for 6 hr or overnight. Do not exceed the maximum temperature limit set for the packing material (OV-17: 350°C; Dexsil 300: >400°C). A good practice for silicone phases is to inject 10 or 20 μl of a silylating agent such as BSTFA to derivatize any free amine or hydroxyl groups that will cause peak tailing and irreversible adsorption of the sample to the column. Prepacked columns purchased from the manufacturers must also be conditioned. After conditioning overnight, cool the oven to room temperature and connect the column to the detector. Check with a leak detector solution for any leaks. Often a connector will need to be tightened after overnight conditioning. Once the column is connected and ascertained to be leakproof, it is ready for analysis.

After the column has been conditioned and made ready for use, column efficiency should be determined and expressed in theoretical plates, which are applicable to all chromatographic systems. Column efficiency is normalized by relating the number of theoretical plates to the length of the column for comparison of columns of varying length. To measure column efficiency, make a solution of 1 μg/μl cholesterol or acenapthene in dichloromethane. Set the column temperature so that the compound elutes in 10 to 20 min. Inject 1 μl, and measure the retention time in minutes from the point of injection to the top of the peak. Next, measure the width of the bottom of the peak. The equation for determining column efficiency is

$$\text{Efficiency} = 16 \times \left(\frac{\text{Retention Time}}{\text{Peak Width}}\right)^2 = N \text{ (theoretical plates)}$$

The equation applies only to GC analysis in the isothermal mode. Good efficiencies for packed GC columns are 1000 to 3000 theoretical plates per meter. Capillary columns have excellent efficiencies ranging up to a total of 300,000 plates, but they are much longer than packed columns, so that the number of plates per meter is only slightly greater. Efficiencies and peak quality should be checked for all columns to assure the quality of chromatography and to determine the life of the column.

Selecting a column packing for drug analysis is generally not a difficult problem, since, of hundreds of packing materials, only a few are sufficient for most drug analyses. Usually, low percentage loadings of the high boiling silicones, such as OV-1, OV-101, SE-30, OV-17, and OV-225, yield optimal results. Occasionally a more polar phase such as FFAP (free fatty acids), EGA, or Carbowax may be needed. The McReynold constants characterize the liquid phase by reflecting the retentions of several standard compounds at specified constant chromatographic parameters. Tables of McReynold constants are found in most catalogs of GC packing materials. Caution must be exercised, however, because McReynold constants apply only to retention and do not indicate degree of peak tailing, irreversible absorption, and the other factors that have to be considered in column selection.

Detectors. Thermal conductivity detectors are not sensitive enough for monitoring most drug levels. The FID, N-FID, and EC detector are discussed below.

Flame Ionization Detector. The eluting carrier gas is mixed with hydrogen gas, and the mixture is ignited over a jet, called the flame tip, in the presence of air or pure oxygen. A large voltage is placed across the flame tip, and a cylindrical ring positioned above and around the flame. Compounds that are mixed with the carrier gas become ionized in the flame. As ionized molecules travel between the flame tip and the collector, a current is produced which is detected and amplified by an electrometer. The FID responds to all organic molecules, and for this reason is called a universal detector. There are some important exceptions to this rule: water, carbon dioxide, nitrogen, and helium give no response in the FID. Furthermore, carbon disulfide gives only a small initial solvent peak when compared with other common solvents such as chloroform and heptane. Therefore carbon disulfide is the recommended sample solvent when practical. Flame ionization detectors have an exceptionally linear response; linearity with respect to mass burned in the flame is over 10,000-fold. Large solvent peaks may represent a problem in GC-FID analysis.

Nitrogen-Sensitive Detector. The N-FID or alkali flame ionization detector is similar to the FID in requiring a hydrogen-air flame. The flame is composed of rubidium sulfate or cesium bromide. For optimization of selectivity the oxidizer flow to the flame is decreased, and the hydrogen flow is maximized for selectivity of nitrogen containing compounds. In the conventional flame detector, the oxidizer is high and the hydrogen flow is maximized for sensitivity. In a properly adjusted N-FID the solvent peak influences a smaller portion of the chromatogram than do solvent peaks obtained with the FID. The N-FID is sensitive to changes in hydrogen and air flow, necessitating the use of accurate gas control valves.

Electron Capture Detector. The EC detector depends on a radioactive foil made of ^{63}Ni or ^{3}H bound to titanium. The radioactive foil emits beta particles that continuously ionize the passing carrier gas, nitrogen or argon-methane. The ionized particles are collected by a pulsing electric field and produce a small standing current, which is measured by an electrometer. If an eluting compound has a sufficiently high electron affinity, such as compounds containing halogen atoms or a nitro group, electrons are captured by the compound, and the standing current decreases. Since the total decrease in current depends on the number of electrons available, the EC detector is easily overloaded with electron absorbing sample and therefore does not have as wide a range of linearity as the FID. Silylation reagents coat the radioactive foil and sharply decrease the number of electrons produced. Therefore silylation can be utilized only if the silylating reagent is completely removed, for instance, by evaporation, before injection onto the column. Many other techniques must be strictly followed to prevent contamination of the detector.

Data Processing. All GC detectors transduce the effect of mass into a usable electrical signal, which is usually small, often as little as 10^{-12} A, and is detected by a very sensitive, yet noise-free, electrometer. The electrometer amplifies the detector signal so that a recorder or data processing system can further convert the signal into a usable or storable data form.

The strip chart recorder is the most common and least expensive data storage device available. Quantitation is usually achieved by measuring peak heights from peak to baseline unless peaks are very wide or asymmetical. Other manual methods of quantitation include disk integration, peak triangulation, cutting and weighing, and planimetry.

Electronic peak integrators have been used for many years with excellent accuracy and precision. The recent generations of gas chromatographs incorporate integrator, data processor, and printer for on-line calculations of concentration, baseline, and retention time. With the design of automatic sampling devices, all aspects of gas chromatography are now automated with the exception of preliminary sample processing, resulting in increased utility of GC analysis for routine use.

Future role of Gas Chromatography in Drug Level Monitoring. Although GC analysis has been somewhat overshadowed by recent HPLC developments, there now seems to be a renewed interest in GC. During 1978–1979 several drug assays were published that are at least comparable to those obtained with HPLC in speed, sensitivity, precision, and specificity for therapeutic drug level monitoring (see monographs on benzodiazepines, theophyllin, lidocaine, etc.). These excellent GC assays are often based on rapid microextraction techniques and the selection of capillary columns, rather than packed columns, and the N-FID, rather than the FID. In particular, capillary GC may be increasingly used for drug assays in biological specimens.

Bibliography

Ahuja, S., Review: Derivatization in Gas Chromatography, *J. Pharm. Sci.*, **65**, 163–182 (1976).

Bobbitt, J. M., A. E. Schwarting, and R. J. Gritta, *Introduction to Chromatography*, Von Nostrand Rheinhold Co., New York, 1968.

Crippen, R., *Identification of Organic Compounds with the Aid of Gas Chromatography*, McGraw-Hill Book Co., New York, Toronto, London, 1973.

David, D. J., *Gas Chromatographic Detectors*, John Wiley and Sons, New York, 1974.

Donaghey, L. F., G. M. Bobba, and D. Jacobs, A Microcomputer System for Real-Time Monitoring and Control of Gas Chromatographs, *J. Chromatogr. Sci.*, **14**, 274–278 (1976).

Eik-Nes, K. B. and E. C. Horning, *Gas Phase Chromatography of Steroids*, Springer-Verlag, Berlin, Heidelberg, New York, 1968.

Finkle, B. S., E. J. Cherry, and D. M. Taylor, A GLC Based System for the Detection of Poisons, Drugs and Human Metabolites Encountered in Forensic Toxicology, *J. Chromatogr. Sci.*, **9**, 393–419 (1971).

Grob, R. L., *Modern Practice of Gas Chromatography*, John Wiley and Sons, New York, London, Sidney, Toronto, 1977.

Haken, J. K., Retention Indices in Gas Chromatography, *Adv. Chromatogr.*, **14**, 367–407 (1976).

McNair, H. M. and E. J. Bonelli, *Basic Gas Chromatography*, 5th edition, Varian Aerograph, Walnut Creek, Calif. 1969.

Mitruka, B. M., *Gas Chromatographic Applications in Microbiology and Medicine*, John Wiley and Sons, New York, London, Sidney, Toronto, 1975.

Moffat, A. C., Use of SE-30 in Drug Analysis (Review), *J. Chromatogr.*, **113**, 69–95 (1975).

Nicholson, J. D., Derivative Formation in the Quantitative Gas Chromatographic Analysis of Pharmaceuticals: Parts I and II, *Analyst,* **103,** 1–28, 192–222 (1977).

Nogare, S. D. and R. S. Juvet, Jr., *Gas-Liquid Chromatography: Theory and Practice,* Interscience Publishers, New York, 1962.

Novak, J., Quantitative Analysis by Gas Chromatography, *Adv. Chromatogr.,* **13,** 1–71 (1975).

Parris, N. A., *Instrumental Liquid Chromatography,* Elsevier Sci. Publ. Co., Amsterdam, 1976.

Riedman, M., Gas Chromatographic Analysis of Drugs and Drug Metabolites in Biological Samples, *Xenobiotica,* **3,** 411–434 (1973).

Scott, R. P. W., Determination of the Optimum Conditions to Effect a Separation by Gas Chromatography, *Adv. Chromatogr.,* **9,** 193–214 (1970).

Sevcik, J., *Detectors in Gas Chromatography,* Elsevier Sci. Publ. Co., Amsterdam, 1976.

Van den Heuvel, W. J., Gas-Liquid Chromatography in Drug Analysis, *Adv. Chromatogr.,* **13,** 265–303 (1975).

Vink, H., Contributions to the Theory of Chromatography, *J. Chromatogr.,* **135,** 1–12 (1977).

Yancey, J. A., *Guide to Stationary Phases for Gas Chromatography,* Analabs, New Haven, Conn., 1977.

4.5. MASS SPECTROMETRY

Mass spectrometry ranks among the most powerful analytical techniques, with important applications in the biomedical area. Its mode of detection is not only highly selective but also sensitive. Under certain conditions, MS can be utilized for the quantitative analysis of a great variety of compounds. Quantitative MS assays in biological fluids are now available for a significant number of the selected drugs (53%, Table III). Pertinent analytical parameters of these assays are summarized in Table IX.

Because of the technical complexity, we can present only a rather scant review of the field. However, there are but a few important parameters that need to be considered in the selection of a suitable MS technique. These include chemical derivatization, modes of sample introduction, ionization, mass spectral detection, and the use of internal standards, preferably stable isotopically labeled drugs (see Table IX).

A schematic of mass spectral analysis is presented in Fig. 6. Mass spectral analysis depends on the formation and separation *in vacuo* of ionized drug molecules. Separation of the ions can be achieved in

Table IX. Mass Spectrometric Assays

Drug	Derivatization	Insertion	Ionization	Selected Ion Monitoring	Stable Isotope
Acetaminophen	Silylation, *O*-methylation	GC	EI, CI	+	D_3
Aminopyrine	None	GC	EI	+	D_3
Amphetamine	Acylation (asymmetric reagent)	GC	EI, CI	+	D_2, D_3
Anticonvulsants	Methylation (CH_2N_2), silylation	GC, direct	CI, API		^{13}C
BCNU	None	Direct	CI	+	D_8
Barbiturates	Flash methylation	GC	EI	+	D_6
Carbamazepine	None	GC	EI	+	
Chloramphenicol	Silylation	GC	EI	+	
Chlorpromazine	None	GC	EI	+	D
Clonazepam	None	GC	CI	+	^{15}N
Clonidine	Flash methylation	GC	EI	+	D_4
Cocaine	None	GC	EI	+	
Cyclophosphamide	Trifluoroacetylation	GC	EI	+	D_4
Diazoxide	Methylation (CH_2N_2)	GC	EI	+	D_3
Digoxin	None	Direct	EI		
Doxorubicin	None; hydrolysis, silylation, and methoxime formation	GC	FD, EI		
Ethambutol	Silylation	GC	CI		D_4
Ethynylestradiol	Silylation	GC	EI		D_2
5-Fluorouracil	Silylation, methylation	GC	EI	+	^{14}C, ^{15}N
Hydralazine		GC	EI		D
Hydrocortisone	Methoxylamine formation silylation	GC	EI	+	^{14}C

Imipramine	Acylation (desipramine)	GC	EI, CI	+	D
		Direct	FI		D_6
Indomethacin	Methylation (metabolite)	GC	EI	+	
Lidocaine	None; propylation (metabolites)	GC	EI	+	D
		Direct	CI		
6-Mercaptopurine	Extractive methylation	GC	EI	+	D_3
Methadone	None	GC	EI, CI	+	D_3, D_5
Methaqualone	Acylation, silylation (metabolite); none	GC	EI	+	
Methyldopa	Ethylation, pentafluoropropionylation	Direct	FI		D_7
		Direct	CI		D
Morphine	Acylation	GC	EI	+	D_3, $^{13}CD_3$
Nicotine	None	GC	EI	+	D_2
		Direct	API		
Phencyclidine	None	GC	EI, CI	+	D_5
		Membrane inlet	EI		
Phenylbutazone	None	Direct	CI	+	D_2
Phenytoin	Methylation, silylation	GC	EI	+	D
Procainamide	None	GC	EI	+	D
Propoxyphene	None	GC	EI	+	D_2/D_7 (d, l)
Propranolol	Acylation	GC	EI	+	D_2 (d, l)
Quinidine	GC direct	EI	+		
		CI			D_2
		GC	EI		D_2, D_3
Tetrahydrocannabinol	None; methylation			+	
Theophylline	Alkylation	GC	EI		D
Tolbutamide	Methylation (CH_2N_2)	Direct	CI		D_2

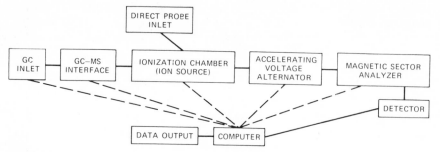

Fig. 6. Schematics of a mass spectrometer system. A single-focusing, magnetic sector mass spectrometer is taken as the example.

various ways, including passage through magnetic and electric fields, quadrupole mass analyzers, or time-of-flight analysis. Mass units are expressed as mass over charge units (m/e), since separation of ions depends on both mass and charge, usually 1 eV. However, doubly charged or multiply charged ions usually also occur and appear at 1/2 or $1/n$ m/e units. Usually, MS analysis is performed on positively charged ions; however, negative ion MS can also be useful and may be considerably more sensitive than positive ion MS under certain conditions. The degree of resolution and the precision of the mass measurement determine whether one achieves resolution of nominal mass units only (e.g., between m/e 300 and 301, low resolution MS) or resolution of ions with the same nominal mass but different empirical formulas (e.g., resolution in m/e 300.01 and 300.02, high resolution MS). All of the MS assays cited in Chapter 5 employ low resolution MS; while high resolution MS may also be useful in drug level monitoring, it is technically more demanding and should be employed for special problems only.

Chemical Derivatization. Chemical derivatization of drugs for MS analysis serves three purposes:

1. Enhancement of the vapor pressure may be necessary, since most MS analyses are performed on compounds in the vapor phase. Reduction of the inlet probe temperature required for analysis may prevent extensive thermolysis of derivatives with high vapor pressures.
2. Derivatizing reagents may direct the fragmentation process during ionization to produce abundant ion fragments at m/e values that are free of interferences from extraneous compounds.
3. Since many MS assays are performed in combination with GC, derivatization may serve to produce suitable GC characteristics.

Usually, a combination of these purposes is applied to the selection of a chemical derivatization mode. (See also Sections 4.2.3 and 4.4.)

Inlet Systems. The major modes of sample introduction are by direct insertion through a vacuum trap or via a GC column through a suitable high vacuum interface. Occasionally, other modes may also be employed, mainly for qualitative purposes [membrane inlet (phencyclidine), HPLC inlet systems, etc.]. The inlet system must be tailored to the ionization mode. First, the required ionization pressure may range from 10^{-5} torr to atmospheric pressure; the GC interface and vacuum pumping systems are designed to achieve the required pressure. The GC carrier gas is normally helium, which can be readily separated from the higher molecular weight drugs by diffusion processes in the GC-MS interface. Second, when introducing complex compound mixtures directly into the ion source, significant ion peaks are likely to appear at essentially every mass unit, unless the formation of fragment ions is suppressed. Therefore quantitative drug analysis in biological samples by direct insertion requires mild ionization modes, such as chemical and field ionization. Mass spectral analysis in conjunction with highly efficient separations (GC-MS) can be performed with electron impact ionization, which often produces many fragment ions.

Ionization. The mode of ionization is a key factor in MS analysis. This area is still rapidly developing and promises to broaden MS applicability to compounds of high molecular weight and very low vapor pressure. The principal modes of ionization in drug level monitoring are electron impact (EI), chemical ionization (CL), field ionization (FI) and field desorption (FD). These modes are listed in the order of decreasing compound fragmentation, based on decreasing energy applied to the ionization process. Atmospheric pressure ionization (API) is another ionization mode causing few fragmentations. The importance of fragment formation to the selection of the insertion mode was discussed in the preceding paragraph. Electron impact utilizes an electron beam, usually with an energy level of 70 eV, which is sufficient to break any chemical bond in organic compounds. Compound fragmentation under electron impact is reproducible, and parent ion (M^+) and fragmentations compose the mass spectrum. In the CI mode the electron beam first ionizes a reactant gas, for example, CH_4, which in turn ionizes the drug by proton charge transfer, a process that is more gentle than EI and causes fewer fragmentations. The fact that the vapor pressure in the ion chamber is higher than that required for EI (1 torr versus 10^{-2} to 10^{-5} torr) is advantageous in the GC-MS mode, where the carrier gas may contain the reactant gas. Mass spectral assays employing direct insertion with CI are available for BCNU, lidocaine, methyldopa, phenylbutazone, and tolbutamide. These assays cannot be performed in the EI mode. Chemical ionization may also be needed if EI does not produce parent ions or abundant fragmentations. Field ionization is based on the ionization of compounds in a high electrostatic field, which conveys still less energy to the compound and therefore causes fewer fragmentations. High sensitivity of any direct insertion MS assay, however, requires extensive sample purification before analysis in order to avoid overloading and interferences (see imipramine and methaqualone monographs). Field desorption ionization occurs at high electrostatic fields with the compound adsorbed to the surface of a specially designed cathode wire. Thus FD ionization partially eliminates the need to vaporize the uncharged drug molecule, which may cause thermolysis prior to ionization. Highly polar drugs of very low vapor pressure yield abundant molecular ions that can serve for drug level monitoring (see doxorubicin monograph). Atmospheric pressure ionization is essentially similar to the electron capture process. Drugs have to possess a relatively high electron affinity for API analysis. Both positive and negative MS is possible, with the negative ion mode showing high sensitivity (anticonvulsants, nicotine).

Detection. Following suitable ion separation (quadrupole filter, magnetic sector), the generated ions are detected by an electrometer, amplified, and recorded. Computerized data analysis is important to achieve optimum utilization of MS data. Normally, the entire m/e mass range is scanned, for instance, by altering the magnetic field, and the mass spectrum recorded (m/e versus ion intensity). This procedure, however, has limited sensitivity, since only a fraction of the analysis time is spent on the ion or ions of interest (e.g., m/e 300). A dramatic sensitivity increase is achieved if the magnetic field is held constant, so that only one preselected ion is focused at the electrometer (m/e 300 alone). If several ions are to be analyzed, rapid peak switching can be obtained by changing the accelerating voltage (accelerating voltage alternator, AVA, Fig. 6). This detection mode, called selected ion monitoring (SIM), or mass fragmentography in earlier reports, is employed in most drug level assays (Table IX) because of its high sensitivity; the specificity of SIM is no different from that of regular scanning MS. Drug assay sensitivity in GC-MS-SIM may be as low as 100 pg/ml serum, but is usually around 1 ng/ml (clonazepan, clonidine, 5-fluorouracil, morphine, nicotine, etc.). An example of a SIM computer record of a GC-MS-SIM assay (morphine) by stable isotope dilution is given in Fig. 7.

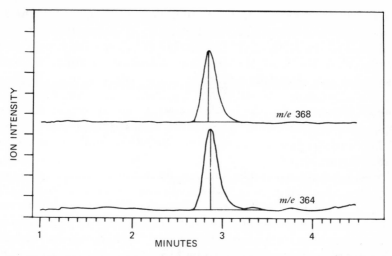

Fig. 7. GC-MS-SIM (selected ion monitoring) record of a rat brain sample 6 hr after administration of 10 mg morphine/kg. The approximate brain level is 20 ng morphine/g. The data were obtained using the assay of Hipps et al. [*J. Pharmacol. Exp. Ther.*, **196,** 642–648 (1976)] as modified in our laboratory (Finn et al., 1976). $N\text{-}^{13}CD_3$-morphine (100 ng) was added to the brain sample before extraction as the internal standard and the analytical carrier. The extracts were derivatized with trifluoroacetic anhydride to give bis(trifluoroacetyl)morphine and analyzed by GC-MS-SIM at *m/e* 364 ($M^+ - [CF_3COO]$ of morphine) and *m/e* 368 ($M - [CF_3COO]$ of $N\text{-}^{13}CD_3$-morphine). The data were plotted on a computer terminal using a smoothing algorithm, baseline adjustment, and scaling (2×) of the *m/e* 364 record for easy inspection. Using the cursor capability of the terminal, an *m/e* 364/368 ratio of 0.317 was obtained, which can be used for quantitation of morphine with a calibration curve. Note that $N\text{-}^{13}CD_3$-morphine has a slightly shorter retention time than morphine (deuterium isotope effect). The solvent peak was eliminated by a GC converter valve. Endogenous brain compounds did not interfere with the assay at this level, attesting to the specificity of GC-MS-SIM.

Quantitation Using Internal Standards. Because of fluctuations in the many processes of MS analysis, proper choice of internal standards is mandatory to achieve high precision. Chemical analogs may be useful as internal standards (carbamazepine, doxorubicin, indomethacin, lidocaine, methaqualone), as discussed for GC analysis. However, drugs labeled with stable isotopes are superior as internal standards, since they behave chemically like the drug to be analyzed but can be separately measured by MS analysis. After addition of a known amount of the labeled drug to the biological specimen and suitable extraction, selected ions of the unlabeled unknown and the labeled internal standard are measured, and the ratio of the relative intensities of these two ions is used to calculate the unknown drug concentration (isotope dilution analysis). Deuterium isotopes may affect the GC behavior of some drugs (see morphine, Fig. 7), a fact that should be considered in the analysis. It is desirable to incorporate several stable isotopic nuclides into the internal standard in order to avoid interferences by naturally occurring stable isotopes (e.g., ^{13}C, ~1% natural abundance). Multilabeled standards are mandatory if the internal standard simultaneously serves as an analytical carrier in higher amounts to avoid irreversible losses during isolation and GC-MS analysis (see monographs on BCNU, clonidine, methadone, methaqualone, etc.).

Drugs labeled with stable isotopes can also be administered to animals or human subjects in order to determine drug kinetics during steady-state conditions under routine maintenance dosing (phenytoin). Furthermore, one isomer of a racemic mixture may be labeled with a stable isotope (pseudoracemic mixture) in order to determine the metabolic fates of both isomers *in vivo* by MS analysis (e.g., propoxyphene, propranolol). Biological stable isotope effects, however, may be rather large for deuterium, and the isotopic label has to be placed in a metabolically inert site of the drug molecule.

Occasionally, ^{14}C can be substituted for stable isotopes, since it has a rather long decay half-life (5-fluorouracil, hydrocortisone). Sample amounts injected have to be small in order to avoid radioactive contamination.

Future Developments. Following a very active period in the development of MS analysis in drug level monitoring, progress has recently slowed because of the advent of technically less demanding alternative techniques. However, the use of stable isotopes in special research problems will continue to provide many applications for MS assays in drug level monitoring. Furthermore, routine analysis may become feasible if specialized MS detectors for limited purposes are available at reasonable costs.

Bibliography

Budzikiewicz, H., C. Djerassi, and D. H. Williams, *Structure Elucidation of Natural Products by Mass Spectrometry*, Holden-Day, San Francisco, London, Amsterdam, 1964.

Carrington, R. and A. Frigerio, Review: Mass Fragmentography in Drug Research, *Drug Metab. Rev.*, **6**, 243–282 (1974).

Costa, E. and B. Holmstedt, Gas Chromatography—Mass Spectrometry in Neurobiology, in *Advances in Biochemical Psychopharmacology*, Vol. 7, Raven Press, New York, 1973.

Frigerio, A., *Essential Aspects of Mass Spectrometry*, Spectrum Publ. Co., Flushing, N.Y., 1974.

Frigerio, A., *Recent Developments in Mass Spectrometry in Biochemistry and Medicine*, Vol. 7, Plenum Press, New York, London, 1978.

Frigerio, A. and N. Castagnoli, *Advances in Mass Spectrometry in Biochemistry and Medicine*, Vol. I, Spectrum Publ. Co., Flushing, N.Y., 1976.

Gross, M. L., *High Performance Mass Spectrometry*, ACS Symposium Series 70, American Chemical Society, Washington, D.C., 1978.

McFadden, W., *Techniques of Combined Gas Chromatography/Mass Spectrometry: Applications in Organic Analysis*, Wiley-Interscience, New York, London, Sydney, Toronto, 1973.

McLafferty, F. W., *Interpretation of Mass Spectra*, W. A. Benjamin, London, 1973.

Porter, Q. N. and J. Baldas, *Mass Spectrometry of Heterocyclic Compounds*, Wiley-Interscience, New York, London, Sidney, Toronto, 1971.

Tolgyessy, J., T. Braun, and T. Kyrs, *Isotope Dilution Analysis*, Pergamon Press, Oxford, 1972.

Waller, G. R., *Biochemical Applications of Mass Spectrometry*, Wiley-Interscience, New York, London, Sydney, Toronto, 1972.

Watson, J. T., *Introduction to Mass Spectrometry: Biomedical, Environmental and Forensic Applications*, Raven Press, New York, 1976.

Williams, D., and I. Howe, *Principles of Organic Mass Spectrometry*, McGraw-Hill Book Co., New York, Toronto, London, 1973.

4.6. THIN-LAYER CHROMATOGRAPHY

Quantitative TLC procedures for the analysis of drug levels in biological fluids are available for 32% of the selected drugs. Many of the reports on TLC assays have been published recently, indicating that TLC is currently an important method in drug level monitoring. Table X lists the drugs that can be assayed by TLC and summarizes pertinent information on the mode of detection after separation.

Thin-layer chromatography is a separation method in which uniform thin layers of sorbent material or other selected media serve as the stationary phase. In principle, TLC separations follow the same rules as HPLC; however, the mobile phase is drawn into the sorbent layer by capillary action, a process called development of the TLC plate. The sorbent layer is usually coated onto a support plate of glass or plastic material. The most common sorbents are silica gel, alumina, kieselguhr (diatomaceous earth), and cellulose. Nearly all of the quantitative TLC assays listed in Table X utilize silica gel plates. These plates often

Table X. Thin-Layer Chromatography (TLC)

Drug	Reactions	Detection
Aminopyrine	(1) HClO₄	Fluorescence scanning
	(2) Elution	UV
Anticonvulsants	—	UV scanning
Barbiturates	(1) Elution	UV or fluorescence in NaOH
	(2) Dansylation	Fluorescence scanning
Carbamazepine	—	UV scanning
Chlordiazepoxide	H₂SO₄	Fluorescence scanning
Chloroquine	(1) Elution	Fluorimetry
	(2) —	UV densitometry
Chlorpromazine	(1) —	UV densitometry
	(2) Dansylation of N-dealkylated metabolites, elution	Fluorescence
Clofibrate	9-Bromomethylacridine	Fluorescence
Clonazepam	(1) NO₂ → NH₂, Bratton-Marshall	Colorimetry scanning
	(2) —	UV scanning
Cocaine	Color reactions	Semiquantitative
Diazepam	—	UV scanning
Digoxin	Xanthydrol (2-deoxy sugars)	Colorimetry
Diphenhydramine	Color reactions	Semiquantitative
Doxorubicine	—	Fluorescence scanning
Imipramine	Oxidation	UV scanning
Methadone	Iodoplatinate	Semiquantitative
Metronidazole	—	UV scanning
Oxazepam	(1) —	UV scanning
	(2) HCl hydrolysis, Bratton-Marshall	Colorimetry
Penicillins	Acid or base treatment	Fluorescence scanning
Propranolol	—	UV and fluorescence scanning
Clonidine	—	UV and fluorescence scanning
Reserpine	Oxidation under acid conditions	Fluorescence scanning
Salicylic acid	Ferric chloride	Colorimetric scanning
Succinylcholine	Elution spray reagents	Semiquantitative
Tetrahydrocannabinol	2-p-Chloro-5-sulfo-phenyl-3-phenylindone	Fluorescence scanning

contain a binder, such as calcium sulfate hemihydrate, and fluorescent substances. Sodium fluorescein-impregnated TLC plates fluoresce with a greenish yellow color when exposed to 254 nm UV light. Any 254 nm UV absorbing substance on the plate quenches the fluorescence and appears as a dark spot under the UV lamp. Binder and fluorescence indicator may interfere with some assays, necessitating the preparation of pure silica gel plates.

Separation of compound mixtures by silica gel TLC is similar to that achieved by silica gel HPLC. However, there are several technical differences that are important in quantitative drug analysis. First, polar compounds stay close to the origin of the silica gel TLC plate and do not interfere with the analysis after plate development. In contrast, polar compounds elute from the silica gel HPLC column after the usually more lipophilic drug, and the succeeding analysis has to wait for all compound to elute from the column. Second, several samples can be run simultaneously on a TLC plate, which usually has a size of 20 × 20 cm. Third, analytical TLC plates have a thickness of the sorbent layer of only 100 to 250 μ; therefore plate overloading, which results in broad tailing spots not suitable for quantitative analysis, may readily occur. Biological samples require more clean-up procedures for TLC than for HPLC.

The quantitative analysis of separated compounds on a developed TLC plate is versatile. The compound spots can be scraped off the plate, and the drug eluted from the sorbent with suitable solvents, most commonly methanol, and analyzed by any applicable analytical technique. In such a case, TLC merely serves as a chromatographic technique in the preparation of the biological sample for analysis. A few examples have been included in Table X (chloroquine, barbiturates). On the other hand, quantitation can also occur directly on the plate, either semiquantitatively by visual inspection of the compound spots or more precisely by direct spectroscopic methods (densitometry).

Densitometry is conducted by scanning the TLC plate with UV or visible light and measuring the intensity of the reflected light with a spectrophotometer. Light absorbing compounds on the plate reduce the intensity of the reflected light and appear as peaks in the densitometry record. Ultraviolet scanning densitometry can be quite precise (5% C.V.) and sensitive. The UV-TLC assay of propranolol is sufficiently sensitive to measure therapeutic plasma concentrations (>20 ng/ml). Furthermore, some TLC assays with UV scanning include the quantitative detection of several drug metabolites (chlorpromazine, diazepam). A TLC record of diazepam and its metabolites extracted from rat liver microsomal preparations is shown in Fig. 8. Greatly increased sensitivity is achieved with fluorescence densitometry. Some drugs possess high native fluorescence yields and can be analyzed in picomole quantities (doxorubicin, propranolol, quinidine).

Frequently, spray reagents are utilized to produce a colored or fluorescent TLC spot. There are numerous color reagents for the qualitative and semiquantitative detection of most drugs (cocaine,

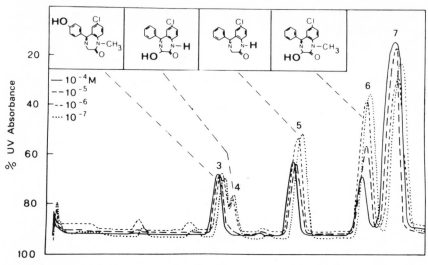

Fig. 8. UV-densitometry of TLC plates after separation of diazepam and its metabolites (silica gel plates, solvent system: chloroform-ethanol-acetic acid, 95 : 5 : 0.5, v/v [Schwandt et al., *N.S. Arch. Pharmacol.*, **294**, 91–98 (1976)]. Diazepam was incubated with a 19,000 g hepatic microsomal preparation from phenobarbital-pretreated rats along with increasing concentrations of the enzyme inhibitor metyrapone (10^{-7} to 10^{-4} M). The incubations were extracted into ether, and the ethereal residues chromatographed and scanned at 320 nm directly on the plates (scans are superimposed). Rapid quantitation of several components down to 0.1 μg/ sample can be achieved.

diphenhydramine, methadone, succinylcholine). The spray reagent also may induce increased UV absorbance (imipramine) or fluorescence (chlordiazepoxide, penicillins, reserpine) for subsequent densitometry. Alternatively, drugs may be derivatized with a fluorescent reagent before TLC separation, most notably dansyl chloride (barbiturates, chlorpromazine metabolites).

We can conclude that quantitative TLC is a versatile, sensitive, and specific technique that can replace HPLC or GC assays in research and clinical laboratories, if the proper instrumentation is available. Systematic technological developments are likely to increase the current level of assay sensitivity by an order of magnitude, thereby making TLC analysis the preferred technique in some biomedical applications. The development of high performance thin-layer chromatography (HPTLC) with very thin layers and small particle size offers the potential of greatly improved separation and sensitivity over conventional TLC.

Bibliography

Kirchner, J. G. and E. S. Perry, *Thin Layer Chromatography*, John Wiley and Sons, New York, 1978.

Scott, R. M., *Clinical Analysis by Thin-Layer Chromatography Techniques*, Ann Arbor-Humphrey Sci. Publ., 1969.

Stahl, E., *Thin-Layer Chromatography; a Laboratory Handbook*, Springer-Verlag, Berlin, New York, and Academic Press, New York, 1965, (translation of *Dünnschicht-Chromatographie*).

Touchstone, J. C., *Quantitative Thin Layer Chromatography*, John Wiley and Sons, New York, 1973.

Touchstone, J. C. and M. F. Dobbins, *Practice of Thin Layer Chromatography*, John Wiley and Sons, New York, 1978.

Zlatkis, A. and R. E. Kaiser, *HPTLC: High Performance Thin-Layer Chromatography*, Elsevier Sci. Publ. Co., Amsterdam, Oxford, New York, 1977.

4.7. SPECTROSCOPIC METHODS

4.7.1. Ultraviolet and Visible Spectroscopy. Ultraviolet and visible light occupy a fraction of the broad spectrum of electromagnetic radiation. Photons interact with organic molecules by exciting electrons to higher energy levels. Electrons that form saturated bonds (σ) absorb photons of high energy in the vacuum UV range between 120 and 180 nm wavelengths. Nonbonding electrons (n), for instance, the unshared electrons of the heteroatoms nitrogen, oxygen, and sulfur, and electrons in unsaturated bonds (π) absorb UV light above 180 nm. Chemical functions containing n or π electrons are called chromophores or auxochromes (e.g., OH) and maximally absorb in the range of 180 to 200 nm. However, if several chromophores are connected via one saturated bond each, they are conjugated, and photons of lower energy are capable of exciting such a chromophore system. The longer the conjugated system, the longer the wavelength of maximum absorption (bathochromic effect of conjugation). An example is given below:

Chromophore: C=O   //\C=O   //\/\C=O   //\/\/\C=O
Absorption maximum (nm):   ~180–190   ~240   ~280   ~340
(Exact location depends on the complete chemical structure.)

The UV range extends from 180 to 400 nm, while the visible range has wavelengths between 400 and 700 nm. A series of chromophores has to be conjugated in order to shift the absorption maximum into the visible range. It is therefore not surprising that the majority of endogenous substrates and most drugs do not absorb in the visible range. For instance, serum which contains thousands of compounds is normally pale yellow, but absorbs strongly UV light below 300 nm. We can postulate, then, that the selectivity of UV measurements increases with increasing absorption maximum wavelength. The fact that colored compounds can often be measured directly in serum without analytical work-up procedures is frequently exploited in diagnostic medicine (e.g., bromsulfphthalein liver function tests).

For most compounds the intensity of light absorbance is proportional to the compound concentration in solution over some specified range (Beer's law). The following relationships apply:

$$A = \log\left(\frac{I^0}{I}\right), \quad A = \epsilon \cdot l \cdot c$$

where A = absorbance, I^0 = intensity of incident light, I = intensity of transmitted light, ϵ = extinction coefficient (1 cm cuvette, at 1 mol/l), l = path length of absorbing solution (cm), c = compound concentration. Unsaturated ketones usually have ϵ values between 10,000 and 20,000, a range that is sufficient for sensitive UV analysis with a lower detection limit of 0.1 to 1 μg/ml solution. The solvent choice depends on the absorption maximum of the compound and the minimum usable wavelength of the solvent (e.g., 190 nm for water, 204 nm for ethanol, 237 nm for chloroform).

Table XI contains the spectrophotometric drug assays in biological fluids. The listed assays are often nonspecific for the particular drug in the presence of its metabolites, since drug biotransformation normally introduces auxochromic functions that have a limited effect on the location of the absorption maximum. A considerable increase in UV assay selectivity can be achieved by differential UV analysis of bases or acids at two different pH values. The ionized and the nonionized drug usually show different UV spectra, and the difference in absorption intensity at a selected wavelength is proportional to the drug concentration (barbiturates, phenylbutazone, salicylic acid) (Table XI).

4.7.2. Colorimetry. Most drugs can be derivatized or chemically reacted to form a colored product that can be detected in the presence of noncolored compounds in biological samples. This procedure is called colorimetry. The specificity of colorimetric assays is determined by the selectivity of the chemical color reaction. Like the spectrophotometric methods (Section 4.7.1), colorimetry is frequently nonspecific for the parent drug in the presence of its metabolites. Furthermore, similar

Table XI. Ultraviolet Spectroscopy (UV)

Drug	Reactions/procedure
Acetaminophen	In NaOH
Acetazolamide	In HCl
Anticonvulsants	In NaOH
Barbiturates	Differential UV absorbance in NaOH and NH_4Cl
Carbamazepine	HCl $\rightarrow$ 9-methylacridine
Chlorpromazine	Oxidation with Co(III)
Clofibrate	At 226 nm
Diatrizoate sodium	At 238 nm
Diazoxide	In Na_2CO_3 solution
Diphenhydramine	In HCl (after distillation)
6-Mercaptopurine	Column chromatography
Methadone	Oxidation with ceric sulfate to benzophenone
Metronidazole	In ethanol-water
Penicillins	Acid or base treatment (amoxicillin)
Phenylbutazone	Dual wavelength at various pH values, oxidation to azobenzene
Phenytoin	Permanganate oxidation to benzophenone
Propoxyphene	Steam distillation after chemical degradation
Salicylic acid	Differential UV absorbance at varying pH
Theophylline	In NaOH or in HCl
Warfarin	In NaOH

drugs and endogenous substrates with the same functional chemical groups are likely to interfere. Sensitivity is usually no better than 0.1 μg drug/ml biological fluid.

Table XII lists the colorimetric drug assays suitable for biological specimens, along with the type of chemical color reaction. The Bratton-Marshall reaction forms the basis for one of the most frequent colorimetric assays. It is applied to the assay of primary aromatic amines (chlorodiazepoxide after hydrolysis, procainamide, sulfonamides), but can also be adapted to aromatic nitro compounds after quantitative reduction of the corresponding aromatic amines (e.g., chloramphenicol, nitrazepam, a clonazepam analog). Phenyl-substituted drugs can be nitrated and similarly treated (anticonvulsants, e.g., phenytoin). Moreover, some drugs release nitrous acid upon acid treatment, which is another component of the Bratton-Marshall reaction (BCNU, metronidazole). Finally, aromatic compounds capable of coupling to diazonium ions to form azo dyes are equally detectable (phenols, 6-mercaptopurine). The Bratton-Marshall reaction of sulfon-

Table XII. Colorimetry (Col)

Drug	Reactions/procedures
Acetaminophen	HCl → p-aminophenol, + various reagents
Amphetamine	Various reagents
Anticonvulsants	Aromatic nitration, NO_2 → NH_2, Bratton-Marshall
BCNU	HCl → HNO_2, Bratton-Marshall
Chloramphenicol	(1) + isoniazid → yellow
	(2) NO_2 → NH_2, Bratton-Marshall
Chlordiazepoxide	HCl → 2-amino-6-chlorobenzophenone, Bratton-Marshall
Chloroquine	Eosin ion-pair extraction, various reagents
Chlorpromazine	Oxidation with Fe(III) in acids → free radicals
Cyclophosphamide	γ-(p-Nitrobenzyl)pyridine (alkylating species)
Diatrizoate sodium	Ashing, discoloration of ceric ammonium sulfate solution
Diphenydramine	Bromthymol blue ion-pair extraction
Ethambutol	Bromthymol blue ion-pair extraction
Hydralazine	Hydrazone formation
Hydrocortisone	Porter-Silber reaction
Isoniazid	Various reagents
Melphalan	γ-(4-Nitrobenzyl)pyridine (alkylating species)
Meperidine	Methyl orange ion-pair extraction
6-Mercaptopurine	Reduction, Bratton-Marshall
Metronidazole	HCl → HNO_2, Bratton-Marshall
Nitrofurantoin	(1) Hydrolysis → 2-nitro-2-furaldehyde → phenylhydrazone
	(2) Nitromethane-hyamine ion-pair extraction
Phenytoin	Aromatic nitration, NO_2 → NH_2, Bratton-Marshall
Procainamide	Bratton-Marshall (N-acetylprocainamide after acid hydrolysis to procainamide)
Salicylic acid	Iron complexes (Trinder reagent)
Succinylcholine	Ion-pair extractions
Sulfonamides	(1) Bratton-Marshall
	(2) 9-Chloroacridine
Tolbutamide	Dinitrofluorobenzene (carboxytolbutamide)

amides is shown in Fig. 9 as an example; this colorimetric assay, albeit nonspecific, is still used in many clinical laboratories with adequate results for therapeutic monitoring of sulfonamide serum concentrations.

Another widely applicable, yet rather nonspecific colorimetric method consists of ion complex formation with dyes, followed by organic ion-pair extraction and spectrometric analysis of the organic phase (chloroquine, diphenhydramine, ethambutol, meperidine, nitrofurantoin, succinylcholine).

The colorimetric assay of salicylate by Trinder is also nonspecific in the presence of some of the salicylate metabolites. Nevertheless, this assay is clinically useful, since metabolite concentrations are insignificant in the plasma following high doses of salicylates administered in the treatment of rheumatoid arthritis or ingested accidentally by children.

4.7.3. Fluorescence. To understand luminescence phenomena a brief discussion of the molecular events of photon excitation and emission is necessary. Following photon excitation of an organic compound (ground state S_0 to excited singlet state S_2), the gained energy can be dissipated by several mechanisms, including heat and light (luminesce). Light can be emitted by organic molecules either as fluorescence or as phosphorescence. The excited singlet state S_2 (all electrons spin-paired) rapidly falls to its lowest vibrational level through collisions, followed by conversion to a comparable energy level of the next lower singlet state S_1 (internal conversion). After additional vibrational relaxation to the lowest excited singlet state S_1, the molecule can drop back to S_0, producing heat or one photon (fluorescence). It also may undergo intersystem crossing to the corresponding triplet energy state T_1, drop to the lowest T_1 energy state by vibrational relaxation, and emit one photon when converting from T_1 to

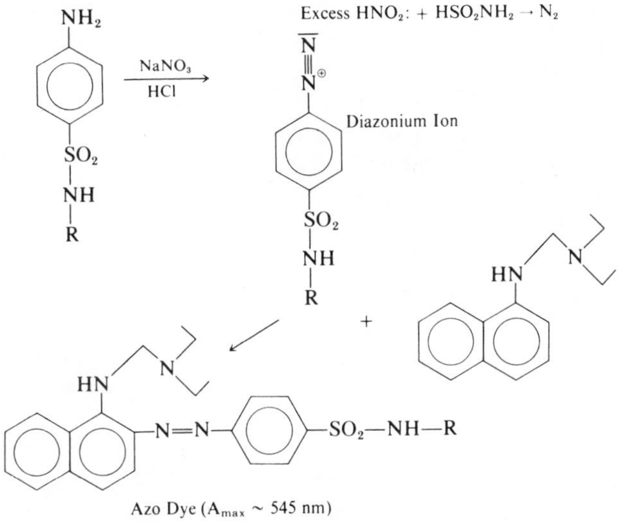

Fig. 9. Chemical structures of the Bratton-Marshall reaction: example, sulfonamides.

S_0 (phosphorescence). The lifetime of the S_1 state is on the order of 10^{-7} to 10^{-9} sec, while the T_1 state has a lifetime of 10^{-6} to 10 sec. Because of the vibrational relaxation, the emitted photon always has a lower energy level (longer wavelength) than the exciting photon. Phosphorescence is rarely used in drug analysis and will not be further discussed. The phosphorimetric assay of sulfonamides at 77°K represents one example of its application (see sulfonamides monograph).

The fluorescence assays for the determination of drugs in biological fluids are summarized in Table XIII. Many drugs do not produce a sufficiently high fluorescence yield to permit their sensitive detection. Therefore chemical reactions, which frequently involve oxidations, serve to introduce an intense fluorescence for a major portion of the drugs listed.

Fluorescence analysis is more selective than UV spectroscopy, since more structural requirements have to be fulfilled to produce a high fluorescence yield (ϕ). For example, the trienone shown in section 4.7.1 has to be embedded into a rigid frame, such as a steroid moiety (see spironolactone monograph). Fluorescence can be manipulated by the solvent, which may favor either the n $\rightarrow$ π^* transition with high fluorescence yield (H_2O, CH_2OH for the trienone) or the $\pi \rightarrow \pi^*$ transition with practically no fluorescence (heptane). Furthermore, fluorescence of bases and acids may be highly pH dependent (chloroquine, salicylate, barbiturates), and the biological fluorescence background may be determined at a pH where the drug does not fluoresce (warfarin). Fluorescence analysis is considerably more sensitive than UV assays because the fluorescent light can be measured above a negligible background, while UV analysis depends on the difference measurement between incident and transmitted light intensity. Fluorescence assays can be sensitive in the range of 1 to 10 ng drug/ml biological fluid, if applied in conjunction with organic solvent extractions (e.g., propranolol, spironolactone). However, drug metabolites and endogenous substrates often display similar fluorescence, thereby limiting the specificity of detection.

4.7.4. Future Trends. The spectroscopic methods are simple and rapid; however, they suffer from lack of specificity. With the exception of the Trinder assay of salicylates and the Bratton-Marshall reaction for sulfonamides, none of these techniques is currently used in therapeutic drug level monitoring. The utility of spectroscopic methods may increase if rapid prepurification steps that compensate for the lack of specificity become available. Extraction of biological samples with microcolumns of high efficiency may indeed provide such a prepurifica-

Table XIII. Fluorescence (Fluor)

Drug	Reactions/procedures
Amphetamine	Aldehydes, ketones, fluorescamine
Chloramphenicol	Zn/HCl: $NO_2 \rightarrow NH_2$ + fluorescamine
Chlordiazepoxide	(1) HCl → 2-amino-6-chlorobenzophenone + fluorescamine
	(2) Photochemical reaction
Chloroquine	At pH 9.5
Doxorubicin	TLC separations
Hydrochlorothiazide	
Hydrocortisone	H_2SO_4 reaction
Indomethacin	In heptane
Isoniazid	Various reactions
Lysergide	In HCl
Melphalan	
Meperidine	CH_2O/H_2SO_4 reaction
6-Mercaptopurine	Permanganate oxidation
Methotrexate	Permanganate oxidation
Methyldopa	Oxidation to dihydroxyindoles
Morphine	Oxidation with $K_3Fe(CN)_6$ to pseudomorphine
Penicillins	Acid or base treatment → diketopiperazines (?)
Phenytoin	Permanganate oxidation to benzophenone + H_2SO_4
Procainamide	(1) In alkaline medium (*N*-acetylprocainamide in acid medium)
	(2) Fluorescamine
Propoxyphene	7-Nitrobenzo-2,1,3-oxadiazole
Propranolol	In HCl
Quinidine	In H_2SO_4 (dilute)
Salicylic acid	At pH 7
Spironolactone	H_2SO_4 → trienones
Sulfonamides	(1) Native fluorescence (weak)
	(2) Fluorescamine
	(3) 9-Chloroacridine fluorescence quenching
Tetracycline	Ternary metal chelate-barbiturate-tetracycline complex
Warfarin	(After TLC separation), at alkaline pH, background at low pH

tion procedure. A significant portion of the selected drugs (67%) can be assayed at therapeutic levels by UV, fluorescence, and colorimetry.

Bibliography

Ultraviolet and Visible Spectroscopy

Dyer, J. R., *Applications of Absorption Spectroscopy of Organic Compounds*, Prentice-Hall, Englewood Cliffs, N.J., 1965.

Rabalais, J. W., *Principles of Ultraviolet Photoelectron Spectroscopy*, John Wiley and Sons, New York, 1977.

Williams, D. H. and I. Fleming, *Spectroscopic Methods in Organic Chemistry*, 2nd edition, McGraw-Hill, London, 1973.

Fluorescence

Bartos, J. and M. Pesez, *Colorimetric and Fluorimetric Analysis of Steroids*, Academic Press, London, New York, 1976.

Elevitch, F. R., *Fluorometric Techniques in Clinical Chemistry*, Little, Brown, and Co., Boston, 1973.

Guilbault, G. G., *Fluorescence: Theory, Instrumentation, and Practice*, Edward Arnold, London, and Marcel Dekker, New York, 1967.

Guilbault, G. G., *Practical Fluorescence: Theory, Methods, and Techniques*, Marcel Dekker, New York, 1973.

White, C. E. and R. J. Argauer, *Fluorescence Analysis: a Practical Approach*, Marcel Dekker, New York, 1970.

Winefordner, J. D., S. G. Schulman, and T. C. O'Haver, *Luminescence Spectrometry in Analytical Chemistry*, Wiley-Interscience, New York, 1972.

Colorimetry

Pesez, M. and J. Bartos, *Colorimetric and Fluorimetric Analysis of Organic Compounds and Drugs*, Marcel Dekker, New York, 1974.

Thomas, L. C. and G. J. Chamberlin, *Colorimetric Chemical Analytical Methods*, 8th edition, John Wiley and Sons, New York, 1974.

Wright, W. D., *The Measurement of Colour*, 3rd edition, Van Nostrand, Princeton, N.J., 1964.

4.8. POLAROGRAPHY

Polarography is based on microelectrolysis, with the generated current proportional to the drug concentration. The general term "voltammetry" is used for all current-voltage recording methods with microelectrodes, while polarography is normally performed with the dropping mercury electrode and the saturated colomel electrode as the auxiliary electrode. The drug potentials are usually given relative to the saturated calomel electrode. Dissolved oxygen is reduced to H_2O_2 at -0.1 V and therefore has to be removed from the sample before polarographic analysis of many drugs. This can be achieved by bubbling nitrogen through the sample. The sensitivity of regular polarography can be as good as 10^{-5} to 10^{-6} M for suitable samples, which is sufficient for some drugs at therapeutic serum levels. A more sensitive technique is pulse polarography; a square wave voltage pulse is applied to the electrode during the end of the mercury drop life at the electrode. The current is measured during the second half of the pulse when the charging current is small. Differential pulse polarography, another variation, employs pulses of constant amplitude superimposed on a slowly increasing linear voltage ramp. The current is measured just before and near the end of the pulse, and the difference plotted. The detection limit of pulse polarography is 10^{-6} to 10^{-7} M, and of differential pulse polarography 10^{-7} to $10^{-8}M$. Because of its superior sensitivity, most of the drugs listed in Table XIV are assayed by differential polarography.

Polarography is selective for chemical functions that can undergo redox reactions under the potentials of the dropping mercury electrode. For instance, the following functions can be readily detected in biological sample extracts: organic nitro groups (chloramphenicol), organic nitrates (isosorbide dinitrate), azomethines (clonazepam), and acetophenone analogs (haloperidol). The same chemical functions may also be present in the drug metabolites, which therefore interfere with the assay of the parent drug. Extensively metabolized drugs (chlordiazepoxide, clonazepam, metronidazole) have to be separated from their metabolites before analysis, for example, by TLC. Some drugs can be determined directly in serum without extractions, depending on the drug concentration and redox potential (penicillins after ultrafiltration, acetaminophen using differential pulse voltammetry at a carbon paste electrode).

Table XIV. Drugs That Can Be Assayed by Polarography

Acetaminophen	Clonazepam
Anticonvulsants	Haloperidol
Chloramphenicol	Isosorbide dinitrate
Chlordiazepoxide	Metronidazole
Chlorpromazine	Oxazepam
(N-oxide and S-oxide metabolites)	Penicillins

The most promising application of electrochemical techniques appears to be their use as flow cell detectors in HPLC analysis. The sensitivity of such combined methods is in the picomole range for suitable drugs. The list in Table XIV gives an indication of which type of drugs may be amenable to analysis by HPLC combined with electrochemical detectors.

Bibliography

Christian, G. D., *Analytical Chemistry*, 2nd edition, John Wiley and Sons, New York, 1977.

Meitos, L. and P. Zuman, *Handbook Series in Organic Electrochemistry*, Vols. I and II, CRC Press, West Palm Beach, Fla., 1977.

Zuman, P., *The Elucidation of Organic Electrode Processes*, Academic Press, New York, 1969.

4.9. COMPETITIVE PROTEIN BINDING ASSAYS

The saturable, high affinity binding of a drug to proteins represents the basis for competitive binding assays. The nature of the protein can serve as a criterion for classifying these assays into immunoassays, which employ drug-specific antibodies and account for the major portion of protein binding assays, and into assays utilizing other proteins, for example, drug target enzymes, receptors, and carrier proteins. Immunoassays are further subdivided according to the analytical method by which the fraction of free and protein-bound drug is measured: radioimmunoassay, enzyme immunoassay, or other immunoassays of lesser current importance in drug analysis (spin, nephelometric, or fluorescence immunoassay).

The competitive binding principle is depicted in Fig. 10. On the assumption of a simple bimolecular reaction between ligand (S) and protein (P), the following expression can be derived from the law of mass action:

$$\frac{B}{F} = K(P^0 - B)$$

where K is the equilibrium constant in the direction of S-P complex formation, P° is the molar concentration of protein binding sites, and B and F are the concentrations of bound and free ligand S. It follows that B/F decreases with increasing B, and the binding of the labeled ligand S to the protein also decreases with increasing concentrations of the unlabeled ligand. The analytical procedure then involves the quantitative

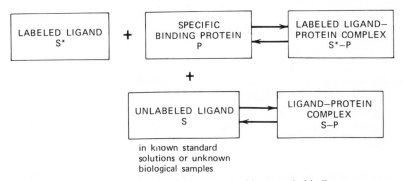

Fig. 10. Fundamental reactions of a competitive protein binding assay.

detection of the bound and/or free fraction of the labeled ligand S* and construction of a standard curve with increasing known drug concentrations. There are numerous variations of the tracer, incubation, and detection in competitive protein binding assays. However, a few principles can be applied to the evaluation of all such drug assays.

1. The major determinant of assay sensitivity and specificity is the nature of the ligand-protein binding interaction. The affinity, or dissociation constant, limits the assay sensitivity. Usually, the affinity constant should be greater than $10^9 \, M^{-1}$ for highly sensitive drug assays in serum which contains many drug binding proteins with low affinity. Extraction and separation of the drug from non-specific serum protein binding may be necessary to achieve maximum assay sensitivity. Furthermore, the cross-affinity of the binding protein to other compounds in the sample determines assay specificity for a drug. Many of the proteins selected for competitive binding assays are highly specific; yet drug metabolites and endogenous substrate analogs may be chemically very similar to the drug of interest and can interfere with the drug-protein binding process. It is therefore mandatory to carefully test for assay interferences by drug analogs which may be present in the assay solution. Frequently, inactive drug metabolites may be present at much higher concentrations than is the active parent drug (see morphine monograph), and a cross-affinity of the protein to a drug metabolite of only 1% may cause significant assay interference.
2. The ability to detect the free and the protein-bound labeled ligand also limits assay sensitivity. Frequently, the labeled ligand is present at trace concentrations relative to the unlabeled drug ligand, requiring extreme sensitivity of detection. For example, a

^{14}C-labeled tracer (^{14}C displacement of one ^{12}C atom) has a specific radioactivity of ~50 mCi/mmol, while a highly enriched ^{3}H tracer has a specific activity of ~20 Ci/mmol and can be detected at a 100-fold lower concentration.

3. The labeled ligand or the mode of detection can contribute to assay interferences if the labeled ligand and the drug possess different binding characteristics or the detection mode is subject to error by factors unrelated to the binding reaction (e.g., enzymatic reactions).

4. The incubation mode of the binding reaction (i.e., equilibrium analysis or nonequilibrium methods), the methods of detecting and/or separating bound and free labeled ligand, and the biological sample preparation all participate in the assay results; however, they are not nearly as significant as the nature of the binding protein and the label.

5. Some binding assays require physical separation of bound and free labeled ligand before analysis, because the analytical method does not differentiate between bound and free label (radioactive tracers). These assays are called heterogeneous. In contrast, homogeneous protein binding assays employ labels that are different in the protein-bound and free forms and do not require separation.

6. The data obtained from a competitive binding assay do not provide any clue as to potential assay interferences, which can often be detected with chromatographic techniques by observing the location and shape of the chromatographic peak. If competitive binding assays are employed in clinical laboratories for therapeutic drug level monitoring, it is therefore necessary to document that assay interferences in a large number of randomly selected patient samples are negligible. Moreover, the availability of reference assays of proven specificity is desirable.

Close to one half of the selected drugs (45%) can be analyzed in biological fluids by competitive protein binding assays, mostly radioimmunoassays. The following sections discuss the immunoassays and methods based on proteins other than immunoproteins.

4.9.1. Radioimmunoassay. Radioimmunoassay (RIA) procedures are widely used for drug analysis in biological samples (Table XV). The ligand S* is a radioactively labeled tracer drug which binds to a drug-specific antibody as the protein. The nature of the antibody is responsible to a large extent for the quality of the RIA method, and antibody production therefore represents an important factor in the assay

Table XV. Radioimmunoassays

Drug	Tracer	Comment
Amphetamine		Stereoselective for d-amphetamine
Atropine	^{14}C	Assays of varying specificity
Barbiturates	^{14}C	Group specific
Chlordiazepoxide	^{14}C	Specific in presence of metabolites
Chlorpromazine	^{3}H	N-Desmethylchlorpromazine cross-reacts
Clonazepam	^{3}H, ^{125}I	Specific in presence of metabolites
Cocaine	^{125}I	Benzoylecgonine cross-reacts
Dexamethasone	^{3}H	
Digoxin	^{3}H, ^{125}I	Various methods
Doxorubicin	^{125}I	Doxorubicinol cross-reacts, specific with HPLC separation
Ethynylestradiol	^{3}H	
5-Fluorouracil	^{3}H	FUdR-directed antibody
Gentamicin	^{3}H, ^{125}I	Variable cross-reactivity with other aminoglycosides
Haloperidol		
Hydrocortisone	^{125}I	Elevated temperature minimizes binding to hydrocortisone binding globulin in serum
Imipramine	^{3}H	Desipramine, amitriptyline, nor-triptyline cross-react
Isoniazid	^{3}H	Specific in presence of metabolites
Lysergide	^{3}H	Varying cross-affinity to metabolites
Methadone		
Methotrexate		Various procedures
Morphine	^{3}H	Various procedures
Nicotine		Nicotine and metabolite assays
Penicillins		Penicilloic acid trace analysis
Phenytoin	^{14}C, ^{3}H, ^{125}I	Metabolite cross-affinity tracer dependent
Propranolol		Stereoselective for active l-form
Reserpine	^{3}H	
Tetrahydrocannabinol		Requires chromatography for specificity

development. Drugs of low molecular weight (<1000) are normally inactive as antigens and do not produce antibodies. However, when coupled to a suitable protein via a chemical spacer bridge, drugs act as immunologically active haptens. The antibodies raised against a drug hapten-protein conjugate also recognize the unbound drug; good antibodies have affinity constants of $>10^9$ M^{-1} toward the drug hapten.

There are many ingenious methods to couple a drug to a protein (often bovine serum albumin, BSA); one is ester formation of a hydroxylated drug with succinic anhydride, followed by amide linkage to the protein:

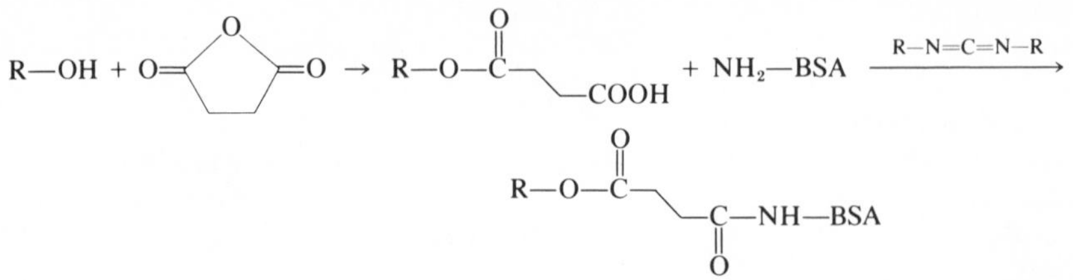

The spacer bridge should allow the drug to protrude from the protein surface to maximize antibody recognition of the hapten. The site of conjugation of the drug to the protein cannot participate in determining the structure of the antibody, and we can frequently predict antibody specificity for a drug in the presence of its metabolites by knowing the site of conjugation. For example, cocaine antibodies have been raised against a conjugate coupled via the COOH function of benzoylecgonine and therefore do not differentiate between cocaine and the pharmacologically inactive benzoylecgonine. Similarly, morphine-specific antibodies raised against a C-3–OH morphine-protein conjugate also bind morphine C-3–*O*-glucuronide, while antibodies against a C-6–OH or a C-1 morphine-protein conjugate do not cross-react with the C-3–*O*-glucuronide. Some antibodies are group selective, for example, for barbiturates and penicilloic acids. However, the degree of antibody selectivity is usually remarkable, allowing the specific detection of many drugs in the presence of closely related metabolites in small serum samples (10 to 100 μl) with a sensitivity as good as 1 ng drug/ml. The antibody may be stereoselective for one chosen isomer of racemic drugs (amphetamine, propranolol). The RIA procedures are frequently performed with the unextracted serum sample; inclusion of chromatographic separations can increase the specificity and sensitivity by eliminating chemically similar substances and nonspecific binding to serum proteins (doxorubicin, tetrahydrocannabinol).

Next to the quality of the antibody, we have to consider the radioactive tracer for RIA method evaluation. Because of increasing specific radioactivity, ^{14}C, ^{3}H, and ^{125}I allow increasing maximum assay sensitivity. For instance, ^{14}C tracers do not possess sufficiently high specific activity for RIAs in the picogram range. While the beta emitters ^{3}H and ^{14}C are measured by liquid scintillation counting, ^{125}I as a gamma emitter can be determined in a gamma counter, which is faster and more readily automated. However, ^{125}I has to be introduced into the drug molecule by reaction with a suitable derivative, such as tyroxine, which can be readily iodinated. Thus the labeled tracer is chemically different from the unlabeled drug; a discussion of the factors involved in selecting a suitable radioactive tracer is included in the digoxin monograph. Cross-affinity of a phenytoin-directed antibody to phenytoin metabolites varies with the use of a ^{3}H and a ^{125}I tracer, indicating that these two tracers bind differently to the antibody. Nevertheless, assay results obtained with ^{3}H-RIA and ^{125}I-RIA are usually identical (see digoxin monograph).

The incubation of antibody with tracer and unknown drug is normally designed to achieve binding equilibrium. Occasionally, sequential saturation RIA is used to increase assay specificity. Sequential saturation RIA depends on a sufficiently slow off-rate of the drug from the antibody binding site. Extraneous compounds binding to the same site with lower affinity dissociate from the binding site during the labeling incubation, whereas the tightly bound drug does not (digoxin).

The procedure used to separate free and antibody-bound labeled tracer can also affect RIA results. Free tracer may be absorbed to dextran-coated charcoal, or bound tracer precipitated with the antibody by ammonium sulfate. Immunoprecipitation in a double-antibody RIA is also frequently used, to name just a few examples of separation techniques. Analysis time can be considerably shortened by using antibodies coated onto glass beads or the surface of the incubation tube (solid phase RIA); the liquid can simply be decanted after incubation to remove the unbound tracer.

We can conclude that RIA techniques are valuable in drug level monitoring and may represent the only assay available at present for a number of drugs. There are two drawbacks, however, one of which is specific to radioimmunoassays, and the other common to all immunoassays. These are, first, the ubiquitous use of radioactive tracers, causing environmental concern in spite of the rather low level of radioactivity needed for the RIA, and, second, the unpredictable quality of the antibody. Specificity of the antibody may vary even in the same immunized animal (sheep, rabbit) when collected at different times during immunization. Moreover, antibodies may not be uniform, but rather heterogeneous. Clinical use of RIA methods is hampered by the changing specificity of antibodies with changing lot numbers or manufacturers. Therefore the validity of the RIA for a drug assay in serum should be reestablished for each antibody lot, using a rigid set of criteria. Unfortunately, this is not common practice; we again refer to the digoxin monograph, which summarizes recent studies on the varying specificities of currently used digoxin RIA methods in the presence of spironolactone metabolites.

The field of drug immunoassays would greatly benefit from uniform, generally available, and well-defined antibodies. Possibly, monoclonal antibodies can meet these criteria. Such antibodies are obtained by harvesting spleen cells, instead of serum, from the immunized animal and then fusing these cells with myeloma cells. The generated hybrid cells retain the ability to produce antibodies and can be grown indefinitely without the growth limitations of differentiated cells. The difficulty remains of selecting for and cloning the hybridoma cell that produces the most desirable antibody.

4.9.2. Enzyme Immunoassay. Immunoassays involving enzyme-labeled antigens, haptens, or antibodies have been recently developed for a large number of macromolecules and small molecular weight substances serving as haptens. Enzyme immunoassays (EIAs) are increasingly used in clinical laboratories for diagnostic tests and for therapeutic drug level monitoring. Table XVI summarizes the published drug assays and some details of the EIA methods. The advantageous features of the EIA include speed, simplicity, ready automation, and versatility, which render it suitable for routine clinical applications and rapid drug screening. Initial lack of sensitivity of the EIA, relative to the RIA, has been largely overcome; for example, the EIA assay of progesterone is sensitive to 15 pg/sample, using β-galactosidase as the enzyme label and a heterogeneous EIA (Dray et al., 1975). Thus the sensitivity and specificity of the EIA are comparable to those of the RIA, while the EIA offers advantages over the RIA in avoiding radioactive tracers and in speed of analysis. Many of the articles cited in the individual monographs (see Table XVI and Chapter 5) include comparison studies performed with the EIA, RIA, and specific physiochemical assays. The results indicate that the EIA and RIA are usually equivalent and specific for the test drug. However, the enzyme label introduces additional potential assay interferences by several mechanisms, including differential binding of free drug and enzyme label to the antibody, presence of enzyme inhibitors, and presence of enzyme activity in the test sample. These factors have to be eliminated or minimized for clinically useful EIA methods.

There are an unlimited number of EIA systems, based on the nature of the immunoreaction and the specific enzyme label. For drug haptens, both heterogeneous and homogeneous EIAs are suitable. The nomenclature is rather confusing in this rapidly developing field; for our purposes

Table XVI. Enzyme Immunoassay

Drug[a]	Enzyme	Substrate-cosubstrate	Detection	Method
Amphetamine	Lysozyme	Mucopolysaccharides (→ cell lysis)	UV	Homogeneous[b]
Anticonvulsants	G-6-P-dehydrogenase[c]	G-6-P, NAD^+	UV	Homogeneous[b]
Carbamazepine				
Ethosuximide				
Phenobarbital				
Phenytoin				
Primidone				
Barbiturates	Lysozyme	Mucopolysaccharides	UV	Homogeneous[b]
Cocaine	Lysozyme	Mucopolysaccharides	UV	Homogeneous[b]
(benzoylecgonine)				
Digoxin	(1) G-6-P-dehydrogenase	G-6-P, NAD^+	UV	Homogeneous[b]
	(2) Peroxidase	Chromogen + H_2O_2	Visible	Heterogeneous[d]
Gentamicin	(1) G-6-P-dehydrogenase	G-6-P, NAD^+	UV	Homogeneous[b]
	(2) Peroxidase	Chromogen + H_2O_2	Visible	Solid phase antibody[d]
	(3) β-Galactosidase	Sisomycin-umbelliferyl-d-galactoside	Fluorescence	Homogeneous[b]
Hydrocortisone	(1) β-Galactosidase	o-Nitrophenyl-β-d-galactopyranoside	UV	Double antibody[a]
	(2) Alkaline phosphatase		UV	Heterogeneous[d]
Lidocaine	G-6-P-dehydrogenase	G-6-P, NAD^+	UV	Homogeneous[b]
Morphine	(1) Lysozyme	Mucopolysaccharides	UV	Homogeneous[b]
	(2) Malate dehydrogenase	Oxaloacetate + NADH	UV	Homogeneous[b]
Procainamide	G-6-P-dehydrogenase	G-6-P, NAD^+	UV	Homogeneous[b]
Tetrahydrocannabinal	Malate dehydrogenase	Oxaloacetate, NADH	UV	Homogeneous[b]
Theophylline	(1) Lysozyme	Mucopolysaccharide (→ cell lysis)	UV	Homogeneous[b]
	(2) Malate dehydrogenase	Oxaloacetate + NADH	UV	Homogeneous[b]

[a] Additional drugs that can be assayed by EMIT include benzodiazepines and methotexate. No literature data are available as yet.

[b] EMIT®: enzyme multiplied immunoassay.

[c] G-6-P-dehydrogenase: glucose-6-phosphate dehydrogenase.

[d] ELISA: enzyme linked immunosorbent assay.

89

we can differentiate between the heterogeneous EIA or enzyme linked immunosorbent assay (ELISA), which requires separation of free and antibody-bound enzyme label, and the homogeneous EIA without separation. The latter is also referred to as the "enzyme multiplied immunoassay technique" (EMIT®, registered trade name of the Syra Corp., Palo Alto, Calif.). The common enzymes that serve as the label are lysozyme, glucose-6-phosphate dehydrogenase, peroxidases, β-galactosidase, malate dehydrogenase, and alkaline phosphatase (Table XVI), each of which offers some procedural advantages and disadvantages. The substrates or cosubstrates are usually detected by UV, visible light spectroscopy, or fluorescence.

The principle of the EMIT assay procedure is outlined in Fig. 11. The drug hapten, for example, digoxin, is covalently bound to the enzyme, G-6-P dehydrogenase, which converts G-6-P to gluconolactone-6-phosphate and NAD+ to NADH. The rate of NADH formation is measured spectrophotometrically at 340 nm and is proportionate to the free enzyme-drug conjugate under the assay conditions. Upon binding to a digoxin-specific antibody, the enzyme activity is suppressed, resulting in decreased production of NADH. Further addition of test serum containing free digoxin reduces the binding and therefore the inhibition of the enzyme label by competitive displacement. By the use of appropriate calibration sera, standard curves can be constructed for the quantitation of unknown drug sera. The G-6-P dehydrogenase of bacterial origin utilizes NAD^+, rather than $NADP^+$, which is the cosubstrate for the mammalian enzyme; this cosubstrate selectivity minimizes assay interferences by sera containing native G-6-P dehydrogenase. However, cosubstrate specificity is not absolute; moreover, enzyme label inhibition upon antibody binding is not complete. Consequently, there exists a baseline production of NADH which limits assay sensitivity. This problem was particularly difficult to overcome for the digoxin assay (therapeutic range 0.5 to 2 ng/ml serum); yet the currently available EMIT digoxin assay yields good correlation of results with other clinical digoxin assays. The current advantage of the EMIT assay over other EIA techniques for therapeutic drug level monitoring lies in its documented reliability under clinical conditions and its commercial availability for a large number of the drugs that are routinely measured in clinical laboratories (Table XVI). Furthermore, automation of the EMIT using the centrifugal analyzer can reduce assay speed and reagent costs considerably. The required serum sample volume is usually below 200 μl, and results can be available in less than 1 hr.

The ELISA principle avoids the problem of less than complete inactivation of the antibody-bound enzyme label; rather, the drug-enzyme

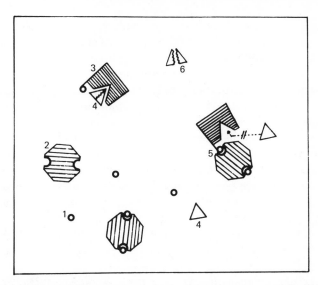

Fig. 11. Principles of a homogeneous enzyme immunoassay: example, EMIT® (Syva, Palo Alto, Calif.). Bacterial glucose-6-phosphate dehydrogenase, utilizing NADPH, is one of the enzymes used in EMIT tests. The test drug molecules are conjugated with the enzyme without affecting its catalytic activity, while binding to a drug antibody suppresses the enzyme activity of the conjugate. Competition for antibody binding sites causes the enzyme-drug conjugate to remain unbound. The resultant enzyme activity is directly related to the concentration of free drug molecules in the sample. (1) Free drug molecule. (2) Drug-specific antibody. (3) Enzyme-drug conjugate, catalyzing substrate reaction. (4) Enzyme substrate. (5) Enzyme-drug conjugate–antibody complex with minimal catalytic activity. (6) Substrate products.

conjugate binds in a way to preserve enzyme activity. Following separation of free and bound enzyme label, the catalytic activity of either fraction can be determined spectrophotometrically. This procedure provides a potentially higher assay sensitivity because of low or absent extraneous substrate conversion (progesterone, Dray et al., 1975). However, the separation represents one more step over those needed for EMIT and makes automation more difficult. The commercially available digoxin ELISA method (Enzymun-Test®, Boehringer Mannheim, G.m.b.H.) is based on a solid phase EIA which simplifies the separation of free and bound enzyme label. It employs test tubes coated with digoxin-specific antibodies and a digoxin-horseradish peroxidase conjugate. Free digoxin in the test serum and in the enzyme label compete for the available antibody binding sites. When equilibrium of the reaction is reached, the incubation mixture is decanted and the antibody-bound enzyme activity measured colorimetrically with H_2O_2 and a chromogen. The ELISA assay of digoxin yields clinical results comparable to those obtained by other immunoassays; similar Enzymun-Tests are not available for other drugs of clinical interest.

Among many EIA variations, we find the homogeneous fluorescence immunoassay of gentamicin of particular interest. Here the drug is bound to the substrate, umbelliferyl-β-D-galactoside, rather than to the enzyme, β-galactosidase (Table XVI; sisomycin is a close analog of gentamicin with high antibody cross-affinity). The competitive binding reaction now occurs between free gentamicin and the gentamicin-labeled substrate, which is completely protected against the enzyme in the bound state. Thus residual enzyme activity after binding in EMIT is eliminated. Moreover, the nonfluorescent substrate develops an intense fluorescence after enzymatic liberation of umbelliferone, providing excellent assay sensitivity with only a few microliters of serum. This gentamicin EIA gives an indication of the versatility of the EIA and the potential future development of the method in drug level monitoring.

4.9.3. Other Immunoassays. We have included several additional immunoassay techniques in Chapter 5 in order to demonstrate the versatility of this analytical approach to drug level monitoring. Some of these methods already are of importance or have the potential for broad applicability to drug assays. Table XVII contains the following immunoassay variants:

Fluorescence immunoassay
Spin immunoassay
Nephelometric immunoassay
Hemagglutination-inhibition immunoassay

Details of the immunoassay procedure are given in Chapter 5 under the respective drug listed in Table XVII. Here we will briefly discuss the analytical potential and major drawbacks of each method. None of the four immunoassays requires separation steps after incubation of hapten and antibody, although the nephelometric and the hemagglutination-inhibition tests depend on precipitation reactions and are therefore not homogenous immunoassays. All assays are rapid and relatively simple to perform.

The fluorescence immunoassays of gentamicin are based on changes in the fluorescent properties of fluoresceine caused by binding to a macromolecule; either fluorescence quenching or fluorescence polarization induction is measured, the latter requiring a specialized fluorimeter. Fluoresceine is covalently linked to gentamicin, and the complex binds to a gentamicin-specific antibody as the macromolecule.

Table XVII. Further Immunoassays

Drug	Tracer	Detection	Procedure
Gentamicin	Fluoresceine labeled gentamicin	(1) Fluorescence polarization induction (2) Fluorescence quenching	Homogeneous fluorescence immunoassay
Hydrocortisone	Fluoresceine labeled hydrocortisone	Fluorescence polarization induction	Homogeneous fluorescence immunoassay
Morphine	(1) Spin-labeled morphine	Electron spin resonance (ESR)	Homogeneous spin immunoassay
	(2) Morphine sensitized erythrocytes	Hemagglutination	Hemagglutination-inhibition immunoassay
Phenytoin	Spin-labeled phenytoin	ESR	Homogeneous spin immunoassay
Theophylline	Theophylline labeled antigen	Light scattering by Ag-Ab precipitate	Nephelometric immunoassay

Fluorescence immunoassays are rather sensitive; the major disadvantage may be the potential for assay interference by nonspecific fluorescence quenching or background fluorescence by other compounds in the biological sample.

The electron spin resonance (ESR) signal of a free radical label also undergoes a measurable change upon binding to a macromolecule. While the rapidly rotating free radical in solution gives rise to sharp ESR signals, the relatively slow motion of the same radical bound to an antibody abolishes any measurable ESR peak by substantial peak broadening. Therefore no separation of bound and free label is needed. The drugs of interest, for example, phenytoin and morphine, are covalently bound to a functional group containing a free radical, usually a stable nitroxyl radical. Spin immunoassays are rapid and have been mainly used in drug screening programs for semiquantitative analysis (morphine: FRAT®, free radical assay technique, Syra Corp.). However, quantitative assays are also possible, and the sensitivity can be in the low nanomolar range if appropriate ESR equipment is available. Furthermore, the degree of drug-plasma protein binding can be rapidly determined with the corresponding spin-labeled drug. However, spurious results are obtained if the spin label alters the protein binding behavior of the drug (phenytoin). A further problem with spin immunoassays arises from potential chemical reactions of the free radical label with endogenous substances. For instance, ascorbic acid (vitamin C) reduces the nitroxyl radical and may cause false-positive results in urine samples that are screened for morphine content.

The nephelometric assay of theophylline offers great potential for clinically applicable assays of other drugs as well, because of its simplicity and speed. The assay principle is simple; a theophylline-labeled protein antigen and a theophylline-specific antibody form an antibody-antigen complex that is large enough to scatter light. The dispersed light is quantitatively measured in a nephelometer (a specially constructed spectrophotometer). Any hapten (theophylline) in the sample solution prevents the immunoprecipitation in a competitive fashion. We are looking forward to the implementation and successful clinical application of this assay technique to a variety of suitable drugs.

The classical hemagglutination-inhibition immunoassay, while rather sensitive, represents only a semiquantitative tool with insufficient precision for therapeutic drug level monitoring. In the hands of experienced immunologists, however, it can be a highly sensitive tool.

4.9.4. Other Competitive Protein Binding Assays. Any protein, or any other macromolecule, can be used in competitive binding assays if it binds the drug with sufficiently high affinity and specificity. The target enzyme or receptor of a drug often presents itself for drug binding assays, since the drug in question has already been selected over less active congeners for high target affinity and specificity. Table XVIII lists the drug assays along with the corresponding target proteins, which are discussed in Chapter 5. These assays employ the enzymes Na-K-ATPase and dihydrofolate reductase-NADPH, a soluble estrogen receptor preparation, and the hydrocortisone carrier protein, transcortin, as the binding proteins. With the use of ^{3}H-labeled tracers, the assay procedures are identical to the radioimmunoassay procedures, except for substitution of the antibody with a drug target protein. The ability of a substance to bind to the target protein usually implies pharmacological activity; therefore drug target protein binding assays tend to measure all active components in the biological sample, exogenous or endogenous. Chromatographic separations are required if the parent drug has to be measured in the presence of active metabolites and endogenous substrates (ethynylestradiol). Similar considerations apply to the hydrocortisone assay with transcortin, although not all glucocorticosteroids bind to the carrier protein in plasma (e.g., dexamethasone). In contrast, the binding of methotrexate to dihydrofolate reductase is several orders of magnitude greater than the binding of its natural substrate, dihydrofolate. Addition of the cofactor NADPH further increases the binding affinity of methotrexate, providing specific detection of the drug in the presence of dihydrofolate without chromatographic separations. However, binding affinities may vary considerably, depending on the source of the enzyme.

All of the cardiac glycosides bind to the target enzyme Na-K-ATPase. Their relative binding affinities correlate somewhat with their respective intrinsic activities, a circumstance that may be advantageous under special assay conditions. The drug binding affinities of

Table XVIII. Competitive Protein Binding Assays, except for Immunoassays

Drug	Tracer	Protein
Digoxin	^{3}H-ouabain	Na-K-ATPase
Ethynylestradiol	^{3}H-ethynylestradiol	Estrogen receptor in rabbit uterine cytosol
Hydrocortisone	^{3}H-hydrocortisone	Cortisol binding globulin in serum (transcortin)
Methotrexate	^{3}H-methotrexate	Dihydrofolate reductase-NADPH

these proteins are comparable to those of antibodies in radioimmunoassays. Enzyme or receptor binding assays can be important tools in pharmacological research and in the clinical laboratory. At UCSF we utilized the competitive protein binding assay of methotrexate until very recently. The major drawbacks of the methotrexate assay are the limited dynamic serum concentration range and the variable nature of the enzyme when obtained from different sources. In general, immunoassays and receptor assays may be used as interchangeable or complementary test methods.

Bibliography

Ayscough, P. B., *Electron Spin Resonance in Chemistry*, Methuen and Co., London, 1967.

Bastiani, R. and J. Chang, *Performance Evaluation of the EMIT® Antiepileptic Drug Assays*, Summary Report, Syva, Palo Alto, Calif., 1978.

Brattin, W. J. and I. Sunshine, Immunological Assays for Drugs in Biological Samples, *Am. J. Med. Technol.*, **39**, 223–230 (1973).

Dray, F., J.-M. Andrieu, and F. Renaud, Enzyme Immunoassay of Progesterone at the Picogram Level Using β-Galactosidase as Label, *Biochim. Biophys. Acta*, **403**, 131–138 (1975).

Feldman, G., P. Druet, J. Bignon, and S. Avrameas, *Immuno Enzymatic Techniques*, Inserm Symposium 2, North-Holland Publ. Co./Amer. Elsevier Publ. Co., 1976.

Gupta, D., *Radioimmunoassay of Steroid Hormones*, Verlag Chemie, Weinheim, 1975.

Jaffe, B. M. and H. R. Behrman, *Methods of Hormone Radioimmunoassay*, Academic Press, New York, 1974.

Mule, S. J., *Immunoassays for Drugs Subject to Abuse*, CRC Press, West Palm Beach, Fla., 1974.

Odell, W. D. and W. H. Daughaday, *Principles of Competitive Protein Binding Assays*, J. B. Lippincott Co., Philadelphia, Toronto, 1971.

Parker, C. W., *Radioimmunoassay of Biologically Active Compounds*, Prentice-Hall, Englewood Cliffs, N.J., 1976.

Ransom, J. P., *Practical Competitive Binding Methods*, C. V. Mosby, St. Louis, Mo., 1976.

Ritcie, R. F., *Automate Immunoanalysis*, Marcel Dekker, New York, 1978.

Soini, E. and I. Hemmilä, Fluoroimmunoassay: Present Status and Key Problems, *Clin. Chem.*, **25**, 353–361 (1979).

Thorell, J. I. and S. M. Larson, *Radioimmunoassay and Related Techniques: Methodology and Clinical Applications*, C. V. Mosby, St. Louis, Mo., 1978.

Vogt, W., *Enzymimmunoassay*, Georg Thieme Verlag, Stuttgart, 1978.

Wisdom, G. B., Enzyme Immunoassay, *Clin. Chem.*, **22**, 1243–1255 (1976).

4.10. ENZYMATIC ASSAYS

The interactions of a number of drugs with specific enzymes can be utilized as the basis for an enzymatic drug assay in a biological specimen (Table XIX). The assay procedures, enzymes, and detection modes are quite varied, but usually involve the measurement of reaction rates by UV and visible spectroscopy or radioactivity analysis. For instance, methotrexate inhibits the enzyme dihydrofolate reductase, which can be determined by measuring the disappearance rate of the UV absorption maximum at 340 nm of NADPH. This enzymatic assay is similar in its selectivity to the competitive protein binding assay with dihydrofolate reduction described in the preceding section; however, the enzymatic assay is more cumbersome and less sensitive than the competitive protein binding assay. The digoxin assay employing Na-K-ATPase can also be performed as a protein binding assay (see preceding section) or as an enzymatic assay. Cardiac glycosides suppress the uptake of ^{86}Rb into red blood cells by inhibiting Na-K-ATPase. The extent of enzyme inhibition reflects the intrinsic activity of the cardiac glycoside.

Several assays employ bacterial enzymes (R factors) that inactivate specific antibiotics and thus mediate bacterial resistance against these antibiotics (chloramphenicol, gentamicin). The assay end point is the complete inactivation of the antibiotic by acetylation or adenylation; the products can be isolated and their radioactivity determined, if

Table XIX. Enzymatic Assays

Drug	Enzyme	Substrate/cosubstrate	Detection
Acetazolamide	Carbonic anhydrase	H_2CO_3	
Amphetamine	N-Methyltransferase	S-Adenosyl-L [methyl-^{3}H] methionine	^{3}H
Chloramphenicol	R factor (acetyltransferase)	^{14}C-acetyl-CoA	^{14}C
Digoxin	Na-K-ATPase		^{86}Rb uptake by red blood cells
Gentamicin	R factors:		
	(1) Adenyltransferase	γ-^{32}P-ATP	^{32}P
	(2) N-Acetyltransferase	^{14}C-acetyl-CoA	Colorimetry (SH) or ^{14}C
Methotrexate	Dihydrofolate reductase	Dihydrofolic acid/NADPH	UV

radioactive cofactors are utilized. Alternatively, the sulfhydryl group of CoA can be assayed colorimetrically in the gentamicin assay, since cosubstrate consumption is stoichiometric. Such an approach requires a high enzyme specificity. The amphetamine assay is also based on complete product formation (*N*-methylamphetamine) as the end point; however, the enzyme is rather nonspecific, and extensive purification of the radioactive product is necessary.

Enzymatic assays are moderately useful as drug assays in biological fluids, but they compete against a number of other analytical techniques. Lack of high quality enzyme sources may limit the utility of enzymatic assays.

Bibliography

Bergmeyer, H. U., *Methods of Enzymatic Analysis*, Verlag Chemie, Weinheim, and Academic Press, New York, 1974.

Guilbault, G. G., *Handbook of Enzymatic Methods of Analysis*, Marcel Dekker, New York, 1976.

4.11. MICROBIOLOGICAL ASSAYS

The analysis of antibiotics in serum or urine samples was one of the earliest drug level assays of therapeutic value. With a few exceptions the sole assay method of choice was the microbiological assay. Chapter 5 contains original references to microbiological assays of the following drugs:

Chloramphenicol	Gentamicin	Nitrofurantoin
Ethambutol	Ioniazid	Penicillins
5-Fluorouracil	Methotrexate	Tetracyclines

Microbiological assays are also useful for the determination of vitamins and amino acids. In principle, the ability of a compound to inhibit (or promote) bacterial growth is measured using two major analytical designs, the agar diffusion assay and the tube assay. A discussion of microbiological assay techniques is beyond the scope of this book, and we refer the reader to the textbooks listed in the bibliography. We can, however, discuss briefly some of the advantages and drawbacks of these methods.

Above all, a microbiological assay measures the concentration of the biologically active species in serum. Thus differences between the fluorescence assay and the microbiological assay of penicillins are indicative of the presence of fluorigenic, inactive penicillin metabolites, that is, penicilloic acids. The value of microbiological assays is diminished, however, if active (or toxic) metabolites are formed or if multiple antibiotics are administered. It is unlikely that the test organism mimics precisely the drug sensitivity of the infecting organism, and the assay results will not truly represent the antibiotic serum activity directed against the infection organism. Drug level monitoring of the aminoglycosides (e.g., gentamicin) is also concerned with the toxic side effects, such as renal toxicity and ototoxicity. If mixtures of antibiotics are present, the rather nonspecific microbiological assays are not capable of assessing the potential for toxic side effects of gentamicin. The specificity of microbiological assays can be improved either by introducing a chromatographic separation before analysis or by employing multiple-resistant test organisms that are sensitive only to the antibiotic in question. However, the clinical use of multiple-resistant bacteria is hazardous and should be avoided, except for specially engineered test organisms that can survive only under certain laboratory conditions.

The limited accuracy of microbiological assays of antibiotics and the low reproducibility among different clinical laboratories (see gentamicin monograph) contribute to the current trend to employ more specific assay methods in therapeutic antibiotic level monitoring.

Bibliography

Board, R. G. and D. W. Lovelock, *Some Methods for Microbiological Assay*, Academic Press, New York, San Francisco, London, 1974.

Hewitt, W., *Microbiological Assay: An Introduction to Quantitative Principles and Evaluation*, Academic Press, New York, San Francisco, London, 1977.

Kavanagh, F., *Analytical Microbiology*, Vols. 1 and 2, Academic Press, New York, San Francisco, London, 1963 and 1972.

4.12. VARIOUS ASSAY TECHNIQUES

Many more assay systems are potentially applicable to drug analysis in biological samples; some of these are included in Table XX and discussed in further detail in Chapter 5. Three assay categories are of general interest: biological assays (atropin, warfarin, cyclophosphamide), isotope derivatization procedures (morphine), and luminometry (gentamicin).

Biological assays are valuable because they truly represent the drug activity in the sample, regardless of the formation of drug metabolites

Table XX. Various Assay Techniques

Drug	Assay Procedure
Atropine	Biological assay: mouse eye pupil dilation (10 ng atropin measurable)
Cyclophosphamide	Mammalian cell culture (Walker 256 rat carcino-sarcoma), toxicity assay
Gentamicin	Microbiological assay using the firefly bioluminescence system
Lithium	Atomic absorption and emission spectrometry
Morphine	Isotope derivatization procedure (^{3}H-dansyl-chloride) with TLC separation
Sulfonamide	Phosphorimetry (at 77° K)
Warfarin	Biological assay: plasma prothrombin time

of unknown activity. The sensitivity of biological assays is often limited, but can be in the low nanogram range (atropine). Determination of the prothrombin time is not a drug (warfarin) level assay, but an indicator of the principal drug effect in the patient. The mammalian cell culture assay of cyclophosphamide is essentially equivalent to microbiological assays of antibiotics.

Derivatization of a drug with radioactive reagents, followed by isolation and measurement of the product, can be a highly sensitive and specific tool. Such an assay may be performed either as a single-isotope assay (see morphine monograph) or as a double-isotope dilution procedure. The latter may involve the addition of a ^{14}C-labeled tracer (e.g., morphine—N—^{14}CH$_3$) to the biological sample containing the unknown, unlabeled morphine. The mixture is then derivatized with ^{3}H-dansyl chloride, and the ^{14}C/^{3}H ratio of the product determined by liquid scintillation counting. The higher the concentration of unlabeled morphine in the sample, the lower is the ^{14}C/^{3}H ratio of the reaction product. The ^{14}C-isotope dilution can be quantitatively measured using a calibration curve. Isotope dilution assays require reagents of high specific radioactivity and extensive purification steps; therefore they are primarily used as reference assays, if no other procedures are available.

Luminometry is a technique based on the chemoluminescence of a light generating system. One of the reactions used in luminometric assays is based on the oxidation of a mollusk luciferin by horseradish peroxidase:

$$\text{Pholad luciferin} + O_2 \xrightarrow[\text{(oxidase)}]{\text{peroxidase}} \text{light}$$
$$\text{(glycoprotein)}$$

Luminometry is three orders of magnitude more sensitive than fluorescence; the detection is simple (phototube) and linear over a wide concentration range. The technique is applicable to the assay of a large variety of enzymes and substrates of biomedical interest. The major reactions that can be coupled to chemoluminescent systems are (1) formation of hydrogen peroxide; (2) interconversions of oxidized and reduced forms of pyridine nucleotides; (3) production or utilization of ATP; and (4) enzyme immunoassays utilizing peroxidase-labeled antibodies (the Pholad luciferin system).

The gentamicin assay is in principle a microbiological assay with ATP concentrations as the analytical end point. The ATP levels correlate with the number of bacteria in the incubation mixture and can be used to monitor bacterial growth. The ATP levels are measured by luminometry, using the firefly bioluminescence system. The extreme sensitivity of luminometry offers many potential applications in drug level monitoring, and this technique may replace other detection systems in the future.

Bibliography

DeLuca, M. A., *Bioluminescence and Chemoluminescence*, Academic Press, New York, 1978.

Glick, D., The Future of Clinical Microchemical Analysis, *Clin. Chem.*, **24**, 189–192 (1978).

Nodine, J. H., P. E. Siegler, and Y. H. Moyer, *Animal and Clinical Pharmacologic Techniques in Drug Evaluation*, Vols. I–VI, Year Book Medical Publ., Chicago, 1964.

Tölgyessy, J., T. Braun, and T. Kyrs, *Isotope Dilution Analysis*, International Series of Monographs in Analytical Chemistry, Vol. 49, Pergamon Press, Oxford, 1972.

Winefordner, J. D., S. G. Schulman, and T. C. O'Haver, *Luminescence Spectrometry in Analytical Chemistry*, Wiley-Interscience, New York, 1972.

4.13. ASSAY EVALUATION

The important performance characteristics of a drug assay are precision, bias, accuracy, sensitivity, selectivity, range, time of analysis, and applicability to biological samples. The precision is usually expressed as the standard deviation of the assay:

$$S = \sqrt{\frac{\sum_1^N (X_i - \overline{X})^2}{n - 1}}$$

The coefficient of variation (C.V.), or relative standard deviation, is also frequently used:

$$\% \text{ C.V.} = \left(\frac{S}{\overline{X}}\right) 100 \qquad (n \geqslant 6)$$

The sample number should not be smaller than 6. The bias (B) of an assay method is determined by assaying spiked serum samples and comparing the results with the known true value. The results are averaged ($\overline{X}$); the standard deviation (S_x) and the standard error of the average of the mean ($S_x/\sqrt{n}$) are determined. The significance of the bias is established by setting a confidence limit:

$$\hat{B} - t\left(\frac{S_x}{\sqrt{n}}\right) < B < \hat{B} + t\left(\frac{S_x}{\sqrt{n}}\right)$$

(B = true bias). The t value is taken from Student's t table; it is $t = 2.447$ for $n = 6$ and $P = .975$. Precision and bias together determine the accuracy of the method. Each publication that includes analytical methods should discuss accuracy and the other characteristics given above. For drug assays in biological samples, selectivity is a particularly important parameter, since many closely related compounds may be present in the sample, most notably drug metabolites. Spiking serum samples with known amounts of the pure drug is therefore not sufficient to establish assay selectivity; rather, serum samples have to be obtained after ingestion of the drug by the patient. Assay results then have to be compared to those obtained with an assay of proven relia-bility and specificity. One useful parameter describing the degree of correlation between the two assay results is the Pearson correlation coefficient r:

$$r = \frac{(X_i - \overline{X}) (Y_i - \overline{Y})}{n \cdot S_x \cdot S_y}$$

The maximum value of r is 1, indicating exact correlation between the two variables; $r = 0$ means complete independence. However, the degree of selectivity of the reference method is often not precisely known because of the complexity of the biological sample.

Assays suitable for therapeutic drug level monitoring have to satisfy special requirements based on the nature of the sample material and the special purpose of the assay. Clinical serum samples are randomly obtained from patients with a broad spectrum of diseases. These patients commonly ingest several drugs in addition to the test drug. Selectivity criteria therefore have to be rather stringent, and the assay has to be tested in blank sera obtained from patients with various diseases, treated with a large number of drugs (but not the test drug). Furthermore, control sera should be spiked with rather high concentrations of pure drugs that are likely to be taken concomitantly in order to test for interferences at peak drug levels. Table XXI gives an example of the specificity test of a quinidine HPLC assay. The disease states should include hyperbilirubinemia and hyperlipidemia as potential causes of assay interference. Absence of false-positive results for the test drug under all these conditions indicates assay specificity with respect

Table XXI. List of Drugs Tested for Intereference of the Quinidine HPLC-UV Assay of Powers and Sadée [*Clin. Chem.* **24**, 299–302 (1978)]
Pure drug substances and sera from patients receiving the indicated drugs, but not quinidine, were tested and did not show any interferences, except for primaqine and quinine.

Drug[a]	Retention (min)	Drug	Retention (min)
N-Acetylprocainamide[b]		Methotrexate	1.1
Acetylsalicylic acid[b]	2.0	Morphine[b]	
Amikacin		Phenacetin	
ε-Aminocaproic acid		Phenobartital	
Caffeine[b]		Phenylbutazone	
Chlordiazepoxide	2.9	Phenytoin[b]	
Chlorpheniramine		Primaquine	2.7
Desimipramine		Procaine	1.8
Diazoxide		Procaineamide	
Digitoxin		Propranolol	1.2
Digoxin[b]		Quinine	2.7
Ephedrine		Salicylate	2.0
Ethchlorvynol		Sulfisoxazole	
5-Fluorouracil	1.3	Theobromine	
Hydrocortisone[b]		Theophylline[b]	
Indomethacin	7.8	Tranexamine acid	
Isoniazid	1.3	Warfarin[b]	5.6

[a] Quinidine has a retention time of 2.7 min. No peak appeared within 6 min in the case of the compounds for which no retention time is listed.
[b] These drugs and their metabolites were present in patient sera.

to other drugs and endogenous substrates under clinical conditions. Assay specificity in the presence of metabolites of the test drug should be determined by comparison to a reference assay of known specificity. By the use of standard statistical methods, the linear regression line is calculated of Y (new assay method) on X (reference method) in a scatter diagram, and the intercept and slope are determined. If the X/Y ratio deviates from unity or the regression line does not pass through the origin, the new method is biased with respect to the comparison method. A typical scatter diagram, along with a statistical evaluation of the comparison between two quinidine assays, is given in Fig. 12.

Next to assay specificity, clinical quality control procedures are concerned with assay precision. The within-run precision can be determined by measuring drug serum levels in duplicate in a number of samples (~ 20). The differences between the duplicates are used to calculate the standard deviation. The between-run precision is measured on separate days with 20 replicate samples at a low, an intermediate, and a high drug serum concentration. From these three sets of 20 replicate samples, the between-run standard deviation is calculated for each drug level.

Once a drug assay is established for clinical use, patient samples are measured, along with a suitable calibration curve and at least two inde-

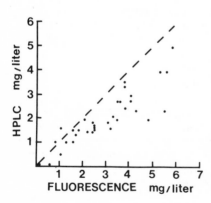

Fig. 12. Comparison of results for quinidine serum concentrations by direct HPLC-UV assay and a single-extraction fluorescence assay [taken from Powers and Sadée, *Clin. Chem.*, **24**, 299–302 (1978)]. The broken line represents the line of perfect correlation with a slope of 1. Correlation and regression statistics (X: fluorescence, Y: HPLC, $n = 35$; in μg quinidine/ml serum):

$$X = 2.837, \quad Y = 1.926, \quad X - Y = 0.911 \, (n = 35)$$
$$S_X = 1.573, S_Y = 1.109, S_{X-Y} = 0.79$$

$r = 0.883$ (correlation coefficient); $S = 0.848$ (standard deviation, duplicates); regression line, slope = 0.622, intercept = 0.160 μg/ml; standard error of estimate = 0.529. These results demonstrate a bias of results; fluorimetry gave 50 ± 7% (SEM) higher values than the HPLC method, with a poor correlation ($r = .88$) not explicable by the known reproducibility error of both assays (corrected $r = .92$). The assay discrepancy is caused by fluorescent metabolites of quinidine which are not measured with the HPLC assay.

pendently prepared control samples. A large number of identical control samples are stored under conditions where the spiked drug is stable. Patient samples are run in duplicate. Assay results have to be rejected if the difference between duplicate assay results on the patient sample is greater than a carefully selected relative range that is different for each drug assay, and if both control samples deviate from the previously determined mean by more than twice the between-run standard deviation. Regulations for clinical laboratory procedures may vary among states and countries.

Further important characteristics of clinical drug assays are the time of analysis, reliability, ease of operation, and instrumentation required. The assay should be as simple as possible with a minimum number of analytical manipulations in order to reduce the frequency of operator mistakes during analysis.

Requirements for assay sensitivity and selectivity are dictated by the purpose for which the assay results will be used (see individual drug monographs). Clinical drug level assays do not have to be the most selective and certainly not the most sensitive techniques, unless the therapeutic drug concentration range is very low. For example, the nonspecific, relatively insensitive colorimetric Trinder assay of salicylates is quite suitable for clinical applications, since salicylate serum levels are routinely monitored only after high, anti-inflammatory doses or overdoses; plasma levels of salicylate metabolites are negligible under these conditions.

The time required to perform the drug assay and report the data to the therapist should be short enough to permit individual dosage adjustment, before the next dose is given. The frequency of drug administration is usually influenced by the drug's biological half-life; the shorter the half-life, the shorter the dosing interval should be. Modern analytical techniques can usually be performed in less than 1 hr, and the rate limiting factor in obtaining the drug level data may be not the assay procedure, but rather the organization of blood drawing, distribution of blood samples, data processing, and so on.

Bibliography

Christian, G. D., *Analytical Chemistry*, 2nd edition, John Wiley and Sons, New York, 1977.

Description of Analytical Methods and Results: Instruction to Authors, *Clin. Chem.*, **25**, 3-4 (1979).

Hamilton, L. F. and S. G. Simpson, *Calculations of Analytical Chemistry*, McGraw-Hill Book Co., New York, Toronto, London, 1960.

International Federation of Quality Control in Clinical Chemistry. Part 2: Assessment of Analytical Methods for Routine Use, *Clin. Chim. Acta*, **63**, F1–F17 (1976).

Mandel, J., *The Statistical Analysis of Experimental Data*, Interscience Publishers, 1964.

Pecsok, R. L., L. D. Shields, T. Cairns, and I. G. McWilliams, *Modern Methods of Chemical Analysis*, 2nd edition, John Wiley and Sons, New York, 1976.

4.14. CLINICAL PHARMACOKINETICS LABORATORY

With the clinical use of drug level data, the CPL has to conform to the regulations applicable to all clinical chemistry laboratories. These regulations include licensure, quality control, and general laboratory rules and may vary in different locations. Detailed information cannot be given here. However, the successful operation of a CPL depends not only on compliance with all regulations, but also on the selection of suitable assay procedures, the interface between the CPL and the hospital, proper data interpretation, and implementation of subsequent therapeutic decisions.

There are several suitable assay techniques for therapeutic drug level monitoring of the major drugs (Table II). The methods usually include HPLC, GC, colorimetry, RIA, and EIA, most of which give virtually identical results. However, no single method is sufficient for all clinical drug level assays. It is advisable to select two general assay methods, in addition to colorimetry (salicylates, sulfonamides, isoniazid) and flame spectrophotometry (lithium), for example, HPLC and RIA, or GC and EIA. In spite of its broad applicability, HPLC is still rather expensive relative to GC, particularly for maintenance and column utilization; thus HPLC is not absolutely required, although it presently dominates new assay developments. When comparing RIA and EIA, one has to weigh several factors; the equipment requirements are somewhat simpler for EIA and even overlap those for colorimetry. Furthermore, EIA eliminates any radiation hazards. However, many RIA procedures are available for the same drug from different commercial sources, providing a greater choice of antibodies and assay variations than is possible with EIA. Often, the decision will have to be made on the basis of availability of equipment and experience of laboratory personnel. In any case maintenance of three assay systems, for example, HPLC, GC, and RIA, is not readily justified because of the high cost of these procedures.

The interface of the CPL with the hospital is best achieved via a computer terminal that is connected to the central computer of the clinical laboratories. The terminal permits the personnel of the CPL to rapidly enter assay results and to obtain any clinical patient parameters needed for interpretation of the result.

The correct interpretation and utilization of drug level data is the key to the potential benefit of therapeutic drug level monitoring. The mere availability of a clinical drug level assay compels many therapists to order such assays without a firm intent to derive maximum utility from the results (see statistics in the gentamicin monograph). We therefore feel that at the present time a health-care professional with specific training and experience in therapeutic drug level monitoring should be either directly associated with the CPL or available for consultations. At UCSF, we have clinical pharmacologists and clinical pharmacists directly associated with the CPL to ensure proper clinical utilization of drug level data. Each unusual drug level result is accompanied by a written interpretation of the data (but no therapeutic recommendations) for ready use by the therapist. This procedure may be necessary, since therapeutic drug level monitoring in its present scope is a rather young discipline. In the future, clinical conditions requiring drug level data and the application of such data may become sufficiently standardized so that a specialized health-care professional is no longer needed to assist in the day-to-day operation of a CPL.

5

DRUG MONOGRAPHS

The selection criteria for the 102 drugs reviewed in this chapter were enumerated in the Introduction. A few monographs represent drug classes, that is, anticonvulsants, barbiturates, penicillins, sulfonamides, and tetracyclines, rather than individual drugs. Some of the major anticonvulsants are also reviewed in separate individual drug monographs because of their importance in therapeutic drug level monitoring. The monographs are designed to convey and interpret the most important data on drug metabolism, pharmacokinetics, and analytical techniques for each drug, cited mostly from original research articles published before May, 1979. Therefore the monographs should enable one to select a suitable drug assay for a given pharmacological or clinical purpose and, conversely, to critically evaluate drug level data based on the nature of the analytical technique. Moreover, assay development for new drugs is facilitated by ready access to analytical methods applicable to close chemical congeners described here (see also cross-references to chemical analogs of the listed drugs). Finally, the reviews of individual drug disposition, including the formation of active or toxic metabolites, are of value per se as a general information resource to health-care professionals, pharmacologists, and clinical chemists.

ACETAMINOPHEN

4'-Hydroxyacetanilide, paracetamol.
Amphenol, Calpol, Tylenol, Valadol, and so on. Analgesic, antipyretic.

$C_8H_9NO_2$; mol. wt. 151.16; pK_a 9.55

108

a. Therapeutic and Toxic Concentration Range. From 1 to 10 μg acetaminophen/ml plasma (Horwitz and Jatlow, 1977; Prescott et al., 1971). Renal failure and massive, often fatal liver necrosis occur after ingestion of large doses ($>>10$ g) acetaminophen. Toxicity is associated with a prolongation of the normal plasma elimination half-life from 2 hr to over 4 to 10 hr and with the persistence of high acetaminophen levels over several days (Boyer and Rouff, 1971; McJunkin et al., 1976). Ambre and Alexander (1977) note that acetaminophen plasma levels are a more reliable index of the risk of toxic liver injury than is the total dose ingested. Since liver necrosis can be successfully prevented by the administration of thiol reagents, such as cysteamine, L-methionine, N-acetylcysteine, and penicillamine (Prescott et al., 1976), numerous analytical methods for monitoring acetaminophen intoxication have been published. Furthermore, the metabolism of acetaminophen has been extensively studied because of its role in the mechanism of activation of acetaminophen to cytotoxic agents.

b. Metabolism. The major metabolites of acetaminophen are its sulfate and glucuronide conjugates, which are excreted into the urine (Andrews et al., 1976). Further urinary metabolites include the corresponding 3-hydroxy and 3-methoxy conjugates (Andrews et al., 1976) and 3-methylthio conjugates (Klutsch et al., 1978). The C-3-cystein and C-3-mercapturate conjugates are found in the urine in rather small amounts after low doses of acetaminophen, but are considerably increased after high, toxic doses (Andrews et al., 1976; Davis et al., 1976). Microsomal preparations are known to oxidize acetaminophen to active alkylating metabolites (Hinson et al., 1977; Potter et al., 1973) which bind convalently to macromolecular components such as proteins, presumably thereby causing tissue necrosis. Thiol conjugation with active amino acids or oligopeptides (glutathione) traps these active intermediates and protects the liver against injury. However, massive doses of acetaminophen deplete the glutathione stores of the liver and kidney, thereby resulting in severe tissue injury (Davis et al., 1976). The normal plasma elimination half-life of acetaminophen is below 2 hr, but can be considerably longer after massive doses (Prescott et al., 1971).

Acetaminophen is a major metabolite of phenacetin and acetanilide (Baty and Robinson, 1977).

c. Analogous Compounds. Acetanilide, phenacetin.

d. Analytical Techniques

    1. High Performance Liquid Chromatography—(Gotelli et al. (1977)

Discussion. This reverse phase HPLC assay with UV detection at 254 nm requires only 0.1 ml plasma with a sensitivity of 0.5 μg acetaminophen/ml and analysis time of 5 min. The extraction technique is rapid and includes acetoacetanilide as the internal standard. Of 36 other drugs tested, only theophylline interfered with the determination of acetaminophen. The assay is also suitable for the quantitative measurement of phenacetin.

Detailed Method. Plasma (0.5 ml), 0.5 ml of a 50 μg/ml solution of acetoacetanilid in methanol-water (1:10), and 0.5 ml 1 M phosphate buffer (pH 7.0) are mixed and extracted with 7 ml ethyl acetate by shaking for 10 min. After centrifugation the ethyl acetate layer is separated and evaporated under reduced pressure. The residue is dissolved in 50 μl methanol, and 10 to 20 μl is injected onto the column. Standard curves are constructed, using the peak height ratio method.

Analysis is performed on a Model 601 liquid chromatograph with a Model LC-55 variable-wavelength detector (Perkin-Elmer Corp.). The 30 cm $\times$ 4 mm (i.d.) column is packed with octadecyl trichlorosilane (μ-Bondapak C_{18}; Waters Assoc.). The detector is set at 254 nm. The column is eluted with a pH 4.4 phosphate buffer-acetonitrile (81:19) at a flow rate of 3 ml/min at 50°C. The pH 4.4 phosphate buffer is prepared by adding 0.3 ml 1 M potassium dihydrogen sulfate and 0.05 ml 4.4 M phosphoric acid to 1800 ml distilled water.

2. Other HPLC Assays—Black and Sprague (1978), Blair and Rumack (1977), Buckpitt et al. (1977), Horwitz and Jatlow (1977), Knox and Jurand (1977), Minet and Kissinger (1979), Munson et al. (1978), Riggin et al. (1975), Wong et al. (1976)

Discussion. Horwitz and Jatlow (1977) and Knox and Jurand (1977) also employ an octadecyl silica reverse phase column. Horwitz and Jatlow (1977) include an ether extraction step and report no assay interference by theophylline; this may be due to the ether extraction step, rather than the ethyl acetate used by Gotelli et al. (1977). The recovery of acetaminophen is 65% with ether and above 80% with ethyl acetate; however, theophylline might not be extractable into ether under the assay conditions of Horwitz and Jatlow (1977). Also, the low pH (2.7) of the HPLC eluent used by Horwitz and Jatlow might have caused separation of the theophyllin from the acetaminophen. The pH of 2.7 approaches the useful pH limit of the reverse phase column. Additional silylation of the octadecyl silica phase in order to mask free Si-OH functions improves the separation of acetaminophen and its metabolites from endogenous substances in urine (Knox and Jurand, 1977). Black and Sprague (1978) report that the theophyllin assay of Orcutt et al. (1977) (see theophyllin monograph) is directly applicable to the measurement of acetaminophen. Acetaminophen elutes before theophyllin on the C_{-18} reverse phase column; β-hydroxyethyltheophyllin can serve as the internal standard.

Blair and Rumack (1977) describe a microscale HPLC assay for acetaminophen on a cation-exchange column requiring only 1.5 μl plasma. The retention times of acetaminophen and its internal standard, N-butyryl-p-aminophenol, are 32 and 50 min, respectively, a circumstance that limits the applicability of this assay. Another mode of HPLC utilizes 10 μ particle size silica gel columns for the detection of acetaminophen in urine with a sensitivity of 1 $\mu g/ml$ (Wong et al., 1976). Also, pellicular polyamide packings are suitable for acetaminophen analysis in serum (Riggin et al., 1975). Excellent sensitivity of 50 ng acetaminophen/ml plasma is achieved with an electrochemical detector which is particularly sensitive for p-aminophenol as a potential impurity in acetaminophen (Riggin et al., 1975; Minet and Kissinger, 1979; Munson et al., 1978).

Conjugates of acetaminophen with cysteine, N-acetylcysteine, and glutathione can be quantitatively measured in the urine by HPLC (Buckpitt et al., 1977).

3. Gas Chromatography—Dechtiaruk et al. (1976), Evans and Harbison (1977), Grove (1971), Kalra et al. (1977), Prescott (1971), Serfontein et al. (1976), Street (1975).

Discussion. Of the many published GC methods, most employ ether extraction of plasma and urine samples at approximately neutral pH and flame ionization detection of acetaminophen. Phenacetin is simultaneously measured in some studies (Evans and Harbison, 1977; Prescott, 1971). The GC detection of acetaminophen in urine without derivatization allows a sensitivity of 2 $\mu g/ml$ after fluorisil column chromatography prepurification and ether extraction (Grove, 1971). However, underivatized acetaminophen leads to tailing GC peaks and nonlinear standard curves at low concentrations. The following compounds have been suggested as internal standards: N-propionyl-p-aminophenol (Dechtiaruk et al., 1976), p-bromoacetanilide (Evans and Harbison, 1977; Prescott, 1971), and p-chloroacetanilide (Prescott, 1971).

Derivatization methods include alkylation, acylation, and silylation. Prescott (1971) proposes silylation with N,O-bis(trimethylsilyl)acetamide, which forms the O,N-bis(TMS) derivative of acetaminophen and the N-TMS derivative of phenacetin. In the absence of phenacetin, N-trimethylsilylimidazole is used to selectively form the O-TMS derivative of acetaminophen with equally suitable GC properties. Phenacetin can be sensitively analyzed by GC without derivatization. The sensitivity in plasma is 0.05 μg acetaminophen/ml plasma. Sulfates and glucuronides can be measured after incubation with sulfatases and glucuronidases.

Flash methylation with trimethylanilinium hydroxide yields a permethylated derivative with good GC properties allowing a sensitivity of 0.1 μg acetaminophen/ml plasma (Evans and Harbison, 1977). Dechtiaruk et al. (1976) generate an O-heptyl-N-methyl acetaminophen derivative by sequential alkylation. Reaction with heptyl iodide off-column yields the O-heptyl derivative, followed by on-column N-methylation with trimethylanilinium hydroxide. The O-heptyl derivative was selected from a series of homologous alkyl groups in order to minimize interference of the acetaminophen derivative GC peak by endogenous substrates. The sensitivity is 0.5 μg acetaminophen/ml plasma.

Serfontein et al. (1976) utilize a microphase extraction method and formation of the O-butyrate derivative for GC analysis. Street (1975) contends that many of the GC assays are subject to interference by barbiturates. Such interference can be overcome by use of the O-benzoyl derivative of acetaminophen. Subsequent N-trimethylsilylation may be used for additional differentiation of the acetaminophen peak (Street, 1975). The sensitivity of this assay is 2 μg acetaminophen/ml plasma.

The potential interference of acetaminophen GC assays by the endogenous substrate p-hydroxyphenylacetic acid is pointed out by Kalra et al. (1977).

In conclusion, GC analysis provides potentially specific assay methods for acetaminophen in plasma and urine, given suitable extraction and derivatization steps. The assays of Prescott (1971) and Evan and Harbison (1977) are both rapid and sensitive with minimal sample preparation steps.

4. Gas Chromatography-Mass Spectrometry—Baty et al. (1976), Garland et al. (1977)

Discussion. Both GC-MS assays employ trideuteroacetaminophen, trideuterophenacetine (Baty et al., 1976; Garland et al; 1977), and trideuteroacetanilide (Baty et al., 1976) as the respective internal standards, which have been synthesized by reaction of hexadeuteroacetic anhydride with the appropriate aromatic amine. The deuterium exchange is negligible if the samples are kept at a pH below 8. The derivatization involves either silylation with N,O-bis(trimethylsilyl)acetamide (Baty et al., 1976) or O-methylation with diazomethane (Garland et al., 1977). The GC-MS is performed in the selected ion monitoring mode, using isobutane chemical ionization (Garland et al., 1977) and electron impact (Baty et al., 1976). Sensitivities are in the low nanograms per milliliter of plasma range.

5. Ultraviolet-Spectrophotometry—e.g., Kneipl (1974)

Discussion. The UV assays involve organic solvent extractions at neutral or acidic pH and back-extraction into aqueous base, followed by UV absorbance measurement ($UV_{max} \sim 273$ nm). Many drugs, including barbiturates, salycylic acid, theophylline, and oxyphenbutazone, interfere with these assays.

6. Colorimetry—Brodie and Axelrod (1949), Chambers and Jones (1976), Glynn and Kendal (1975), Kendal et al. (1976), Plakogiannis and Saad (1978), Routh et al. (1968), Swale (1977), Welch and Conney (1965), Widdop (1976), Widdop and Goulding (1976), Wiener (1977), Wilkinson (1976) (review).

Discussion. The various colorimetric assays have proved useful for monitoring acetaminophen intoxications. However, none of the procedures achieves the specificity and sensitivity of GC and HPLC.

The earliest colorimetric assay of phenacetin and acetaminophen in plasma was proposed by Brodie and Axelrod (1949). After acid hydrolysis, the formed p-aminophenol is diazotized and coupled to α-naphthol, and the azo dye is measured by spectrophotometry. Other procedures utilize the reaction of p-aminophenol with $NaNO_2$-HCl to form the colored 2-nitro-p-aminophenol (Chambers and Jones, 1976; Glynn and Kendal, 1975; Wiener, 1977) and with vanillin to a stable imine with an absorbance maximum at 395 nm (Plakogiannis and Saad, 1978). p-Aminophenol can also be coupled with phenol in the presence of hypobromite to form an indophenol dye (Welch and Conney, 1965) which may be interfered with by ascorbic acid (Swale, 1977).

Routh et al. (1968) suggest an assay based upon decoloration of diphenylpicrylhydrazyl, a free radical dye, by acetaminophen, which is quite sensitive but lacks specificity. All of the colorimetric assays are sufficiently sensitive to monitor toxic acetaminophen levels. For rapid clinical use, several commercial test kits have been described (Kendal et al., 1976; Widdop, 1976; Widdop and Goulding, 1976).

7. Polarography—Munson and Abdine (1978)

Discussion. The direct determination of acetaminophen at toxic concentrations in plasma (20 to 400 μg/ml) employs differential pulse voltametry. The method is based on the oxidation of the phenolic moiety of acetaminophen at a carbon paste electrode.

8. Comments. The literature volume on acetaminophen metabolism and analysis is large owing to the clinical importance of metabolite-mediated liver and kidney toxicity. It is clear that HPLC methods offer a number of advantages over other analytical techniques. Next to speed, these are sensitivity, specificity, and the ability to perform analyses without derivatization and to measure key metabolites (e.g., the thiol conjugates) after toxic doses.

e. References

Ambre, J. and M. Alexander, Liver Toxicity after Acetaminophen: Inadequacy of the Dose Estimate as an Index of Risk, *J.A.M.A.*, **238**, 500–501 (1977).

Andrews, R. S., C. C. Bond, J. Burnett, A. Saunders, and K. Watson, Isolation and Identification of Paracetamol Metabolites, *J. Int. Med. Res.*, **4** (4. Suppl.), 34–39 (1976).

Baty, J. D. and P. R. Robinson, Acetaminophen Production in Man after Coadministration of Acetanilid and Phenacetin: A Study with Stable Isotopes, *Clin. Pharmacol. Ther.*, **21**, 177–186 (1977).

Baty, J. D., P. R. Robinson, and J. Wharton, A Method for the Estimation of Acetanilide, Paracetamol and Phenacetin in Plasma and Urine Using Mass Fragmentography, *Biomed. Mass. Spectrom.*, **3**, 60–63 (1976).

Black, M. and K. Sprague, Micromethod for Acetaminophen Determination in Serum, *Clin. Chem.*, **24**, 1288–1289 (1978).

Blair, D. and B. H. Rumack, Acetaminophen in Serum and Plasma Estimated by High-Pressure Liquid Chromatography: A Micro-scale Method, *Clin. Chem.*, **23**, 743–745 (1977).

Boyer, T. D. and S. L. Rouff, Acetaminophen-Induced Hepatic Necrosis and Renal Failure, *J.A.M.A.*, **218**, 440–441 (1971).

Brodie, B. B. and J. Axelrod, The Fate of Acetophenetidine (Phenacetin) in Man and Methods for the Estimation of Acetophenetidin and Its Metabolites in Biological Material, *J. Pharm. Exp. Ther.*, **97**, 58–67 (1949).

Buckpitt, A. R., D. E. Rollins, S. D. Nelson, R. B. Franklin, and J. R. Mitchell, Quantitative Determination of the Glutathione, Cysteine, and *N*-Acetyl

Cysteine Conjugates of Acetaminophen by High-Pressure Liquid Chromatography, *Anal. Biochem.*, **83**, 168-177 (1977).

Chambers, R. E. and K. Jones, Comparison of a Gas Chromatographic and Colorimetric Method for the Determination of Plasma Paracetamol, *Ann. Clin. Biochem.*, **13**, 433-434 (1976).

Davis, M., C. J. Simmons, N. G. Harrison, and R. Williams, Paracetamol Overdose in Man: Relationship Between Pattern of Urinary Metabolites and Severity of Liver Damage, *Quart. J. Med.*, **45**, 181-191 (1976).

Dechtiaruk, W. A., G. F. Johnson, and H. M. Solomon, Gas Chromatographic Method for Acetaminophen (*N*-Acetyl-*p*-aminophenol) Based on Sequential Alkylation, *Clin. Chem.*, **22**, 879-883 (1976).

Evans, M. A. and R. D. Harbison, GLC Microanalyses of Phenacetin and Acetaminophen Plasma Levels, *J. Pharm. Sci.*, **66**, 1628-1629 (1977).

Garland, W. A., K. C. Hsiao, E. J. Pantuck, and A. H. Conney, Quantitative Determination of Phenacetin and Its Metabolite Acetaminophen by GLC-Chemical Ionization Mass Spectrometry, *J. Pharm. Sci.*, **66**, 340-344 (1977).

Glynn, J. P. and S. E. Kendal, Paracetamol Measurement, *Lancet* (1), 1147-1148 (1975).

Gotelli, G. R., P. M. Kabra, and L. J. Marton, Determination of Acetaminophen and Phenacetin in Plasma by High-Pressure Liquid Chromatography, *Clin. Chem.*, **23**, 957-959 (1977).

Grove, J., Gas-Liquid Chromatography of *N*-Acetyl-*p*-aminophenol (Paracetamol) in Plasma and Urine, *J. Chromatogr.*, **59**, 289-295 (1971).

Hinson, J. A., S. D. Nelson, and J. R. Mitchell, Studies on the Microsomal Formation of Arylating Metabolites of Acetaminophen and Phenacetin, *Mol. Pharmacol.*, **13**, 625-633 (1977).

Horwitz, R. A. and P. I. Jatlow, Determination of Acetaminophen Concentrations in Serum by High-Pressure Liquid Chromatography, *Clin. Chem.*, **23**, 1596-1598 (1977).

Kalra, J., O. A. Mamer, B. Gregory, and M. H. Gault, Interference by Endogenous *p*-Hydroxyphenylacetic Acid with Estimation of *N*-Acetyl-*p*-aminophenol in Urine by Gas Chromatography, *J. Pharm. Pharmacol.*, **29**, 127-128 (1977).

Kendal, S. E., G. Lloyd-Jones, and C. F. Smith, The Development of a Blood Paracetamol Estimation Kit, *J. Int. Med. Res.*, **4** (4. Suppl.), 83-88 (1976).

Klutch, A., W. Levin, R. L. Chang, F. Vane, and A. H. Conney, Formation of a Thiomethyl Metabolite of Phenacetin and Acetaminophen in Dogs and Man, *Clin. Pharmacol. Ther.*, **24**, 287-293 (1978).

Kneipl, J., A Sensitive, Specific Method for Measuring *N*-Acetyl-*p*-aminophenol (Paracetamol) in Blood, *Clin. Chim. Acta*, **52**, 369-372 (1974).

Knox, J. H. and J. Jurand, Determination of Paracetamol and Its Metabolites in Urine by High-Performance Liquid Chromatography Using Reversed-Phase Bonded Supports, *J. Chromatogr.*, **142**, 651-670 (1977).

McJunkin, B., K. W. Barwick, W. C. Little, and J. B. Winfield, Fatal Massive Hepatic Necrosis Following Acetaminophen Overdose, *J.A.M.A.*, **236**, 1874–1875 (1976).

Miner, D. J. and P. T. Kissinger, Trace Determination of Acetaminophen in Serum, *J. Pharm. Sci.*, **68**, 96–97 (1979).

Munson, J. W. and H. Abdine, Direct Determination of Acetaminophen in Plasma by Differential Pulse Voltametry, *J. Pharm. Sci.*, **67**, 1775–1776 (1978).

Munson, J. W., R. Weierstall, and H. B. Kostenbauder, Determination of Acetaminophen in Plasma by High-Performance Liquid Chromatography with Electrochemical Detection, *J. Chromatogr.*, **145**, 328–331 (1978).

Plakogiannis, F. M. and A. M. Saad, Quantitative Determination of Acetaminophen in Plasma, *J. Pharm. Sci.*, **67**, 581 (1978).

Potter, W. Z., D. C. Davis, J. R. Mitchell, D. J. Jollow, J. R. Gillette, and B. B. Brodie, Acetaminophen-Induced Hepatic Necrosis. 3: Cytochrome P-450-Mediated Covalent Binding *in vitro*, *J. Pharmacol. Exp. Ther.*, **187**, 203–210 (1973).

Prescott, L. F., The Gas-Liquid Chromatographic Estimation of Phenacetin and Paracetamol in Plasma and Urine, *J. Pharm. Pharmacol.*, **23**, 111–115 (1971).

Prescott, L. F., P. Roscoe, N. Wright, and S. S. Brown, Plasma-Paracetamol Half-Life and Hepatic Necrosis in Patients with Paracetamol Overdosage, *Lancet* (1), 519–522 (1971).

Prescott, L. F., G. R. Sutherland, J. Park, I. J. Smith, and A. T. Proudfoot, Cysteamine, Methionine, and Penicillamine in the Treatment of Paracetamol Poisoning, *Lancet* (2), 109–113 (1976).

Riggin, R. M., A. L. Schmidt, and P. T. Kissinger, Determination of Acetaminophen in Pharmaceutical Preparations and Body Fluids by High-Performance Liquid Chromatography with Electrochemical Detection, *J. Pharm. Sci.*, **64**, 680–683 (1975).

Routh, J. I., N. A. Shane, E. G. Arredondo, and W. D. Paul, Determination of *N*-Acetyl-*p*-aminophenol in Plasma, *Clin. Chem.*, **14**, 882–889 (1968).

Serfontein, W. J., L. S. De Villiere, L. S. Villiers, and D. Botha, GLC Determination of Paracetamol and D-Propoxyphene, *S. Afr. J. Med. Sci.*, **41**, 297–304 (1976).

Street, H. V., Estimation and Identification in Blood Plasma of Paracetamol (*n*-Acetyl-*p*-aminophenol) in the Presence of Barbiturates, *J. Chromatogr.*, **109**, 29–36 (1975).

Swale, J., Elimination of Interference Due to Ascorbic Acid When Detecting Paracetamol in Urine, *Lancet* (2), 981 (1977).

Welch, R. M. and A. H. Conney, A Simple Method for the Quantitative Determination of *N*-Acetyl-*p*-aminophenol (APAP) in Urine, *Clin. Chem.*, **11**, 1064–1067 (1965).

Widdop, B., The Paracetamol Test-Kit in Practice, *J. Int. Med. Res.*, 4(4. Suppl.), 89–92 (1976).

Widdop, B. and R. Goulding, Letter: Paracetamol Test Kit, *Lancet* (2), 583–584 (1976).

Wiener, K., Paracetamol Estimation: Comparison of A Quick Colorimetric Method with a Standard Spectrophotometric Method, *Ann. Clin. Biochem.*, **14**, 55–58 (1977).

Wilkinson, G. S., Rapid Determination of Plasma Paracetamol, *Ann. Clin. Biochem.*, **13**, 435–437 (1976).

Wong, L. T., G. Solomonraj, and B. H. Thomas, High-Pressure Liquid Chromatographic Determination of Acetaminophen in Biological Fluids, *J. Pharm. Sci.*, **65**, 1064–1066 (1976).

ACETAZOLAMIDE—by B. Silber

5-Acetamido-1,3,4-thiadiazole-2-sulfonamide.
Diamox®. Carbonic anhydrase inhibitor, diuretic.

$C_4H_6N_4O_3S_2$; mol. wt. 222.25; pK_a 7.2.

a. Therapeutic Concentration Range. Lehmann et al. (1969) found that the minimum effective plasma concentration necessary for lowering intraocular pressure was 10 μg/ml. A single oral dose of 250 mg acetazolamide produces peak plasma levels of 10 to 18 μg/ml (Wallace et al., 1977).

b. Metabolism. Utilizing an enzymatic technique, Maren (1960) reported that approximately 70% of an oral dose of acetazolamide was recovered unchanged in the urine. Metabolites of acetazolamide have thus far not been identified. Wallace et al. (1977) reported that acetazolamide elimination was biphasic with alpha and beta half-lives of approximately 2 and 13 hr, respectively.

c. Analogous Compounds. Chemical analogs: sulfa drugs, diazoxide, hydrochlorothiazide.

d. Analytical Methods

1. Gas Chromatography—Wallace et al. (1977)

Discussion. A GC technique was developed for acetazolamide, utilizing flash methylation by either tetramethylammonium hydroxide or trimethylphenylammonium hydroxide and electron capture detection for quantitation with the internal standard fluoranthene. The method requires solvent extraction. It is sensitive to about 10 ng acetazolamide per sample or 100 ng/ml plasma. No interferences were observed.

Detailed Method. Plasma (50 to 100 μl) is acidified to pH 5.0 with a 0.1 M acetate buffer and extracted with 10 ml ether–dichloromethane–2-propanol (6:4:2); the organic layer is transferred to another test tube and evaporated. If the concentration of acetazolamide exceeds the linear portion of the standard curve, an appropriate dilution of the organic layer is performed. To the residue, internal standard (2.5 μg fluoranthene in ether) is added. The solvent is then evaporated under nitrogen, and 50 μl of either 0.2% tetramethylammonium hydroxide or 0.1 M trimethylammonium hydroxide in methanol is added. Two microliters of this solution is then injected into the chromatograph.

A Varian Aerograph Model 1200 equipped with a ^{63}Ni electron capture detector was utilized under these conditions: 1.8 m × 0.3 cm (o.d.) glass column packed with 3% OV-17 on Gas Chrom Q (100/200 mesh). Carrier gas (95% argon, 5% methane) flow was at 30 ml/min. Injector, column, and detector temperatures were 280, 215, and 340°C, respectively. Under these conditions the retention times for the internal standard and acetazolamide were 2 and 2.5 min, respectively.

2. High Performance Liquid Chromatography—Bayne et al. (1975)

Discussion. Reverse phase HPLC with UV detection yields a sensitive (25 ng/ml) and specific assay for acetazolamide. Selection of the propioamide analog as internal standard increases assay reproducibility. However, the described solvent extraction is rather tedious and time consuming. Six extractions are required.

3. Enzymatic Methods—Maren (1960), Yakatan et al. (1976)

Discussion. Maren (1960) first proposed an enzymatic technique for quantitating acetazolamide concentrations based on carbonic anhydrase inhibition. Yakatan et al. (1976) subsequently simplified and improved the technique. The assay is sufficiently sensitive (200 ng/ml); however, it lacks the specificity of GC and HPLC.

4. Comments. The GC-EC technique described by Wallace et al. (1977) is recommended because of its rapidity and specificity. However, HPLC techniques utilizing reverse phase should be further developed to allow simplified solvent extractions or direct analysis of serum samples.

e. References

Bayne, W. F., G. Rogers, and N. Crisologo, Assay for Acetazolamide in Plasma, *J. Pharm. Sci.*, **64**, 402–404 (1975).

Lehmann, B., E. Linner, and P. J. Winstrand, The Pharmacokinetics of Acetazolamide in Relation to Its Use in the Treatment of Glaucoma and to Its Effects as an Inhibitor of Carbonic Anhydrases, *Adv. Biosci.*, **5**, 197–217 (1969).

Maren, T. H., A Simplified Micromethod for the Determination of Carbonic Anhydrase and Its Inhibitors, *J. Pharmacol. Exp. Ther.*, **130**, 26–29 (1960).

Maren, T. H., E. Mayer, and B. C. Wadsworth, Carbonic Anhydrase Inhibition. I: The Pharmacology of Diamox®, 2-Acetylamino-1,3,4-thiadiazole-5-sulfonamide, *Bull. Johns Hopkins Hosp.*, **95**, 199–243 (1954).

Wallace, S. M., V. P. Shaw, and S. Riegelman, GLC Analysis of Acetazolamide in Blood, Plasma, and Saliva Following Oral Administration, *J. Pharm. Sci.*, **66**, 527–530 (1977).

Yakatan, G. J., C. A. Martin, and R. V. Smith, Enzymatic Determination of Acetazolamide in Human Plasma, *Anal. Chim. Acta*, **84**, 173–177 (1976).

ε-AMINOCAPROIC ACID

6-Aminohexanoic acid.
Amicar®. Antifibrinolytic agent.

$C_6H_{13}NO_2$; mol. wt. 131.17.

a. Therapeutic Concentration Range. From 100 to 400 μg/ml (Keucher et al., 1976). Oral doses of 1 g ε-aminocaproic acid given every hour maintained an effective serum level of about 130 μg/ml in the human subjects tested (McNicol et al., 1962).

b. Metabolism. Approximately 80% of a single oral dose is recovered in the urine within 12 hr, suggesting that metabolism represents a minor pathway of elimination in man (McNicol et al., 1962).

c. Analogous Compounds. Tranexamic acid, amino acids.

d. Analytical Methods

1. High Performance Liquid Chromatography—Adams et al. (1977)

Discussion. The method consists of dansylation and reverse phase HPLC with fluorescense detection, using norleucine as internal standard. Very small ($<10 \mu l$) serum samples can be analyzed. Methionine potentially interferes with the assay; however, its average serum concentration is only $3.2 \mu g/ml$, which is well below the therapeutic range of ϵ-aminocaproic acid.

Detailed Method. A 10-μl volume of serum, 2 μl of the norleucine (4 μg) internal standard, and 40 μl of ethanol are mixed. The protein precipitate is centrifuged, and the supernatant fluid separated and evaporated at 70°C with a slow current of air. Ten microliters of 0.1 M bicarbonate solution is added, followed by 40 μl of a 10% dansyl chloride solution in acetone. The vial is capped and heated at 70°C for 15 min. After cooling, 50 μl of acetone is added, and 1 μl of the final solution chromatographed. Quantitation is by calculation of the drug internal standard peak area ratio.

Analysis is performed on a Series 2/2 HPLC with a Model 204S fluorescence detector (Perkin-Elmer), using a 0.26 × 25 cm column with a 10 μ octyl silane chemically bonded phase (RP8, E. Merck). The column is heated at 60°C and eluted with methanol-water-H_3PO_4 (40:60:0.1, v/v), at a flow rate of 1.5 ml/min. The column effluent is monitored with the detector set at an excitation wavelength of 345 nm and an emission wavelength of 545 nm.

2. Other HPLC Method—Shepherd et al. (1973)

Discussion. This method uses a modified rapid amino acid analyzer system employing postcolumn derivatization with ninhydrin and colorimetric detection. The method is sufficiently sensitive and precise for therapeutic drug level monitoring.

3. Gas Chromatography—Keucher et al. (1976)

Discussion. Keucher et al. (1976) utilize column chromatography preseparation and derivatization with *n*-butanol in HCL and with trifluoroacetic anhydride followed by GC-FID. Tranexamic acid serves as the internal standard. This method is considerably more cumbersome than the HPLC methods.

4. Comments. Other published assay methods include clot-lysis time determination, electrophoresis, and column chromatography (summarized by Adams et al., 1977).

e. References

Adams, R. F., G. J. Schmidt, and F. L. Vandemark, Determination of ε-Aminocaproic Acid in Serum by Reversed-Phase Chromatography with Fluorescence Detection, *Clin. Chem.*, **23**, 1226–1229 (1977).

Keucher, T. R., E. B. Solow, J. Metaxas, and R. L. Campbell, Gas Chromatographic Determination of an Antifibrinolytic Drug, ε-Aminocaproic Acid, *Clin. Chem.*, **22**, 806–809 (1976).

McNicol, G. P., A. P. Fletcher, N. Alkjaersig, and S. Sherry, The Absorption, Distribution, and Excretion of ε-Aminocaproic Acid Following Oral or Intravenous Administration to Man, *J. Lab. Clin. Med.*, **59**, 15–24 (1962).

Shepherd, J. A., D. W. Nibbelink, and L. D. Steginrk, Rapid Chromatographic Technique for the Determination of ε-Aminocaproic Acid in Physiological Fluids, *J. Chromatogr.*, **86**, 173–177 (1973).

AMINOPYRINE

4-Dimethylamino-2,3-dimethyl-1-phenyl-3-pyrazolin-5-one. Pyramidone, and so on. Antipyretic, analgesic.

$C_{13}H_{17}N_3O$; mol. wt. 231.29; pK_a 5.0.

a. Therapeutic Concentration Range. An oral dose of 300 mg aminopyrine yields peak plasma levels of approximately 5 μg/ml in adults (e.g., Kaneo et al., 1973).

b. Metabolism. Aminopyrine is rapidly metabolized by oxidative N-demethylation to 4-methylaminoantipyrine and further to the pharmacologically active 4-aminoantipyrine, which is then acetylated to the inactive 4-acetylaminoantipyrine as the major urinary excretion product (Brodie and Axelrod, 1950; Koizumi et al., 1974). Minor pathways of aminoantipyrine lead to 4-hydroxyantipyrine and its conjugates (Gradnik and Fleischman, 1973b; Schütz, 1975) and 4-formylaminoantipyrine (Noda et al., 1976), which has been implicated in mediating aminopyrine allergy (Shimeno and Yoshimura, 1972). The red urine color following ingestion of aminopyrine is caused by methylrubazoic acid and, to a lesser extent, rubazoic acid (Gradnik and Fleischman, 1973b). Furthermore, the potent carcinogen dimethylnitrosoamine may be formed *in vivo* in the stomach through reaction of aminopyrine with nitrites from foodstuffs (Labar and Sander, 1975), a reaction that can be prevented by ascorbic acid (Mirvish, 1975). Recovery of the known metabolites in the urine represents one third to one half of the dose, of which more than 50% is accounted for by 4-acetylaminoantipyrine and less than 5% by the parent drug (Gradnik and Fleischman, 1973b). The plasma elimination half-life of aminopyrine averages 2.7 hr (Vesell et al., 1976).

Since hepatic oxidative N-demethylation appears to be the predominant mode of elimination of aminopyrine, this drug has become a widely used test compound for liver function in drug metabolism. *In vitro* studies with hepatic microsomal preparations are usually based upon measurements of formaldehyde as the product of N-demethylation (Matsubara et al., 1977), while the $^{14}CO_2$ breath test following doses of ^{14}C-aminopyrine is the common assay technique in the intact animal and in human beings (Caspary, 1978; Hepner and Vesell, 1976). Reduced levels of expired $^{14}CO_2$ occur in liver diseases such as cirrhosis (Bircher et al., 1976) and cancer (Hepner et al., 1976) and in hepatocellular injuries associated with hyperbilirubinemia (Hepner and Vesell, 1977). The assay can also be performed with ^{13}C-aminopyrine and determination of the $^{12}CO_2/^{13}CO_2$ ratios in expired air by a mass analyzer, thereby avoiding radiation exposure (Schneider et al., 1978).

c. Analogous Compounds. Antipyrine.

d. Analytical Methods

1. Gas Chromatography—Windorfer and Röttger (1974)

Discussion. This article describes GC-FID assays for plasma containing aminopyrine, paracetamol, and three anticonvulsants. Docosane serves as the internal standard. Plasma samples are alkalinized and extracted into chloroform, and the extraction residues analyzed by GC, using a temperature program. The sensitivity is 1 μg aminopyrine/ml plasma, using a 0.5 ml sample volume, and the precision is 4%. The long retention time of the internal standard (35 min) and the temperature program render the GC analysis rather time consuming.

Detailed Method. Plasma (0.5 ml), 0.5 ml 0.1 N NaOH, and 7 μl 0.1% docosane solution in acetone are extracted twice, first with 25 ml and then with 10 ml of chloroform. The combined organic layers are evaporated at 60°C, the residue is transferred with 3 ml acetone to a smaller test tube, and the solution is again evaporated. The residue is dissolved in 100 μl acetone, which is evaporated, and again in another volume of 100 μl acetone. Of this final solution 5 μl is injected onto the GC column.

The GC-FID analysis is performed on a Model 546 Packard chromatograph equipped with a 4 m × 0.2 mm (i.d.) glass column packed with 4% DC-200 on Gas Chrom Q (120 mesh). The column temperature program starts at 185°C for 12 min, followed by a temperature increase of 2°C/min up to 240°C. Nitrogen, hydrogen, and air flow rates are 40, 40, and 400 ml/min, respectively.

2. Other GC Assays—Caille et al. (1977), Lavene et al. (1974)

Discussion. The GC assay of lidocaine plasma levels (Caille et al., 1977) utilizes aminopyrine as the internal standard and is therefore applicable also to the determination of aminopyrine. Lavene et al. (1974) report a procedure for the simultaneous detection of aminopyrine, 4-aminoantipyrine, and 4-acetylaminoantipyrine in plasma and urine on a 5% SE-30 column. The GC-FID is performed with a temperature program. With hexobarbital as the internal standard and a 5 ml sample volume, the assay is sensitive to 0.1 μg/ml plasma. The plasma extraction into chloroform is performed at pH 6, rather than the usually higher pH values, in order to allow extraction of both the weak pyrazolone bases and the weak acid hexobarbital.

3. Gas Chromatography-Mass Spectrometry—Goromaru et al. (1978), Kaneo et al. (1973)

Discussion. The GC-MS assay of aminopyrine plasma levels by Kaneo et al. (1973) is performed on a 1% OV-17 column with electron impact MS in the selected ion monitoring mode. Quantitation is achieved by peak area integration. Goromaru et al. (1978) report GC-MS procedures for the detection of aminopyrine, its *N*-demethylated metabolites, 4-acetylaminoantipyrine, and 4-formylantipyrine. Trideuterated internal standards for all of these compounds are synthesized from d_3-antipyrine, which is obtained by reaction of 1-phenyl-3-methyl-5-pyrazolone and d_3-dimethylsulfate. The GC liquid phases are either 1.5% OV-17 or 1% XE-60. Electron impact GC-MS is performed in the selected ion monitoring mode, utilizing the molecular ions of the respective deuterated and nondeuterated pyrazolones. Quantitation of all of these species is possible in plasma and urine by the isotope dilution principle.

4. Thin-Layer Chromatography—Gradnik and Fleischman (1973a), Oehne and Schmid (1972), Schütz (1975)

Discussion. There are many reports on the TLC separation of aminopyrine and its metabolites, mostly using silica gel preparations. Schütz (1975) suggests the Bratton-Marshall azo dye reaction for sensitive on-plate detection of the metabolite 4-aminoantipyrine in the urine. The major urinary metabolite, 4-acetylaminoantipyrine, can be converted to an azo dye following acidic hydrolysis on the plates to aminoantipyrine. Oehne and Schmid (1975) quantitatively detect aminopyrine, 4-aminoantipyrine, and 4-acetylaminoantipyrine directly on the TLC plates with a fluorescence scanner after treatment with 70% HClO₄ at 120°C for 60 min. Their quantitative urinary excretion data for human beings agree well with those obtained by Gradnik and Fleischman (1973b). These authors quantitate aminopyrine and six metabolites, including methyl rubazoic and rubazoic acid, by extracting urine samples and separating extracts on silica gel plates (Gradnik and Fleischmann, 1973a). Excellent separations are achieved on silica gels without gypsum binder. Appropriate bands are scraped off and eluted with CHCl₃-acetone, and UV absorbance is measured at 256 nm.

5. Spectrophotometry—Brodie and Axelrod (1950)

Discussion. Aminopyrine is measurable in plasma samples by UV spectrophotometry at 260 nm. Plasma samples are extracted at alkaline pH into dichloroethane, which is back-extracted into 0.1 *N* HCl. The UV absorbance of the acidic layer is determined.

6. Comments. The commonly neglected metabolites of oxidative dealkylation, formaldehyde and its product, carbon dioxide, have become important tools in assessing hepatic cytochrome P-450 activity under many conditions (see Section b). Rapid, simple, and sensitive assays of aminopyrine in plasma commensurate with current assay standards are not available. The reported GC assays are useful but can be considerably shortened. In view of the existence of many metabolites, HPLC-UV or TLC with UV scanning may become the method of choice.

e. References

Bircher, J., K. Upfer, I. Gikalof, and R. Preisig, Aminopyrine Demethylation Measured by Breath Analysis in Cirrhosis, *Clin. Pharmacol. Ther.*, **20**, 484–492 (1976).

Brodie, B. and J. Axelrod, The Fate of Aminopyrine (Pyramidon) in Man and Methods for the Estimation of Aminopyrine and Its Metabolites in Biological Material, *J. Pharmacol. Exp. Ther.*, **99**, 171–184 (1950).

Caille, G., J. Lelorier, Y. Latour, and J. G. Besner, GLC Determination of Lidocaine in Human Plasma, *J. Pharm. Sci.*, **66**, 1383–1385 (1977).

Caspary, W. F., Use of Breath Analysis in Diagnosis of Liver Function, *Z. Gastroenterol.*, **16**, 188–197 (1978).

Goromaru, T., K. Matsuyama, A. Noda, and S. Iguchi, The Measurement of Plasma Concentration of Aminopyrine and Its Metabolites in Man, *Chem. Pharm. Bull.* (Tokyo), **26**, 33–37 (1978).

Gradnik, B. and L. Fleischman, Quantitative Determination of Aminopyrine and Aminopyrine Metabolites by Thin Layer Chromatographic Separation, *Pharm. Acta Helv.*, **48**, 144–150 (1973a).

Gradnik, B. and L. Fleischman, Quantitative Urinary Excretion of Various Aminopyrine Metabolites in Man, *Pharm. Acta Helv.*, **48**, 181–191 (1973b).

Hepner, G. W. and E. S. Vesell, Aminopyrine Disposition: Studies on Breath, Saliva, and Urine of Normal Subjects and Patients with Liver Disease, *Clin. Pharmacol. Ther.*, **20**, 654–660 (1976).

Hepner, G. W., and E. S. Vesell, Aminopyrine Metabolism in the Presence of Hyperbilirubinemia Due to Cholestasis or Hepatocellular Disease: Combined Use of Laboratory Tests to Study Disease-Induced Aterations in Drug Disposition, *Clin. Pharmacol. Ther.*, **21**, 620–626 (1977).

Hepner, G. W., S. R. Uhlin, A. Lipton, H. A. Harvey, and V. Rohrer, Abnormal Aminopyrine Metabolism in Patients with Hepatic Neoplasm: Detection by Breath Test, *J.A.M.A.*, **236**, 1587–1590 (1976).

Kaneo, Y., T. Goromaru, and S. Iguchi, Microdetermination of Aminopyrine in Human Plasma by Mass Fragmentography, *J. Pharm. Soc. Jap.*, **93**, 258–260 (1973).

Koizumi, T., M. Ueda, and S. Takada, Urine Data Analysis for Pharmacokinetics of Aminopyrine and Its Metabolites in Man, *Chem. Pharm. Bull.* (Tokyo), **22**, 894–906 (1974).

Labar, J., and J. Sander, Carcinogenic *N*-Nitro-Dimethylamine from the Reaction of the Analgesic Amidopyrine and Nitrite Extracted from Foodstuffs, *Z. Krebsforsch.*, **84**, 299–310 (1975).

Lavene, D., M. Guerret, H. Humbert, and J. L. Kiger, Simultaneous Determination by Gas Chromatography of Butalbital, Caffeine, Amidopyrine, and 2 Metabolites of the Latter in Plasma and Urine, *Ann. Pharm. Fr.*, **32**, 505–512 (1974).

Matsubara, T., A. Touchi, and Y. Tochino, Hepatic Aminopyrine *N*-Demethylase System: Further Studies of Assay Procedure, *Jap. J. Pharmacol.*, **27**, 127–136 (1977).

Mirvish, S. S., Blocking the Formation of *N*-Nitroso Compounds with Ascorbic Acid *in vitro* and *in vivo*, *Ann. N.Y. Acad. Sci.*, **258**, 175–180 (1975).

Noda, A., T. Goromaru, N. Tsubone, K. Matsuyama, and S. Iguchi, *In vivo* Formation of 4-Formylaminoantipyrine as a New Metabolite of Aminopyrine, *Chem. Pharm. Bull.* (Tokyo), **24**, 1502–1505 (1976).

Oehne, H. and E. Schmid, Quantitative Determination of Amidopyrine and Its Metabolites in Urine Using Thin-Layer Chromatography and Spectrofluorometry, *Arzneim.-Forsch.*, **22**, 2115–2117 (1972).

Schneider, J. F., D. A. Schoeller, B. Nemchausky, J. L. Boyer, and P. Klein, Validation of $^{13}CO_2$ Breath Analysis as a Measurement of Demethylation of Stable Isotope Labeled Aminopyrine in Man, *Clin. Chim. Acta*, **84**, 153–162 (1978).

Schütz, H., The Chromatographic Reaction Detection of Aminophenazone (Pyramidon) and Its Major Metabolites, *Pharmazie*, **30**, 160–162 (1975).

Shimeno, H. and H. Yoshimura, Possible Implication of an Aldehyde Metabolite in Aminopyrine Allergy, *Xenobiotica*, **2**, 461–468 (1972).

Vesell, E. S., G. T. Passananti, and G. W. Hepner, Interaction Between Antipyrine and Aminopyrine, *Clin. Pharmacol. Ther.*, **20**, 661–669 (1976).

Windorfer, A. and H. J. Röttger, Micro-Method for the Quantitative Determination of Sedatives and Antipyretics (Aminophenazone, Diphenylhydantoin, Paracetamol, Phenobarbital, Primidone) Using Thermo-Gas Chromatography, *Arzneim.-Forsch.*, **24**, 893–895 (1974).

AMPHETAMINE

1-Phenyl-2-aminopropane.
Benzedrine, Phenedrine, and so on. Central nervous system stimulant.

$C_9H_{13}N$; mol. wt. 135.20; pK_a 9.83.

a. Therapeutic Concentration Range. Oral doses of 10 mg amphetamine in human beings yield peak plasma levels of about 20 ng/ml after 1 to 2 hr (Wan et al., 1978).

b. Metabolism. Oxidative deamination of amphetamine to phenylacetone oxime and phenylacetone represents the primary metabolic pathway (Hucker, 1973), which occurs to the extent of 30% of the dose under conditions of maximal urinary clearance, while parahydroxylation accounts for only a minor fraction of the dose (Vree et al., 1971). The renal excretion is highly dependent on the urinary pH with maximal urinary clearance at acidic pH (NH_4Cl treatment) and very little urinary excretion at alkaline pH ($NaHCO_3$ treatment) (Beckett and Rowland, 1965b; Vree et al., 1971; Wan et al., 1978). The *d*-isomer (dexamphetamine) is more active than the *l*-isomer, although its plasma elimination half-life is shorter than that of the *l*-isomer. Half-lives are 15.6 and 25.0 hr (basic urine) and 6.8 and 7.7 hr (acidic urine) for the *d*- and *l*-isomers, respectively (Wan et al., 1978). Minimal urinary excretion of amphetamine under alkaline conditions has to be considered in doping control when assaying amphetamine in urine rather than plasma samples. The complex metabolism of amphetamine has been reviewed by Cho and Wright (1978).

c. Analogous Compounds. Amphetamines (e.g., cypenamine, dimethylamphetamine, fencamfamine, methamphetamine, methylphenidate, phenmetrazine, xylopropamine).

d. Analytical Methods

1. Gas Chromatography-Mass Spectrometry—Matin et al. (1977)

Discussion. The quantitative determination of the optical isomers of amphetamine in plasma and saliva employs derivatization with *N*-pentafluorobenzoyl-S-(−)-prolyl-1-imidazolide. This reaction yields diastereomers of *d*- and *l*-amphetamine which can be separated by GC. The eluting diasteromers are quantitated by methane-chemical ionization MS in the selected ion monitoring mode, using *dl*-1,2-dideuteroamphetamine as internal standard. Linear calibration curves are obtained in the range of 0 to 80 ng amphetamine/ml plasma with a correlation coefficient of 3.5% ($N = 5$) at 50 ng/ml.

Detailed Method. To plasma (1 ml) or saliva (1 ml) containing amphetamine an aqueous solution of *dl*-1,2-dideuterio-1-phenyl-2-

aminopropane-HCl (^{2}H$_2$-amphetamine) (100 ng in 100 μl) is added as the internal standard. Samples are made basic by the addition of 2.5 N NaOH (0.2 ml) and are extracted with hexane (8 ml). The layers are separated by centrifugation, and the hexane layer is then transferred to another test tube. To the hexane extract 1 ml of a solution of N-pentafluorobenzoyl-S-($-$)-prolyl-l-imidazolide (PFBPI) in CH$_2$Cl$_2$ (100 μg/ml) is added. The test tubes are placed in an 80°C bath under a slow stream of nitrogen until the volume of hexane is reduced to 2 to 3 ml. On cooling, 2.5 N NaOH (1 ml) is added, and the test tubes are shaken for 15 min. After centrifugation the hexane layer is transferred to another test tube and evaporated to dryness under a stream of nitrogen in a water bath at 80°C. The residue is reconstituted in 25 to 50 μl of ethyl acetate, and 2 to 6 μl subjected to GC-MS analysis. A Finnigan Model 3200 gas chromatograph-mass spectrometer system is used with methane as the reactant gas. Proteo- and deuterioamphetamine isomers are measured quantitatively as they elute from the gas chromatograph by the mass spectrometer computer system by continuously monitoring the [MH]$^+$ ion at m/e 427 for proteoamphetamine and the [MH]$^+$ ion at m/e 429 for ^{2}H$_2$-amphetamine. The gas chromatograph is equipped with a 5 ft 2 mm (i.d.) glass column packed with 5% OV-275 coated on 100/120 mesh Chromosorb W-AW (Supelco, Inc., Bellefonte, Pa.). The oven and injector port temperatures are maintained at 235 and 280°C, respectively. The interface is maintained at 250°C, and the chemical ionization (c.i.) source is operated at 1 torr with methane as the carrier gas. The retention times of the d- and l-amphetamine derivatives are 19.2 and 13.4 min, respectively, under these conditions. Quantitation is achieved by measuring unlabeled deuterated amphetamine peak height ratios.

2. Other GC-MS Assays—Cho et al. (1973), Gal (1978)

Discussion. In the assay of Cho et al. (1973), *dl*-amphetamine is purified from plasma samples with a double back-extraction, derivatized with trifluoroacetic anhydride, and analyzed by quadrupole GC-MS at 70 eV in the selected ion monitoring mode. The internal standard is *dl*-3-trideuteroamphetamine. The peaks chosen for GC-MS analysis are the abundant ions at m/e 140 and 143 resulting from beta cleavage between C$_1$ and C$_2$ of the side chain. Underivatized amphetamine does not yield suitable ion peaks under electron impact. The d- and l-isomers of amphetamine are not separated by this technique, since the derivatives contain only one chiral center. The sensitivity of the assay is about 0.5 ng/ml plasma, using a 2 ml sample volume.

The GC-MS (electron impact) assay of Gal (1978) involves extraction from urine and derivatization of *dl*-amphetamine with the chiral reagent (*S*)-α-methoxy-α-(trifluoromethyl)phenylacetyl chloride. The diastereomeric *d*- and *l*-amphetamine derivatives are separated on the GC column and quantitated by selected ion monitoring, using trideuteroamphetamine as the internal standard. Urinary excretion of *l*-amphetamine is 10 to 20% higher than that of *d*-amphetamine after administration of racemic amphetamine to rats.

3. Gas Chromatography—Beckett and Rowland (1965a), Campbell (1969)

Discussion. The method of Beckett and Rowland (1965) can detect 0.1 μg *dl*-amphetamine/ml urine. The method uses GC-FID of underivatized amphetamine with *N,N*-dimethylaniline as the internal standard. Further identification of amphetamine in biological samples can be achieved by adding acetone to the ether extract. Following concentration of the solvent mixture and injection on the GC column, the amphetamine peak is shifted from the original retention time because of formation of the acetone imine. Campbell (1969) improved the sensitivity of the GS assay considerably to 2 ng amphetamine/ml plasma, using a 10 ml sample volume and a double back-extraction procedure. Evaporation of samples to dryness has to be avoided in order to prevent evaporation of amphetamine. The GC column has to be preinjected with nicotine to overcome irreversible absorption of amphetamine to the column packing material.

4. Fluorescence—Mehta and Schulman (1974)

Discussion. Although the native fluorescence of amphetamine in dilute acids is rather weak, strong fluorescence can be induced by reaction with a series of reagents, including aldehydes, ketones, and fluorescamine. The reactions with acetylacetone and formaldehyde, to give a 1,4-dihydrolutidine derivative, and with fluorescamine are sufficiently sensitive to detect amphetamine in urine.

5. Colorimetry—e.g., Frings et al. (1971)

Discussion. A series of colorimetric reagents is available to detect amphetamine for drug screening purposes in urine. Sensitivity and specificity are relatively low.

6. Chromatography—Hetland et al. (1972)

Discussion. Adsorption column and thin layer chromatography are described for amphetamine, coupled with TLC detection reagents for drug screening in urine.

7. Radioimmunoassay—Faraj et al. (1976)

Discussion. The antibody was raised by immunization of rabbits with a conjugate of N-(4-aminobutyl)methamphetamine coupled to bovine serum albumin. The specificity of the antibody was thoroughly investigated among a series of amphetamine derivatives. p-Hydroxymetabolites do not interfere because of the coupling to BSA at the amino function. l-Amphetamine is three times less reactive than d-amphetamine. The sensitivity is 5 ng amphetamine/ml plasma.

8. Enzyme Immunoassay—Broughton and Ross (1975)

Discussion. The homogeneous enzyme immunoassay of amphetamine in the urine serves as a drug screening method with a sensitivity of 1 μg amphetamine/ml urine. Amphetamine analogs also react with the antibody. The EMIT method is based on lysozyme acting on mucopolysaccharides in the cell wall of bacteria. The resulting lysis of bacteria causes the release of light absorbing substances into the medium, which can be monitored at 436 nm.

9. Enzymatic Assay—(Kreuz and Axelrod (1974)

Discussion. The assay is based on the transfer of the tritiated methyl group of S-adenosyl-L-[methyl-^{3}H]methionine to amphetamine in the presence of a partially purified N-methyltransferase from rabbit lung. The sensitivity is 10 ng amphetamine/ml plasma, using a 4 ml sample volume. The assay requires extensive purification steps to assure sufficient specificity and sensitivity at therapeutic amphetamine levels. The l-isomer is 30% less reactive with the enzyme than the d-isomer.

10. Comments. The most versatile and specific method for the quantitation of amphetamine in plasma is GC-MS. Derivatization with fluorescamine and detection by HPLC-fluorescence appears to be a likely alternative that warrants further investigation.

e. References

Beckett, A. H. and M. Rowland, Determination and Identification of Amphetamine in Urine, *J. Pharm. Pharmacol.*, **17**, 59–60 (1965a).

Beckett, A. H. and M. Rowland, Urinary Excretion Kinetics of Amphetamine in Man, *J. Pharm. Pharmacol.*, **17**, 628–639 (1965b).

Broughton, A. and D. L. Ross, Drug Screening by Enzymatic Immunoassay with the Centrifugal Analyzer, *Clin Chem.*, **21**, 186–189 (1975).

Campbell, D. B., A Method for the Measurement of Therapeutic Levels of (+)-Amphetamine in Human Plasma, *J. Pharm. Pharmacol.*, **21**, 129–130 (1969).

Cho, A. K. and J. Wright, Pathways of Metabolism of Amphetamine and Related Compounds, *Life Sci.*, **22**, 363–372 (1978).

Cho, A. K., B. Lindeke, B. J. Hodshon, and D. J. Jenden, Deuterium Substituted Amphetamine in a Gas Chomatographic/Mass Spectrometric (GC/MS) Assay for Amphetamine, *Analt. Chem.*, **45**, 570–574 (1973).

Faraj, B. A., Z. H. Israili, N. E. Kight, E. E. Smissman, and T. J. Pazdernik, Specificity of an Antibody Directed Against D-Methamphetamine: Studies with Rigid and Nonrigid Analogs, *J. Med. Chem.*, **19**, 20–25 (1976).

Frings, C. S., C. Queen, and L. B. Foster, Improved Colorimetric Method for Assay of Amphetamines in Urine, *Clin. Chem.*, **17**, 1016–1019 (1971).

Gal, J., Mass Spectra of N-[(S)-α-Methoxy-α-(trifluoromethyl)phenylacetyl] Derivatives of Chiral Amines: Stereochemistry of Amphetamine Metabolism in the Rat, *Biomed. Mass Spectrom.*, **5**, 32–37 (1978).

Hetland, L. B., D. A. Knowlton, and D. Couri, A Method for the Detection of Drugs at Therapeutic Dosages in Human Urine Using Adsorption Column Chromatography and Thin-Layer Chromatography, *Clin. Chim. Acta*, **36**, 473–478 (1972).

Hucker, H. B., Phenylacetone Oxime—an Intermediate in Amphetamine Deamination, *Drug Metab. Dispos.*, **1**, 332–336 (1973).

Kreuz, D. S. and J. Axelrod, Amphetamine in Human Plasma: A Sensitive and Specific Enzymatic Assay, *Science*, **183**, 420–421 (1974).

Matin, S. B., S. H. Wan, and J. B. Knight, Quantitative Determination of Enantiomeric Compounds. I: Simultaneous Measurement of the Optical Isomers of Amphetamine in Human Plasma and Saliva Using Chemical Ionization Mass Spectrometry, *Biomed. Mass Spectrom.*, **4**, 118–121 (1977).

Mehta, A. C. and S. G. Schulman, Comparison of Fluorometric Procedures for Assay of Amphetamine, *J. Pharm. Sci.*, **63**, 1150–1151 (1974).

Vree, T. B., J. P. Gorgels, A. T. Muskens, and J. M. Van Rossum, Deuterium Isotope Effects in the Metabolism of N-Alkylsubstituted Amphetamine in Man, *Clin. Chim. Acta*, **34**, 333–344 (1971).

Wan, S. H., S. B. Matin, and D. L. Azarnoff, Kinetics, Salivary Excretion of Amphetamine Isomers, and Effect of Urinary pH, *Clin. Pharmacol. Ther.*, **23**, 585–590 (1978).

ANTICONVULSANTS

BARBITURATES

Mephobarbital (Mebaral)

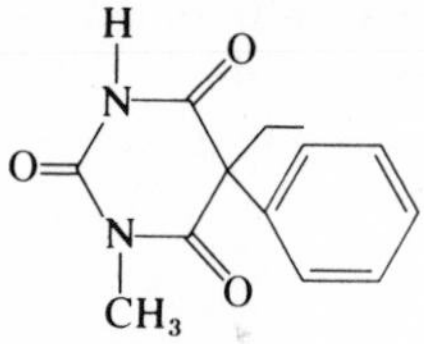

mol. wt. 246.26; pK_a 7.7.

Phenobarbital (Luminal)

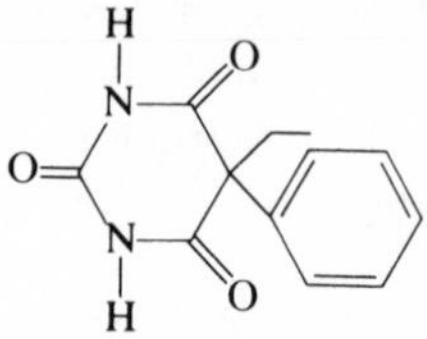

mol. wt. 232.21; pK_a 7.5.

Primidone (Mysoline)

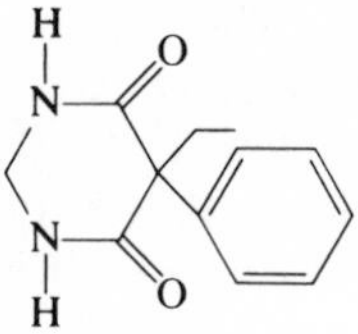

mol. wt. 218.25.

HYDANTOINS

Ethotoin (Nirvanol)

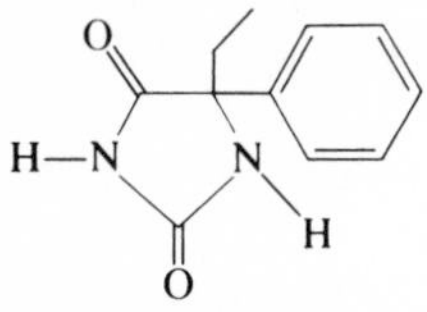

mol. wt. 204.22.

Mephenytoin (Mesantoin)

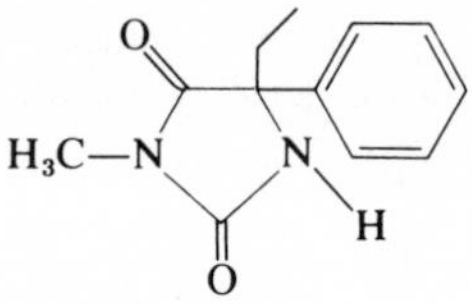

mol. wt. 218.25.

Phenytoin (Dilantin)

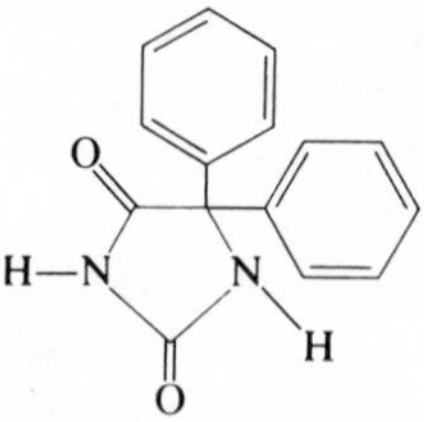

mol. wt. 252.26; pK_a 8.3.

OXAZOLIDONES

Paramethadione (Paradione)

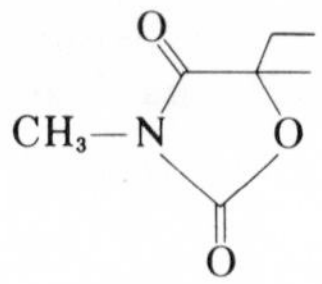

mol. wt. 157.17.

Trimethadione (Tridione)

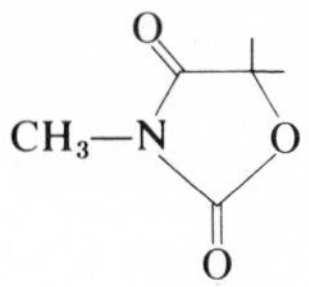

mol. wt. 143.14.

SUCCINIMIDES

Ethosuximide (Zarontin)

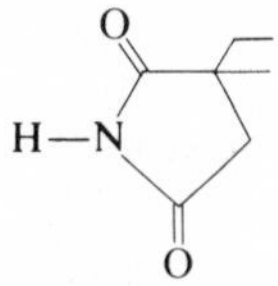

mol. wt. 141.17; pK_a 9.0

Methsuximide (Celontin)

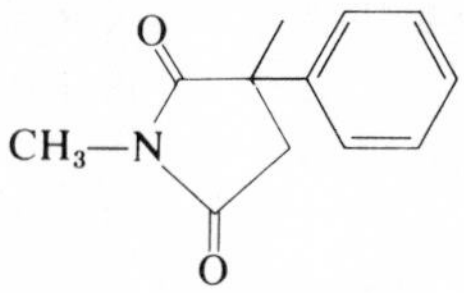

mol. wt. 203.23.

Phensuximide (Milontin)

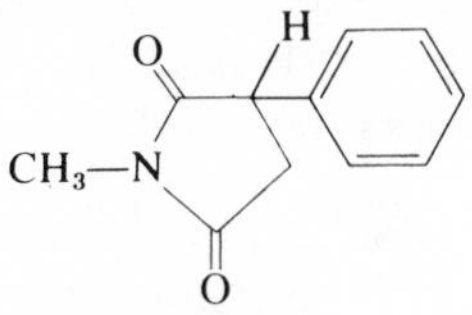

mol. wt. 189.21.

MISCELLANEOUS DRUGS

Carbamazepine

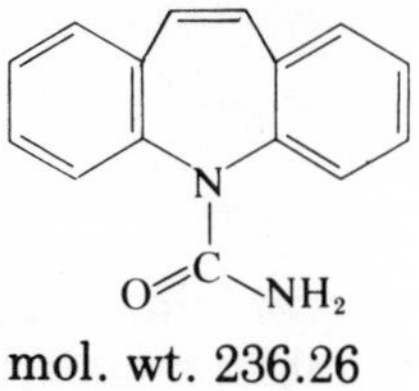

mol. wt. 236.26

Of the anticonvulsant drugs with rather different chemical structures, carbamazepine is frequently included with general anticonvulsant assays in plasma and urine. The barbiturates (including mephobarbital and phenobarbital), primidone, phenytoin, ethosuximide, and carbamazepine, as well as additional anticonvulsants (i.e., benzodiazepines and valproic acid) are also presented in separate individual monographs.

a. Therapeutic Concentration Range (μg/ml, in plasma)

Carbamazepine	4–8
Ethosuximide	40–100
Ethotoin	15–50
Mephenytoin	15–40 (sum of mephenytoin and its metabolite phenylethylhydantoin)
Mephobarbital	Not applicable (therapeutic levels of the metabolite phenobarbital)
Methsuximide	10–40 (as the N-desmethylmethsuximide metabolite)
Paramethadione	Not applicable (active metabolite, ethylmethyloxazolidinedione)
Phenobarbital	10–40
Phensuximide	5–15
Phenytoin	10–20
Primidone	4–12 (and therapeutic levels of its metabolite phenobarbital)
Trimethadione	100 (as its metabolite dimethadione)

(Guelen and van der Kleijn, 1978; Kalman et al., 1976; Leal and Troupin, 1977; Rose et al., 1971; Strong et al., 1974).

b. Metabolism. See individual monographs (carbamazepine, ethosuximide, barbiturates, phenytoin, primidone).

c. Analogous Compounds. Barbiturates, hydantoins, oxazolidones, succinimides.

d. Analytical Methods

1. High Performance Liquid Chromatography—(Kabra et al. (1977)

Discussion. This is a direct method for simultaneously determining five anticonvulsants (phenobarbital, phenytoin, primidone, ethosuximide, carbamazepine) in as little as 25 μl of serum. Proteins are precipitated with an acetonitrile solution containing hexobarbital as the internal standard. The drugs are eluted from a reverse phase column and detected by their absorbance of 195 nm. Each analysis requires less than 14 min at an optimum column temperature of 50°C. The lower limit of detection of these drugs is less than 1.0 μg/ml serum, and analytical recoveries vary from 97 to 107% with C.V.s between 3.9 and 5.0%. Choice of UV absorbance of 195 nm for the eluent detector greatly enhances the sensitivity for ethosuximide, primidone, and carbamazepine, which possess only minimal absorptivity at 254 nm (Adams, 1977). However, the rather short UV wavelength increases the potential for other drug interferences. Of the over 30 drugs tested for possible interference, only ethotoin interferes with the analysis of phenobarbital. Comparison with a gas-liquid chromatographic method gave good correlations for these anticonvulsant drugs ($r > .98$). In addition, background from drug-free serum and plasma sample is negligible (less than 0.1 μg/ml).

Detailed Method. To 25 to 500 μl serum is added an equal volume of acetonitrile containing 50 μg hexobarbital/ml. The mixture is vortex-mixed for ten seconds and centrifuged for 2 min with an Eppendorf® Model 5412 centrifuge. About 20 μl of the supernatant is injected onto the high pressure liquid chromatograph.

The instrument used for analysis is a Model 601 (Perkin-Elmer Corp., Norwalk, Conn. 06856) high pressure liquid chromatograph equipped with a variable-wavelength detector (Perkin-Elmer LC-55) and a temperature-controlled oven set at 50°C. A 30 cm x 3.9 mm (i.d.) μ-Bondapack C_{18} reverse phase column (Waters Assoc, Milford, Mass. 07157) is mounted on the chromatograph. The mobile phase is acetonitrile pH 4.4 phosphate buffer (4.5×10^{-5} M) (19:18, v/v). Concentrations of the drugs are calculated from the response factors of a known standard and the internal standard (25 μg/ml each, except hexobarbital, which is 50 μg/ml) injected onto the column.

2. Other HPLC Assays—Adams (1977) review, Adams and Schmidt (1976), Adams and Vandemark (1976), Atwell et al (1975), Evans (1973), Kabra et al. (1976), Kitazawa and Komura (1976), Soldin and Hill (1976, 1977)

Discussion. Reverse phase HPLC assays similar to the one of Kabra et al. (1977) have also been developed by Adams and Schmidt (1976), Adams and Vandemark (1976), and Soldin and Hill (1976, 1977). The method of Soldin and Hill (1976) employs a pH 8 eluent buffer which may limit the useful life time of the column material. However, because of the small sample volume injected (equivalent to 1.5 μl deproteinized serum), the useful life of the column is over 1200 analyses (Soldin and Hill, 1977). The same five major anticonvulsants determined by Kabra et al. (1977) are also measured in this study at 200 nm, using cyheptamide as the internal standard. Correlation coefficients, obtained by comparing the HPLC assay with a GC method, are excellent except for primidone (r .77). The primidone metabolite phenylethylmalondiamide may not have been separated from primidone under the assay conditions. In 1300 clinical plasma samples, only gentamicin, diazoxide, and mephobarbital were found to interfere with the assay. Soldin and Hill (1977) suggest subsequently the routine use of dual-wavelength UV detection at 254 and 200 nm in order to identify assay interferences. Only 12 out of 2000 phenytoin and phenobarbital assays gave 254/200 ratios different from those of pure sample injections. Additional GC analysis of these 12 samples indicated that the 254 nm, but not the 200 nm, detection was interfered with in 10 sera; the 200 nm detection gave results different from GC analysis for only 2 samples.

Adams and Vandemark (1976) propose a sensitive (0.1 to 0.5 μg/ml plasma) reverse phase HPLC assay for the five major anticonvulsants and methsuximide, based on charcoal absorption from plasma and UV detection at 195 nm. Phenacetin serves as the internal standard, a questionable choice for clinically applied anticonvulsant assays because of the frequent use of this analgesic. Analytical conditions for the determination of less common anticonvulsants are given by Adams and Schmidt (1976).

Earlier HPLC assays employ organic solvent extractions and silica gel columns with 254 nm UV detection of phenobarbital and phenytoin (Evans, 1973; Atwell et al., 1975). Phenobarbital, phenytoin, and carbamazepine can be detected simultaneously by anion-exchange chromatography, using α-naphthol as the internal standard, with a sensitivity of 1 μg drug/ml plasma (Kitazawa and Komura, 1976).

3. Gas Chromatography—(Heipertz et al. (1977)

Discussion. Heipertz et al. (1977) present a simultaneous assay method for plasma levels of ethosuximide, carbamazepine, phenytoin, phenobarbital, primidone, and the primidone metabolite phenylethylmalonediamide, with mephenytoin as the internal standard. The procedure includes a dual ether extraction of the drug from serum before and after addition of ammonium sulfate in order to assure nearly complete recovery yields. Standard curves can therefore be constructed from control solutions without plasma extractions. Gas chromato-graphic analysis of the underivatized anticonvulsants is performed using a linear temperature program and nitrogen-sensitive flame ioniza-tion detection. Choice of a slightly acidic stationary phase (SP-1000, a terephthalic acid-modified Carbowax 20M) minimizes on-column absorption of the acidic anticonvulsants, in particular phenobarbital, as well as the acid-catalyzed thermolysis of carbamazepine to iminostilbene (16 to 22%). During routine analysis of over 800 samples, the procedure gave a low day-to-day variability with C.V.s below 5% for each drug.

Detailed Method. Serum (1 ml) and 4 μl of an aqueous mephenytoin solution (1 mg/ml) are extracted with 5 ml diethyl ether, and after separation of the ethereal layer and addition of 200 mg ammonium sulfate the extraction is repeated with another 5 ml ether. The mixture is then centrifuged, and the two ethereal layers are combined, dried over sodium sulfate, and evaporated under a stream of nitrogen at 45°C. the residue is taken up in 40 μl methanol, and 0.5 μl of this solu-tion injected into the gas chromatograph.

Analysis is performed on a Packard Model 419 (or Pye Model 104) gas chromatograph with a nitrogen-selective thermionic detector con-nected to an Autolab System I integrator. The gas chromatograph is equipped with 1.8 mm (i.d.) silanized glass columns filled with 0.8% SP-1000 on 100/120 Chromosorb 750. Flow rates of the gases argon, hydrogen, and oxygen are 30, 40, and 240 ml/min, respectively. The injection port and detector are maintained at temperatures of 240 and 270°C, respectively, while the column oven temperature is programmed after 40 sec at 180°C to 255°C with a rate of 150°C/min. The detector system consists of a rubidium bromide crystal. Lower limits of detection range from 1 to 3 ng of the various drugs injected onto the column. Peak area ratios are used for construction of standard curves from calibration solutions. The GC analysis lasts for approximately 15 min.

4. Other GC Assays—Abraham and Gresham (1977), Ayers et al. (1977), Cramers et al. (1976), Davis et al. (1975), Dorrity and Lin-noila (1976), Hill and Latham (1977), Larsen et al. (1972), Least et al. (1977), Malkus et al. (1978), Nishina et al. (1976), Papadopoulos et al. (1973), Roger et al. (1973), Serfontein and de Villiers (1977), Toseland et al. (1972), Vandemark and Adams (1976)

Discussion. The listed articles are representative of the numerous studies on the simultaneous GC analysis of several anticonvulsant drugs. Most methods include the major traditional anticonvulsants phenobarbital, phenytoin, and primidone; carbamazepin is measured in addition in only some of these techniques (Cramers et al., 1976; Larsen et al. 1972; Nishina et al. 1976; Roger et al., 1973; Toseland et al., 1972). Succinimides, such as ethosuximide, are codetermined by Abraham and Gresham (1977) and Hill and Latham (1977). The most commonly employed internal GC standard compounds are mephenytoin, p-methylphenyl-phenylhydantoin, and cyheptamide.

The large number of published GC assays with minor modifications attests to the difficulties in obtaining reliable results for all of the major anticonvulsants using the same assay procedure for routine clinical applications. Several investigators perform GC analysis of the underivatized drugs (Cramers et al., 1976; Larsen et al., 1972; Papadopoulos et al., 1973; Toseland et al., 1972). The major problems are column absorption of phenobarbital and thermolysis of carbamazepine (compare Heipertz et al., 1977), which are critically dependent on the chromatographic conditions. Cramers et al. (1976) suggest a highly sensitive (10^{-9} g injected) capillary GC method which requires only 0.1 ml serum for routine analysis. These authors report no chromatographic loss during analysis with OV-225 as the stationary phase.

All other procedures involve a derivatization step, usually by off-column or on-column methylation. Alkylation of all these drugs yields suitable GC properties, which reduce the risk of artifacts due to this method. However, methylation of phenobarbital precludes its specific analysis in the presence of mephobarbital because of the formation of identical derivatives. If mephobarbital is administered therapeutically, alkylation other than methylation has to be utilized (e.g., hexylation; see Serfontein and de Villiers, 1977). Furthermore, the commonly used methylating reagents, tetramethylammonium hydroxide and trimethylphenylammonium hydroxide, cause decomposition of some of the anticonvulsant, particularly phenobarbital, which cleaves to N-methyl-α-phenylbutyramide ("early phenobarbital peak"). This chemical degradation may occur during prolonged contact (~ 10 min) with the derivatizing reagent in solution (Davis et al., 1975) or in the injector port during flash methylation (Nishina et al., 1976). When using flash methylation, phenobarbital degradation can be minimized either by diluting the trimethylphenylammonium hydroxide reagent in methanol from 2 M to 30 mM (Least et al., 1977), or by reducing the basicity of the reagent to a pH between 8 and 10 with appropriate buffers (Serfontein and de Villiers, 1977). Alternatively, methylation can be carried out off-column by heating the quaternary ammonium reagent in an organic solvent such as dimethylformamide (Nishina et al., 1976) or N,N-dimethylacetamide containing methyl iodide (Vandemark and

Adams, 1976), or by evaporating the reagent solution to dryness at elevated temperatures (Malkus et al., 1978; Hill and Latham, 1977). (See also barbiturates monograph.)

Serum samples are commonly extracted with organic solvents, such as chloroform, ether, ethyl acetate, and toluene. Some assays include a back-extraction into aqueous bases (Malkus et al., 1978), which, however, was not considered necessary by most authors. Several procedures include back-extraction of the anticonvulsants from toluene into a small volume of the methanolic quaternary ammonium hydroxide solution, thereby avoiding evaporation steps (Dorrity and Linnoila, 1976; Vandemark and Adams, 1976). The predominantly neutral carbamazepine cannot be readily extracted from organic solvents into alkaline solutions and is excluded from GC analysis.

Nitrogen-sensitive FID analysis often increases the sensitivity over that obtained with conventional FID (e.g., Ayers et al., 1977; Least et al., 1977; Vandemark and Adams, 1976) and reduces serum sample requirements to 0.2 ml and less. The elegant assay of Least et al. (1977) for phenobarbital, primidone, and phenytoin is designed for pediatric clinical use with a sample volume of only 20 μl and a run-to-run precision between 2.8 and 4.4%, using on-column methylation. Unfortunately, carbamazepine and other anticonvulsants were not considered in this study. Employing regular FID and larger sample volumes, Nishina et al. (1976) report similarly low C.V.s (about 5%) for their assays of phenobarbital, primidone, phenytoin, and carbamazepine with off-column methylation.

In summary, several reliable GC assays of the anticonvulsant drugs have been presented, either with or without derivatization prior to analysis. Comprehensive GC assays for all of the major anticonvulsants usually require temperature programming because of differences in drug volatility and polarity. Care has to be taken in selecting and maintaining proper assay conditions in order to avoid erratic results.

5. Mass Spectrometry—Horning et al. (1974)

Discussion. Horning and co-workers (1974) have developed two mass spectrometric techniques for the analysis of anticonvulsants in plasma, urine, and breast, Gas chromatography-mass spectrometry in the chemical ionization mode with methane as the carrier and reactant gas is used for measurements of phenobarbital, primidone, phenylethylmalondiamide, phenytoin, and the succinimides. The drugs are extracted from plasma with ammonium carbonate-ethyl acetate. Phenobarbital is methylated with diazomethane, followed by silylation of the mixture in order to form N,N'-ditrimethylsilyl derivatives of primidone and phenylethylmalondiamide. The succinimides are not derivatized and analyzed as such. A narrow mass range including the respective $M^+ + 1$ ions is repetitively scanned, and mass chromatograms of individual drugs and their internal standards are reconstructed with the aid of a computer. Phenobarbital-2,4,5-^{13}C, phenytoin-2,4,5-^{13}C, and chemical succinimide analogs serve as the internal standards.

Atmospheric pressure ionization mass spectrometry of dichloromethane extracts containing phenobarbital and phenytoin is used with direct injection of the underivatized organic extract into the external ion source. The M^{-1} ions are quantitatively monitored in the negative ion mode with the ^{13}C-labeled internal standards. Both MS procedures are sensitive and specific; however, they require expensive instrumentation, which limits their wide applicability.

      6. Enzyme Immunoassay—Brunk et al. (1976), Finley et al. (1976), Legas and Raisys (1976), Ollerich et al. (1977), Spiehler et al. (1976), Sun and Szafir (1977), Turri (1977), Ven Lente et al. (1977)

Discussion. The "enzyme multiplied inhibition technique" (EMIT, Syva Co.) has found wide applicability for the clinical measurement of the anticonvulsant drugs, mainly because of its speed and small sample volume requirement. Its major disadvantage, when compared to GC and HPLC, is the necessity to use specific antibodies in separate incubations for each individual anticonvulsant drug. If active metabolites are formed, as is the case for primidone leading to phenobarbital, two or even more separate incubations may be needed. However, the ready adaptation of EMIT to automated procedures using the centrifugal analyzer overcomes this disadvantage (Brunk et al., 1976; Finley et al., 1976; Turri, 1977; Van Lente et al., 1977). Coefficients of variation for the determination of phenobarbital, phenytoin, primidone, carbamazepine, and ethosuximide are in the range of 5 to 10% (day-to-day variation). Microscale methods for phenobarbital and phenytoin have been published (Brink et al., 1976), requiring only 3 μl serum for duplicate analysis with a C.V. of 15%. Such procedures can considerably reduce reagent costs for routine anticonvulsant assays.

Correlations of the EMIT technique with other clinically established assays, such as GC, RIA, and spectrophotometry, are good to excellent. Therefore immunoassays can be substituted for conventional techniques without changing the clinical interpretation of results.

The EMIT assay is in principle a competitive protein binding method. The anticonvulsant drug is covalently bound to a bacterial glucose-6-phosphate dehydrogenase, which, in the presence of substrate, converts the cofactor NAD^+ to $NADH$, measurable at 340 nm. Upon binding of the drug-enzyme complex to a drug-specific antibody, the enzyme activity is partially inhibited, as measured spectrophotometrically at 340 nm. The free drug present in serum is competing with the drug-enzyme complex for antibody binding sites, thereby reversing the enzyme inhibition. Increasing drug concentrations yield a displacement curve, measured by UV absorbance, equivalent to that observed with RIA methods. As with other immunoassays, the antibody specificity is critical to the quality of the assay. For example, the EMIT phenobarbital antibody cross-affinity to mephobarbital is 20%, and for other barbiturates, phenytoin, and primidone is below or equal to 1% (Spiehler et al., 1976). The phenytoin antibody possesses a cross-affinity of 0.5% to 5-(p-hydroxyphenyl)-5-phenylhydantoin, the phenytoin metabolite, and negligible cross-affinity to the other anticonvulsants (Spiehler et al., 1976).

7. Radioimmunoassay—e.g., Spiehler et al. (1976)

Discussion. Radioimmunoassay does not readily lend itself to automation. In view of the need for separate incubations for each drug and its active metabolites, RIA techniques are discussed in the individual drug monographs.

8. Thin-Layer Chromatography—e.g., Breyer and Villumsen (1975), Pippinger et al. (1969)

Discussion. Serum and urine assays of the anticonvulsant drugs are well within the range of quantitative analysis using TLC with a UV scanning detector. Such techniques can be quite specific and rapid; however, suitable instrumentation for their routine application is often not available.

9. Ultraviolet Spectrophotometry and Colorimetry—e.g., Fellenberg and Pollard (1976), Huisman (1966), Spiehler et al. (1976)

Discussion. The UV spectrophotometric assay for barbiturates after organic solvent extraction and back-extraction into aqueous base may be useful for clinical assays of phenobarbital (Spiehler et al., 1976), although it is subject to many interferences. Several chemical reactions or derivatizations have been proposed for the various anticonvulsants in order to achieve better specificity in biological assays. Fellenberg and

Pollard (1976) propose a specific assay of phenytoin and carbamazepin in the same serum sample, based on the permanganate oxidation of phenytoin to benzophenone and the acid-catalyzed rearrangement of carbamazepin to 9-methylacridine. The products are measured spectrophotometrically. Serum sample requirements are only 0.1 to 0.2 ml with a sensitivity of 1 μg/ml.

The colorimetric technique of Huisman (1966) is based on the aromatic nitration of the anticonvulsant drugs, followed by TLC separation, extraction from the plate, and reduction of the nitro group. The generated aromatic amines are then estimated by diazotization and azo dye formation according to the Bratton-Marshall method. Huisman also reviews earlier spectrophotometric assays.

10. Various Assays
Differential Pulse Polarography—Brooks et al. (1973). Electron Spin Resonance Immunoassay—Montgomery et al. (1975)

11. Assay Comparisons and Reviews—Abraham (1977), Dixon et al. (1976), Schneider et al. (1975), Siest and Young (1976), Spiehler et al. (1976)

12. Comments. Anticonvulsant serum assays have undoubtedly proved useful in improving the clinical application of these drugs. Current generally applicable techniques include HPLC, GC, and enzyme immunoassays. More assay techniques may be useful for the determination of individual anticonvulsant drugs. A recent review of 112 clinical laboratories, however, has revealed wide interlaboratory variability in anticonvulsant drug assay results (Pippinger et al., 1976), which probably reflects the analytical difficulties discussed in this chapter. Meticulous care has to be given to the selection and routine performance of assay procedures and to vigorous quality control in order to assure the therapeutic benefits expected from drug level data for seizure control.

e. References
Abraham, C. V., Comparison of Six Procedures for the Simultaneous Determination of Antiepileptic Drugs, *Am. J. Med. Technol.*, **43,** 935–938 (1977).

Abraham, C. V. and D. Gresham, Simultaneous Gas Chromatographic Analysis for the Seven Commonly Used Antiepileptic Drugs in Serum, *J. Chromatogr.*, **136,** 332–336 (1977).

Adams, R. F., The Determination of Anticonvulsants in Biological Samples by Use of High-Pressure Liquid Chromatography, *Adv. Chromatogr.*, **15**, 131–168 (1977).

Adams, R. F. and G. Schmidt, Determination of Less Common Anticonvulsants in Serum Using Reversed-Phase High-Pressure Liquid Chromatography, *Chromatogr. Newslett.*, **4**, 8–10 (1976).

Adams, R. F. and F. L. Vandemark, Simultaneous High-Pressure Liquid-Chromatographic Determination of Some Anticonvulsant in Serum, *Clin. Chem.*, **22**, 25–31 (1976).

Atwell, S. H., V. A. Green, and W. G. Haney, Development and Evaluation of Method for the Simultaneous Determination of Phenobarbital and Diphenylhydantoin in Plasma by High-Pressure Liquid Chromatography, *J. Pharm. Sci.*, **64**, 806–809 (1975).

Ayers, G. J., J. H. Goudie, K. Reed, and D. Burnett, Quality Control in the Simultaneous Assay of Anticonvulsants Using an Automated Gas Chromatographic System with a Nitrogen Sensitive Detector, *Clin. Chim. Acta*, **76**, 113–124 (1977).

Breyer, U. and D. Villumsen, Thin-Layer Chromatographic Determination of Barbiturates and Phenytoin in Serum and Blood, *J. Chromatogr.*, **115**, 493–500 (1975).

Brooks, M. A., J. A. de Silva, and M. R. Hackman, The Determination of Phenobarbital and Diphenylhydantoin in Blood by Differential Pulse Polarography, *Anal. Chim. Acta*, **64**, 165–175 (1973).

Brunk, S. D., T. P. Hadjiioannou, S. I. Hadjiioannou, and H. V. Malmstadt, Adaptation of "EMIT" Technique for Serum Phenobarbital and Diphenylhydantoin Assays to the Miniature Centrifugal Analyzer, *Clin. Chem.*, **22**, 905–907 (1976).

Cramers, C. A., E. A. Vermeer, L. G., van Kuik, J. A. Hulsman, and C. A. Meijers, Quantitative Determination of Underivatized Anticonvulsant Drugs by High Resolution Gas Chromatography with Support-Coated Open Tubular Columns, *Clin. Chim. Acta*, **73**, 97–109 (1976).

Davis, H. L., K. J. Falk, and D. G. Bailey, Improved Method for the Simultaneous Determination of Phenobarbital, Primidone, and Diphenylhydantoin in Patient's Serum by Gas-Liquid Chromatography, *J. Chromatogr.*, **107**, 61–66 (1975).

Dixon, P. F. et al., Eds., *High Pressure Liquid Chromatography in Clinical Chemistry*, Academic Press, London, 1976.

Dorrity, F., Jr., and M. Linnoila, Rapid Gas-Chromatographic Measurement of Anticonvulsant Drugs in Serum, *Clin. Chem.*, **22**, 860–862 (1976).

Evans, J. E., Simultaneous Measurement of Diphenylhydantoin and Phenobarbital in Serum by High-Performance Liquid Chromatography, *Anal. Chem.*, **45**, 2428,-2429 (1973).

Fellenberg, A. J. and A. C. Pollard, A Rapid Spectrophotometric Procedure for the Simultaneous Micro Determination of Carbamazepine and 5,5-Diphenylhydantoin in Blood, *Clin. Chim. Acta*, **69**, 429–431 (1976).

Finley, P. R., R. J. Williams, and J. M. Byers III, Assay of Phenytoin: Adaptation of "EMIT" to the Centrifugal Analyzer, *Clin. Chem.*, **22**, 911–914 (1976).

Guelen, P. J. M. and E. van der Kleijn, *Rational Anti-Epileptic Drug Therapy*, Elsevier/North-Holland, Amsterdam, 1978.

Heipertz, R., H. Pilz, and K. Eickhoff, Evaluation of a Rapid Gas-Chromatographic Method for the Simultaneous Quantitative Determination of Ethosuximide, Phenylethylmalonediamide, Carbamazepine, Phenobarbital, Primidone and Diphenylhydantoin in Human Serum, *Clin. Chim. Acta*, **77**, 307–316 (1977).

Hill, R. E. and A. N. Latham, Simultaneous Determination of Anticonvulsant Drugs by Gas-Liquid Chromatography, *J. Chromatogr.*, **131**, 341–346 (1977).

Huisman, J. W., The Estimation of Some Important Anticonvulsant Drugs in Serum, *Clin. Chim. Acta*, **13**, 323–328 (1966).

Horning, M. G. et al., Anticonvulsant Drug Monitoring by GC-MS-COM Techniques, *J. Chromatogr. Sci.*, **12**, 630–635 (1974).

Kabra, P. M., G. Gotelli, R. Stanfill, and L. J. Marton, Simultaneous Measurement of Phenobarbital, Diphenylhydantoin and Primidone by High-Pressure Liquid Chromatography, *Clin. Chem.*, **22**, 824–827 (1976).

Kabra, P. M., B. E. Stafford, and L. J. Marton, Simultaneous Measurement of Phenobarbital, Phenytoin, Primidone, Ethosuximide, and Carbamazepine in Serum by High-Pressure Liquid Chromatography, *Clin. Chem.*, **23**, 1284–1288 (1977).

Kalman, S. M. et al., Blood Concentrations of Antiepileptic Drugs, *Drug Assay Lab. Newslett.* **2**, No. 5–8 (1976).

Kitazawa, S. and T. Komura, High-Speed Liquid Chromatographic Determination of Antiepileptic Drugs in Human Plasma, *Clin. Chim. Acta.* **73**, 31–38 (1976).

Larsen, N.-E., J. Naestoft, and E. Hvidberg, Rapid Routine Determination of Some Anti-epileptic Drugs in Serum by Gas Chromatography, *Clin. Chim. Acta*, **40**, 171–176 (1972).

Leal, K. W. and A. S. Troupin, Clinical Pharmacology of Antiepileptic Drugs: A Summary of Current Information, *Clin. Chem.*, **23**, 1964–1968 (1977).

Least, C. J., Jr., G. F. Johnson, and H. M. Solomon, Microscale Anticonvulsant Assay with Use of Nitrogen-Phosphorus Detector and On-Column Methylation Compared with a Macro-Scale Procedure Involving Flame-Ionization Detection, *Clin. Chem.*, **23**, 593–595 (1977).

Legaz, M. and V. A. Raisys, Correlation of the EMIT Antiepileptic Drug Assay with a Gas-Liquid Chromatographic Method, *Clin. Biochem.*, **9**, 35–38 (1976).

Malkus, H., P. I. Jatlow, and A. Castro, A Practical Routine Gas Chromatographic Determination of Phenobarbital, Primidone, and Diphenylhydantoin, *Clin. Chim. Acta*, **82**, 113–117 (1978).

Montgomery, M. R., J. L. Holtzman, and R. K. Lente, Application of Electron Spin Resonance to Determination of Serum Drug Concentrations, *Clin. Chem.*, **21**, 1323–1328 (1975).

Nishina, T., K. Okoshi, and M. Kitamura, Improved Method for Measurement of Serum Levels of Phenobarbital, Carbamazepine, Primidone, and Diphenylhydantoin by Gas-Liquid Chromatography, *Clin. Chim. Acta*, **73**, 463–468 (1976).

Ollerich, M. K., W. R. Ulpmann, R. Haeckel, and R. Heyer, Determination of Phenobarbital and Phenytoin in Serum by a Mechanized Enzyme Immunoassay (EMIT) in Comparison with a Gas-Liquid Chromatographic Method, *J. Clin. Chem. Clin. Biochem.*, **15**, 353–358 (1977).

Papadopoulos, A. S., E. Maty Baylis, and D. E. Fry, A Rapid Micro-Method for Determining Four Anticonvulsant Drugs by Gas-Liquid Chromatography, *Clin. Chim. Acta*, **48**, 135–141 (1973).

Pippinger, C. E., J. E. Scott, and H. W. Gillen, Thin-Layer Chromatography of Anticonvulsant Drugs, *Clin, Chem.*, **15**, 255 (1969).

Pippinger, C. E., J. Kiffin Penry, B. G. White, D. D. Daly, and R. Buddington, Interlaboratory Variability in Determination of Plasma Antiepileptic Drug Concentrations, *Arch. Neurol*, **33**, 351–355 (1971).

Roger, J.-C., G. Rodgers, and A. Soo, Simultaneous Determination of Carbamazepine ("Tegretol") and Other Anticonvulsants in Human Plasma by Gas-Liquid Chromatography, *Clin. Chem.*, **19**, 590–592 (1973).

Rose, S. W., L. D. Smith, and J. K. Penry, *Blood Level Determinations of Antiepileptic Drugs: Clinical Value and Methods*, U.S. Dept. of Health, Education and Welfare, National Institute of Health, Bethesda, Md., 1971.

Schneider, H. et al., Eds., *Clinical Pharmacology of Antiepileptic Drugs*, Springer Verlag, Berlin, 1975.

Serfontein, W. J. and L. S. de Villiers, Quantitative Gas Chromatographic Analysis of Barbiturates and Hydantoins with Quaternary Ammonium Hydroxides, *J. Chromatogr.*, **130**, 342–345 (1977).

Siest, G. and D. S. Young, Eds., *Drug Interference and Drug Measurement in Clinical Chemistry*, Karger Verlag, Basel, 1976.

Soldin, S. J. and G. H. Hill, Rapid Micromethod for Measuring Anticonvulsant Drugs in Serum by High-Performance Liquid Chromatography, *Clin. Chem.*, **22**, 856–859 (1976).

Soldin, S. J. and J. G. Hill, Routine Dual-Wavelength Analysis of Anticonvulsant Drugs by High-Performance Liquid Chromatography, *Clin. Chem.*, **23**, 2352–2353 (1977).

Spiehler, V. et al., Radioimmunoassay, Enzyme Immunoassay, Spectrophotometry and Gas-Liquid Chromatography Compared for Determination of Phenobarbital and Diphenylhydantoin, *Clin. Chem.*, **22**, 749–753 (1976).

Strong, J. M., T. Abe, E. L. Gibbs, and A. J. Atkinson, Jr., Plasma Levels of Methsuximide and *N*-Desmethylsuximide During Methsuximide Therapy, *Neurology*, **24**, 250–255 (1974).

Sun, L. and I. Szafir, Comparison of Enzyme-Immunoassay and Gas Chromatography for Determination of Carbamazepin and Ethosuximide in Human Serum, *Clin. Chem.*, **23**, 1753–1756 (1977).

Toseland, P. A., J. Grove, and D. J. Berry, An Isothermal GLC Determination of the Plasma Levels of Carbamazepine, Diphenylhydantoin, Phenobarbitone, and Primidone, *Clin. Chim. Acta*, **38**, 321–328 (1972).

Turri, J. J., Enzyme Immunoassay of Phenobarbital, Phenytoin and Primidone with the ABA 100 Biochromatic Analyzer, *Clin. Chem.*, **23**, 1510–1512 (1977).

Vandemark, F. L. and R. F. Adams, Ultramicro Gas-Chromatographic Analysis for Anticonvulsants, with Use of a Nitrogen-Selective Detector, *Clin. Chem.*, **22**, 1062–1065 (1976).

Van Lente, F., D. Warkentin, and T. Ohno, Phenytoin and Phenobarbital Assay by Use of EMIT and the ABA-100, *Clin. Chem.*, **23**, 761–762 (1977).

ATROPINE—by G. M. Wientjes

dl-Hyoscyamine; *dl*-tropyl tropate. Anticholinergic agent.

$C_{17}H_{23}NO_3$; mol. wt. 289.38; pK_a 9.7

a. Therapeutic Concentration Range. Atropine disappears rapidly from plasma into tissues. Serum levels after therapeutic doses of atropine are in the low nanogram range (Wurzburger et al., 1977).

b. Metabolism. Following administration of ^{14}C-labeled atropine to human beings, the half-lives of ^{14}C excreted into the urine were 2 and 13 to 38 hr in the fast and slow elimination phases, respectively (Kalser and McLain, 1970). Metabolites accounted for the major portion of the excreted ^{14}C activity.

c. **Analogous Compounds.** Tropane alkaloids: benzoyltropine, cocaine, homatropine, scopolamine.

d. Analytical Methods

1. Radioimmunoassay—Wurzburger et al. (1977)

Discussion. Bovine serum albumin coupled to atropine via the phenyl ring of the tropic acid moiety serves as an antigen to produce atropine-specific antibodies (Wurzburger et al., 1977). The assay is specific for atropine with a detection limit of 6 ng/ml using 10 μl plasma, and of 1 ng/ml using 50 μl plasma.

Detailed Method. Fifty microliters of plasma is incubated overnight at 4°C with 500 μl of an appropriate antiserum dilution in 0.01 M phosphate buffer (pH 7.4) in saline. Antibody-bound atropine is separated from free atropine by addition of an equal volume of saturated pH 7.4 ammonium sulfate solution. The mixture is incubated for 30 min and centrifuged at 2500 g for 30 min. The precipitate is washed once with 0.6 ml of a 50% saturated pH 7.4 $(NH_4)_2SO_4$ solution, and again centrifuged for 30 min. The precipitate containing bound atropine is dissolved in 0.5 ml water and transferred to a counting vial containing 6 ml Riafluor® (New England Nuclear, Boston, Mass.). The ^{14}C activity is then determined in a liquid scintillation counter.

2. Other Radioimmunoassay—Fasth et al. (1975)

Discussion. The antibody in the procedure of Fasth et al. (1975) was raised against BSA coupled to atropine via the primary aliphatic OH of the tropic acid moiety as the antigen. This antibody is therefore less specific for atropine in the presence of benzoyltropine, homatropine, and atropine glucuronide as a potential metabolite of atropine.

3. High Performance Liquid Chromatography—Verpoorte and Svendsen (1976)

Discussion. Tropane alkaloids can be separated on silica gel columns. Ultraviolet detection at wavelengths greater than 254 nm is used with poor sensitivity (>1 μg atropine). This technique is therefore not suitable for biological samples until more sensitive detection methods have been developed (e.g., at 200 to 220 nm).

4. Biological Assay—Tønnesen (1948)

Discussion. Direct instillation of atropine containing tissue extract into mouse eye causes a measurable pupil dilation with as little as 10 ng atropine or its pharmacological equivalent.

5. Comments. Only the radioimmunoassay of Wurzburger et al. (1977) is presently sensitive and specific enough for quantitation of atropine at therapeutic plasma levels.

e. References

Fasth, A., J. Sollenberg, and B. Sorbo, Production and Characterization of Antibodies to Atropine, *Acta Pharm. Suec.*, **12**, 311–322 (1975).

Kalser, S. C. and P. McLain, Atropine Metabolism in Man, *Clin. Pharmacol. Ther.*, **11**, 214–227 (1970).

Tønnesen, M., Chemical and Biological Methods for Determination of Small Concentrations of Atropine and Allied Alkaloids in Forensic Analyses, *Acta Pharmacol.*, **4**, 186–198 (1948).

Verpoorte, L. and A. Svensen, High Performance Liquid Chromatography of Some Tropane Alkaloids, *J. Chromatogr.*, **120**, 203–205 (1976).

Wurzburger, R. J., R. L. Miller, H. G. Boxenbaum, and S. Spector, Radio Immunoassay of Atropine in Plasma, *J. Pharmacol. Ther.*, **203**, 435–441 (1977).

BARBITURATES

AMOBARBITAL

5-Ethyl-5-isoamylbarbituric acid.
Eunoctal, Somnal, and so on. Intermediate-duration sedative, hypnotic.

$C_{11}H_{18}N_2O_3$, mol. wt. 226.27; pK_a 7.7.

HEXOBARBITAL

5-(1-Cyclohexen-1-yl)-1,5-dimethylbarbituric acid.
Evipan, Somnalert. Short-duration sedative, hypnotic.

$C_{12}H_{16}N_2O_3$; mol. wt. 236.26; pK_a 8.19.

MEPHOBARBITAL

5-Ethyl-1-methyl-5-phenylbarbituric acid.
Prominal. Sedative, long-acting hypnotic, anticonvulsant.

$C_{13}H_{14}N_2O_3$; mol. wt. 246.26; pK_a 7.7.

PENTOBARBITAL SODIUM

Sodium 5-ethyl-5-(1-methylbutyl)barbiturate.
Nembutal. Sedative, short-acting hypnotic, anticonvulsant.

$C_{11}H_{17}N_2O_3Na$; mol. wt. 248.26; pK_a 8.0.

PHENOBARBITAL

5-Ethyl-5-phenylbarbituric acid.
Luminal. Long-acting sedative, hypnotic, anticonvulsant.

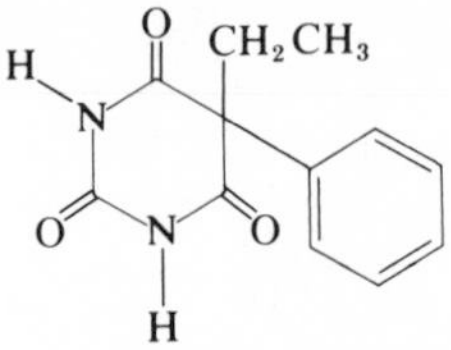

$C_{12}H_{12}N_2O_3$; mol. wt. 232.23; pK_a 7.5 and 10.8.

THIOPENTAL SODIUM

Sodium 5-ethyl-5-(1-methylbutyl)-2-thiobarbiturate.
Pentothal Sodium. Short-acting intravenous anesthetic.

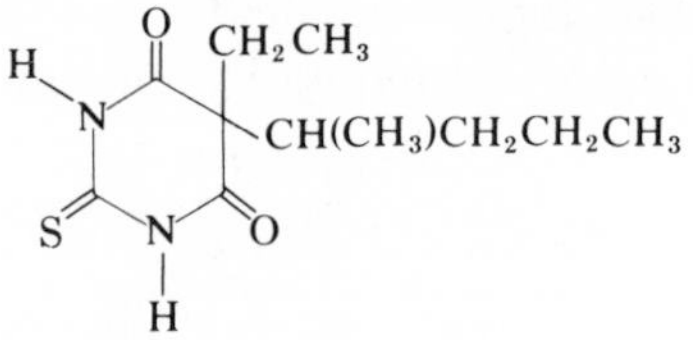

$C_{11}H_{17}N_2NaO_2S$; mol. wt. 264.33; pK_a 7.5.

a. Therapeutic Concentration Range. Most barbiturates used in sedative and hypnotic therapy possess therapeutic concentration ranges between 1 and 5 μg/ml plasma (Breimer, 1974; Winek, 1970); levels above 10 μg/ml are often associated with overdoses (Winek, 1970). Phenobarbital and its prodrug, mephobarbital, are primarily used as anticonvulsants with therapeutic phenobarbital plasma levels of 10 to 30 μg/ml (Leal and Troupin, 1977).

b. Metabolism. Most barbiturates are extensively metabolized in the body prior to urinary excretion of the products, except for barbital and phenobarbital, which are predominantly (barbital) or to a significant extent (phenobarbital) excreted in unchanged form. During long-term treatment with phenobarbital in human beings, the sum of unchanged drug, its *p*-hydroxyphenyl metabolite, and the corresponding glucuronide in the urine accounts for approximately 40% of the dose, the major part of which is contributed by unchanged phenobarbital (Whyte and Dekaban, 1977; Kallberg et al., 1975). An additional important metabolite of phenobarbital and mephobarbital is the 5-(3,4-dihydroxy-1,5-cyclohexadien-1-yl) oxidation product (Harvey et al., 1972). Mephobarbital is mainly converted to phenobarbital by oxidative demethylation, resulting in therapeutic phenobarbital levels (Leal and Troupin, 1977).

Side chain oxidation in C-5 represents the major metabolic pathway of the barbiturates yielding less active or inactive metabolites (Breimer, 1974). Amobarbital, pentobarbital, and hexobarbital form 3'-hydroxy metabolites (Breimer, 1974; Draffan et al., 1973; Holtzman and Thompson, 1975). 3'-OH-Hexobarbital is further oxidized to 3'-keto-hexobarbital as a major product. The N-methyl substituents of hexobarbital and mephobarbital are subject to oxidative N-dealkylation, which does not abolish pharmacological activity (Breimer, 1974; Leal and Troupin, 1977).

Several of the listed barbiturates, (i.e., hexobarbital, mephobarbital, pentobarbital, and thiopental) exist in two stereoisomeric forms. The metabolic and pharmacologic properties of the R- and S-forms may be quite different. For example, the plasma elimination half-lives of (+)-hexobarbital and (−)-hexobarbital are 4.6 and 1.4 hr, respectively, while the racemic mixture is eliminated with a $t_{1/2}$ of 4.4 hr (Breimer and van Rossum, 1973). The half-life of the racemic mixture is mainly determined by the (+)-isomer, which also determines the duration of hypnotic effects. Holtzman and Thompson (1975) have shown that two different microsomal enzymes may be responsible for the metabolism of R-(+)- and S-(−)-pentobarbital catalyzing the formation of diastereomeric 3'-OH metabolites.

The duration of the anesthetic effect of the highly lipophilic thiopental is controlled chiefly by redistribution of the drug from the central nervous system to other tissues, rather than by metabolism (Brodie, 1952; Saidman and Eger, 1973). The terminal log-linear elimination half-life of thiopental may be rather long; however, it occurs only at subtherapeutic levels. The intravenous anesthetic hexobarbital, which is also lipophilic because of the N-methyl substitution, lacks a sufficiently high affinity to adipose tissues and therefore does not exhibit a marked tissue distribution effect. Rather, its duration of effect is mainly determined by drug metabolism (Breimer, 1974).

Plasma elimination half-lives (hr) of the listed barbiturates are as follows:

Amobarbital	14–42	(Kadar et al., 1973)
Hexobarbital	3–7	(Breimer, 1974)
Mephobarbital	24–45	(Leal and Troupin, 1977)

Pentobarbital	23–30	(Smith et al., 1973)
Phenobarbital	84–108	(Leal and Troupin, 1977)
Thiopental sodium	>8	(Brodie, 1952)

c. Analogous Compounds. Other barbiturates; metabolic phenobarbital precursors (N,N-dimethoxymethylphenobarbital, primidone); glutethimide, methyprylon; anticonvulsants.

d. Analytical Methods. Mephobarbital and phenobarbital assays are discussed in the anticonvulsants monograph. Other barbiturate assays are listed here.

1. Gas Chromatography—Greely (1974)

Discussion. This paper presents a specific and sensitive (0.5 μg/ml plasma) GC-FID analysis of 14 common barbiturates, which are separated on the GC column using a temperature program. Derivatization is achieved in quantitative yield with alkyl iodides in N,N-dimethylacetamide, methanol, and tetramethylammonium hydroxide. The butyl derivatives possess GC characteristics superior to those of the methyl and ethyl derivatives and permit differentiation between N-H and N-methylated barbiturates, for example, mephobarbital and hexobarbital. This procedure circumvents the common problems encountered with flash alkylation of barbiturates.

Detailed Method. Whole blood (1 ml), 0.5 ml 1 M phosphoric acid, and 5 ml toluene are shaken for 2 min and centrifuged (extraction according to Kananen et al., 1972). The organic layer is separated and evaporated to dryness under nitrogen. To the residue are added 40 μl N,N-dimethylacetamide, 10 mM tetramethylammonium hydroxide in 5 μl methanol, and 10 μl 1-iodobutane. Alkylation is complete within less than 10 min. One microliter of the final solution is injected into a gas chromatograph.

Analysis is performed on a Model 402 GC-FID (Hewlett Packard, Palo Alto, Calif.) equipped with a 180 cm by 2 mm (i.d.) glass column packed with 1.5% SP-2250 on Super W (Pacific Analytical Consultants, Palo Alto, Calif.). The columns are treated by injection of Silyl 8 (Pierce Chemical Co., Rockford, Ill.) at 200°C after every 8 to 10 analytical injections to prevent a loss in performance. The temperature program is run at 7.5°C/min from 170 to 260°C after an isothermal period of 4 min at 170°C.

2. **Other GC Assays**—Dünges and Bergheim-Irps (1978), Ehrnebo et al. (1973), Hooper et al. (1975), Kallberg et al. (1975), Kelly et al. (1977), Lehane et al. (1976), Marigo et al. (1977), Perchalski and Wilder (1973), van Meter and Gillen (1973), Whyte and Dekaban (1977), Wu (1974)

Discussion. The GC analysis of barbiturates without prior derivatization is feasible; however, irreversible absorption on the column leads to trailing GC peaks and nonlinear standard curves. Such irreversible absorption can be minimized by conditioning the column by silylation prior to analysis (Marigo et al., 1977). Derivatization by alkylation affords better chromatographic behavior and has been adopted by most investigators. On-column flashmethylation with trimethylphenylammonium hydroxide is the most common procedure (Ehrnebo et al., 1973; Whyte and Dekaban, 1977; Wu, 1974). For example, Ehrnebo et al. (1973) extract amobarbital and pentobarbital from plasma at pH 5.5 into ether. The ethereal layer is back-extracted with a small volume of a trimethylphenylammonium hydroxide solution, a small aliquot of which is analyzed by GC-FID. A less common barbiturate, for example, vinbarbital, serves as the internal standard. Methylation results in the same reaction product from mephobarbital and phenobarbital; therefore Hooper et al. (1975) employ on-column butylation with tetrabutylammonium hydroxide for these drugs.

Phenobarbital yields two products upon treatment with alkaline on-column methylating reagents, the *N*-dimethyl derivative and a decomposition product that elutes earlier from the GC column. This product was first thought to be a permethylated 2-ethyl-2-phenylmalondiamide (e.g., van Meter and Gillen, 1973); however, Kelley et al. (1977) determined its structure as *N*-methyl-2-phenylbutyramide by GC-MS analysis. This reaction of phenobarbital may result in erroneous results for plasma level determinations, and several approaches have been taken in order to obtain reproducible data. Selection of an internal standard with equivalent reaction products (e.g., *p*-methylphenobarbital) cancels out phenobarbital decomposition (Whyte and Dekaban, 1977). Use of dilute solutions of trimethylphenylammonium hydroxide, added immediately prior to GC analysis to prevent hydrolysis in solution (van Meter and Gillen, 1973), or judicious choice of the reaction solvents (Kelley et al., 1977) suppresses phenobarbital decomposition. Conversely, Perchalski and Wilder (1973) obtain complete conversion of phenobarbital to *N*-methyl-2-phenylbutyramide with high reactant concentration and GC inlet temperature and use this product for quantitation of phenobarbital. Similar considerations apply to the measurement of the major phenobarbital metabolite, p-hydroxyphenobarbital (Kallberg et al., 1975; Whyte and Dekaban, 1977).

Increased sensitivity and specificity of the GC detection of barbiturates can be achieved with nitrogen-sensitive flame ionization detection (Lehane et al., 1976), which permits determination of as little as 1 ng barbiturate/25 μl blood (Dünges and Bergheim-Irps, 1978).

3. Gas Chromatography—Mass Spectrometry—Draffan et al. (1973)

Discussion. Amobarbital, its 3'-OH metabolites, and other barbiturates can be sensitively analyzed by electron impact (23 eV) GC-MS in the selected ion monitoring mode. Detection limit is between 0.01 and 0.05 μg/ml, using a 0.1 ml plasma sample. The barbiturates are flash-methylated and detected at the abundant ions at m/e 169 and 184, arising from the barbiturate moiety and part of the side chain at C-5. Appropriate internal standards are prepared by reaction of the barbiturate with dimethyl sulfate-d_6, yielding perdeuterated permethylated products.

4. High Performance Liquid Chromatography—Blackman and Jordan (1978), Dünges et al. (1974)

Discussion. The HPLC analysis of anticonvulsant barbiturates is discussed in the anticonvulsants monograph. The assay of Blackman and Jordan (1978) measures plasma levels of the anesthetic thiopental, using reverse phase (Partisil 10/25 ODS column) HPLC with UV detection at 290 nm. Plasma is directly injected, together with quinoline, as the internal standard onto a precolumn connected to the analytical column. The precolumn, containing the same reverse phase packing, protects the analytical column and can be readily exchanged. The assay is sensitive to 0.5 μg thiopental/ml plasma, using a 10 μl sample volume.

After derivatization with dansyl chloride in ethyl acetate-solid potassium carbonate, barbiturates can be detected in picomole amounts by reverse phase HPLC with fluorescence detection (Dünges et al., 1974). Only 20 μl serum is required for the detection of therapeutic barbiturate serum levels.

5. Ultraviolet Spectrophotometry—Goldbaum (1952), Gupta et al. (1973), Jatlow (1973), Sinnema (1967), Wallace (1969), Williams and Zak (1959)

Discussion. Spectrophotometric assays of barbiturate serum levels are derived from the method of Goldbaum (1952). Barbiturates with no *N*-alkyl substituents are divalent N-H acids. The monoanion possesses a UV absorbance maximum at 240 nm, while the maximum of the dianion, which predominates above pH 11, is shifted to 260 nm (Sinnema, 1967). In the assay of Goldbaum (1952), the UV absorbance is read at 260 nm in 0.45 *N* NaOH (pH > 12) and in 0.45 *N* NaOH containing 16% NH_4Cl (pH 10.5); the difference between the two readings is directly proportional to the barbiturate concentration with a detection limit of 1 μg/ml plasma. *N*-Methyl- and thiobarbiturates cannot be analyzed by this procedure. Further publications describe minor variations of the Goldbaum assay (Sinnema, 1967; Wallace, 1969), including automatic analysis (Williams and Zak, 1959). Despite a relative degree of specificity, interfering substances, such as phenytoin, have been observed in patient samples (Gupta et al., 1973; Jatlow, 1973). However, the UV assay of Wallace (1969) is specific for phenobarbital in the presence of phenytoin, which can also be determined in the same sample (see anticonvulsants monograph).

6. Thin-Layer Chromatography—Hsiung et al. (1974), Itiaba et al. (1970), Kananen (1972)

Discussion. Several TLC assays with varying degrees of specificity and sensitivity have been published. Methods utilizing UV-TLC compare well with specific GC assays (Kananen, 1972). Itiaba et al. (1970) employ instant TLC (fiberglass sheets impregnated with silica gel) and rhodamine B for qualitative barbiturate analysis. Hsiung et al. (1974) extract the barbiturates from the instant TLC plates and measure the fluorescence of barbiturate dianions in NaOH with an excitation of 265 nm and an emission of 410 nm. Fluorescence analysis is one to two orders of magnitude more sensitive than UV absorbance measurements. The TLC-fluorescence assay of amobarbital and penobarbital is specific in the presence of their respective metabolites (Hsiung et al., 1974).

7. Radioimmunoassay—Flynn and Spector (1972), Spector and Flynn (1971), Viswanathan et al. (1977)

Discussion. Antibodies have been raised against a 5-allyl-5-(1-carboxyisopropyl)barbituric acid conjugate with bovine gamma globulin (Spector and Flynn, 1971). The procedure uses ¹⁴C-pentobarbital as the tracer and ammonium sulfate precipitation to separate bound and free tracer. The assay sensitivity is 500 pg/sample and does not differentiate between barbiturates with varying substituents in C-5 (Flynn and Spector, 1972). However, *N*-methyl barbiturates (hexobarbital, mephobarbital) and thiopental have little or no cross-affinity to the antibody. Other anticonvulsants or hypnotics demonstrate negligible cross-affinity. Viswanathan et al. (1977) point out that the major phenobarbital metabolite, *p*-hydroxyphenobarbital, does interfere with the phenobarbital assay. They suggest a selective solvent extraction of phenobarbital to eliminate the *p*-hydroxy metabolite interference.

8. Enzyme Immunoassay—Belfield et al. (1977), Finley et al. (1977), Walberg (1974)

Discussion. The phenobarbital EMIT assay compares well with GC and UV assays (Belfield et al., 1977; Walberg, 1974) and has been automated using the centrifugal analyzer (Belfield et al., 1977; Finley et al., 1977). (For further details see anticonvulsants monograph.) The antibody recognizes the barbiturate moiety of a number of common barbiturates, that is, amobarbital, butobarbital, pentobarbital, phenobarbital, and secobarbital; therefore the barbiturate EMIT assay can be used as a rapid screening method for barbiturate intoxication. Assay sensitivity for the listed barbiturates ranges from 1 to 3 μg/ml urine (Walberg, 1974).

9. Comments. The spectrophotometric assay for barbiturates has recently been replaced as the major technique in clinical laboratories by either GC methods or HPLC and EMIT (anticonvulsants). Therapeutic drug level measurements are common for the anticonvulsants mephobarbital and phenobarbital, while barbiturates mainly used as hypnotics and sedatives are of interest to the toxicologist (overdose) and pharmacologist. Different assays may be suitable for the broad spectrum of studies requiring plasma level determinations. No convenient method is presently available to study the separate biological fates of asymmetric barbiturates that are given as the racemic mixture.

e. References

Belfield, A., A. M. Duncan, and P. C. Reavey, Enzyme Multiplied Immunoassay of Phenobarbital by LKP Reaction Rate Analyzer: A Comparison with Analysis by Gas-Liquid Chromatography, *Ann. Clin. Biochem.*, **14**, 218–222 (1977).

Blackman, G. L. and G. J. Jordan, Analysis of Thiopentone in Human Plasma by High-Performance Liquid Chromatography, *J. Chromatogr.* **145**, 492–495 (1978).

Breimer, D. D., Pharmacokinetics of Hypnotic Drugs, Thesis, Nijmegen, 1974.

Breimer, D. D. and J. M. van Rossum, Pharmacokinetics of (+)-, (−)-, and (±)-Hexobarbitone in Man after Oral Administration, *J. Pharm. Pharmacol.*, **25**, 762–764 (1973).

Brodie, B.B., Physiological Disposition and Chemical Fate of Thiobarbiturates in the Body, *Fed. Proc.*, **11**, 632–639 (1952).

Draffan, G. H., R. A. Clare, and F. T. Williams, Determination of Barbiturates and Their Metabolites in Small Plasma Samples by Gas Chromatography—Mass Spectrometry: Amylobarbitone and 3-Hydroxyamylobarbitone, *J. Chromatogr.*, **75**, 45–53 (1973).

Dünges, W. and E. Bergheim-Irps, Microtechniques for the Gas Chromatographic Determination of Barbiturates in Small Blood Samples, *J. Chromatogr.*, **145**, 265–274 (1978).

Dünges, W., G. Naundorf, and N. Seiler, High Pressure Liquid Chromatographic Analysis of Barbiturates in the Picomole Range by Fluorimetry of Their DANS-Derivatives, *J. Chromatogr. Sci.*, **12**, 655–657 (1974).

Ehrnebo, M., S. Agurell, and L. O. Boreus, Gas Chromatographic Determination of Therapeutic Levels of Amobarbital and Pentobarbital in Plasma, *Eur. J. Clin. Pharmacol.*, **4**, 191–195 (1973).

Finley, P. R., R. J. Williams, D. F. Lichti, and J. M. Byers, Assay of Phenobarbital: Adaptation of "EMIT" to the Centrifugal Analyzer, *Clin. Chem.*, **23**, 738–740 (1977).

Flynn, E. J. and S. Spector, Determination of Barbiturate Derivatives by Radioimmunoassay, *J. Pharmacol. Exp. Ther.*, **181**, 547–554 (1972).

Goldbaum, L. R., Determination of Barbiturates: Ultraviolet Spectrophotometric Method with Differentiation of Several Barbiturates, *Anal. Chem.*, **24**, 1604–1607 (1952).

Greely, R. H., New Approach to Derivatization and Gas-Chromatographic Analysis of Barbiturates, *Clin. Chem.*, **20**, 192–194 (1974).

Gupta, R. N., P. M. Keane, R. G. Cooper, M. S. Greaves, and G. Owen, Diphenylhydantoin Interference with Spectrophotometric Barbiturate Estimations, *Clin. Chem.*, **19**, 433–434 (1973).

Harvey, D. J. et al., Detection of a 5-(3,4-Dihydroxy-1,5-cyclohexadien-1-yl) Metabolite of Phenobarbital and Mephobarbital in Rat, Guinea Pig, and Human, *Res. Commun. Chem. Pathol. Pharmacol.*, **3**, 557–565 (1972).

Holtzman, J. L. and J. A. Thompson, Metabolism of *R*-(+)- and *S*-(−)-Pentobarbital by Hepatic Microsomes from Male Rats, *Drug Metab. Dispos.*, **3**, 113–117 (1975).

Hooper, W. D., D. K. Dubetz, M. J. Eadie, and J. H. Tyrer, Simultaneous Assay of Methylphenobarbitone and Phenobarbitone Using Gas-Liquid Chromatography with On-Column Butylation, *J. Chromatogr.*, **110**, 206–209 (1975).

Hsiung, M. W., K. Itiaba, and J. C. Crawhall, A Spectrophotofluorimetric Method for the Estimation of Barbiturates Separated by Instant Thin-Layer Chromatography, *Clin. Biochem.*, **7**, 45–51 (1974).

Itiaba, K., J. C. Crawhall, and C. M. Sin, A Rapid High Resolution Chromatographic Method for Serum Barbiturates: Instant Thin-Layer Chromatography, *Clin. Biochem.*, **3**, 287–293 (1970).

Jatlow, P., Ultraviolet Spectrophotometric Analysis of Barbiturates: Evaluation of Potential Interferences, *Am. J. Clin. Pathol.*, **59**, 167–173 (1973).

Kadar, D., T. Inaba, L. Endrenuyi, G. E. Johnson, and W. Kalow, Comparative Drug Elimination Capacity in Man—Glutethimide, Amobarbital, Antipyrine, and Sulfinpyrazone, *Clin. Pharmacol. Ther.*, **14**, 552–560 (1973).

Kallberg, N. et al., Quantitation of Phenobarbital and Its Metabolites in Human Urine. *Eur. J. Clin. Pharmacol.*, **9**, 161–168 (1975).

Kananen, G., R. Osiewicz, and I. Sunshine, Barbiturate Analysis—a Current Assessment, *J. Chromatogr. Sci.*, **10**, 283–287 (1972).

Kelly, R. C., J. C. Valentour, and I. Sunshine, Solvent Suppression of the Decomposition of Phenobarbital During On-Column Methylation with Trimethylanilinium Hydroxide, *J. Chromatogr.*, **138**, 413–422 (1977).

Leal, K. W. and A. S. Troupin, Clinical Pharmacology of Antiepileptic Drugs: A Summary of Current Information, *Clin. Chem.*, **23**, 1964–1968 (1977).

Lehane, D. R., P. Menyharth, G. Lum, and A. L. Levy, Therapeutic Drug Monitoring: Measurements of Antiepileptic and Barbiturate Drug Levels in Blood by Gas Chromatography with Nitrogen-Sensitive Detector, *Ann. Clin. Lab. Sci.*, **6**, 404–410 (1976).

Marigo, M., S. D. Ferrara, and L. Tedeschi, A Sensitive Method for Gas-Chromatographic Assay of Barbiturates in Body Fluids, *Arch. Toxicol.*, **37**, 107–112 (1977).

Perchalski, R. J. and B. J. Wilder, GC Assay of Phenobarbital, *Clin. Chem.*, **19**, 788–789 (1973).

Saidman, L. J. and E. I. Eger, Uptake and Distribution of Thiopental after Oral, Rectal, and Intramuscular Administration: Effect of Hepatic Metabolism and Injection Site Blood Flow, *Clin. Pharmacol. Ther.*, **14**, 12–20 (1973).

Sinnema, Y. A., The Identification and Determination of Barbiturates in Serum, *Z. Klin. Chem. Klin. Biochem.*, **5**, 21–25 (1967).

Smith, R. B., L. W. Dittert, W. O. Griffin, Jr., and J. T. Doluisio, Pharmacokinetics of Pentobarbital After Intravenous and Oral Administration, *J. Pharmacokin. Biopharm.*, **1**, 5–16 (1973).

Spector, S. and E. J. Flynn, Barbiturates—Radioimmunoassay, *Science*, **174,** 1036–1038 (1971).

Van Meter, J. C. and H. W. Gillen, A Source of Variation in the Gas-Chromatographic Assay of Phenobarbital Treated with Trimethylanilinium Hydroxide, *Clin. Chem.*, **19,** 359 (1973).

Viswanathan, C. T., H. E. Booker, and P. G. Welling, Interference by *p*-Hydroxyphenobarbital in the ^{125}I-Radioimmunoassay of Serum and Urinary Phenobarbital, *Clin. Chem.*, **23,** 873–876 (1977).

Walberg, C. B., Correlation of the EMIT urine Barbiturate Assay with a Spectrophotometric Serum Barbiturate Assay in Suspected Overdose. *Clin. Chem.*, **20,** 305–306 (1974).

Wallace, J., Simultaneous Spectrophotometric Determination of Phenytoin and Phenobarbitone in Biological Specimens, *Clin. Chem.*, **15,** 323–330 (1969).

Whyte, M. P., and A. S. Dekaban, Metabolic Fate of Phenobarbital: A Quantitative Study of *p*-Hydroxyphenobarbital Elimination in Man, *Drug Metab. Dispos.*, **5,** 63–70 (1977).

Williams, L. A. and B. Zak, Determination of Barbiturates by Automatic Differential Spectrophotometry, *Clin. Chim. Acta*, **4,** 170–174 (1959).

Winek, C. L., Laboratory Criteria for the Adequacy of Treatment and Significance of Blood Levels, *Clin. Toxicol.*, **3,** 511–549 (1970).

Wu, A., Concerning On-Column Methylation of Phenobarbital, *Clin. Chem.*, **20,** 630 (1974).

BCNU—by Z. Hosein

1,3-Bis(2-chloroethyl)-1-nitrosourea.
Carmustine. Antineoplastic alkylating agent.

$$Cl-CH_2-CH_2-\underset{\underset{NO}{|}}{N}-\overset{\overset{O}{\|}}{C}-NH-CH_2-CH_2-Cl$$

$C_5H_9Cl_2N_3O_2$; mol. wt. 214.

a. Therapeutic Concentration Range. Therapeutic levels of BCNU in the plasma of patients are usually less than 1 $\mu g/ml$ (Weinkam et al., 1978).

b. Metabolism. Tissue uptake and metabolism of BCNU occur rapidly. The *in vitro* half-life of BCNU in plasma incubations is 15 min (Loo and Dion, 1965). Approximately 80% of the radioactively labeled drug appears in the urine within 24 hr as degradation product. The mechanism of action is thought to be alkylation of DNA and carbamoylation of protein.

c. Analogous Compounds. Nitrosureas: CCNU, methyl-CCNU, PCNU.

d. Analytical Methods

    1. Chemical Ionization Mass Spectrometry—Weinkam et al. (1978)

Discussion. Deuterium-labeled BCNU as the internal standard is added to the sample (e.g., blood, plasma), followed by a hexane-ether extraction. The organic extract residue is analyzed by direct insertion chemical ionization (isobutane) MS. Selected ion monitoring of the protonated molecular ions of the drug and internal standard yields ion intensity ratios, from which the concentration of the drug is calculated. Masses of interest are detected without significant interference from other compounds in the sample. The sensitivity for plasma samples is about 100 ng BCNU/ml.

Detailed Method. The internal standard is the octadeuterio analog of BCNU, labeled on the four methylene carbons (BCNU-$^2H_8$). Glass-distilled hexane was used for extraction. A solution of a known amount of BCNU-$^2H_8$ in 1 ml hexane-diethyl ether (1:1) and 0.5 ml plasma are mixed for 20 sec. The organic layer is pipetted into a conical vial and stored at $-70°C$ until analysis, at which time the hexane-ether is evaporated under nitrogen at 30°C. Any traces of water are removed by adding and again evaporating 200 μl absolute alcohol. The plasma residue is dissolved in dichloromethane and deposited on the ceramic tip of the direct insertion probe. The probe is inserted into the mass tip of the direct insertion probe. The probe is inserted into the mass spectrometer and slowly heated to volatilize BCNU. Data acquisition cycles require 3 min per sample.

    Mass spectral analysis is performed with a Model 3200 mass spectrometer (Finnigan Corp.), equipped with a dual electron impact, chemical ionization source operated in chemical ionization mode. Isobutane at 80 Pa (0.6 torr) is used as the reagent gas at an ion source temperature of 120°C. Mass spectrometer parameters are set to monitor the protonated molecular ions (MH$^+$) *m/e* 214 and 216 (BCNU-$^{35}Cl_2$ and BCNU-$^{35}Cl^{37}Cl$) and *m/e* 222 and 224 (BCNU-$^2H_8$-$^{35}Cl_2$ and BCNU-$^{35}Cl^{37}Cl$).

The mass spectrometer is interfaced to a Nova 830 computer (Data General Corp.) system. Data output is provided by a Model 4000 terminal and hard copy device (Tektronix Corp.). Peak height ratios are measured, and BCNU concentrations obtained from a standard curve plot of peak height ratios versus BCNU concentrations.

2. Colorimetry—Loo and Dion (1965)

Discussion. In hydrochloric acid solution BCNU decomposes, liberating nitrous acid, which is determined colorimetrically by the Bratton-Marshall reaction. Plasma samples are extracted with a 2:1 mixture of ether and water before analysis. The sensitivity of the procedure is 0.5 μg/ml with a precision of 2%. Although this method is adequately sensitive, it lacks the specificity of the technique described by Weinkam et al. (1978), since nitrosamides, nitrosamines, and other nitrosoureas also give this reaction.

3. Polarography—Bartosek et al. (1978)

Discussion. This assay is sensitive to well below 1 μg/ml plasma, employing differential pulse polarography. Plasma samples are extracted into n-pentane, and the extract residue is measured in 0.2 M citric acid solutions. The specificity of this assay is questionable.

4. Comments. Few analytical techniques have been published for BCNU because of its chemical instability. Analysis by chemical ionization MS is, at present, the method of choice, because of its specificity as well as its sensitivity. Addition of isotopically labeled BCNU as the internal standard automatically accounts for chemical degradation and extraction losses during the analytical procedure, which could present a major analytical problem with other assay techniques.

e. References

Bartosek, I., S. Daniel, and S. Sykora, Differential Pulse Polarographic Determination of Submicrogram Quantities of Carmustine and Related Compounds in Biological Samples, *J. Pharm. Sci.*, **67**, 1160–1163 (1978).

Loo, T. L. and R. L. Dion, Colorimetric Method for the Determination of 1,3-Bis(2-chloroethyl)-1-nitrosourea, *J. Pharm. Sci.*, **54**, 809–811 (1965).

Weinkam, R. J., J. H. C. Wen, D. E. Furst, and V. A. Levin, Analysis for 1,3-Bis(2-chloroethyl)-1-nitrosourea by Chemical Ionization Mass Spectrometry. *Clin. Chem.*, **24**, 45–49 (1978).

CARBAMAZEPINE

5H-Dibenz[b,f]azepine-5-carboxamide.
Tegretol. Anticonvulsant, therapy of trigeminal neuralgia.

$C_{15}H_{12}N_2O$; mol. wt. 236.26.

a. Therapeutic Concentration Range. From 5 to 10 μg/ml serum (Bertilsson, 1978; Leal and Troupin, 1977).

b. Metabolism. The major metabolites of carbamazepine are the 10,11-dihydro-10,11-diol (Baker et al., 1973) and the 10,11-epoxide (Frigerio et al., 1972), which possesses anticonvulsant activity comparable to that of the parent drug (Bertilsson, 1978; Morselli and Frigerio, 1975). The dihydrodiol is rapidly cleared by the kidneys and represents the major metabolite in urine. The active 10,11-epoxide is present in plasma at concentrations between 5 and 81% of carbamazepine during long-term treatment. Enzyme inducers, such as phenobarbital and phenytoin, increase the concentration of the 10,11-epoxide relative to carbamazepine (Bertilsson, 1978). The half-life of carbamazepine plasma elimination after single doses ranges between 18 and 65 hr and decreases to between 10 and 20 hr after multiple dosing because of autoinduction of the oxidative hepatic enzymes responsible for carbamazepine metabolism (Bertilsson, 1978).

c. Analogous Compounds. Tricyclic antidepressants; iminobenzyl derivatives. Note, however, that carbamazepine behaves as a neutral compound over a wide pH range.

d. Analytical Methods. The major assay methods for the detection of carbamazepine and other anticonvulsants in serum, for example, HPLC, GC, and enzyme immunoassay, are discussed in detail in the anticonvulsants monograph. Additional assays designed only for the detection of carbamazepine and its metabolites are listed below.

1. **High-Performance Liquid Chromatography**—Adams et al. (1977), Eichelbaum and Bertilsson (1975), Kabra and Marton (1976), Mihaly et al. (1977), Westenberg and de Zeeuw (1977).

Discussion. All of these assays require organic solvent extraction of plasma or serum prior to HPLC analysis. Differences in the analytical conditions cause remarkable differences in the sensitivity of carbamazepine and 10,11-epoxide detections. Chromatography on small particle silica gel affords excellent sensitivity with UV detection at 254 nm (Eichelbaum and Bertilsson, 1975) or 250 nm (Westenberg and de Zeeuw, 1977). Westenberg and de Zeeuw (1977) claim a detectable limit of 2 ng carbamazepine/ml plasma, using a 1 ml sample. Detection of the 10,11-epoxide is 4 (250 nm) to 10 (254 nm) times less sensitive since the lack of conjugation in C-10,11 causes a bathochromic shift of the UV absorbance maximum. Nitrazepam (Westenberg and de Zeeuw, 1977) and 10,11-dihydrocarbamazepine (Eichelbaum and Bertilsson, 1975) serve as internal standards. Choice of the anticonvulsant nitrazepam as the internal standard is questionable for a clinical carbamazepine assay.

Kabra and Marton (1976) and Adams et al. (1977) utilize reverse phase chromatography on silica beads chemically bonded with octadecyl trichlorosilane. The internal standard is 5-(4-methylphenyl)-5-phenylhydantoin, which is frequently used in anticonvulsant assays. Ultraviolet detection at 254 nm and a sample volume of 2 ml plasma afford a relatively insensitive detection limit at 0.25 µg carbamazepine/ml (Kabra and Marton, 1976), compared to normal phase HPLC. Adams et al. (1977) achieve a much greater sensitivity with UV detection at 195 nm. With a sample of only 50 µl plasma, their assay is sensitive to 0.2 µg/ml for carbamazepine and its 10,11-epoxide metabolite. In contrast to the higher UV absorbance of carbamazepine at 254 nm, the 10,11-epoxide has a slightly higher absorbance at 195 nm. Some of the common anticonvulsants and barbiturates may interfere with the measurement of the 10,11-epoxide, but not carbamazepine itself (Adams et al., 1977).

2. **Gas Chromatography**—Chambers and Cooke (1977), Drummer et al (1976), Kupferberg (1972), Lensmeyer (1977), Perchalski (1976), Schwertner et al. (1978), Sheehan and Beam (1975)

Discussion. Carbamazepine and its two major metabolites are unstable under GC conditions. The thermolytic product of carbamazepine is iminostilbene, while the 10,11-dihydro-10,11-diol rearranges to 9-acridine carboxyaldehyde (Frigerio et al., 1973; Baker et al., 1973). Nevertheless, Sheehan and Beam (1975) successfully use a GC assay of carbamazepine without derivatization by measuring the sum of GC peaks representing the parent drug and iminostilbene with heptobarbital as the internal standard. When primidone is coadministered with carbamazepine, its metabolite phenylethylmalonamide is not resolved from iminostilbene by GC on OV-17 as the liquid stationary phase; rather, OV-225 should be used in order to avoid assay interferences.

Other investigators attempt to minimize on-column thermolysis of carbamazepine by derivatization with either trimethylsilyl reagents (Kupferberg, 1972; Lensmeyer, 1977) or dimethylformamide dimethylacetal (Perchalski et al., 1976). Other GC assays for carbamazepine plasma levels have been published by Chambers and Cooke (1977) and Drummer et al. (1976).

Gas chromatography with electron capture can be used after formation of the *N*-pentafluorobenzamide derivative with pentafluorobenzoyl chloride (Schwertner et al., 1978). The assay can be performed with 0.5 ml serum and has a C.V. of less than 2% with 10-methoxycarbamazepine as the internal standard.

3. Gas Chromatography-Mass Spectrometry—Frigerio et al. (1973), Palmer et al. (1973), Trager et al. (1978)

Discussion. These GC-MS assays are quite sensitive in the selected ion monitoring mode for the detection of carbamazepine in plasma (50 ng/ml, Palmer et al., 1973). 10,11-Dihydrocarbamazepine can serve as an internal standard for the assay without derivatization, since it undergoes thermolysis and electron impact fragmentations similar to those of carbamazepine (Palmer et al., 1973). The GC-MS assay of Trager et al. (1978) involves the addition of ^{13}C-carbamazepine and ^{13}C-carbamazepine-10,11-epoxide as the internal standards to plasma samples, and MS analysis in the chemical ionization mode.

Frigerio et al. (1973) analyze the GC-MS behavior of carbamazepine and its metabolites, in particular, on-column thermolysis and electron-impact-induced reactions.

4. Thin-Layer Chromatography—Breyer (1975), Farber et al (1974), Hundt and Clark (1975)

Discussion. Quantitative TLC analysis is performed by UV densitometry directly on the plates following organic solvent extractions of plasma and urine. These assays are sufficiently sensitive and specific to analyze therapeutic carbamazepine levels in plasma and can be applied to the simultaneous detection of the two major carbamazepine metabolites (Hundt and Clark, 1975).

5. Enzyme Immunoassay—Kleine (1978), Lacher et al. (1979), Mihaly et al. (1977)

Discussion. The EMIT assay for anticonvulsant drugs has already been described. Kleine (1978) proposes a mechanized microliter system that significantly lowers the reagents requirements and thereby the cost of the assay. Comparison of the carbamazepine EMIT assay with specific GC and HPLC assays reveals high correlation coefficeints ($r \leq .97$), indicating that the active epoxide is not recognized by the antibody (Mihaly et al., 1977). Hemoglobin interference with the EMIT assay on a centrifugal analyzer precludes analysis of severely hemolyzed samples (Lacher et al., 1979).

6. Spectrophotometry—Beyer et al. (1971), Fellenberg and Pollard (1976)

Discussion. Spectrophotometric assays are usually based on the acid-catalyzed rearrangement of carbamazepine to 9-methylacridine, which exhibits a strong UV maximum at 258 nm (Beyer et al., 1971). A microprocedure (Fellenberg and Pollard, 1976) using only 0.1 ml plasma is senstive to 1 μg carbamazepine/ml plasma. The specificity of this assay for unchanged carbamazepine is questionable, however, since its metabolites may undergo similar rearrangements.

7. Assay Comparison—Mihaly et al. (1977)

8. Comments. Clinical assays of carbamazepine plasma levels are widely used because of the well-established relationship between plasma levels and therapeutic effects. The presence of an active metabolite (i.e., the 10,11-epoxide) and the thermochemical instability of carbamazepine make HPLC-UV the assay method of choice for most biomedical applications.

e. References

Adams, R. F., G. J. Schmidt, and F. L. Vandemark, A Micromethod for the Determination of Carbamazepine and 10,11-Epoxide Metabolite in Serum and Urine by Reverse-Phase Liquid Chromatography, *Chromatogr. Newslett.*, **5,** 11–13 (1977).

Baker, K. M. et al., 10,11-Dihydro-10,11-dihydroxy-5H-dibenz(b,f)azepine-5-carboxamide, a Metabolite of Carbamazepine Isolated from Human and Rat Urine, *J. Med. Chem.*, **16,** 703–705 (1973).

Bertilsson, L., Clinical Pharmacokinetics of Carbamazepine, *Clin. Pharmacokin.*, **3,** 128–143 (1978).

Beyer, K.-H., O. Bredenstein, and G. Schenck, Isolierung und Identifizierung eines Carbamazepin-Reaktionsproduktes, *Arzneim.-Forsch.*, **21,** 1033–1034 (1971).

Breyer, U., Rapid and Accurate Determination of the Level of Carbamazepine in Serum by Ultraviolet Reflectance Photometry on Thin-Layer Chromatograms, *J. Chromatogr.*, **108,** 370–374 (1975).

Chambers, R. E. and M. Cooke, Simple and Reliable Gas Chromatographic Assay for the Determination of Carbamazepine in Plasma, *J. Chromatogr.*, **144,** 257–262 (1977).

Drummer, O., P. Morris, and F. Vajda, Plasma Carbamazepine Determinations; A Simple Gas Chromatographic Method, *Clin. Exp. Pharmacol. Physiol.*, **3,** 497–501 (1976).

Eichelbaum, M., and L. Bertilsson, Determination of Carbamazepine and Its Epoxide Metabolite in Plasma by High-Speed Liquid Chromatography, *J. Chromatogr.*, **103,** 135–140 (1975).

Farber, D. B., T. Manin, and W. A. Veld, A Thin-Layer Chromatographic Method for Determining Carbamazepine in Blood, *J. Chromatogr.*, **93,** 238–342 (1974).

Fellenberg, A. J. and A. C. Pollard, A Rapid and Sensitive Spectrophotometric Procedure for the Micro Determination of Carbamazepine in Blood, *Clin. Chim. Acta*, **69,** 423–428 (1976).

Frigerio, A. et al., Mass Spectrometric Characterization of Carbamazepine-10,11-epoxide, A Carbamazepine Metabolite Isolated from Human Urine, *J. Pharm. Sci.*, **61,** 1144–1147 (1972).

Frigerio, A., K. M. Baker, and P. L. Morselli, Gas Chromatographic-Mass Spectrometric Studies on Carbamazepine, *Adv. Biochem. Psychopharmacol.*, **7,** 125–134 (1973).

Hundt, H. K. and E. C. Clark, Thin-Layer Chromatographic Method for Determining Carbamazepine and Two of Its Metabolites in Serum, *J. Chromatogr.*, **107,** 149–154 (1975).

Kabra, P. M. and L. J. Marton, Determination of Carbamazepine in Blood or Plasma by High-Pressure Liquid Chromatography, *Clin. Chem.*, **22,** 1070–1072 (1976).

Kleine, T. D., Lowering the Cost of the Enzyme Immunoassay (EMIT) for Carbamazepine by Its Adaptation to a Mechanized Microliter System, *Clin. Chim. Acta*, **82**, 193–195 (1978).

Kupferberg, H. J., GLC Determination of Carbamazepine in Plasma, *J. Pharm. Sci.*, **61**, 284–286 (1972).

Lacher, D. A., R. Valdes, Jr., and J. Savory, Enzyme Immunoassay of Carbamazepine with a Centrifugal Analyzer, *Clin. Chem.*, **25**, 295–298 (1979).

Leal, K. W. and A. S. Troupin, Clinical Pharmacology of Anti-epileptic Drugs: A Summary of Current Information, *Clin. Chem.*, **23**, 1964–1968 (1977).

Lensmeyer, G. L., Isothermal Gas Chromatographic Method for the Rapid Determination of Carbamazepine ("Tegretol") as Its TMS Derivative, *Clin. Toxicol.*, **11**, 443–454 (1977).

Mihaly, G. W., J. A. Phillips, W. J. Louis, and F. J. Vajda, Measurement of Carbamazepine and Its Epoxide Metabolite by High-Performance Liquid Chromatography, and a Comparison of Assay Techniques for the Analysis of Carbamazepine, *Clin. Chem.*, **23**, 2283–2287 (1977).

Morselli, P. L. and A. Frigerio, Metabolism and Pharmacokinetics of Carbamazepine, *Drug Metab. Rev.*, **4**, 97–113 (1975).

Palmer, L., L. Bertilsson, P. Collste, and M. Rawlins, Quantitative Determination of Carbamazepine in Plasma by Mass Fragmentography, *Clin. Pharmacol. Ther.*, **14**, 827–832 (1973).

Perchalski, R. J., B. D. Andresen, and B. J. Wilder, Reaction of Carbamazepine with Dimethylformamide Dimethylacetal (for Gas Chromatography), *Clin. Chem.*, **22**, 1129–1130 (1976).

Schwertner, H. A., H. E. Hamilton, and J. E. Wallace, Analysis for Carbamazepine in Serum by Electron-Capture Gas Chromatography, *Clin. Chem.*, **24**, 895–899 (1978).

Sheehan, M. and R. E. Beam, GLC Determination of Underivatized Carbamazepine in Whole Blood, *J. Pharm. Sci.*, **64**, 2004–2006 (1975).

Trager, W. F., R. H. Levy, I. H. Patel, and J. N. Neal, Simultaneous Analysis of Carbamazepine and Carbamazepine-10,11-epoxide by GC/CI/MS, Stable Isotope Methodology, *Anal. Lett.*, **11**, 119–133 (1978).

Westenberg, H. G., R. A. de Zeeuw, E. van der Kleijn, and T. T. Oei, Relationship Between Carbamazepine Concentrations in Plasma and Saliva in Man as Determined by Liquid Chromatography, *Clin. Chim. Acta*, **79**, 155–161 (1977).

CHLORAMPHENICOL

D-(−)-Threo-2,2-dichloro-N-[β-hydroxy-α-(hydroxymethyl)-p-nitrophenethyl]acetamide.
Chloromycetin, Paraxin, and so on. Antibiotic.

$$O_2N-\underset{HO\ H}{\overset{H\ \ NH-\overset{\overset{O}{\|}}{C}-CHCl_2}{\underset{|}{\bigcirc}-\underset{|}{\overset{|}{C}}-\underset{|}{\overset{|}{C}}-CH_2OH}}$$

$C_{11}H_{12}Cl_2N_2O_5$; mol. wt. 323.14.

a. Therapeutic Concentration Range. An oral dose of 2 g chloramphenicol gives peak plasma concentrations ranging from 20 to 40 μg/ml (Resnik et al., 1966). High plasma levels may be associated with bone marrow toxicity.

b. Metabolism. The drug is primarily inactivated in the liver by glucuronidation (e.g., Thies and Fischer, 1978). Further metabolic pathways include cleavage of the amide bond (Nakagawa et al., 1975). Chloramphenicol metabolites are considered to be largely inactive (Thies and Fischer, 1978). The plasma elimination half-life of chloramphenicol ranges from 1.5 to 3.5 hr (Goodman and Gilman, 1975).

c. Analogous Compounds. Aromatic nitro derivatives (e.g., nitrazepam); thiamphenicol.

d. Analytical Methods

1. High Performance Liquid Chromatography—Thies and Fischer (1978)

Discussion. Chloramphenicol is extracted together with the internal standard mephenesin into an organic solvent and analyzed by reversed phase HPLC and UV detection at 278 nm. The method is sensitive to 0.1 μg chloramphenicol/ml plasma, serum, urine, or cerebrospinal fluid, using 0.1 ml of the biological sample. Its specificity for chloramphenicol in the presence of metabolites is verified by MS. Chloramphenicol succinate, the common form used for intravenous administration, is separated from chloramphenicol on the HPLC column. Other antibiotics do not interfere, and the authors conclude that this assay is suitable for clinical applications.

Detailed Method. The experimental sample, 0.1 to 0.2 ml serum, plasma, cerebrospinal fluid, or urine, and 50 μl methanolic solution of 10 μg mephenesin are diluted with tris(hydroxymethyl)aminomethane (0.8 mol/l, pH 10.4) buffer to produce a volume of 1.0 ml. The solution is extracted with 10 ml diethyl ether by shaking vigorously for 10 min. The organic layer is separated by centrifugation, and 8 ml of it is transferred to a centrifuge tube and evaporated at room temperature under a stream of nitrogen. (Heat should be avoided during this solvent evaporation.) Acid-washed tubes or disposable glass tubes have been used successfully, but detergent contamination may interfere with the assay. For chromatography, 20 to 50 μl methanol is added to the tube, and a 5 to 10 μl sample is injected into the chromatograph. Peak height measurements are used to construct standard curves.

The chromatograph consists of a Model 6000 A pump and a U6-K sample injector (both from Waters Assoc., Milford, Mass. 01756). A variable-wavelength detector, Model SF770 (Schoeffel Instruments, Westwood, N.J. 07675), is used. The Partisil 10-ODS (Whatman, Clifton, N.J. 071014) column is eluted with methananol-water (30:70, v/v) at a flow rate of 1.5 ml/min.

2. Other HPLC Methods—Peng et al. (1978), Wal et al. (1978)

Discussion. The procedure of Peng et al. (1978) does not require a solvent extraction prior to HPLC analysis by reverse phase at 280 nm. The sensitivity is about 2.5 μg chloramphenicol/ml, which is sufficient for clinical purposes. This procedure therefore should be the method of choice for therapeutic drug level monitoring, if sufficient specificity can be demonstrated in subsequent studies.

Wal et al. (1978) achieve a sensitivity of 0.25 μg chloramphenicol/ml plasma (0.5 ml sample) using ethyl acetate extraction, a similar reverse phase column, and UV detection at 254 nm. External standards are employed for quantitation.

3. Gas Chromatography—Least et al. (1977), Resnick et al. (1966)

Discussion. Resnick et al. (1966) propose a procedure involving solvent extraction, derivatization with a silylation reagent (hexamethyldisilazane and trimethylchlorosilane), and GC with electron capture detection, the monochloro analog of chloramphenicol serving as the internal standard. An improved organic extraction step enables Least et al. (1977) to employ a flame ionization detector with sufficient sensitivity for clinical purposes. Internal standards are thiamphenicol and the acetyl analog of chloramphenicol. The requirement of 0.5 ml serum, however, decreases the assay utility for monitoring chloramphenicol therapy in infants and children. Gas chromatographic methods are specific for unchanged chloramphenicol.

4. Colorimetry—Glazko et al. (1949), Kakemi et al. (1962)

Discussion. The procedure of Kakemi et al. (1962) utilizes the rather specific reaction of chloramphenicol with isoniazid to form a yellow color. Glazko et al. (1949) use the Bratton-Marshall azo dye reaction, following reduction of the nitro group to the primary aromatic amine. This assay is subject to interference by a number of agents, including primary aromatic amines and nitro derivatives.

5. Fluorescence—Clarenburg and Rao (1977)

Discussion. Chloramphenicol is extracted from biological fluids into an organic solvent, reduced with zinc dust in HCl to the amine, and reacted with fluorescamine. Fluorescence analysis of the product allows detection of chloramphenicol down to 60 ng/ml plasma without interference by the glucuronide metabolite.

6. Enzymatic Assay—Lietman et al. (1976)

Discussion. The assay is based on the enzymatic acetylation of chloramphenicol by an R-factor-mediated enzyme with ^{14}C-acetyl-coA as the cofactor. Only 10 μl of plasma is needed for analysis of therapeutic chloramphenicol concentrations. The assay is specific for chloramphenicol in the presence of its glucuronide, succinate, and l-threo derivatives, as well as other antibiotics. Results obtained from clinical samples correlate well with GC data, although GC assays do not differentiate between (–)- and (+)-threochloramphenicol.

7. Polarography—Fossdal and Jacobsen (1971)

Discussion. The direct electrochemical determination of chloramphenicol in milk samples is described within a concentration range of 3 to 60 μg/ml. The drug undergoes a two-electron reduction of the nitro group.

8. Microbiological Assays—e.g, Joslyn and Galbraith (1949)

Discussion. This turbidometric assay measures the ability of chloramphenicol to inhibit the growth of a test organism. It lacks the sensitivity and specificity of chemical analytical methods.

9. Assay Comparison and Mass Spectrometry—Pickering et al. (1979)

Discussion. The report of Pickering et al. (1979) compares three assay methods—radioenzymatic, GC, and GC-MS—all of which gave satisfactory results. The GC-MS is performed at 70 eV electron impact ionization, using selected ion monitoring at m/e 225 of silanized chloramphenicol. D-(–)-Threo-N-[β-hydroxy-α-(hydroxymethyl)-p-nitrophenylethyl]acetamide serves as the internal standard.

10. Comments. Both GC and HPLC are clinically useful techniques; HPLC is more rapid than GC.

References

Clarenburg, R. and V. R. Rao, A Fluorimetric Method to Assay Chloramphenicol, Drug Metab. Dispos., **5**, 246–252 (1977).

Fossdal, K. and E. Jacobsen, Polarographic Determination of Chloramphenicol, *Anal. Chim. Acta*, **56**, 105–115 (1971).

Glazko, A. J., L. M. Wolf, and W. A. Dill, Biochemical Studies on Chloramphenicol (Chloromycetin): I: Colorimetric Methods for the Determination of Chloramphenicol and Related Nitro Compounds, *Arch. Biochem.*, **23**, 411–418 (1949).

Goodman, L. S. and A. Gilman, *The Pharmacological Basis of Therapeutics*, The Macmillan Co., New York, 1975, p. 1183.

Joslyn, D. A. and M. Galbraith, A. Turbidometric Method for the Assay of Antibiotics, *Bacteriology*, **54**(G57), 26–27 (1947).

Kakemi, K., T. Arita, and S. Ohaski, Absorption and Excretion of Drugs. IV: Determination of Chloramphenicol in the Blood, *J. Pharm. Soc. Jap.*, **82**, 342–345 (1962).

Least, C. J., N. J. Wiegand, G. F. Johnson, and H. M. Solomon, Quantitative Gas-Chromatographic Flame-Ionization Method for Chloramphenicol in Human Serum, *Clin. Chem.*, **23**, 220–222 (1977).

Lietman, P. S., T. J. White, and W. V. Shaw, Chloramphenicol: An Enzymological Microassay, *Antimicrob. Agents Chemother.*, **10**, 347–353 (1976).

Nakagawa, T., M. Masada, and T. Uno, Gas Chromatographic Determination and Gas Chromatographic-Mass Spectrometric Analysis of Chloramphenicol, Thiamphenicol and Their Metabolites, *J. Chromatogr.*, **111**, 355–364 (1975).

Peng, G. W., M. A. F. Gadalla, and W. L. Chiou, Rapid and Micro High-Pressure Liquid Chromatographic Determination of Chloramphenicol in Plasma, *J. Pharm. Sci.*, **67**, 1036–1038 (1978).

Pickering, L. K., J. L. Hoecker, W. G. Kramer, J. G. Liehr, and R. M. Caprioli, Assays for Chloramphenicol Compared: Gas Chromatographic with Electron Capture, and Gas Chromatographic-Mass Spectrometric, *Clin. Chem.*, **25**, 300–305 (1979).

Resnik, G. L., D. Corbin, and D. H. Sandberg, Determination of Serum Chloramphenicol Utilizing Gas-Liquid Chromatography and Electron Capture Spectrometry, *Anal. Chem.*, **38**, 582–585 (1966).

Thies, R. L. and L. J. Fischer, High-Performance Liquid-Chromatographic Assay for Chloramphenicol in Biological Fluids, *Clin. Chem.*, **24**, 778–781 (1978).

Wal, J. M., J. C. Peleran, and G. Bories, Dosage Sensible et Rapide du Chloramphenicol dans le Serum par Chromatographie Liquide Haute Pression, *J. Chromatogr.*, **145**, 502–506 (1978).

CHLORDIAZEPOXIDE—by D. Smith

7-Chloro-*N*-methyl-5-phenyl-3H-1,4-benzodiazepin-2-amine 4-oxide. Librium®. Minor tranquilizer.

$C_{16}H_{14}ClN_3O$; mol. wt. 299.75; pK_a 4.8.

a. Therapeutic Concentration Range. From 1.0 to 3.0 µg/ml. Toxicity occurs above 5.5 µg/ml, and lethal concentrations exceed 20 µg/ml (Winek, 1976).

b. Metabolism. Chlordiazepoxide is mainly metabolized to desmethylchlordiazepoxide, which undergoes hydrolytic deamination to demoxepam. This major pathway is followed by reduction of the $N_4 \rightarrow O$ to *N*-desmethyldiazepam, hydroxylation to oxazepam, and glucuronidation (Dixon et al., 1976; Schallek et al., 1972). Except for the glucuronide, all of these metabolites are pharmacologically active (Dixon et al., 1976; Schallek et al., 1972). The plasma elimination half-life of chlordiazepoxide is 20 to 24 hr (Koechlin et al., 1965); that of demoxepam, 14 to 95 hr (Schwartz et al., 1971). Demoxepam may therefore accumulate to a greater extent than the parent drug during multiple dosing.

c. Analogous Compounds. Benzodiazepines: chlorazepate, clonazepam, diazepam, flurazepam, medazepam, nitrazepam, oxazepam.

d. Analytical Methods

  1. Radioimmunoassay—Dixon et al. (1975)

Discussion. The antiserum was obtained from rabbits following immunization with bovine serum albumin coupled to chlordiazepoxide via the 5-phenyl ring. The assay uses ^{14}C-chlordiazepoxide (200 μCi/mg) as the tracer and is sensitive to 20 ng/ml plasma. Chlordiazepoxide can be analyzed directly without prior extractions. The antibody shows minimal cross-reactivity with demoxepam, N-desmethyldiazepam, diazepam, and clonazepam (less than 1%). N-Desmethylchlordiazepoxide shows about a 5% cross-reactivity, which may result in a 10% overestimation of plasma levels 36 to 48 hr after a single dose. The assay is not interfered with by this metabolite under multiple-dosing, steady-state conditions. Results compare well with those obtained by a fluorescence assay (Koechlin and D'Arconte, 1963).

Detailed Method. Plasma samples (20 to 100 μl) are added to tubes containing 0.1 ml of 0.001 N HCL. Then 0.1 ml (4000 cpm) of a chlordiazepoxide-2-^{14}C solution is added to each tube, followed by 0.1 ml of the antiserum solution. The volume in each tube is brought to 1 ml with 0.01 M sodium phosphate buffer (pH 7.4). After mixing on a vortex, each tube is immersed in an ice water bath for 1 hr. An equal volume (1 ml) of saturated ammonium sulfate is then added to precipitate globulin-bound chlordiazepoxide-2-^{14}C. After thorough mixing on a vortex and standing at 4° for 15 min, the tubes are centrifuged at 3000 rpm for 30 min at 4°. The supernatant containing unbound chlordiazepoxide-2-^{14}C is decanted into a counting vial, and 10 ml of a toluene scintillator is added. The vial is then vortexed for 10 sec to extract the radioactive material into the organic phase, and each sample is counted by liquid scintillation. Calibration curves are obtained with 2 to 100 ng chlordiazepoxide per sample.

  2. Colorimetry—Beyer and Sadée (1969), Frings and Cohen (1971)

Discussion. Chlordiazepoxide is extracted into chloroform, and back-extracted into 6 N HCL, followed by acid hydrolysis to 2-amino-6-chlorobenzophenone. The amine is diazotized and coupled with N-(1-

naphthyl)ethylenediamine in the Bratton-Marshall reaction and read at 550 nm. The assay procedure does not differentiate between chlordiazepoxide and its metabolites and is interfered with by primary aromatic amines such as sulfonamides and procainamide. Optimal conditions for the application of the Bratton-Marshall reaction to the measurement of benzodiazepines in biological samples have been developed by Beyer and Sadée (1969), resulting in a sensitivity of less than 1 μg/ml plasma or urine.

3. Fluorimetry—Koechlin and D'Arconte (1963), Stewart and Williamson (1976)

Discussion. The fluorimetric assay of Stewart and Williamson (1976) for chlordiazepoxide is based on a fluorophor which is formed with fluorescamine after acid hydrolysis of the drug. Plasma and urine samples both require heptane extractions for specific analysis of chlordiazepoxide in the presence of its metabolites, with a sensitivity of 0.25 ug/ml. Primary amines, such as amphetamine, may interfere. Koechlin and D'Arconte (1963) have developed a differential extraction method for chlordiazepoxide and its metabolite demoxepam, followed by a photochemical reaction to fluorescent products. The assay is sufficiently sensitive for therapeutic levels of chlordiazepoxide.

4. Gas Chromatography—Sun and Hoffman (1978), Zingales (1971)

Discussion. Chlordiazepoxide is extracted from plasma into *n*-heptane and 1.5% isoamyl alcohol and analyzed in its intact from by GC with electron capture detection (Zingales, 1971). Advantages of this method include sensitivity (lower limit 0.5 ng) and specificity with respect to metabolites. The GC conditions have to be carefully selected to avoid thermolytic degradation of chlordiazepoxide. Alternatively, chlordiazepoxide can be derivatized by flash methylation to a product with suitable GC properties, providing an assay sensitivity of 45 ng/ml serum (0.4 ml specimen) (Sun and Hoffman, 1978).

5. High Performance Liquid Chromatography—Greizerstein and Wojtowicz (1977), Stroijny et al. (1978)

Discussion. Blood samples are extracted with heptane and 1.5% isoamyl alcohol under alkaline conditions and analyzed for chlordiazepoxide and its *N*-desmethyl metabolite, using reverse phase HPLC and UV absorbance at 254 nm (Greizerstein and Wojtowicz, 1977). Chlorpromazine serves as the internal standard. The assay is sensitive to 0.1 μg chlordiazepoxide/ml, using only 50 μl plasma. It requires a time consuming solvent gradient elution of the column.

A similar HPLC assay for chlordiazepoxide and several of its metabolites has been reported by Stroijny et al. (1978) with a sensitivity of 50 to 100 mg ng/ml, using a 1 ml plasma specimen.

6. Thin-Layer Chromatography—Sun (1978)

Discussion. Chlordiazepoxide, demoxepam, and desmethylchlordiazepoxide are extracted from plasma into ether and separated by silica gel TLC. The plates are then treated with a sulfuric acid spray to convert the three compounds to fluorescent products, and the generated fluorescence is measured by fluorescence densitometry. The sensitivity of the assay is 0.05 μg chlordiazepoxide/ml plasma and 0.01 μg/ml for its two major plasma metabolites. Similar techniques may be applicable to other benzodiazepines as well.

7. Differential Pulse Polarography—Hackman et al. (1974)

Discussion. Chlordiazepoxide and its metabolites are extracted from plasma into diethyl ether and separated by TLC. Bands corresponding to chlordiazepoxide, its desmethyl metabolite, and demoxepam are eluted with methanol and quantitated by differential pulse polarography. This method is sensitive (lower limit 50 ng/ml) and specific.

8. Comments.

At present, the radioimmunoassay is the method of choice for the determination of chlordiazepoxide in a large number of plasma samples. The HPLC assay could be further improved, for example, with the adoption of isocratic solvent elution.

e. References

Beyer, K.-H. and W. Sadée, Spectrophotometrische Bestimmung von 5-Phenyl-1,4-benzodiazepin—Derivation and Untersuchungen über den Metabolismus des Nitrazepams, *Arzneim.-Forsch.* **19**, 1929–1931 (1969).

Dixon, W. R., J. Earley, and E. Postma, Radioimmunoassay of Chlordiazepoxide in Plasma, *J. Pharm. Sci.*, **64**, 937–939 (1975).

Dixon, W. R. et al., N-Desmethyldiazepam: A New Metabolite of Chlordiazepoxide in Man, *Clin. Pharmacol. Ther.*, **20**, 450–457 (1976).

Frings, C. S. and P. S. Cohen, Rapid Colorimetric Method for the Quantitative

Determination of Librium (Chlordiazepoxide Hydrochloride) in Serum, *Am. J. Clin. Pathol.*, **56**, 216–219 (1971).

Greizerstein, H. B. and C. Wojtowicz, Simultaneous Determination of Chlordiazepoxide and Its *N*-Desmethyl Metabolite in 50 μl Blood Samples by High Pressure Liquid Chromatography, *Anal. Chem.*, **49**, 2235–2236 (1977).

Hackman, M. R., M. A. Brooks, and J. A. F. de Silva, Determination of Chlordiazepoxide Hydrochloride (Librium) and Its Major Metabolites in Plasma by Differential Pulse Polarography, *Anal. Chem.*, **46**, 1075–1082 (1974).

Koechlin, B. A. and L. D'Arconte, Determination of Chlordiazepoxide (Librium) and of a Metabolite of Lactam Character in Plasma of Humans, Dogs, and Rats by a Specific Spectrofluorometric Micro Method, *Anal. Biochem.*, **5**, 195–207 (1963).

Koechlin, B. A., M. A. Schwartz, G. Krol, and W. Oberhansli, The Metabolic Fate of C^{14}-Labeled Chlordiazepoxide in Man, in the Dog, and in the Rat, *J. Pharmacol. Exp. Ther.*, **148**, 399–411 (1965).

Schallek, W., W. Schlosser, and L. O. Randall, Recent Developments in the Pharmacology of the Benzodiazepines, *Adv. Pharmacol. Chemother.*, **10**, 119–183 (1972).

Schwartz, M. A., E. Postma, and Z. Gaut, Biological Half-Life of Chlordiazepoxide and Its Metabolite, Demoxepam, in Man, *J. Pharm. Sci.*, **60**, 1500–1503 (1971).

Stewart, J. T. and J. L. Williamson, Fluorometric Determination of Chlordiazepoxide in Dosage Forms and Biological Fluids with Fluorescamine, *Anal. Chem.*, **48**, 1182–1185 (1976).

Strojny, N., C. V. Puglisi, and J. A. F. de Silva, Determination of Chlordiazepoxide and Its Metabolites in Plasma by High Pressure Liquid Chromatography, *Anal. Lett.*, **11**, 135–160 (1978).

Sun, S. R., Quantitative Determination of Chlordiazepoxide and Its Metabolites in Serum by Fluorescence TLC-Densitometry, *J. Pharm. Sci.*, **67**, 639–641 (1978).

Sun, S. R. and D. J. Hoffman, Rapid GLC Determination of Chlordiazepoxide and Metabolite in Serum Using On-Column Methylation, *J. Pharm. Sci.*, **67**, 1647–1648 (1978).

Winek, C. L., Tabulation of Therapeutic, Toxic, and Lethal Concentrations of Drugs and Chemicals in Blood, *Clin. Chem.*, **22**, 832–836 (1976).

Zingales, I. A., Determination of Chlordiazepoxide Plasma Concentrations by Electron Capture Gas-Liquid Chromatography, *J. Chromatogr.*, **61**, 237–252 (1971).

CHLOROQUINE

7-Chloro-4-(4-diethylamino-1-methylbutylamino)quinoline.
Aralen, Nivaquine, Resochin, and so on. Antimalarial; treatment of rheumatoid arthritis.

$$CH_3$$
$$NH—CH—CH_2CH_2CH_2N(CH_2CH_3)_2$$

$C_{18}H_{26}ClN_3$; mol. wt. 319.89; pK_a8.3 and 10.2

a. Therapeutic Concentration Range. None defined. Daily oral doses of 310 mg chloroquine over a period of 14 days give steadily increasing chloroquine plasma concentrations up to 120 ng/ml (McChesney et al., 1967).

b. Metabolism. Oxidative N-dealkylation to N-desethylchloroquine represents the major metabolic pathway (Fletcher et al., 1975; McChesney et al., 1966), which accounts for 23% of the detectable urinary excretion (McChesney et al., 1967). Further metabolites include bis(desethyl)chloroquine, 4'-hydroxychloroquine, 4'-carboxychloroquine, and 4-amino-7-chloroquinoline, all of which account for less than 3% of the total dose. Cumulative urinary recovery over 77 days accounts for 55% of the dose, mostly as unchanged chloroquine (70% of amount excreted) (McChesney et al., 1967). Plasma elimination of chloroquine is at least biphasic with half-lives of 3 and 18 days in the alpha and beta phases, respectively (McChesney et al., 1967). Therefore it may take several weeks after initiation of chloroquine therapy before steady-state plasma levels are reached. The very high tissue affinity of chloroquine, which binds to several sites, including nucleic acids (Morris et al., 1970), is primarily responsible for this long half-life. Binding to red blood cells is particularly avid, and the erythrocyte/plasma concentration ratio is > 100 (Williams and Fanimo, 1975). Detection of chloroquine may still be possible several months or even years after the last dose (McChesney et al., 1967), indicating that the terminal log-linear chloroquine elimination half-life may be considerably longer than 18 days.

c. Analogous Compounds. Aminoquinoline antimalarials: hydroxychloroquine, iodochlorhydroxyquine, mepacrine, primaquine, quinacrine, quinine; quinidine.

d. Analytical Methods

1. **Fluorescence**—Vogel and Königk (1975)

Discussion. Following solvent extractions, the fluorescence of chloroquine and its major metabolite, desethylchloroquine, is determined at pH 9.5. The method does not differentiate between chloroquine and its metabolite and is sensitive to 5 ng/ml plasma for both compounds. Analytical recoveries are 80% for chloroquine and 76% for desethylchloroquine. Other fluorescent metabolites of chloroquine are present at concentrations too low for significant assay interference. Quinine, but not quinacrine, interferes with the fluorescence assay.

Detailed Method. By shaking for 30 min, 1 to 5 ml plasma and an equal volume of 1 N NaOH are extracted with 30 ml heptane. Twenty milliliters of the organic layer is then transferred to another flask containing 3 ml 0.1 N HCl. The solution is shaken for 5 min and centrifuged, and the heptane layer is removed by aspiration. Then 2 ml of the aqueous layer is transferred to a cuvette containing 0.5 ml 0.4 N NaOH and 0.5 ml 3 M borate buffer (pH 9.5).

The fluorescence is measured on a fluorescence spectrophotometer (Perkin-Elmer MPF-2A) with an excitation wavelength of 340 nm and an emission wavelength of 385 nm.

2. Other Fluorescence Assays—Rubin et al. (1965), Schulman and Young (1974)

Discussion. Schulman and Young (1974) have investigated the effects of varying pH on the fluorescence yield of chloroquine. The doubly protonated species (pK_a 8.3) produces a 100-fold higher fluorescence yield than the monocation (pK_a 10.2) and the neutral molecule. However, at pH values above 10, significant quenching, which is probably caused by N—H proton abstraction in the fluorescent state, occurs. Thus fluorescence analysis at pH 9.5 to 10 is more than 10-fold more sensitive than at pH 13, the level used by previous workers (e.g., Rubin et al., 1965). Earlier fluorescence methods also include selective solvent extraction steps that separate chloroquine from desethylchloroquine (e.g., Rubin et al., 1965).

3. Colorimetry—Dondero and Mariappan (1974), Lelijveld and Kortmann (1970)

Discussion. Various field tests of urinary chloroquine have been designed to check the regularity of chloroquine intake. As an example, the Dill-Glatzko eosin color test is based on chloroform extraction of chloroquine from urine in the presence of eosine (Lelijveld and

Kortmann, 1970). The nonionized, light yellow eosine forms a chloro-form-soluble ion pair with chloroquine. The ionized eosin in the ion pair is deeply colored and intensely fluorescent, and chloroquine concentrations can be estimated by visual comparison with standards. However, the lack of inherent specificity of the test may yield misleading results (Dondero and Mariappan, 1974).

4. Gas Chromatography—Murayama and Nakajima (1977), Viala et al. (1975)

Discussion. The GC-FID assay of Viala et al. (1975) employs 3% OV-17 as the liquid phase and medazepam as the internal standard. The sensitivity is 0.25 µg chloroquine/ml blood. Considerably greater sensitivity can be achieved with electron capture detection (few ng/ml urine) (Murayama and Nakajiama, 1977). Following extraction of K_2CO_3 saturated urine with hexane, GC-EC analysis is also performed on 3% OV-17. The internal standard is O-ethyl-O-(p-nitrophenyl)phenylphosphonothioate.

5. Thin-Layer Chromatography—McChesney et al. (1966), Vogel and Königk (1975)

Discussion. Urine samples containing chloroquine are extracted into heptane at alkaline pH and back-extracted into 0.1 N HCl, and the aqueous phase is applied to silica gel plates with a fluorescence indicator (F_{254}) (Vogel and Königk, 1975). Ultraviolet densitometry of the plates provides a sensitivity of 5 µg chloroquine/ml urine. The method is specific for chloroquine in the presence of quinine, which interfers with the fluorescence assay described by the same authors (see Section d1).

McChesney et al. (1966) propose a similar extraction and TLC procedure. After development of the TLC plates, bands corresponding to chloroquine and four of its fluorescent metabolites are scraped off. The compounds are then eluted, and their fluorescence is determined. This assay is specific and sensitive to the low nanograms per milliliter plasma range.

6. Comments. Fluorescence and GC-EC assays of chloroquine are suitable for most biomedical applications. Special consideration should be given to the very long half-life of chloroquine and the very high erythrocyte/plasma concentration ratio, which may not remain constant *in vitro*.

e. References

Dondero T. J., Jr., and M. Mariappan, Disappointing Results in Malaysia with the Dill-Glazko Eosine Color Tests for Chloroquine in Urine, *Trans. R. Soc. Trop. Med. Hyg.*, **68**, 338–339 (1974).

Fletcher K. A., J. D. Baty, D. A. Evans, and H. M. Gilles, Studies on the Metabolism of Chloroquine in Rhesus Monkey and Human Subjects, *Trans. R. Soc. Trop. Med. Hyg.*, **69**, 6, (1975).

Lelijveld, J. and H. Kortmann, The Eosin Color Test of Dill and Glazko: A Simple Field Test to Detect Chloroquine in Urine., *Bull. World Health*, **42**, 447–479 (1970).

McChesney, E. W., W. D. Conway, W. F. Banks, Jr., J. E. Roger, and J. M. Shekosky, Studies of the Metabolism of Some Compounds of the 4-Amino-7-chloroquinoline Series, *J. Pharmacol. Exp. Ther.*, **151**, 482–493 (1966).

McChesney, E. W., M. J. Fasco, and W. F. Banks, Jr. The Metabolism of Chloroquine in Man During and After Repeated Oral Dosage, *J. Pharmacol. Exp. Ther.*, **158**, 323–331 (1967).

Morris , C. R., L. V. Andrew, L. P. Whichard, and D. J. Holbrook, Jr., The Binding of Antimalarial Aminoquinolines to Nucleic Acids and Polynucleotides, *Mol. Pharmacol.*, **6**, 240–250 (1970).

Murayama, K.,and A. Nakajima, Determination of Chloroquine in Urine by Gas Chromatography, *J. Pharm. Soc. Jap.*, **97**, 445–449 (1977).

Rubin, M., N. Zvaifler, H. N. Bernstein, and A. Mansour, *Proceedings of the 2nd International Pharmacological Meeting*, Pergamon Press, Oxford, 1965, p. 467.

Schulman, S. G. and J. F. Young, A Modified Fluorimetric Determination of Chloroquine in Biological Samples, *Anal. Chim. Acta*, **70**, 229–232 (1974).

Viala, A., J. P. Cano, and A. Durand, Determination of Chloroquine in Biological Material by Gas-Liquid Chromatography, *J. Chromatogr.*, **111**, 299–303 (1975).

Vogel, C. W. and E. Königk, Spectrofluorometric and Spectrodensitometric Determination of Chloroquine in Plasma and Urine, *Trop. Med. Parasitol.*, **26**, 278–280 (1975).

Williams, S. G. and O. Fanimo, Malaria Studies *in vitro*. IV: Chloroquine Resistance and the Intracellular pH of Erythrocytes Parasitised with *Plasmodium berghei*, *Ann. Trop. Med. Parasitol.*, **69**, 301–309 (1975).

CHLORPROMAZINE—by Z. Hosein

2-Chloro-*N*,*N*-dimethyl-10H-phenothiazine-10-propamine.
Thorazine. Antipsychotic and antiemetic.

$C_{17}H_{19}ClN_2S$; mol. wt. 318.88; pK_a 9.2.

a. Therapeutic Concentration Range. No clear relationship between plasma levels of chlorpromazine and therapeutic results has as yet been defined (Alfredsson et al., 1976). Clinical improvement has been observed in psychotic patients with plasma levels of 50 to 300 ng/ml, upon administration of 400 to 800 mg/day (Rivera-Calimlin et al., 1976).

b. Metabolism. Chlorpromazine is extensively metabolized in the liver. The major metabolic pathways are hydroxylation in C-3 and C-7 with subsequent conjugation with glucuronic acid, formation of sulfoxides, and *N*-demethylation (Hammar et al., 1968). Monodes-methylchlorpromazine (norchlorpromazine) and 7-hydroxychlor-promazine are known to be pharmacologically active (Alfredsson et al., 1976). The $t_{1/2}$ of chlorpromazine is 6 hr or less in plasma, but metabolites are excreted in the urine for 2 to 6 weeks following cessation of medication. This indicates a longer terminal log-linear half-life of chlor-promazine, which might be detectable with more sensitive assay methods.

c. Analogous Compounds. Phenothiazines: fluphenazine, mesori-dazine, perphenazine, prochlorperazine, promazine, promethazine, thioridazine, trifluoperazine.

d. Analytical methods

Gas Chromatography with Electron Capture Detection — Curry (1968, 1974)

Discussion. The GC-EC method is suitable for the determination of chlorpromazine and its sulfoxide and demethylated metabolites in plasma. The assay involves extraction with an *n*-heptane–isoamyl alcohol solvent, concentration by successive back-extractions into aqueous and organic solvents, and analysis of the concentrated extract by GC, using a ^{63}Ni ionization detector. A sensitivity of 10 ng chlorpromazine/ml plasma with a S.D. of ±12.3 at 210 ng/ml (n = 7) was attained, using external standards. The plasma blank showed no interference in the concentration range of 10 to 100 ng/ml.

Detailed Method. To 1 to 5 ml heparinized or oxalated plasma are added 1 ml 1 N NaOH and 10 ml n-heptane containing 1.5% isoamyl alcohol. The tube is shaken mechanically for 30 min and centrifuged. A 9 ml aliquot of the organic layer is transferred to another tube containing 2 ml 0.05 N HCl. The tube is shaken for 10 min and centrifuged. A 1.8 ml aliquot of the aqueous layer is again transferred, and the solution is made alkaline with 0.2 ml aqueous 1 N NH$_3$ and extracted with 100 μl toluene containing 15% isoamyl alcohol. The aqueous layer is removed by aspiration, and 1 to 10 μl of the organic layer injected into the gas chromatograph. Standard solutions of chlorpromazine, the sulfoxide, and demethylated derivatives are prepared by dissolving the respective hydrochlorides in water. The stock solutions are stable when stored at 4°C in the dark. Neutralization of 1 ml samples of these solutions with 1 ml 1.0 N ammonia solution and extraction with 1 ml toluene containing 15% isoamyl alcohol give suitable chromatographic standard solutions.

Gas chromatography is performed with a Dye Model 104 instrument equipped with a 14 mCi ^{63}Ni ionization detector. A coiled, 9 ft glass column (i.d. 3.5 mm) of 3% OV-17 on Chromosorb WHP (80/100 mesh) is operated at 265°C. Dry nitrogen is the carrier gas at a flow rate of 80 ml/min. The inlet and detector temperatures are 275°C.

2. Gas Chromatography-Mass Spectrometry—Alfredsson et al. (1976), Craig et al. (1974), Hammar et al. (1968)

Discussion In the selected ion monitoring mode GC-MS (electron impact) is a very sensitive tool for the detection of chlorpromazine and its metabolites in biological samples, where it can exceed the specificity and sensitivity of GC-EC (Craig et al., 1974). The report by Hammar et al. (1968) is one of the earliest examples of the application of this technique ("mass fragmentography") to drug analysis. Deuterated internal standards enhance the precision of this method (Alfredsson et al., 1976).

3. High Performance Liquid Chromatography— Watson and Stewart (1977)

Discussion. Tricyclic drugs, including chlorpromazine, and their metabolites are assayed in urine by HPLC. Although chlorpromazine and some of its metabolites are separated under the conditions used, a more extensive study with emphasis on specificity is required.

4. Thin-Layer Chromatography—Chan et al. (1974)

Discussion. Chlorpromazine and 17 metabolites in human plasma and urine are assayed by direct UV spectrodensitometry of the thin-layer chromatograms. The compounds are identified and quantitated by comparison with reference samples. Reproducibility is ±10%; the sensitivity is not specified.

5. Thin-Layer Chromatography-Fluorescence—Kaul et al. (1970, 1976)

Discussion. N-Demethylated chlorpromazine metabolites are measured by fluorescence analysis as their dansyl derivatives after TLC separation (Kaul et al., 1970). A similar method, involving quaternization with 9-bromomethylacridine and photolysis, is proposed for the quantitation of chlorpromazine and its sulfoxide (Kaul et al., 1975). Although adequately precise and sensitive, these techniques are laborious.

6. Ultraviolet Spectrophotometry—Wallace and Biggs (1971).

Discussion. Phenothiazines and their sulfoxide metabolites are readily oxidized by cobalt(III) and can be measured by their UV absorbance maximum at ~275 nm. Linearity is achieved in the range of 0.5 to 50 μg/ml; however, the method is nonspecific.

7. Colorimetry—Forrest and Forrest (1960)

Discussion. Phenothiazines are readily oxidized in acidic medium to stable free radicals. The color reagent, consisting of 5% ferric chloride, 20% perchloric acid, and 50% nitric acid (5:45:50), reacts with phenothiazines in the urine, forming a range of colors from light pinkish orange to violet to purple, depending on the dose administered. This method is used by some clinical laboratories as a routine screening procedure for phenothiazines.

8. Radioimmunoassay—Hubbard et al. (1978), Kawashima et al. (1975), Midha et al. (1979)

Discussion. The chlorpromazine-bovine serum albumin antibody obtained by Kawashima et al. (1975) is rather specific for chlorpromazine in the presence of its known metabolites. The method is sensitive to 10 pg chlorpromazine, using 10 μl plasma, without extraction. A thorough evaluation of the specificity of this assay for chlorpromazine in human plasma is needed.

Hubbard et al.. (1978) have synthesized three chlorpromazine haptens and produced antibodies against well-defined hapten-BSA conjugates. The antiserum obtained by immunizing rabbits with a conjugate of *N*-(2-carboxyethyl)desmethylchlorpromazine is specific for chlorpromazine and its minor active metabolite, *N*-desmethylchlorpromazine (Midha et al., 1979). Other chlorpromazine metabolites and tested psychotropic drugs do not cross-react significantly with the antibody. The sensitivity is below 34 ng chlorpromazine assayed in 200 μl plasma, using ^{3}H-chlorpromazine as the tracer.

9. Polarography—Beckett et al. (1974)

Discussion. Cathode-ray polarography is used to measure the N-oxide, N-oxide-sulfoxide, and sulfoxide metabolites of chlorpromazine after their separation from urine, plasma, and microsomal samples. Concentrations as low as 20 ng/ml can be measured.

10. Comments. The choice of method depends on the metabolites to be assayed. The GC-EC assay appears to be the most versatile method at present.

e. References

Alfredsson, G., B. Wode-Helgodt, and G. Sedvall, A Mass Fragmentographic Method for the Determination of Chlorpromazine and Two of Its Active Metabolites in Human Plasma and CSF, *Psychopharmacology*, **48**, 123–131 (1976).

Beckett, A. H., E. E. Essien, and F. Smyth, A Polarographic Method for the Determination of the N-Oxide, N-Oxide-Sulphoxide, and Sulphoxide Metabolites of Chlorpromazine, *J. Pharm. Pharmac.*, **26**, 399–407 (1974).

Chan, T. L., G. Sakalis, and S. Gershon, Quantitation of Chlorpromazine and Its Metabolites in Human Plasma and Urine by Direct Spectrodensitometry of Thin Layer Chromatograms, *Adv. Biochem. Psychopharmacol.*, **9**, 335–345 (1974).

Craig, J. C., W. A. Garland, L. D. Gruenke, L. R. Kray, and K. A. N. Walker, The Use of Combined Gas Chromatography-Mass Spectrometry Techniques for the Identification of Hydroxylated and Dihydroxylated Metabolites of Phenothiazine Drugs, *Adv. Biochem. Psychopharmacol.*, **9**, 405–411 (1974).

Curry, S. H., Determination of Nanogram Quantities of Chlorpromazine and Some of Its Metabolites in Plasma Using Gas-Liquid Chromatography with an Electron Capture Detector, *Anal. Chem.*, **40**, 1251–1255 (1968).

Curry, S. H., Chlorpromazine Analysis by Gas Chromatography with an Electron Capture Detector, *Adv. Biochem. Psychopharmacol.*, **9**, 593–602 (1974).

Forrest, I. S. and F. M. Forrest, Urine Color Test for the Detection of Phenothiazine Compounds, *Clin. Chem.*, **6**, 11–15 (1960).

Hammar, C. G., B. Holmstedt, and R. Ryhage, Mass Fragmentography Identification of Chlorpromazine and Its Metabolites in Human Blood by a New Method, *Anal. Biochem.*, **25**, 532–548 (1968).

Hubbard, J. W., K. K. Midha, I. J. McGilveray, and J. K. Cooper, Radioimmunoassay for Psychotropic Drugs. I: Synthesis and Properties of Haptens for Chlorpromazine *J. Pharm. Sci.*, **67**, 1563–1571 (1978).

Kaul, P. N., M. W. Conway, M. L. Clark, and J. Huffine, Chlorpromazine Metabolism. I: Quantitative Fluorometric Method for 11 Chlorpromazine Metabolites, *J. Pharm. Sci.*, **59**, 1745–1749 (1970).

Kaul, P. N., L. R. Whitfield, and M. L. Clark, Chlorpromazine Metabolism. VII: New Quantitative Fluorometric Determination of Chlorpromazine and Its Sulfoxide, *J. Pharm. Sci.*, **65**, 689–694 (1976).

Kawashima, K., R. Dixon, and S. Spector, Development of Radioimmunoassay for Chlorpromazine, *Eur. J. Pharmacol.*, **32**, 195–202 (1975).

Midha, K. K., J. C. K. Loo, J. W. Hubbard, M. L. Rowe, and I. J. McGilveray, Radioimmunoassay for Chlorpromazine in Plasma, *Clin. Chem.*, **25**, 166–168 (1979).

Rivera-Calimlin, L., H. Nasrallah, J. Strauss, and L. Lasagna, Clinical Response and Plasma Levels: Effects of Dose, Dosage Schedules and Drug Interactions on Plasma Chlorpromazine Levels, *Am. J. Psychiatr.*, **133**, 646–651 (1976).

Wallace, J. E. and J. D. Biggs, Determination of Phenothiazine Compounds in Biologic Specimens by UV Spectrophotometry, *J. Pharm. Sci.*, **60**, 1346–1350 (1971).

Watson, I. D. and M. J. Stewart, Assay of Tricyclic Structured Drugs and Their Metabolites in Urine by High Performance Liquid Chromatography, *J. Chromatogr.*, **134**, 182–186 (1977).

CLOFIBRATE—by G. Rosner

Ethyl-*p*-chlorophenoxyisobutyric acid.
Atromid-S. Antihypercholesteremic and antihypertriglyceridemic.

$C_{12}H_{15}ClO_3$; mol. wt. 242.71; pK_a 3.0 (free acid).

a. Therapeutic Concentration Range. From 80 to 200 μg of the free carboxylic acid/ml plasma (Berlin, 1975; Sedeghat, 1974; Thorp, 1962).

b. Metabolism. Clofibrate is well absorbed from the gastrointestinal tract and rapidly and quantitatively hydrolyzed to its corresponding free acid, chlorophenoxyisobutyrate (clofribinic acid), which is the active form. It is 96% bound to serum albumin and largely excreted into the urine as the glucuronide with a half-life of 12 hr (Gugler et al., 1975).

c. Analogous Compounds. Lipophilic carboxylic acids

d. Analytical Methods

1. High Performance Liquid Chromatography—Bjornsson et al. (1977)

Discussion. This is a rapid, sensitive, and specific reverse phase method with UV detection of underivatized clofibrinic acid at 235 nm, applicable to plasma, saliva, and urine following a two-step organic extraction. Sensitivity is to 0.5 μg/ml, with a C.V. of 1 to 6%. No interferences are observed in control samples.

Detailed Method. Aqueous samples (0.1 to 1.0 ml plasma, 1.0 ml saliva, or 1.0 ml urine, diluted 1:100) are shaken with 100 μl of an internal standard solution, 0.5 ml 0.5 N H$_2$SO$_4$, and 5 ml toluene and centrifuged, and the lower aqueous phase is frozen in a dry ice-acetone bath. The organic phase is decanted into another tube containing 50 μl 0.2 N NaOH, vigorously shaken, and again centrifuged. The aqueous phase is drawn into a syringe containing 5% glacial acetic acid in water, and the mixture is injected into the HPLC column. Urine samples are heated for 30 min with HCl prior to extraction in order to liberate clofibrinic acid from its glucuronide. Concentrations of clofibrinic acid are determined using calibration curves of clofibrinic acid/internal standard peak height ratios versus concentrations of clofibrinic acid standards.

A Varian (Model 8500) HPLC fitted with a MicroPak CH-10 reverse phase column, 25 × 6.3 mm (i.d.), is used with a Varian Variscan variable-wavelength UV detector at 235 nm. A mixture of 5% acetic acid and 42% acetonitrile (v/v) in water is used at a flow rate of 70 ml/hr.

2. **Gas Chromatography**—Berlin (1975), Chuong and Tuong (1975), Crouthamel and Cenedella (1975), Gugler and Jensen (1976), Horning et al. (1972), Sedeghat et al. (1974)

Discussion. Several GLC methods have been described, but many of these involve extensive steps, such as TLC (Crouthamel and Cenedella, 1975; Sedeghat et al., 1974), column chromatography (Horning et al., 1972), or multiple solvent extractions (Berlin, 1975; Chuong and Tuong, 1975). The free carboxylic acid function of clofibrinic acid has to be derivatized for GC. Some methods use diazomethane (Berlin, 1975; Chuong and Tuong, 1975; Houin et al., 1975), which should be avoided because of its toxicity, if other alkylating agents are applicable. Gugler and Jensen (1976) have developed a rapid method employing GC-FID for plasma and urine samples. The assay involves extraction into toluene and back-extraction of clofibrinic acid and its internal standard, 2-naphthoic acid, into the methylating reagent, trimethylphenylammonium hydroxide. Sensitivity extends to 1 μg/ml with a C.V. of 3.3% for plasma and 2.7% for urine. Recovery studies yield 71 ± 2% (n = 10). No endogenous or other compounds are shown to interfere.

3. **Spectrophotometry**—Barrett and Thorp (1967)

Discussion. This method measures clofibrinic acid at its UV absorption maximum of 226 nm after solvent extraction from acidified plasma or urine. It is rapid and convenient, but lacks specificity.

4. **Comments.** High performance liquid chromatography offers a simple, accurate, and rapid technique for the measurement of clofibrinic acid in biological samples.

e. References
Barrett, A. M. and J. M. Thorp, Studies on the Mode of Action of Clofibrate: Effects of Hormone-Induced Changes in Plasma Free Fatty Acids, Cholesterol, Phospholipids and Total Esterified Fatty Acids in Rats and Dogs, *Brit. J. Pharm. Chemother.*, **32**, 381–391 (1968).

Berlin, A., Quantitative Gas Chromatographic Determination of Clofibrinic Acid in Plasma, *J. Pharm. Pharmacol.*, **13**, 465–473 (1975).

Bjornsson, T. D., T. F. Blaschke, and P. J. Meffin, High-Pressure Liquid

Chromatographic Analysis of Drugs in Biological Fluid. IV: Determination of Clofibrinic Acid, *J. Chromatogr.* **137**, 145–152 (1977).

Bridgeman, J. F., S. M. Rosen, and J. M. Thorp, Complications During Clofibrate Treatment of Nephrotic-Syndrome Hyperlipoproteinaemia, *Lancet*, *II*, 506–509 (1972).

Chuong, T. C. and A. Tuong, Quantitative Determination of *p*-Chlorophenoxyisobutyric Acid in Blood Plasma by Gas-Liquid Chromatography, *J. Chromatogr.*, **106**, 97–102 (1975).

Crouthamel, W. G. and R. J. Cenedella, Clofibrate Pharmacokinetics: Effect of Elevation of Plasma Free Fatty Acids, *Pharmacology*, **13**, 465–473 (1975).

Gugler, L. and C. Jensen, A Rapid Gas Chromatographic Method for the Determination of Chlorophenoxyisobutyric Acid in Plasma and Urine, *J. Chromatogr.*, **117**, 175–179 (1976).

Gugler, R., D. W. Shoeman, D. H. Huffman, J. B. Cohlmia, and D. L. Azarnoff, Pharmacokinetics of Drugs in Patients with the Nephrotic Syndrome, *J. Clin. Invest.*, **55**, 1182–1189 (1975).

Horning, M. G., et al., Effects of Ethyl-*p*-chlorophenoxyisobutyrate on Biliary Secretion of Bile Acids, Cholesterol and Phosphatidyl Choline, *Lipids*, **7**, 114–120 (1972).

Houin, G., J. J. Thébault, P. d'Athis, J.-P. Tillemont, and J. L. Beaumont, A GLC Method for Estimation of Chlorophenoxyisobutyric Acid in Plasma: Pharmacokinetics of a Single Oral Dose of Clofibrate in Man, *Eur. J. Clin. Pharmacol.*, **8**, 433–437 (1975).

Sedeghat, A., H. Nakamura, and E. H. Ahrens, Determination of Clofibrate in Biological Fluids by Thin-Layer and Gas-Liquid Chromatography, *J. Lipid Res.*, **15**, 352–355 (1974).

Thorp, J. M., Experimental Evaluation of an Orally Active Combination of Androsterone with Ethyl Chlorophenoxyisobutyrate, *Lancet*, **I**, 1323–1326 (1962).

CLONAZEPAM

5-(*o*-Chlorophenyl)-1,3-dihydro-7-nitro-2H-1,4-benzodiazepin-2-one. Clonopin, Rivotril. Anticonvulsant.

$C_{15}H_{10}ClN_3O_3$; mol. wt. 315.72; pK_a 1.5 and 10.5.

a. Therapeutic Concentration Range. From 5 to 50 ng clonazepam/ml plasma (Brown, 1976; Leal and Troupin, 1977).

b. Metabolism. The major mode of clonazepam inactivation and elimination occurs by metabolic reduction of the 7-nitro group to 7-aminoclonazepam, followed by acetylation to 7-acetamidoclonazepam (Min and Garland, 1977; de Silva et al., 1974). The 7-amino metabolite reaches steady-state plasma levels equal to those of clonazepam during maintenance therapy (usually 3 × 2 mg/day) (Min and Garland, 1977; Naestoft and Larsen, 1974; Sjö et al., 1975), while 7-acetamidoclonazepam is at least 5 times less concentrated (Naestoft and Larsen, 1974). 3-Hydroxylation of the 1,4-benzodiazepine moiety represents only a minor pathway of clonazepam metabolism (de Silva et al., 1974), which is similar to that of nitrazepam (Beyer and Sadée, 1969). There is no correlation between clonazepam side effects and plasma levels of clonazepam or its 7-amino metabolite, although a potential association between 7-aminoclonazepam and clonazepam withdrawal symptoms has been suggested (Sjö et al., 1975). The plasma elimination half-life of clonazepam shows a large interindividual variation, and $t_{1/2}$ ranges between 22 and 32 hr (Brown, 1976), 13 and 34 hr (de Boer et al., 1978), and 19 and 60 hr (Berlin and Dahlström, 1975) have been reported.

c. Analogous Compounds. Benzodiazepines (see diazepam monograph), in particular, the potent anticonvulsant 7-nitrobenzodiazepines, flunitrazepam and nitrazepam; aromatic nitro compounds (e.g., chloramphenicol, metronidazole, nitrofurantoin).

d. Analytical Methods

1. Gas Chromatography—de Silva et al. (1976)

Discussion. De Silva et al. (1976) review the GC assays of a series of benzodiazepines in blood. Electron capture detection is needed to achieve the requisite sensitivity for the low plasma levels. Although some benzodiazepines can be analyzed by GC-EC without derivatization, clonazepam has to be converted to the *N*-1-methyl analog by reaction with methyl iodide in basic solution in order to avoid adsorption losses on the column at low concentrations. Linearity of response is further increased by priming the column with control blood extracts and by injecting an external benzodiazepine standard solution for every two or three sample injections to deactivate column adsorption sites.

It is necessary to precipitate blood proteins along with lipid contaminations, if high sensitivity is to be achieved. Classical precipitation methods (alcohols, acids) result in poor recovery of benzodiazepines because of coprecipitation. Mild protein denaturation by heating the sample in a 1 M borate buffer (pH 9) gives excellent recoveries for clonazepam (>90%). Solvent extraction is accomplished with benzene-methylene chloride (9:1). The overall extraction recovery is 85 ± 5%, and nitrazepam serves as the internal standard. Sensitivity of the assay is 1 ng clonazepam/ml blood.

Detailed Method. Whole blood (1 ml), internal standard solution containing 10 ng nitrazepam, and 2 ml 1 M borate buffer (pH 9.0) are thoroughly mixed and heated briefly three times for 12, 6, and 6 sec, respectively, in a boiling water bath, followed by thorough mixing after each heating period. The mixture is then extracted with 6 ml benzene-methylene chloride (9:1) by shaking for 5 min. After centrifuging at 8 to 10°C, 5.5 ml of the organic layer is transferred to another test tube, and 50 μl CH_3I and 0.5 ml of a freshly prepared 0.025 M solution of tetrabutylammonium hydrogen sulfate in 0.1 N NaOH are added. After shaking for 10 min and centrifuging, the aqueous layer is removed and the remaining organic layer washed with 2 ml 0.1 N HCl and 2 ml water, to remove the tetrabutylammonium reagent. The organic layer is then transferred to another tube and evaporated to dryness at 60°C under a stream of N_2. The residue is dissolved in 100 μl benzene-acetone-methanol (80:15:5), and 10 μl injected onto the column.

The GC analysis is performed on a MicroTek MT-220 gas chromatograph (Tracor Instruments, Austin, Tex.), equipped with a 15 mCi ^{63}Ni electron capture detector. The column is a U-shaped 4 ft x 4 mm (i.d.) borosilicate glass column containing 3% OV-17 on 60/80 mesh Gas Chrom Q (Applied Science, State College, Pa.). Argon-methane (9:1) serves as the organic carrier gas at a flow rate of 75 ml/min with a detection purge flow of 20 ml/min.

Injection port, detector, and column are heated to 290, 325, and 240°C, respectively. Careful conditioning of the column is required. The column is first primed with an appropriate blood extract, carried through the procedure. Linearity of response has to be established daily by repetitively injecting 10 μl external standard solution until constant detector response is established. The standard solution consists of nitrazepam in benzene-acetone-methanol (80:15:5, 2 to 10 ng/100 μl). Peak areas are determined by measuring peak height times width at half-height. Calibration curves are constructed with the clonazepam/nitrazepam peak area ratios.

2. Other GC Assays—Cano et al. (1977), de Boer et al. (1978), de Silva et al. (1974), Edelbroek and De Wolff (1978), Gerna and Morselli (1976), Kangas (1977), Naestoft and Larsen (1974)

Discussion. All of these GC assays incorporate various organic solvent extraction procedures and electron capture detection with a sensitivity limit between 0.5 ng clonazepam/ml plasma (Cano et al., 1977; de Silva et al., 1974) and 3 to 5 ng/ml (Naestoft and Larsen, 1974; Gerna and Morselli, 1976). The GC analysis of clonazepam is performed either following methylation to avoid nonlinear response at very low levels (see Section d 1), after acid hydrolysis to the more volatile 2-amino-2'-chloro-5-nitrobenzophenone (Cano et al., 1977; Kangas, 1977; de Silva et al., 1974) or directly with the unchanged drug, yielding lower sensitivity because of column adsorption problems (Gerna and Morselli, 1976; Naestoft and Larsen, 1974). However, the GC assay of underivatized clonazepam (and/or nitrazepam) using an OV-17-coated open tubular capillary column and a single extraction can measure as little as 1 ng clonazepam/ml plasma (de Boer et al., 1978).

Similarly, Edelbroek and De Wolff (1978) describe a highly sensitive assay (1 ng/ml) of underivatized clonazepam on a properly conditioned packed column (3% OV-17 on Supelcoport, 60/70 mesh). The assay employs a solid injection system, requiring only 100 μl serum for analysis.

Hydrolysis of clonazepam results in the corresponding benzophenone with superior GC properties due to its high thermostability and vapor pressure. In addition, the earlier GC-EC titanium tritide detectors have an upper temperature limit of 225°C, which is sufficient for benzophenones, but too low for the sensitive direct analysis of benzodiazepines. With the use of the [63]Ni detector, which is stable at high temperatures (400°C), direct benzodiazepine analysis is feasible, thereby avoiding potential 3-OH metabolite assay interferences which give the same benzophenones as the parent drug. However, direct clonazepam analysis and the assay of benzophenone after H_2SO_4 hydrolysis yield similar results, indicating that 3-hydroxylation of clonazepam is only a minor pathway (Kangas, 1977). Furthermore, clonazepam can be selectively extracted into organic solvents in the presence of its more polar 3-hydroxy metabolites (de Silva, 1974). It appears, therefore, that the benzophenone method is applicable to the specific analysis of clonazepam, but not of certain other benzodiazepines with substantial 3-OH metabolites (e.g., diazepam).

Naestoft and Larsen (1974) measure plasma levels of clonazepam, 7-aminoclonazepam, and 7-acetamidoclonazepam following extraction of the rather weak base clonazepam at acidic pH and of the stronger basic metabolites at pH 7.4. The assay of other nitrobenzodiazepines (e.g., nitrazepam and flunitrazepam) is described by de Boer et al. (1978), Cano et al. (1977), Kangas (1977), and de Silva et al. (1974).

3. Gas Chromatography-Mass Spectrometry—Min and Garland (1977)

Discussion. Gas chromatography-mass spectrometry using chemical ionization with ammonia as the reactant gas is capable of measuring clonazepam and 7-aminoclonazepam with a sensitivity of 1 to 2 ng/ml plasma. The corresponding ^{15}N-labeled internal standards have been synthesized using $K^{15}NO_3$ as the nitrating agent, followed by reduction to 7-aminoclonazepam. The respective MH^+ ions are measured in the selected ion monitoring mode. This isotope dilution method automatically accounts for extraction losses and column adsorption and therefore is performed with the underivatized compounds.

4. High Performance Liquid Chromatography—Perchalski and Wilder (1978)

Discussion. Clonazepam, diazepam, and desmethyldiazepam can be detected with a sensitivity of 5 to 10 ng per sample by HPLC on a fully porous 5 μ silica gel column (Partisil 5, Reeve Angel, Clifton, N.J.). The assay includes a single extraction step and uses 4,5-dihydrodiazepam as the internal standard. A fixed-wavelength (254 nm) UV detector is used for monitoring the HPLC eluent; however, wavelength detection at the major UV absorbance maximum of benzodiazepines at 230 nm may increase the sensitivity. Other benzodiazepines are also measurable under suitable chromatographic conditions.

5. Thin-Layer Chromatography—Ebel and Schütz (1977), Haefelfinger (1979), Wad and Hanifl (1977)

Discussion. The qualitative detection of clonazepam and its reduced metabolites on TLC plates is based on the Bratton-Marshall azo dye reaction (Ebel and Schütz, 1977). The 7-nitro group has to be reduced prior to diazotation to allow detection of clonazepam and nitrazepam. Wad and Hanifl (1977) report a TLC method for the measurement of clonazepam, diazepam, and their metabolites in the presence of each other. The 7-amino metabolites of the 7-nitrobenzodiazepines can be measured with high sensitivity (~1 ng/ml serum) by using the Bratton-Marshall reaction on-plate, followed by colorimetric densitometry (Haefelfinger, 1979).

6. Polarography—Kobiela-Kryzanowska (1976), de Silva et al. (1974)

Discussion. Nitrobenzodiazepines can be readily detected by polarography after acute poisoning (Kobiela-Kryzanowska, 1976). Differential pulse polarography of benzodiazepines is based on the azomethine ($>C_5{=}N_4{-}$) reduction. In conjuction with TLC separations, it has been used by de Silva et al. (1974) to quantitate clonazepam and its metabolites in urine.

7. Radioimmunoassay—Dixon and Crews (1977), Dixon et al. (1977), Spiegel et al. (1978)

Discussion. Clonazepam antibodies have been raised against bovine serum albumin conjugates of 3-hemisuccinyloxyclonazepam (Dixon et al., 1977). The 7-amino and 7-acetamido metabolite, as well as other anticonvulsants, do not cross-react significantly with this antibody. With ^{3}H-clonazepam as the tracer, the RIA is sensitive to 5 ng clonazepam/ml plasma, using 0.1 ml samples (Dixon et al., 1977). Correlation with a GC-EC assay is good. Introduction of a ^{125}I tracer greatly enhances the sensitivity and speed of this RIA, allowing quantitation of clonazepam in only 10 μl plasma samples at 5 ng/ml (Dixon and Crews, 1977). The tracer can be synthesized from 3-aminoclonazepam by acylation with ^{125}I-N-succinimidyl-3-(4-hydroxyphenyl)propionate. Clinical application of this assay is satisfactory, showing no interferences by other anticonvulsants (Spiegel et al., 1978). Clonazepam stability in stored plasma samples is good (Spiegel et al., 1978).

8. Comments. Clonazepam has become an important anticonvulsant drug, with a concomitant increase of clinical drug level determinations. The methods GC-EC, HPLC, and RIA are all suitable for the clinical clonazepam assay. Capillary GC-EC appears to be an excellent choice because of the speed and sensitivity of analysis. Unfortunately, use of capillary columns in clinical laboratories is rather uncommon.

e. References

Berlin, A and H. Dahlström, Pharmacokinetics of the Anticonvulsant Drug Clonazepam Evaluated from Single Oral and Intravenous Doses and by Repeated Oral Administration, *Eur. J. Clin. Pharmacol.*, **9**, 155–159 (1975).

Beyer, K.-H., and W. Sadée, Spektrophotometrische Bestimmung von Benzodiazepinderivaten und Untersuchungen über den Metabolismus des Nitrazepams, *Arzneim.-Forsch.*, **19**, 1929–1931 (1969).

Brown, T. R., Clonazepam: A Review of a New Anticonvulsant Drug, *Arch. Neurol.*, **33**, 326–332 (1976).

Cano, J. P., J. Guintrand, C. Aubert, and A. Viala, Determination of Flunitrazepam, Desmethylflunitrazepam and Clonazepam in Plasma by Gas Liquid Chromatography with an Internal Standard, *Arzneim.-Forsch.*, **27**, 338–342 (1977).

De Boer, A. G., J. Rost-Kaiser, H. Bracht, and D. D. Breimer, Assay of Underivatized Nitrazepam and Clonazepam in Plasma by Capillary Gas Chromatography Applied to Pharmacokinetic and Bioavailability Studies in Humans, *J. Chromatogr.*, **145**, 105–114 (1978).

De Silva, J. A., C. V. Puglisi, and N. Munno, Determination of Clonazepam and Flunitrazepam in Blood and Urine by Electron-Capture GLC, *J. Pharm. Sci.*, **63**, 520–527 (1974).

De Silva, J. A., I. Bekersky, C. V. Puglisi, M. A. Brooks, and R. E. Weinfeld, Determination of 1,4-Benzodiazepines and -diazepin-2-ones in Blood by Electron-Capture Gas-Liquid Chromatography, *Anal. Chem.*, **48**, 10–19 (1976).

Dixon, R. and T. Crews, A ^{125}I-Radioimmunoassay for the Determination of the Anticonvulsant Agent Clonazepam Directly in Plasma, *Res. Commun. Chem. Pathol. Pharmacol.*, **18**, 477–486 (1977).

Dixon, W. R., R. L. Young, R. Ning, and A. Liebman, Radioimmunoassay of the Anticonvulsant Agent Clonazepam, *J. Pharm. Sci.*, **66**, 235–237 (1977).

Ebel, S., and H. Schütz, Studies on the Detection of Clonazepam and Its Main Metabolites, Considering in Particular Thin-Layer Chromatography, *Arzneim.-Forsch.*, **27**, 325–337 (1977).

Edelbroek, P. M. and F. A. De Wolff, Improved Micromethod for Determination of Underivatized Clonazepam in Serum by Gas Chromatography, *Clin. Chem.*, **24**, 1774–1777 (1978).

Gerna, M. and P. L. Morselli, A Simple and Sensitive Gas Chromatographic Method for the Determination of Clonazepam in Human Plasma, *J. Chromatogr.*, **116**, 445–450 (1976).

Haefelfinger, P., Determination of the 7-Amino Metabolites of the 7-Nitrobenzodiazepines in Human Plasma by Thin-layer Chromatography, *J. High Res. Chromatogr. Commun.*, **1**, 39–42 (1979).

Kangas, L., Comparison of Two Gas-Liquid Chromatographic Methods for the Determination of Nitrazepam in Plasma, *J. Chromatogr.*, **136**, 259–270 (1977).

Kobiela-Krzyzanowska, A., Polarographic Determination in Whole Blood of Nitro Derivatives of Benzodiazepine in Acute Poisoning, *Pharmazie*, **31**, 649–650 (1976).

Leal, K. W. and A. S. Troupin, Clinical Pharmacology of Anti-epileptic Drugs: A Summary of Current Information, *Clin. Chem.*, **23**, 1964–1968 (1977).

Min, B. H. and W. A. Garland, Determination of Clonazepam and Its 7-Amino Metabolite in Plasma and Blood by Gas Chromatography-Chemical Ionization Mass Spectrometry, *J. Chromatogr.*, **139**, 121–133 (1977).

Naestoft, J. and N. E. Larsen, Quantitative Determination of Clonazepam and Its Metabolites in Human Plasma by Gas Chromatography, *J. Chromatogr.*, **93**, 113–122 (1974).

Perchalski, R. J. and B. J. Wilder, Determination of Benzodiazepine Anticonvulsants in Plasma by High-Performance Liquid Chromatography, *Anal. Chem.*, **50**, 554–557 (1978).

Sjö, O., E. F. Hvidberg, J. Naestoft, and M. Lund, Pharmacokinetics and Side Effects of Clonazepam and Its 7-Amino Metabolite in Man, *Eur. J. Clin. Pharmacol.*, **8**, 249–254 (1975).

Spiegel, H. E., J. Symington, and A. Savulich, The Stability and Reliability of Radioimmunoassays for Clonazepam, Diphenylhydantoin and Phenobarbital in Blood, Serum or Plasma, *Res. Commun. Chem. Pathol. Pharmacol.*, **19**, 271–280 (1978).

Wad, N. T. and E. J. Hanifl, Simplified Thin-Layer Chromatographic Method for the Simultaneous Determination of Clonazepam, Diazepam and Their Metabolites in Serum, *J. Chromatogr.*, **143**, 214–218 (1977).

CLONIDINE—by L. Lambert

2-(2,6-Dichloroanilino)-2-imidazoline.
Catapress. Antihypertensive.

$C_9H_9Cl_2N_3$; mol. wt. 230.10; pK_a 8.2.

a. Therapeutic Concentration Range. Davies et al. (1977) found effective blood pressure reduction in normotensive men occurring with plasma concentrations up to 1.5 to 2.0 ng/ml. Substantial reduction of blood pressure is seen in human beings following doses as small as 0.3 to 1.5 mg/day given orally (Dollery et al., 1976). The side effects of clonidine, sedation and dry mouth, also correlated directly with plasma levels and reached a maximum at 1.5 to 2 ng/ml for sedation and 1 ng/ml for dry mouth. At plasma levels above 2 ng/ml the antihypertensive effect is attenuated in normotensive men. The therapeutic range has yet to be studied in hypertensive subjects.

b. Metabolism. Ehrhardt (1972) showed that clonidine is hydroxylated at the C_4 position to give 4-OH-clonidine in the rat. This metabolite is found in the brain at a concentration 100 times less than that of the parent compound. Although pharmacologically active when injected intraventricularly, 4-hydroxyclonidine is therefore unlikely to contribute to the pharmacological effects of the parent drug. Clonidine has a mean terminal half-life of about 13 hr with a range of 5.2 to 23.4 hr (Davies et al.,1977; Dollery et al., 1976; Rehbinder and Deckers, 1969).

c. Analogous Compounds. Alpha adrenergic acting imidazoline derivatives; naphazoline, tetrahydrozoline, xylometazoline; alpha adrenergic blockers; phentolamine, tolazoline.

d. Analytical Methods

1. Gas Chromatography-Selective Ion Monitoring Mass Spectrometry—(Davies et al. (1977)

Discussion. This assay requires flash derivatization with trimethylphenylammonium hydroxide and GC-MS-SIM with deuterated clonidine as the internal standard. Following alkaline extraction from plasma, the drug is back-extracted into HCl, dried, and redissolved in the alkylating reagent for flash methylation in the GC injector port. Calibration is linear for clonidine over the range 0 to 25 ng/ml. Sensitivity of detection is 0.15 ng/ml $\pm$ 10% (S.D.). No interferences are reported.

Detailed Method. To 4 ml aliquots of plasma (or diluted urine, 1:4) is added 400 ng C-4,5-d_4-clonidine in 100 μl methanol as the internal standard and analytical carrier. Clonidine is extracted from alkalinized plasma (10% aqueous Na_2CO_3) or diluted urine (pH 10 to 11) with ethyl acetate. The drug is back-extracted into 0.1 N HCl, the pH is adjusted to 11, and the clonidine is then extracted with diethyl ether. After drying with nitrogen gas, the ether extract is redissolved in 20 μl trimethylphenylammonium hydroxide solution (TMPA, Pierce Chem.) for flash methylation.

The GC analysis is done using a 5 ft glass column of 3% OV-17 on 100/120 Gas Chrom Q, with helium as the carrier gas at 30 ml/min. The injection port temperature is 260°C, and the column temperature

195°C. The quadrupole mass spectrometer (Finnigan 3200; ionization voltage 25 eV) was set to monitor ions at m/e 228 for clonidine-d and m/e 257 for clonidine. The retention time is 3.5 min. Peak height ratios are determined using an interactive Model 6000 data system.

2. Other GC-MS Methods—Dollery et al. (1976)

Discussion. This method is similar to that described by Davies et al. (1977), except that the GC-MS analysis is performed without derivatization, thereby lowering the sensitivity. Precision at 1.0 ng/ml was ± 13% (S.D.) and at 0.5 ng/ml was ±16.5%.

3. Gas Chromatography-Electron Capture Detection—Cho and Curry (1969), Chu et al. (1979), Edlund and Paalzow (1977)

Discussion. The method described by Edlund and Paalzow (1977) is sensitive, with a minimum detectable quantity of 3.3 pg pure clonidine. The smallest amount of clonidine detectable in plasma (100 to 200 μl) with a precision less than 20% (S.D.) is 200 pg; therefore this GC-EC assay of clonidine is less sensitive than GC-MS-SIM (Davies et al., 1977) in biological samples. The method requires formation of a pentafluorobenzyl derivative and a chemical analog as the internal standard. The GC-EC assay described by Cho and Curry (1969) is still less sensitive; the minimum detectable sample of clonidine is 2 to 3 ng. However, Chu et al. (1979) have recently overcome the sensitivity problems with GC-EC by removal of excess derivatizing reagent (perfluoroacylation) and interfering compounds. They report a sensitivity of 25 pg clonidine/ml, using a 4 ml plasma sample.

4. Gas Chromatography-Flame Ionization Detection—Timmermans et al. (1977)

Discussion. A method for the determination of clonidine in rat brain is presented. The minimum detectable tissue concentration is 10 ng/g tissue, using a nitrogen-sensitive FID. Overall sensitivity is considerably below that of GC-EC.

5. Comments. Both the GC-MS-SIM assay (Davies et al., 1977) and the GC-EC procedure (Edlund and Paalzow, 1977) are suitable for the detection of therapeutic clonidine plasma levels.

e. References

Cho, A. K. and S. H. Curry, The Physiological Disposition of 2-(2,6-Dichloroanilino)-2-imidazoline (St-155), *Biochem. Pharmacol.*, **18**, 511–520 (1969).

Chu, L.-C., W. F. Bayne, F. T. Tao, L. G. Schmitt, and J. E. Shaw, Determination of Submicro Quantities of Clonidine in Biological Fluids, *J. Pharm. Sci.*, **68**, 72–74 (1979).

Davies, D. S. et al., Pharmacokinetics and Concentration-Effect Relationships of Intravenous and Oral Clonidine, *Clin. Pharmacol. Ther.*, **21**, 593–601 (1977).

Dollery, C. T. et al., Clinical Pharmacology and Pharmacokinetics of Clonidine, *Clin. Pharmacol. Ther.*, **19**, 11–17 (1976).

Edlund, P. O. and L. K. Paalzow, Quantitative Gas-Liquid Chromatographic Determination of Clonidine in Plasma, *Acta Pharmacol. Toxicol.*, **40**, 145–152 (1977).

Ehrhardt, J. D., Metabolisme du Catapressan Action Hypotensive du 4-Hydroxycatapressan, *Thérapie*, **27**, 947–954 (1972).

Rehbinder, D., and W. Deckers, Untersuchungen zur Pharmakokinetik und zum Metabolismus des 2-(2,6-Dichlorphenylamino)-2-imidazolin Hydrochlorids (St-155), *Arzneim-Forsch.*, **19**, 169–176 (1969).

Timmermans, P. B., A. Brands, and P. A. Van Zwieten, Gas-Liquid Chromatographic Determination of Clonidine and Some Analogues in Rat Brain Tissue, *J. Chromatogr.*, **144**, 215–222 (1977).

COCAINE

2β-Carbomethoxy-3β-benzoxytropane.
Topical local anesthesic. Drug of abuse.

$C_{11}H_{21}NO_4$; mol. wt. 303.35; pK_a 8.4.

a. Pharmacologic Concentration Range. Oral or intranasal administration of 2 mg cocaine/kg in human beings causes peak plasma concentrations of about 200 ng/ml, which correlate with a subjective experience of euphoria (Javaid et al., 1978b; Van Dyke et al., 1978).

b. Metabolism. Cocaine is rapidly metabolized, with the inactive metabolites benzoylecgonine and methylecgonine as the principal products. Serum cholinesterase catalyzes the reaction to methylecgonine as the major metabolite of cocaine in urine (Inaba et al., 1978). The half-life of cocaine is 30 to 70 min (Javaid et al., 1978b; Jatlow and Bailey, 1975; Van Dyke et al., 1978).

c. Analogous Compounds. Tropane alkaloids: atropine, benzoylecgonine, homatropine, scopolamine.

d. Analytical Methods

1. Gas-Liquid Chromatography—Kogan et al. (1977)

Discussion. Cocaine plasma concentrations are measured by GC and nitrogen-sensitive FID with a sensitivity of 10 ng/ml, using 0.5 ml sample volumes and chlorproethazine as internal standard. Benzoylecgonine is determined with GC-EC after extraction and derivatization with pentafluorobenzyl bromide. The sensitivity is 5 ng benzoylecgonine/ml plasma with chlorproethazine as the internal standard. The simultaneous determination of cocaine and benzoylecgonine in urine is accomplished by GC-FID, using scopolamine as the internal standard. Benzoylecgonine is silanized prior to GC analysis. Sensitivity for cocaine and bezoylecgonine is 0.5 and 1.0 μg/ml urine, respectively. The C.V. ranges between 0.9 and 2.2%.

Detailed Method (Determination of Cocaine in Plasma). Forty microliters of a methanolic internal standard solution (chlorproethazine) is added to 0.5 ml plasma and then made basic with 0.5 ml pH 9.5 sodium carbonate-bicarbonate buffer. The mixture is extracted with 10 ml of 2% (v/v) isoamyl alcohol in heptane, and the organic phase transferred to another tube and back-extracted with 1 ml 0.1 N H_2SO_4. The organic phase is discarded, and the aqueous phase washed once with 3 ml of the extracting organic solvent. The organic phase is then made basic by saturation with a solid mixture of sodium carbonate-sodium bicarbonate (1.14:1, w/w) and extracted into 2 ml benzene. The benzene layer is transferred to another tube and evaporated to dryness at room temperature. The final extract is dissolved in 20 μl methanol, and 1.0 to 2.5 μl is injected into the gas chromatograph.

Analysis is performed on a Perkin-Elmer 900 gas chromatograph equipped with a rubidium bead nitrogen-phosphorus detector. Column packing is 3% OV-22 on 80/100 Supelcoport. Column, injector, and manifold temperatures are 250, 285, and 300°C, respectively. Helium carrier gas, air, and hydrogen flow rates are 30, 120, and 3 ml/min, respectively.

2. Other GC Methods—Jatlow and Bailey (1975), Javaid et al. (1978a), von Minden and D'Amato (1977), Wallace et al. (1975)

Discussion. The methods of von Minden et al. (1977) and Wallace et al. (1975) use regular FID detection and are applicable only to urine samples because of their relatively low sensitivity. Both cocaine and benzoylecgonine, after n-propylation (von Minden et al., 1977) or methylation (Wallace et al., 1975) of the COOH function, can be detected at levels of about 0.2 μg/ml. The internal standards are n-pentylbenzoylecgonine (von Minden et al., 1977) and butylanthraquinone (Wallace et al., 1975). The methylation reagent (H_2SO_4—CH_3OH, 1:2) of Wallace et al. (1975) converts benzoylecgonine to cocaine. Deletion of the methylation step permits analysis of cocaine alone. The difference between the two measurements represents the benzoylecgonine level. The nitrogen-sensitive FID-GC technique of Jatlow and Bailey (1975) is sensitive to 5 to 10 ng cocaine/ml plasma, using a 2 ml sample volume and propylbenzoylecgonine as the internal standard. It preceded the more generally applicable method of Kogan et al. (1977).

Javaid et al. (1978a) utilize GC-EC for the sensitive analysis of cocaine in plasma (10 ng/ml). The drug is extracted from plasma at pH 8.9 into cyclohexane, reduced with lithium aluminum hydride, and acylated with pentafluoropropionic anhydride. Analytical recovery is determined with ³H-cocaine added at the beginning of the procedure. The GC standard curves are constructed with external standards, limiting the precision of the method. The organic extraction step, using a highly lipophilic solvent, apparently excludes the metabolites methylecgonine and benzoylecgonine from the final GC-EC analysis, although both compounds should yield the same derivatization product as cocaine after reduction and acylation. Thus cocaine plasma levels obtained by this procedure correlate with those obtained by an established GC–N–FID or GC–MS assay (Lin et al., 1977).

3. Gas Chromatography-Mass Spectrometry—Jindal and Vestergaard (1978), Karasek et al. (1976), Lin et al. (1977)

Discussion Mass spectrometry in the selected ion monitoring mode should provide a sensitive assay for cocaine in biological samples. However, extensive fragmentation occurs under electron impact. Karasek et al. (1976) propose plasma chromatography at atmospheric pressure, using ⁶³Ni ionization. Generated positive and negative ions can then be detected according to their relative mobility, producing ion mobility spectra at trace quantities. The application of this technique to cocaine plasma level determinations remains to be demonstrated.

Lin et al. (1977) propose a GC-MS electron impact ionization assay for cocaine in plasma. Fragment ions at both *m/e* 82 and *m/e* 182 appear to be suitable for selected ion monitoring of cocaine. External standard curves are used for quantitation, and results as low as 10 ng cocaine/ml plasma are reported. However, no statistical data are given on the assay performance.

Jindal and Vestergaard (1978) have introduced N-CD$_3$-labeled cocaine and benzoylecgonine as the internal standards for a GC-MS electron impact assay. By measuring the molecular ions of the deuterated and unlabeled compounds, this assay is sensitive to 2 ng cocaine/ml and 5 ng benzoylecgonine/ml urine.

4. High Performance Liquid Chromatography—Jatlow et al. (1978)

Discussion. Reverse phase HPLC analysis on an octadecylsilica-coated packing is sensitive for cocaine and benzoylecgonine at rather low UV wavelengths (200 and 230 nm). The absorptivity at 200 nm is 2.5 times higher than that of the UV absorbance maximum of cocaine at 230 nm. Urine samples (5 ml) are washed with an organic solvent at an acidic pH and then extracted with chloroform-ethanol at a mildly basic pH. The sensitivity for cocaine and benzoylecgonine is 0.1 μg/ml, using benzoylecgonine ethyl ester as the internal standard. The HPLC step also separates the potential metabolites norcocaine and benzoylnorecgonine from cocaine and benzoylecgonine.

5. Thin-Layer Chromatography—Meola and Brown (1975), Wallace et al. (1975)

Discussion. These techniques utilize colorimetric TLC spray reagents to visualize cocaine and are applied for qualitative urine tests only.

6. Radioimmunoassay—Kaul et al. (1976), Mule et al., (1977)

Discussion. Kaul et al. (1976) use an antibody directed against ecgonine as the haptene. Nevertheless, the antibody shows twice as much affinity to benzoylecgonine as to ecgonine, while it is considerably less reactive with cocaine and other tropane metabolites. The ^{125}I-RIA of Mule et al. (1977) is based on an antibody raised against benzoylecgonine by coupling to antigenic protein via the free COOH function. Therefore cocaine is not differentiated by this antibody and actually has twice the affinity of benzoylecgonine. The sensitivity is 2 ng/ml urine, which contains considerably more benzoylecgonine than cocaine.

7. Enzyme Immunoassay—e.g., Van Dyke et al. (1977)

Discussion. The "enzyme multiplied inhibition technique" (EMIT, Syva Corp.) measures urinary benzoylecgonine with a sensitivity of 1 μg/ml.. The assay is based on the enzyme lysozyme, which acts on mucopolysaccharides in the cell wall of bacteria and causes cell lysis and release of UV absorbing material into the test solution. The method is primarily suggested as a screening test for illicit cocaine use.

8. Comments. Most assays are designed for urine analysis of benzoylecgonine, a cocaine metabolite, as a screening test for illicit cocaine use. The method of Jatlow et al. (1977) seems adequate for plasma cocaine determinations with sufficient sensitivity to follow plasma levels over several hours after a pharmacologic dose. Serum cholinesterase presents an assay problem, since it catalyzes the hydrolysis of cocaine to methylecgonine (Inaba et al., 1978). The addition of a saturated solution of sodium fluoride to serum samples inactivates this enzyme and prevents *in vitro* decomposition of cocaine in serum or plasma (Van Dyke et al., 1978).

e. References

Inaba, T., D. J. Stewart, and W. Kalow, Metabolism of Cocaine in Man, *Clin. Pharmacol. Ther.*, **23**, 547–552 (1978).

Jatlow, P. I. and D. N. Bailey, Gas-Chromatographic Analysis for Cocaine in Human Plasma, with Use of a Nitrogen Detector, *Clin. Chem.*, **22**, 1918–1921 (1975).

Jatlow, P. I., C. van Dyke, P. Barash, and R. Byck, Measurement of Benzoylecgonine and Cocaine in Urine: Separation of Various Cocaine Metabolites Using Reversed-Phase High-Performance Liquid Chromatography, *J. Chromatogr.*, **152**, 115–121 (1978).

Javaid, J. I., H. Dekirmenjian, J. M. Davis, and C. R. Schuster, Determination of Cocaine in Human Urine, Plasma, and Red Blood Cells by Gas-Liquid Chromatography, *J. Chromatogr.*, **152**, 105–113 (1978a).

Javaid, J. I., M. W. Fischman, C. R. Schuster, H. Dekirmenjian, and J. M. Davis, Cocaine Plasma Concentration; Relation to Physiological and Subjective Effects in Humans, *Science*, **202**, 227–228 (1978b).

Jindal, S. P. and P. Vestergaard, Quantitation of Cocaine and Its Principal Metabolite, Benzoylecgonine, by GLC-Mass Spectrometry Using Stable Isotope labeled Analogs as Internal Standards, *J. Pharm. Sci.*, **67**, 811–813 (1978).

Karasek, F. W., H. H. Hill, and S. H. Kim, Plasma Chromatography of Heroin and Cocaine with Mass Identified Mobility Spectra, *J. Chromatogr.*, **117**, 327–336 (1976).

Kaul, B., J. Millian, and B. Davidow, The Development of Radioimmunoassay for Detection of Cocaine Metabolites, *J. Pharmacol. Exp. Ther.*, **199**, 171-178 (1976).

Kogan, M. J., K. G. Verebey, A. C. DePace, R. B. Resnick, and S. J. Mule, Quantitative Determination of Benzoylecgonine and Cocaine in Human Biofluids by Gas-Liquid Chromatography, *Anal. Chem.*, **49**, 1965-1969 (1977).

Lin, R.-L., N. Narasimhachari, and J. M. Davis, Determination of Cocaine in Plasma Samples by GC-MS-SIM: Annual Conference on Mass Spectrometry and Allied Topics, *Am. Soc. Mass Spectrom.*, **25**, 583-585 (1977).

Meola, J. M. and H. H. Brown, Detection of Benzoylecgonine in Urine, Drug Screening by TLC, *Clin. Chem.*, **21**, 945 (1975).

Mule, S. J., D. Jukofsky, M, M. Kogan, A. de Pace, and K. Verebey, Evaluation of the Radioimmunoassay for Benzoylecgonine (A Cocaine Metabolite) in Human Urine, *Clin. Chem.*, **23**, 795-801 (1977).

Van Dyke, C., R. Byck, P. G. Barash, and P. Jatlow, Urinary Excretion of Immunologically Reactive Metabolites after Intranasal Administration of Cocaine, as Followed by Enzyme Immunoassay, *Clin. Chem.*, **23**, 241-244 (1977).

Van Dyke, C., P. Jatlow, J. Ungerer, P. G. Barash, and R. Byck, Oral Cocaine: Plasma Concentrations and Central Effects, *Science*, **200**, 211-213 (1978).

Von Minden, D. and N. A. D'Amato, Simultaneous Determination of Cocaine and Benzoylecgonine in Urine by Gas-Liquid Chromatography, *Anal. Chem.*, **49**, 1974-1977 (1977).

Wallace, J. E., H. E. Hamilton, H. Schwertner, and D. E. King, TLC Analysis of Cocaine and Benzoylecgonine in Urine, *J. Chromatogr.*, **114**, 433-441 (1975).

Wallace, J. E. et al., GLC Determination of Cocaine and Benzoylecgonine in Urine, *Anal. Chem.*, **48**, 34-38 (1976).

CYCLOPHOSPHAMIDE—by G. M. Wientjes

2-[Bis(2-chloroethyl)amino]tetrahydro-2H-1,3,2-oxazaphosphorine 2-oxide.
Cytoxan, Endoxan, Sendoxan. Alkylating antineoplastic agent.

$C_7H_{15}Cl_2N_2O_2P$; mol. wt. 261.10.

a. Therapeutic Concentration Range. From 10 to 150 ng cyclophosphamide/ml are observed following therapeutic doses. The range of plasma concentrations of the alkylating active metabolites, expressed as alkylating activity, is 2 to 20 nmol/ml (Bagley et al., 1973).

b. Metabolism. Cyclophosphamide is converted by liver microsomal enzymes to active metabolites such as aldophosphamide and phosphoramide mustard (Jardine et al., 1976; Hill et al., 1972; Struck et al., 1975). The plasma half-life of cyclophosphamide is 5.6 to 8.4 hr (Bagley et al., 1973) in cancer patients.

c. Analogous Compounds. Ifosfamide, trofosfamide; NH-Lost; chlorambucil, melphalan, mechlorethamine.

d. Analytical Methods

1. Gas Chromatography—(Pantarotto et al. (1974)

Discussion. This assay involves an ether extraction under basic conditions, acylation of cyclophosphamide with trifluoroacetic anhydride, and electron capture or flame ionization detection. Isophosphamide is used as the internal standard. The minimum detectable amount of cyclophosphamide is 25 pg per injection, using EC detection; FID is several orders of magnitude less sensitive and is used for cyclophosphamide concentrations in the micrograms per milliliter range. Sensitivity limits for the GC assay in plasma are not specified.

Detailed Method. To 0.1 ml water, serum, or urine are added (depending on the expected concentration range) either 100 ng or 5 μg isophosphamide and 0.9 ml 0.1 N NaOH. The mixture is extracted twice with 5 ml ether, and the combined ether layers are evaporated. The extraction recovery is about 80%. The dry residue is dissolved in 150 μl trifluoroacetic anhydride and ethyl acetate (1:2, v/v) and heated at 70° for 20 min in a stoppered test tube. The samples are dried under nitrogen and reconstituted in 100 μl ethyl acetate, and 1 μl aliquots are injected onto the column. Calibration curves are obtained with peak height ratios.

The GC equipment consists of a Carlo Erba Fractovap G1 Gaschromatograph, with FID and EC detector (^{63}Ni). A glass column, 1.5 m x 4 mm (i.d.), is used, packed with 100/120 mesh Chromosorb Q coated with 3% SE-30. The column, injector port, FID, and EC detector are kept at 200, 250, 250, and 280°C, respectively. The carrier gas is nitrogen at 35 ml/min.

2. Gas Chromatography-Mass Spectrometry—Jardine et al. (1976)

Discussion. The GC-MS method with chemical ionization is applied to the detection of cyclophosphamide, nornitrogen mustard, and the active metabolite phosphoramide mustard. the corresponding deuterated compounds are synthesized from tetradeuterated nornitrogen mustard and serve as internal standards. Cyclophosphamide and nornitrogen mustard are extracted from urine into organic solutions, derivatized with trifluoroacetic anhydride, and analyzed by GC-MS in the selected ion monitoring mode. Phosphoramide mustard is separated from urine and plasma, using a XAD-2 column, and derivatized with diazomethane to give mono-, di- and trimethyl products. The GC-MS analysis is performed monitoring ion peaks of the di- and trimethyl derivatives. Tetradeuterated internal standards give ion peaks at m/e values that are 4 units higher; standard curves can be produced by measuring the normotopic/isotopic peak height ratios. The method seems to be sensitive to about 1 μg/ml urine or plasma.

3. Field Desorption Mass Spectrometry—Schulten (1976)

Discussion. Schulten (1976) reports on the qualitative determination of cyclophosphamide and its metabolites with field desorption MS. Preliminary investigations are also presented on the use of this technique for quantitation of cyclophosphamide with hexadeuterated cyclophosphamide as the internal standard.

4. Colorimetry of Alkylating Metabolites—Friedman and Boger (1961)

Discussion. The colorimetric detection of active cyclophosphamide metabolites is based on the reaction of alkylating agents with γ-(p-nitrobenzyl)pyridine. Less than 1 μg of active mustard derivative is detectable. Cyclophosphamide itself is considered to be an inactive prodrug form and does not react with this reagent.

5. Tissue Culture Cytotoxicity Assay—Weaver et al. (1978)

Discussion. Active cyclophosphamide metabolites can be quantitatively determined by measuring their cytotoxicities in cell cultures with the highly sensitive Walker-256 rat carcinosarcoma cell line. The sensitivity is two orders of magnitude greater for pure compounds than that of the chemical alkylating test, providing a lower detection limit for active metabolites of 100 ng/ml plasma or urine. Cyclophosphamide cannot be measured by the cell culture assay, since it has to be activated by microsomal cytochrome P-450 enzymes, which are virutally absent in most cultured cell lines.

6. Comments. The GC-EC method of Pantarotto et al. (1974) is useful for quantitation of the parent drug at levels found in clinical situations, while GC-MS (Jardine et al., 1976) is presently the method of choice for the detection of some of the metabolites, such as phosphoramide mustard. It should be noted that the chemically labile 4-hydroxycyclophosphamide readily and reversibly reacts with sulfhydryl groups and is covalently bound to serum albumin (Voelcker et al., 1976). Care has to be taken to avoid this reaction *in vitro* during the analytical work-up.

e. References

Bagley, C. M., F. W. Bostick, and V. T. De Vita, Clinical Pharmacology of Cyclophosphamide, *Cancer Res.*, **33**, 226–233 (1973).

Friedman, O. M. and E. Boger, Colorimetric Estimation of Nitrogen Mustards in Aqueous Media, *Anal. Chem.*, **33**, 906–910 (1961).

Hill, D. L., W. R. Laster, and R. F. Struck, Enzymatic Metabolism of Cyclophosphamide and Nicotine and Production of a Toxic Cyclophosphamide Metabolite, *Cancer Res.*, **32**, 658–665 (1972).

Jardine, I., R. Brundrett, M. Colvin, and C. Fenselau, Approaches to the Pharmacokinetics of Cyclophosphamide (NSC-26271): Quantitation of Metabolites, *Cancer Treat. Rep.*, **60**, 403–408 (1976).

Pantarotto, C. et al., Quantitative GLC Determination of Cyclophosphamide and Isophosphamide in Biological Specimens, *J. Pharm. Sci.*, **63**, 1554–1558 (1974).

Schulten, H.-R., Qualitative and Quantitative investigations of Cyclophosphamide, (NSC-26271), Cyclophosphamide Metabolites, and Related Compounds by Field-Desorption Mass Spectrometry, *Cancer Treat. Rep.*, **60**, 501–507 (1976).

Struck, R. F., M. C. Kirk, M. H. Witt, and W. R. Laster, Isolation and Mass Spectral Identification of Blood Metabolites of Cyclophosphamide: Evidence for Phosphoramide Mustard as the Biologically Active Metabolite, *Biomed. Mass Spectrom.*, **2**, 46–52 (1975).

Voelcker, G., T. Wagner, and H.-J. Hohorst, Identification and Pharmacokinetics of Cyclophosphamide (NSC-26271) Metabolites *in vivo*, *Cancer Treat. Rep.*, **60**, 415–422 (1976).

Weaver, F. A., A. R. Torkelson, W. A. Zygmunt, and H. P. Browder, Tissue Culture Cytotoxicity Assay for Cyclophosphamide Metabolites in Rat Body Fluids, *J. Pharm. Sci.*, **67**, 1009–1012 (1978).

DEXAMETHASONE—by Z. Hosein

9α-Fluoro-11β-17,21-trihydroxy-16α-methylpregna-1,4-dien-3,20-dione. Calonat, Decasone, Decacortin, Decadron, Deconil. Anti-inflammatory and immunosuppressive glucocorticoid.

$C_{22}H_{29}FO_5$; mol. wt. 392.45.

a. Therapeutic Concentration Range. Therapeutic plasma levels of dexamethasone have not been clearly defined. Pharmacologic activity of dexamethasone is present in the low nanograms per milliliter range (English et al., 1975). The intrinsic biological potency is 17 times that of hydrocortisone (Meikle and Tyler, 1977).

b. Metabolism. The two principal metabolites are 6-hydroxydexamethasone and 20-dihydroxydexamethasone (English et al., 1975). The plasma elimination half-life ranges from 2.5 to 6.5 hr (Haque et al., 1972; Meikle et al., 1973).

c. Analogous Compounds. Anti-inflammatory and immunosuppressive steroids (e.g., beclomethasone, betamethasone, fludrocortisone, hydrocortisone, methylprednisolone, prednisolone, prednisone, triamcinolone).

d. Analytical Methods

1. Radioimmunoassay—English et al. (1975)

Discussion. A radioimmunoassay is applied to determine dexamethasone levels in human plasma and urine samples, using C-1,2-³H-dexamethasone, specific activity 22 Ci/mmol. The antibody is raised in rabbits to dexamethasone-21-succinate conjugated to bovine serum albumin. The sensitivity is 100 pg/0.1 ml plasma with a mean blank value of 0.38 ± 0.02 ng/ml. Between-batch precision is 5% with plasma and urine samples containing 5000 pg/ml, stored at $-20°C$. Less than 1% cross-reactivity with endogenous steroids (e.g., cortisol, corticosterone, progesterone, testosterone, aldosterone) is observed. The two major metabolites, 6-hydroxydexamethasone and 20-dihydroxydexamethasone, give insignificant cross-reactivity.

Detailed Method. The antiserum is diluted 1:1500 in 0.1 M phosphate buffer, pH 7.0, containing 0.9% NaCl and 0.1% sodium azide. Urine samples are diluted 1:10 with the phosphate buffer and assayed like plasma samples. Plasma (0.1 ml), antiserum (0.1 ml), and ^{3}H-dexamethasone are incubated at 4°C for 2 hr. Dextran-coated charcoal (0.1% solution, 0.1 ml) is added, and the mixture is incubated for 30 min at 4°C and centrifuged. The bound ^{3}H-dexamethasone fraction is determined by liquid scintillation counting.

Aliquots (0.5 ml) of the supernatant are separated and counted in a Packard Model 2425 liquid scintillation counter, using toluene–Triton-X100 (2:1, v/v) containing 0.6% butyl PBD as the scintillation fluid.

2. Other Radioimmunoassays—Hichens and Hogans (1974), Meikle et al. (1973)

Discussion. The assay of Hichens and Hogans (1974) is equivalent to that of English et al. (1975), except for different antibody specificity. The double-antibody RIA of Meikle et al. (1973) uses antibodies raised against dexamethasone-3-carboxymethyloxim–bovine serum albumin and employs paper chromatography to circumvent the cross-affinity of the antibody to hydrocortisone (0.4%).

3. High Performance Liquid Chromatography—De Paolis et al. (1977), Tsuei et al. (1978)

Discussion. An HPLC method using silica gel adsorption and UV detection has been described (De Paolis et al., 1977) for the determination of dexamethasone in milk. Detection limits are estimated at 5 ppb (= 5 ng/ml). The applicability of a similar HPLC procedure to plasma and urine samples has been demonstrated by Tsuei et al. (1978), who achieve a sensitivity of 15 ng/ml.

4. Comments.
The radioimmunoassay procedure is the method of choice to determine dexamethasone in plasma and urine samples. Combined with a chromatographic separation (e.g., Meikle et al., 1973), it provides sufficient specificity for a broad range of applications in the presence of endogenous steroids and drug metabolites.

e. References

De Paolis, A. M., G. Schnabel, S. E. Katz, and J. D. Rosen, Determination of Dexamethasone in Milk by High Pressure Liquid Chromatography, *J. Assoc. Off. Anal. Chem.*, **60**, 210–212 (1977).

English, J., J. Chakraborty, V. Marks, and A. Parke, A Radioimmunoassay Procedure for Dexamethasone: Plasma and Urine Levels in Man, *Eur. J. Clin. Pharmacol.*, **9**, 239–244 (1975).

Haque, N. et al., Studies on Dexamethasone Metabolism in Man: Effect of Diphenylhydantoin, *J. Clin. Endocrinol. Metab.*, **34**, 44–50 (1972).

Hichens, M. and A. F. Hogans, Radioimmunoassay for Dexamethasone in Plasma, *Clin. Chem.*, **20**, 266–271 (1974).

Meikle, A. W. and F. H. Tyler, Potency and Duration of Action of Glucocorticoids: Effects of Hydrocortisone, Prednisone and Dexamethasone on Human Pituitary Adrenal Function, *Am. J. Med.*, **63**, 200–207 (1977).

Meikle, A. W., L. G. Lagerquist, and F. H. Tyler, A Plasma Dexamethasone Radioimmunoassay, *Steroids*, **22**, 193 (1973).

Tsuei, S. E., J. J. Ashley, R. G. Moore, and W. G. McBride, Quantitation of Dexamethasone in Biological Fluids Using High-Performance Liquid Chromatography, *J. Chromatogr.*, **145**, 213–220 (1978).

DIATRIZOATE SODIUM

3,5-Bis(acetylamino)-2,4,6-triiodobenzoic acid sodium salt.
Hypaque, Renografin. Diagnostic radiation contrast medium.

$C_{11}H_8I_3N_2NaO_4$; mol. wt. 635.92.

a. Therapeutic Concentration Range. Plasma levels of diatrizoate are $\gg$ 100 μg/ml to achieve diagnostic purposes.

b. Metabolism. Diatrizoate is mainly eliminated by glomerular filtration as measured with ^{131}I-labeled tracers (Golman and Scient, 1976). Ultraviolet spectrophotometric measurements affirm that little dissociation of iodine from the diatrizoate moiety occurs *in vivo* (Purkiss et al., 1968). The plasma elimination half-life of diatrizoate, normally about 4 hr, is very long in patients with renal failure, requiring hemodialysis to prevent side effects (Bahlmann and Krüskemper,

1973). Taenzer et al. (1973) suggest a multicompartmental model to describe the pharmacokinetics of diatrizoate, including a compartment with long drug retention. Clinical thyroid function tests based on iodine determinations may be invalidated for several months after organic iodine radiopaque administration because of the long retention of residual amounts in the body (Jones and Shultz, 1967). Chromatographic separation of thyroxine can be used to avoid such interference (Jones and Shultz, 1967). The inorganic iodide concentrations in contrast media range from 0.6 to 8.3 mg % (USP maximum levels are 20 mg %) (Lang et al., 1974), a circumstance that may contribute to some of the side effects observed with these agents (Talner et al., 1971).

   c. **Analogous Compounds.** Iodinated organic compounds (e.g., benzoates): acetrizoate, amidotrizoate, iodipamide, iodopyracet, iothalamate, metrizamide; thyroid hormones.

   d. **Analytical Methods**

   1. **Ultraviolet Spectrophotometry**—Purkiss et al. (1968)

Discussion. Diatrizoate can be measured in urine and plasma at its UV absorbance maximum of 238 nm after protein precipitation and dilution. Lack of significant interference by endogenous substrates is due to the rather high diagnostic levels of diatrizoate, allowing its measurement after 1:100 (plasma) to 1:10,000 (urine) dilutions. Removal of iodine from the benzene ring of diatrizoate does not change the UV absorbance at 238 nm, and UV absorbance is therefore not a direct indicator of radiodensity. However, since the UV method yields the same result as a chemical assay of total iodine concentrations *in vivo* in urine samples, it can be assumed that the UV assay measures the intact diatrizoate moiety and therefore is useful in establishing its kinetics of disposition.

Detailed Method. Proteins in heparinized plasma are precipitated by the method of Somogyi and filtered (Whatman No. 40). The filtrates are diluted 1:100 to 1:500, depending on the expected concentration of diatrizoate. A blank plasma sample obtained prior to diatrizoate administration is treated in the same fashion. Appropriate dilutions of plasma (and urine) minimize interferences by endogenous substrates at 238 nm. Plasma-free filtrates show a blank absorbance of 0.02 O.D. at a dilution of 1:100.

   The UV absorption is measured against standards in 1 cm matched silica cuvettes in a Unicam SP 500 spectrophotometer at a wavelength of 238 nm.

2. Colorimetry of Iodine Content—Benotti and Benotti (1963), Zak and Boyle (1952)

Discussion. Assays of the organic iodine content of plasma samples include the determination of total plasma iodine, protein-bound iodine, and butanol-extractable iodine as thyroid function tests. The total plasma iodine assay is also applicable to the measurement of iodinated radiopaque compounds. The procedure includes an ashing step, for example, wet ashing with chlorate (Zak and Boyle, 1952). The generated iodine can then be detected by a variety of methods, including titration with thiosulfate, UV spectrophotometry, and polarography (Zak and Boyle, 1952). Benotti and Benotti (1963) propose an automated colorimetric iodine assay with excellent sensitivity based on decoloration of a yellow ceric ammonium sulfate solution.

3. Comments. Both UV and iodine content assays are suitable for the analysis of radiopaque agents in biological samples.

e. References

Bahlmann, J. and H. L. Krüskemper, Elimination of Iodine-Containing Contrast Media by Haemodialysis, *Nephron*, **10**, 250–255 (1973).

Benotti, J. and N. Benotti, Protein-Bound Iodine: Total Iodine and Butanol-Extractable Iodine by Partial Automation, *Clin. Chem.*, **9**, 408–416 (1963).

Golman, K. and C. Scient, Metrizamide in Experimental Urography. V: Renal Excretion Mechanism of a Non-ionic Contrast Medium in Rabbit and Cat, *Invest. Radiol.*, **11**, 187–194 (1976).

Jones, J. E. and J. S. Shultz, Determination of the Thyroxine Iodine Content of Serum Contaminated by Organic Iodines: A Method Based on Gel Filtration, *J. Clin. Endocrinol. Metab.*, **27**, 877–884 (1967).

Lang, J. H., E. C. Lasser, L. B. Talner, S. Lyon, and M. Coel, Inorganic Iodine in Contrast Media, *Invest. Radiol.*, **9**, 51–55 (1974).

Purkiss, P., R. D. Lane, W. R. Cattell, I. K. Fry, and A. G. Spencer, Estimation of Sodium Diatrizoate by Absorption Spectrophotometry, *Invest. Radiol.*, **3**, 271–274 (1968).

Taenzer, V., P. Koeppe, K. F. Samwer, and G. H. Kolb, Comparative Pharmacokinetics of Sodium- and Methylglucamine Diatrizoate in Orography, *Eur. J. Clin. Pharmacol.*, **6**, 137–140 (1973).

Talner, L. B., J. H. Lang, R. C. Brasch, and E. C. Lasser, Elevated Salivary Iodine and Salivary Gland Enlargement Due to Iodinated Contrast Media, *Am. J. Roentgenol. Radium Ther. Nucl. Med.*, **112**, 380–382 (1971).

Zak, B. and A. J. Boyle, A Simple Method for the Determination of Organic Bound Iodine, *J. Am. Pharm. Assoc.*, **41**, 260–262 (1952).

DIAZEPAM—by G. Rosner

7-Chloro-1,3-dihydro-1-methyl-5-phenyl-2-H-1,4-benzodiazepin-2-one. Valium. Minor tranquillizer.

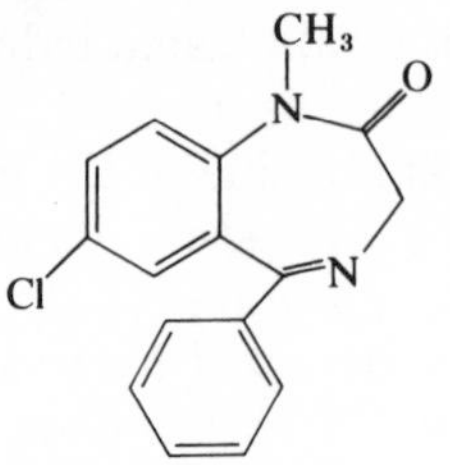

$C_{16}H_{13}ClN_2O$; mol. wt. 284.75; pK_a 3.3.

a. Therapeutic Concentration Range. From 0.1 to 1.0 μg diazepam 1 ml plasma and 0.03 to 2.7 μg/ml of the active metabolite, desmethyldiazepam, are observed during chronic treatment (Zingales, 1973). Definitive therapeutic levels are not well established.

b. Metabolism. Diazepam is metabolized by N-demethylation to the active N-desmethyldiazepam as the major metabolite in plasma. Further metabolic oxidation in C_3 and glucuronidation lead to oxazepam glucuronide as the major excretion product in the urine (Berlin et al., 1972). Diazepam and desmethyldiazepam have plasma half-lives of 26 to 53 hr (Berlin et al., 1972). Desmethyldiazepam accumulates to a greater extent than diazepam during long-term treatment because its half-life is longer than that of the parent drug (de Silva et al., 1966).

c. Analogous Compounds. Benzodiazepines: chlorazepate, chlordiazepoxide. clonazepam, flurazepam, medazepam, nitrazepam, oxazepam.

d. Analytical Methods

1. Gas Chromatography—Rutherford (1977)

Discussion. This is a rapid micromethod (20 to 100 μl plasma) with a simple extraction (requiring neither sample transfer nor evaporation steps), for the simultaneous measurement of diazepam and desmethyldiazepam. After adding extraction solvent containing the internal standard, the sample is injected into the column of a gas chromatograph equipped with a ^{63}Ni electron capture detector. Sensitivity is 20 ng/ml for both diazepam and desmethyldiazepam in plasma and can be increased to 5 ng/ml if needed. The precision is good (4 to 5% S.D.). Interference from other drugs, including benzodiazepines and their metabolites, is minimal except for chlorazepate and chlordiazepoxide (high doses only), which also form the metabolite desmethyldiazepam. This method offers advantages in analysis time, sample volume, and ease of extraction over many other methods involving GC with electron capture detection.

Detailed Method. One hundred microliters of internal standard (4.0 μg/ml prazepam in *n*-butyl acetate) is added to 100 μl plasma, vortexed for 30 sec, and centrifuged at 1700 g for 5 min. Then 2 μl of the upper organic phase is withdrawn by a syringe and injected onto the column of the gas chromatograph. Concentrations of diazepam are assessed by comparing the resulting drug/prazepam peak height ratio to a standard curve.

A Pye 104 Model 84 gas chromatograph equipped with a 10 mCi ^{63}Ni electron capture detector is used. The operating parameters are carrier gas (argon) flow rate, 80 ml/min; column oven temperature, 250°; and detector oven temperature, 270°. The column, coiled glass 2 m x 4 mm (i.d.), is first treated with 2% dimethyldichlorosilane in toluene, then washed in ethanol, and dried. It is packed with 3% OV-17 on a 80/100 mesh support phase. The column is periodically treated with injections of dipalmitoylphosphatidylcholine in ethanol to improve the detection of desmethyldiazepam.

2. Other GC methods—Baselt et al. (1977), Berlin et al. (1972), de Silva et al. (1976), Vandemark and Adams (1977)

Discussion. The GC-EC methods of Berlin et al. (1972) and de Silva et al. (1976) are equivalent to, but more cumbersone than, the method of Rutherford (1977). Although GC methods using other forms of detection (FID: Baselt et al., 1977; nitrogen-sensitive FID: Vandemark and Adams, 1977) have been published, they are less sensitive than EC detection.

3. High Performance Liquid Chromatography—Bugge (1976), Kabra et al. (1978)

Discussion. The method of Kabra et al. (1978) requires 2 ml blood for the determination of diazepam and desmethyldiazepam by isocratic reverse phase HPLC with prazepam as the internal standard. Ultraviolet detection at 240 nm yields a sensitivity of 0.04 μg/ml for diazepam and 0.03 μg/ml for desmethyldiazepam. Bugge (1976) proposes a normal phase silica gel adsorption HPLC method comparable in sensitivity to the reverse phase procedure.

4. Thin-Layer Chromatography—Sun (1978), Wad et al. (1976)

Discussion. Diazepam and its metabolites are measured directly on an unstained thin-layer plate by densitometry at their absorption maximum (232 nm), following toluene extraction from plasma (Wad et al., 1976). Sensitivity is 0.025 μg/ml. The TLC method involves one extraction and no derivatization, but takes 2.5 hr for development.

Sun (1978) describes a highly sensitive (5 to 18 ng/ml serum) fluorescence-densitometric assay of diazepam and its metabolites desmethyldiazepam and oxazepam. Fluorescence is generated by treating the silica gel plates with a sulfuric acid spray. This procedure is applicable to other benzodiazepines as well.

5. Radioimmunoassay—Peskar and Spector (1973)

Discussion. Two separate antibodies, one against diazepam alone, and the other recognizing desmethyldiazepam as well, have been raised against the two analogs coupled to antigenic protein via the 5-phenyl ring. The RIA is capable of detecting 0.02 μg/ml of the drugs in as little as 50 μl serum. This method thus shows excellent sensitivity but lacks the specificity of chromatographic procedures with respect to metabolites.

6. Assay Comparison and Reviews—Clifford and Smith (1974), Hailey (1974)

7. Comments. Probably GC-EC will continue to be the superior method, since the HPLC procedures lack sensitivity. Methods involving hydrolysis of the benzodiazepine moiety to benzophenones should be avoided because of lack of specificity.

e. References
Baselt, R. C., C. B. Stewart, and S. J. Franch, Toxicological Determination of Benzodiazepines in Biological Fluids and Tissues by Flame-Ionization Gas Chromatography, *J. Anal. Toxicol.*, **1**, 10–15 (1977).

Berlin, A. et al., Determination of Bioavailability of Diazepam in Various

Formulations from Steady State Plasma Concentration Data, *Clin. Pharmacol. Ther.*, **13**, 733–744 (1972).

Bugge, A., Quantitative High-Performance Liquid Chromatography of Diazepam in Blood, *J. Chromatogr.*, **128**, 111–116 (1976).

Clifford, J. M. and W. F. Smith, The Determination of Some 1,4-Benzodiazepines and Their Metabolites in Body Fluids: A Review, *Analyst*, **99**, 241–272 (1974).

De Silva, J. A. F., B. A. Koechlin, and G. Bader, Blood Level Distribution Patterns of Diazepam and Its Major Metabolite in Man, *J. Pharm. Sci.*, **55**, 692–697 (1966).

De Silva, J. A. F., I. Bekersky, C. V. Puglisi, M. A. Brooks, and R. E. Weinfeld, Determination of 1,4-Benzodiazepines and -diazepin-2-ones in Blood by Electron-Capture Gas-Liquid Chromatography, *Anal. Chem.*, **48**, 10–15 (1976).

Hailey, D. M., Chromatography of the 1,4-Benzodiazepines, *J. Chromatogr.*, **98**, 527–568 (1974).

Kabra, P. M., G. L. Stevens, and L. J. Marton, High Pressure Liquid Chromatography of Diazepam, Oxazepam and N-Desmethyldiazepam in Human Blood, *J. Chromatogr.*, **150**, 355–360 (1978).

Peskar, B. and S. Spector, Quantitative Determination of Diazepam in Blood by Radioimmunoassay, *J. Pharmacol. Exp. Ther.*, **186**, 167–172 (1973).

Rutherford, D. M., Rapid Micro-Method for the Measurement of Diazepam and Desmethyldiazepam in Blood Plasma by Gas-Liquid Chromatography, *J. Chromatogr.*, **137**, 439–448 (1977).

Sun, S. R., Fluorescence-TLC Densitometric Determination of Diazepam and Other 1,4-Benzodiazepines in Serum, *J. Pharm. Sci.*, **67**, 1413–1415 (1978).

Vandemark, F. L. and R. F. Adams, Determination of Some Benzodiazepines in Serum Utilizing Gas Chromatography and a Nitrogen Detector, *Chromatogr. Newslett.*, **5**, 4 (1977).

Wad, N., H. Rosenmund, and E. Hanifl, A Simplified Quantitative Method for the Simultaneous Determination of Diazepam and Its Metabolites in Serum by Thin-Layer Chromatography, *J. Chromatogr.*, **128**, 231–234 (1976).

Zingales, I. A., Diazepam Metabolism During Chronic Medication: Unbound Fraction in Plasma, Erythrocytes and Urine, *J. Chromatogr.*, **75**, 55–78 (1973).

DIAZOXIDE—by L.-L. Tsai

3-Methyl-7-chloro-1,2,4-benzothiadiazine 1,1-dioxide.
Eudemine, Proglicem, Hyperstat, Hypertonalum, Mutabase. Antihypertensive agent, antihypoglycemic agent.

$C_8H_7ClN_2O_2S$; mol. wt. 230.7; pK_a 8.5.

a. Therapeutic Concentration Range. Diazoxide, when used in the management of acute hypertensive crises, reaches a plasma level of about 20 μg/ml 10 min after rapid intravenous injection of 300 mg (Sellers et al., 1969). The maintenance concentration range is 15 to 50 μg/ml for children treated for hypoglycemia. Blood levels greater than 100 μg/ml may produce hyperglycemia and acidosis (Pruitt et al., 1973).

b. Metabolism. Two metabolites of diazoxide have been identified: 3-hydroxymethyl- and 3-carboxydiazoxide. They are present in plasma at relatively low concentrations (Pruitt et al., 1974). The plasma half-life of diazoxide is 24 to 36 hr in adults and slightly shorter in children. The volume of distribution is 33% of body weight in children (Pruitt et al., 1973).

c. Analogous Compounds. Chlorthalidone, chlorothiazide, furosemide, hydrochlorothiazide.

d. Analytical Methods

1. Gas Chromatography-Mass Spectrometry—Sadée et al. (1973)

Discussion. This method uses 3-trideuteromethyldiazoxide (d_3-diazoxide) as the internal standard, solvent extraction, and *N*-methylation with diazomethane prior to GC. Alkaline alkylating reagents, as well as protic GC solvents, have to be avoided in order to prevent on-column deuterium exchange of d_3-diazoxide. The method is specific and sensitive (1 ng diazoxide/0.1 ml plasma or urine). Comparison of the GC-MS assay to a UV assay (Symchowicz et al., 1967) showed discrepancies at lower levels, which may be caused by the high, erratic UV plasma blank or by the presence of metabolites.

Detailed Method. Plasma or urine (0.1 ml) is added to 0.5 ml 0.1 *M* phosphate buffer (pH 7) containing d_3-diazoxide internal standard (100 ng to 1 μg) in order to afford a measurement range of 0.1 to 100 μg/ml of diazoxide. The sample is extracted with 2.5 ml ethyl ether, and 2 ml of the extract transferred and evaporated to dryness. One milliliter of ethereal diazomethane is added to the tube and left for 30 min at room temperature. The ether is evaporated to dryness, the residue dissolved in 50 μl ether, and 1 to 2 μl injected onto the column. The retention time of diazoxide, 1.3 min, is 3 sec longer than that of d_3-diazoxide. The molecular ion intensities of nondeuterated and deuterated diazoxide at *m/e* 244 and 247 are used to calculate peak height ratios. The ratio 244/247 is multiplied by 0.95 to correct for the 95% deuterium label enrichment of the standard. Both corrected peak areas and peak height ratios plotted against known diazoxide concentrations gave a linear standard curve with a slope of 1 and an intercept of 0.

Analysis is performed on a Varian 2700 gas chromatograph connected to a Varian CH-7 mass spectrometer at 70 eV equipped with a custom-desgined four-channel accelerating voltage alternator for selected ion monitoring.

2. Ultraviolet Spectrophotometry—Symchowicz et al. (1967)

Discussion. This assay requires extraction of whole blood with ethyl acetate, back-extraction of the organic layer into aqueous Na_2CO_3; and UV detection of diazoxide at 280 nm. To identify the material read at 280 nm, TLC is employed to verify the presence of diazoxide in plasma. Although this method lacks specificity, it is useful for routine analysis with minimal interference by metabolites at high concentrations in plasma.

3. Other UV Method—Pruitt et al. (1974)

Discussion. By the use of ethyl acetate extraction with suitable pH selection, the 3-carboxy metabolite is separated from the parent drug and 3-hydroxymethyldiazoxide, which are resolved from each other by TLC. All derivatives are assayed by a modification of the UV method of Symchowicz et al. (1967). One half of the urinary excretion products is accounted for by unchanged diazoxide following administration of ^{14}C-diazoxide to human beings.

4. High Performance Liquid Chromatography—Pruitt et al. (1974)

Discussion. Pruitt et al. (1974) suggest that HPLC, in preliminary studies, has also proved to be capable of separating the parent drug from its metabolites (Z. H. Israili, personal communication).

5. Comments. The UV assays are suitable for most biomedical applications, while the GC-MS procedure provides higher sensitivity and selectivity.

e. References

Pruitt, A. W., P. G. Dayton, and J. H. Patterson, Disposition of Diazoxide in Children, *Clin. Pharmacol. Ther.*, 14, 73–82 (1973).

Pruitt, A. W., B. A. Faraj, and P. G. Dayton, Metabolism of Diazoxide in Man and Experimental Animals, *J. Pharmacol. Exp. Ther.*, 188, 248–256 (1974).

Sadée, W., J. Segal, and C. Finn, Diazoxide Urine and Plasma Levels in Humans by Stable-Isotope Dilution-Mass Fragmentography, *J. Pharmacokin. Biopharm.*, 1, 295–305 (1973).

Sellers, E. M. and J. Koch-Wesser, Protein Binding and Vascular Activity of Diazoxide, *N. Engl. J. Med.*, 281, 1141–1145 (1969).

Symchowicz, S. et al., Diazoxide Blood Levels in Man, *J. Pharm. Sci.*, 56, 912–914 (1967).

DIGOXIN

Digoxigenin trisdigitoxoside.
Digacin, Lanacardin, Lanicor, Lanoxin and so on, Cardiac glycoside.

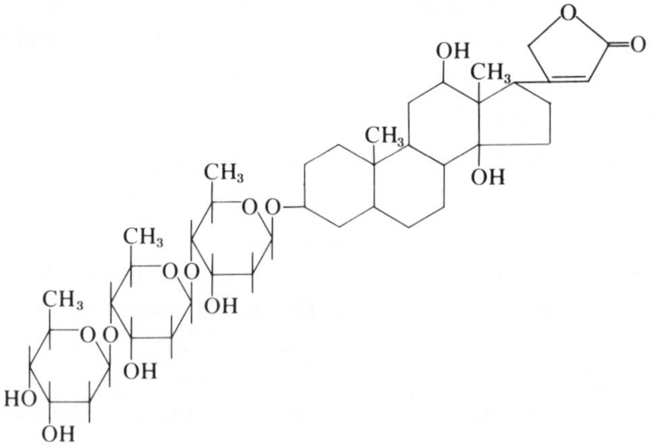

$C_{41}H_{64}O_{14}$; mol. wt. 780.92.

a. Therapeutic Concentration Range. From 0.5 to 2 ng/ml serum (Beller et al., 1971; Butler, 1972; Smith et al, 1969). The therapeutic range may be higher in infants (Butler, 1972). Digoxin toxicity may already occur at therapeutic serum levels and is frequently observed above 2 ng/ml, often with an ominous prognosis (Beller et al., 1971). Prediction of a proper dosage regimen by pharmacokinetic methods alone, using patient size and renal function, resulted in unacceptably large individual deviations from the expected target serum levels (Peck et al., 1973). This argues for measurement of digoxin concentrations and their use for feedback dosage adjustment (Peck et al., 1973). However, Ingelfinger and Goldman (1976) maintained that the usefulness of the serum digitalis concentration as a test for digitalis toxicity is not established. Nevertheless, the digoxin serum level is probably the most frequently requested drug level test today. Its therapeutic utility has been widely accepted.

b. Metabolism. Digoxin is very slowly metabolized and mostly excreted unchanged by glomerular filtration with an average half-life of 1.6 days (Beller et al., 1971). The $t_{1/2}$ can be considerably longer, however, in patients with renal impairments. 20(22)-Dihydrodigoxin is the major metabolite of digoxin and accounted for an average of 13% of the glycosides in a methylene chloride extract of urine (Clark and Kalman, 1974). In some patients this metabolite may be excreted in amounts equal to those of unchanged digoxin. Hydrolysis of the sugar moieties to give the bis- and monodigitoxosides and digoxigenin represents a minor pathway (Clark and Kalman, 1974).

c. Analogous Compounds. Cardiac glycosides (e.g. digitoxin, β-methyldigoxin, ouabain); steroids (e.g., spironolactone, bearing a spiro-γ-lactone ring in C_{17}).

d. Analytical Methods

1. **Radioimmunoassay**—Smith et al. (1969); modified in the Clinical Pharmacokinetics Laboratory, UCSF.

Discussion. The original RIA procedure of Smith et al. (1969) with dextran-coated charcoal separation of unbound 3H-digoxin has been changed into a sequential saturation assay, a modification that may increase its specificity. The assay is sensitive to 0.3 ng digoxin/ml serum. Cross-affinity of the digoxin antibody in this assay system is ~5% with dihydrodigoxin and digitoxin. Spironolactone administration does not interfere with the assay (Silber et al., 1979). Radiodiagnostic agents such as radioactive gallium may cause an underestimation of the digoxin levels, which occasionally results in "negative" digoxin levels. Depending on the decay kinetics of the radioactive nuclides, recounting scintillation vials after an appropriate time corrects for this error. Data obtained from sera with high bilirubin have to be quench corrected.

Detailed Method. Serum (50 μl) is added to 150 μl phosphate-buffered saline (pH 7.4; 0.1 M Na$_2$HPO$_4$, 0.8% NaCl, 1 g/l NaNO$_3$). Then 300 μl diluted digoxin antibody (Antibodies Inc., Davis, Calif., lot P10610) is added to this solution, which is mixed and then incubated for 45 min at 25°C. Fifty microliters of a 10 ng/ml solution of a ^{3}H-digoxin (New England Nuclear, Boston, Mass., lot 772-067) is added to the incubate, and the resulting solution is mixed and then incubated again for 15 min. Free and bound digoxin are separated by the addition of 0.5 ml dextran-coated charcoal (20 g/l activated charcoal and 200 mg/l dextran, both from Sigma Chemical Co., St. Louis, Mo.). The suspension is incubated for 5 min and then cen rifuged for 10 minutes (1800 g). Ten milliliters Oxifluor scintillation fluid (New England Nuclear, Boston, Mass.) is added to the supernatant, which is then counted for 2 min in a Searle Mark III scintillation counter (Searle Analytical Inc., Des Plaines, Ill.). Percent digoxin bound is converted to digoxin concentration by reference to a standard curve, prepared daily. The C.V. is 5% and 10% at digoxin concentrations of 3.0 and 1.0 ng/ml, respectively.

2. Other RIAs

Discussion. Either ^{3}H-digoxin or ^{125}I-digoxin derivatives, such as 3-O-succinyldigoxigenin–^{125}I-tyrosine or ^{125}I-tyrosine—methylester of digoxin, are used as the tracer (^{3}H-RIA: Berk et al., 1974; Line et al., 1973; Wagner et al., 1977; ^{125}I-RIA; Drewes and Pileggi, 1974; Greenwood et al., 1974, 1977; Larson et al., 1977; Walsh et al., 1977). The ^{125}I-tyrosine—methylester of digoxin is preferred over the other ^{125}I tracer, since it retains the digitososide moiety of digoxin. Nevertheless, the fact that the ^{125}I derivatives are chemically different from digoxin may cause assay problems, for instance, potentially in patients with elevated thyroid binding globulin, which could bind the ^{125}I containing moiety of the tracer. Such problems can be avoided by using the ^{3}H-digoxin tracer, although gamma counting of ^{125}I is more rapid than ^{3}H liquid scintillation counting. Weiler and Zenk (1979) review the ^{125}I-digoxin tracers currently available and propose that a newly synthesized ^{125}Ityramine–digoxin derivative is superior because of high stability, low unspecific binding, and improved immunological behavior. Usually ³H-RIA and ¹²⁵I-RIA give equivalent results (Pippin and Markus, 1976). Methods to separate free and bound tracer include double-antibody precipitation (Berk et al., 1974), solid phase antibody (Line et al., 1973), dextran-coated charcoal absorption (Wagner et al., 1977), ammonium sulfate precipitation (Drewes and Pileggi, 1974), gel equilibration (Greenwood et al., 1974), and antibodies linked to magnetizable particles (Greenwood et al., 1977). The digoxin RIA can be fully automated (Greenwood et al., 1977). Wagner et al, (1977) describe a ³H-RIA with a sensitivity of 0.05 ng/ml serum.

The major factor in determining assay quality appears to be the antibody, with tracer and assay procedure as contributing factors. Methodologic sources of assay errors have been extensively studied (e.g., Holtzman et al., 1974; Kuno-Sakai and Sakai, 1975). The contribution of individual sera variability to digoxin assay errors has been well recognized (Anggard et al., 1972). Kramer et al. (1978) suggested that the relatively cardioinactive dihydrodigoxin metabolite might interfere with some digoxin RIAs in patients with high metabolite formation. Digitoxin has also been recognized as a potential source of error (Kuno-Sakai et al., 1975). Conflicting reports have appeared on potential assay interference by spironolactone administration (Lichey et al., 1977; Müller et al., 1978: Ravel, 1975). We have recently determined that the major variable in spironolactone assay interference is the specificity of the antibody (Silber et al., 1979). Furthermore, the assay interference is probably caused by an unknown polar metabolite of spironolactone and not by its major active metabolite, canrenone. Therefore specificity testing of digoxin RIAs has to be performed on sera of patients who have ingested spironolactone in order to account for all spirolactone metabolites. The detailed method presented in Section d1 is specific in the presence of spironolactone metabolites and suitable for a small sample load (5 to 15 sera per day).

Digoxin RIA reliability deserves continual attention, as recently demonstrated by Bergdahl et al. (1979). Several of the tested, commercially available RIA kits gave significantly higher digoxin serum levels than a ⁸⁶Rb uptake-inhibition technique. The free fatty acid content of the serum sample represents a factor that may influence the assay results under certain conditions (O'Leary et al., 1979).

3. Enzyme Immunoassay (EMIT)—Brunk and Malmstadt (1977), Drost et al. (1977), Rosenthal et al. (1976)

Discussion. The principle of this homogeneous enzyme multiplicating immuno test is shown by the following reactions:

1. G-6-PD–digoxin + digoxin + antibody $\rightleftharpoons$ digoxin-antibody + G-6-PD–digoxin–antibody.

2. Glucose 6-phosphate $\xrightarrow{\text{free G-6-PD–digoxin}}$ gluconolactone 6-phosphate

$$NAD^+ \qquad\qquad NADH$$

(G-6-PD–digoxin: digoxin chemically bonded to glucose 6-phosphate dehydrogenase. The bacterial G-6-PD requires NAD^+ rather than $NADP^+$.)

Unlabeled digoxin in the patient serum and enzyme-labeled digoxin compete for the digoxin antibody binding site, which partially inactivates the G-6-PD enzyme activity. In the presence of excess substrate and cofactor, the amount of generated NADH is directly proportional to the amount of free G-6-PD–digoxin and can be measured by UV absorption. The less than complete inhibition of the G-6-PD–digoxin upon binding represents a problem as do potential interferences with the enzymatic reaction by serum components. Advantages over RIA include lack of separation steps, UV detection and ready automation (Brunk and Malmstadt, 1977; Ericksen and Anderson, 1978). Correlation between EMIT and RIA is good (e.g. $r = .90$: Drost et al., 1977). The EMIT procedure shows sufficient sensitivity (0.5 ng digoxin/ml serum) with good day-to-day reproducibility (at 1.33 ng/ml: S.D. = 0.11, C.V. = 8.42%, $n = 10$).

4. Enzyme Immunoassay (ELISA)—Munz et al., (1978), Borner and Rietbrock, (1978)

Discussion. The heterogeneous enzyme-linked immunosorbant assay differs from EMIT in that the enzyme-digoxin complex which is not inactivated by binding to the digoxin antibody, is coated onto the glass surface of the test vial. Following the incubation period the digoxin containing serum mixture is decanted, and the digoxin-enzyme-antibody complex is allowed to react with the substrate to produce a UV absorbing product. The advantage over EMIT lies in complete differentiation of bound and free enzyme activity, while the required separation step is a disadvantage of ELISA. The ELISA method has a sensitivity of 0.3 ng digoxin/ml serum with an intra-assay precision of 8 to 14% (Borner and Rietbrock, 1978). Correlation with a solid phase

^{125}I-RIA is good. Spironolactone administration interferes with the ELISA assay.

5. Na-K-ATPase Displacement Assay.—Brooker and Jeliffe (1970), Marcus et al. (1975).

Discussion. This assay utilizes displacement of ^{3}H-ouabain binding to membrane ATPase and, therefore, may reflect the biologic activity of cardiac glycosides in serum more closely than do immunoassays. It is inferior to immunoassays as an analytical technique, however, because of its lower reproducibility.

6. Rubidium-86 Uptake by Red Blood Cells.—Lader et al. (1972), Lowenstein and Corrill, (1966)

Discussion. The method is dependent on the inhibition by cardiac glycosides of ^{86}Rb flux across the human red cell membrane *in vitro*. The main advantages of the immunoassays over the Rb uptake method are simplicity and speed (Lader et al., 1972).

7. Gas Chromatography.—Clark and Kalman (1973), Watson and Kalman (1971)

Discussion. The GC assay of digoxin in plasma (Watson and Kalman, 1971) requires extensive prepurification steps and derivatization with heptafluorobutyric anhydride. This reagent hydrolyzes digoxin to digoxigenin and forms a heptafluorobuturate derivative with sensitive electron capture properties. ^{3}H-Digoxin serves as recovery tracer, and digitoxigenin heptafluorobuturyrate as internal GC standard. Irreversible absorption on-colmn of the derivatized genins represents a serious problem of this assay that is difficult to overcome. Dihydrodigoxin can be separated from digoxin and analyzed in urine by an equivalent procedure (Clark and Kalman, 1973).

8. Mass Spectrometry.—Greenwood et al. (1975)

Discussion. Mass Spectrometry is utilized to measure urinary dihydrodigoxin in the presence of digoxin. Representative ion signals obtained from direct insertion electron impact ionization are integrated for dihydrodigoxin and digoxin during a temperature program of the insertion probe. The average dihydrodigoxin excretion is 16.4% relative to digoxin in this study. Absolute amounts of digoxin in the urine were measured by ^{125}I-RIA.

9. Thin-Layer Chromatography and Colorimetry.—Jeliffe (1966)

Following extraction of a 2 l urine sample and separation of digoxin from its internal standard, digitoxin, the cardiac glycosides are quantitated by a reaction of the 2-deoxy sugar, digitoxose, with xanthydrol.

10. Liquid Chromatography.—e.g., Nachtmann et al. (1976)

Discussion. No HPLC technique is capable as yet of detecting digoxin serum levels. Nachtmann et al. (1976) used derivatization with 4-nitrobenzoylchloride to enhance UV adsorptivity and chromatographic behavior. It should be possible, however, to employ a flourescence label of digoxin, thus making HPLC serum digoxin assays feasible.

11. Assay Comparisons and Reviews.—Bodem and Gilfrich (1973), Butler (1972)

12. Comments. The very low therapeutic serum concentrations of digoxin represent the major problem with this assay. While most assays using control sera are specific, accurate, and precise, this is not true for randomly selected clinical samples for which no blank serum without digoxin is available. A rapid, chemically specific assay for digoxin using physicochemical measurements is needed either as a direct clinical procedure or as a reference method for quality control studies of current immunoassays. Perhaps HPLC with fluorescence detection after derivatization may be such a method.

e. References

Beller, G. A., et al., Digitalis Intoxication: A Prospective Clinical Study with Serum Level Correlations, *N. Engl. J. Med.*, **284**, 989–997 (1971)

Bergdahl, B. et al., Four Kits for Plasma Digoxin Radioimmunoassay Compared, *Clin. Chem.*, **25**, 305-308 (1979).

Berk, L. S., J. L. Lewis, and J. C. Nelson, One-Hour Radioimmunoassay of Serum Drug Concentrations, as Exemplified by Digoxin and Gentamicin, *Clin. Chem.*, **20**, 1159–1164 (1974).

Bodem, G. and H. J. Gilfrich, Methoden zur Bestimmung von Digoxin und Digitoxin in Blut und ihre klinische Bedeutung, *Klin. Wochenschr.*, **51**, 57–62 (1973).

Borner, K. and N. Rietbrock, Bestimmung von Digoxin im Serum Vergleich von

Radioimmunoassay und heterogenem Enzymimmunoassay, *J. Clin. Chem. Clin. Biochem.*, **16**, 335-342 (1978).

Brooker, G. and R. W. Jeliffe, Serum Digoxin Levels in Toxic and Nontoxic Patients by Enzymatic Isotope Displacement, *Clin. Res.*, **18**, 299 (1970).

Brunk, S. D. and H. V. Malmstadt, Adaptation of the EMIT Serum Digoxin Assay to a Mini-Disc Centrifugal Analyzer, *Clin. Chem.*, **23**, 1054-1056 (1977).

Butler, V. P. , Assays of Digitalis in the Blood, *Prog. Cardiovasc. Dis.*, **14**, 571-600 (1972).

Clark, D. R. and S. M. Kalman, Dihydrodigoxin: A Common Metabolite of Digoxin in Man, *Drug Metab. Dispos.*, **2**, 148-150 (1974).

Drewes, P. A. and V. J. Pileggi, Faster and Easier Radioimmunoassay of Digoxin, *Clin. Chem.*, **20**, 343-347 (1974).

Drost, R. H., T. A. Plomp, A. J. Tennisen, A. H. J. Maas, and R. A. A. Maes, A Comparative Study of the Homogeneous Enzyme Immunoassay (EMIT) and Two Radioimmunoassays (RIA's) for Digoxin, *Clin. Chim. Acta*, **79**, 557-567 (1977).

Eriksen, P. B. and O. Andersen, Homogeneous Enzyme Immunoassay of Serum Digoxin with Use of a Bichromatic Analyzer, *Clin. Chem.*, **25**, 169-171 (1978).

Greenwood, H., M. Howard, and J. Landon, A Rapid, Simple Assay for Digoxin, *J. Clin. Pathol.*, **27**, 490-494 (1974).

Greenwood, H., W. Snedden, R. P. Hayward, and J. Landon, The Measurement of Urinary Digoxin and Dihydrodigoxin by Radioimmunoassay and by Mass Spectroscopy, *Clin. Chim. Acta*, **62**, 213-224 (1975).

Greenwood, H., J. Landon, and G. C. Forrest, Radioimmunoassay for Digoxin with a Fully Automated Continuous-Flow System, *Clin. Chem.*, **23**, 1868-1872 (1977).

Holtzman, Y. L., R. B. Shafer, and R. E. Erickson, Methodological Causes of Discrepancies in Radioimmunoassay for Digoxin in Human Serum, *Clin. Chem.*, **20**, 1194-1198 (1974).

Ingelfinger, J. A. and P. Goldman, The Serum Digitalis Concentration—Does it Diagnose Digitalis Toxicity? *N. Engl. J. Med.*, **294**, 867-899 (1976).

Jelliffe, R. W., A Chemical Determination of Urinary Digitoxin and Digoxin in Man, *J. Lab. Clin. Med.*, **67**, 694-708 (1966).

Kramer, W. G., N. L. Kinnear, and H. K. Morgan, Variability Among Commercially Available Digoxin Radioimmunoassay Kits in Cross Reactivity to Dihydrodigoxin, *Clin. Chem.*, **24**, 155-157 (1978).

Kuno-Sakai, H. and H. Sakai, Effects on Radioimmunoassay of Digoxin of Varying Incubation Periods for Antigen-Antibody Reaction and Varying Periods of Adsorption by Dextran-Coated Charcoal, *Clin. Chem.*, **21**, 227-229 (1975).

Lader, S., A. Bye, and P. Marsden, The Measurement of Plasma Digoxin

Concentration: A Comparison of Two Methods, *Eur. J. Clin. Pharmacol.*, **5**, 22–27 (1972).

Larson, J. H., H. Beckala, and H. A. Homburger, Intercomparison of Four ^{125}I-Labeled Digoxin RIA Kits, *Clin. Chem.*, **23**, 1792–1793 (1977).

Lichey, J., R. Schröder, and N. Rietbrock, The Effect of Oral Spironolactone and Intravenous Canrenoate-K on the Digoxin Radioimmunoassay, *Int. J. Clin. Pharmacol. Biopharm.*, **15**, 557–559 (1977).

Line, W. F. et al., Solid-Phase Radioimmunoassay for Digoxin, *Clin. Chem.*, **19**, 1361–1365 (1973).

Lowenstein, J. M. and E. M. Corrill, An Improved Method for Measuring Plasma and Tissue Concentrations of Digitalis Glycosides, *J. Lab. Clin. Med.*, **67**, 1048–1052 (1966).

Marcus, F. J., J. N. Ryan, and M. G. Stafford, The Reactivity of Derivatives of Digoxin and Digitoxin as Measured by the Na-K-ATPase Displacement Assay and by Radioimmunoassay, *J. Lab., Clin. Med.*, **85**, 610–620 (1975).

Müller, H., H. Bräuer, and B. Resch, Cross Reactivity of Digitoxin and Spironolactone in Two Radioimmunoassays for Serum Digoxin, *Clin. Chem.*, **24**, 706–709 (1978).

Munz, E., A. Kessler, P. U. Koller, and E. W. Busch, Erfahrungen mit Enzymun—Test® Digoxin, *Labt. Mediz.*, in press (1978).

Nachtmann, F., H. Spitzy, and R. W. Frei, Ultraviolet Derivatization of Digitalis Glycosides as 4-Nitrobenzoates for Liquid Chromatographic Trace Analysis, *Anal. Chem.*, **48**, 1576–1579 (1976).

O'Leary, T. D., L. A. Howe, and T. D. Geary, Improvement in a Radioimmunoassay for Digoxin, *Clin. Chem.*, **25**, 332–334 (1979).

Peck, C. C. et al., Computer-Assisted Digoxin Therapy, *N. Engl. J. Med.*, **289**, 441–446 (1973).

Pippin, S. L. and F. J. Marcus, Digoxin Immunoassay with Use of [³H]Digoxin vs. [125I]Tyrosine-methyl-ester of Digoxin, *Clin. Chem.*, **22**, 286–287 (1976).

Ravel, R., Negligible Interference by Spironolactone and Prednisone in Digoxin Radioimmunoassay, *Clin. Chem.*, **21**, 1801–1803 (1975).

Rosenthal, A. F., M. G. Vargas, and C. S. Klass, Evaluation of Enzyme-Multiplied Immunoassay Technique (EMIT) for Determination of Serum Digoxin, *Clin. Chem.*, **22**, 1899–1902 (1976).

Silber, B., L. B. Sheiner, J. L. Powers, M. E. Winter, and W. Sadée, Spironolactone Associated Digoxin Radioimmunoassay Interference, *Clin. Chem.*, **25**, 48-50 (1979).

Smith, T. W., V. C. Butler, and E. Haber, Determination of Therapeutic and Toxic Serum Digoxin Concentrations by Radioimmunoassay, *N. Engl. J. Med.*, **281**, 1212–1216 (1969).

Wagner, J. G., M. R. Hallmark, E. Sakmar, and J. W. Ayres, Sensitive Radioimmunoassay for Digoxin in Plasma and Urine, *Steroids*, **29**, 787–806 (1977).

Walsh, P. R. and J. F. Crawford, and C. D. Hawker, Measurement of Digoxin by Radioimmunoassay, *Ann. Clin. Lab. Sci.*, **7**, 79–87 (1977).

Watson, E. and S. M. Kalman, Assay of Digoxin in Plasma by Gas Chromatography, *J. Chromatogr.*, **56**, 209–218 (1971).

Weiler, E. W. and M. H. Zenk, Iodinated Digoxin Derivatives with Improved Reactivity and Stability, for Use in Radioimmunoassay, *Clin. Chem.*, **25**, 44–47 (1979).

DIPHENHYDRAMINE.—by M. Raeder-Schikorr

2-Diphenylmethoxy-N,N-dimethylethylamine.
Allergin, Bagodyl, Benodin, Dibenfrin, Syntedril. Antihistaminic agent.

$C_{17}H_{21}NO$; mol. wt. 255.35; pK_a 8.98

a. Therapeutic Concentration Range. Unknown. Doses vary between 10 and 400 mg daily, given orally or parenterally.

b. Metabolism. Only a small amount of unchanged diphenhydramine is excreted in urine. The major metabolite in monkeys is a deaminated carboxylic acid derivative, (diphenylmethoxy)acetic acid, and its glutamate conjugate. Several other metabolic pathways also occur, including *N*-demethylation, *N*-oxidation, and cleavage of the ether bond (Drach and Howell, 1968).

c. Analogous Compounds. Brompheniramine, chlorpheniramine, hydroxyzine, meclizine, tripolidine.

d. Analytical Methods

1. Gas Chromatography—Reiss (1970)

Discussion. Separation of seven antihistaminic drugs in pharmaceutical preparations is described, employing a gas chromatograph with dual-column direct injection system and two separate detectors. The two chromatograms show complete separation of each of the antihistaminic peaks. This method can be employed in the presence of the most commonly accompanying drugs, such as aspirin, salicylamide, phenacetin, and caffeine, without showing interferences.

Detailed Method. Four microliters of each aqueous solution is injected into the gas chromatograph (F & M Model 5750), equipped with two columns: (1) 4 ft x ¼ cm (o.d.) glass, containing 2% SE-30 + 2% Carbowax 20M on 80/100 mesh Diatoport S; (2) 4 ft x ¼ cm (o.d.) glass, containing 10% silicone oil DC-200 on 60/80 mesh Diaport S. The following conditions are used: oven temperature, 210°C; injector, 240°C; detector, 230°C; helium, hydrogen and air flow rates of 50, 40, and 420 ml/min, respectively.

2. Colorimetry—Matsui et al.. (1969)

Discussion. Aqueous solutions of diphenhydramine are shaken with bromthymol blue, and the formed ion pair is extracted into an organic phase. The concentration of the resulting ion pair is determined spectrophotometrically at 410 nm. Important parameters for the accuracy and sensitivity of the assay are proper choice of pH and dye concentration. Mixtures of pharmaceutical amines can be measured following TLC separation. Concentrations as low as 5μg/ml show satisfactory absorbance.

3. High Performance Liquid Chromatography—Frei et al. (1979)

Discussion. Tertiary amines, (i.e., chlorpheniramine and brompheniramine) can be readily separated on a reverse phase HPLC column (Micro-Pack CN-10, Varian). The assay of Frei et al. (1979) is of interest because of the rather unusual detection mode, which is based on fluorescent ion-pair formation between tertiary amine drugs and dimethoxyanthracene sulfonate. A dynamic post-column micro-extraction principle is then used to isolate the ion/pairs from the excess reagent in an auto analyzer. The method is sensitive to the low nanograms of drug per milliliter of urine level and should be readily applicable to diphenhydramine.

4. Thin-Layer Chromatography—Fike and Sunshine (1965)

Discussion. Twenty-five antihistaminic drugs are identified by TLC, using three different solvent systems and two spray reagents (Mandelin's and Draggendorff's reagent). Diphenhydramine can be determined in alkaline or neutral extracts of urine, blood, or tissue in concentrations as low as 3 μg/ml.

5. Distillation and Ultraviolet Spectrophotometry—Tompsett (1969)

Discussion. Diphenhydramine can be separated from urine samples by distillation in the presence of sodium bicarbonate. The collected distillate is made acidic and examined spectrophotometrically. The distillation procedure does not cause any chemical changes, and there is little interference by endogenous urinary components. However, this method lacks the sensitivity required for quantitative analysis.

6. Comments. No suitable methods are available at present for the specific detection of diphenylhydramine in biological samples. Either HPLC or GC procedures should ultimately be feasible for this purpose.

e. References

Drach, J. C. and J. P. Howell, Identification of Diphenhydramine Urinary Metabolites in the Rhesus Monkey, *Biochem, Pharmacol.*, **17**, 2125–2136 (1968).

Fike, W. W. and I. Sunshine, Identification of Antihistamines in Extracts of Biological Materials Using Thin-Layer Chromatography, *Anal. Chem.*, **37**, 127–129 (1965).

Frie, R. W., J. F. Lawrence, U. A. Th. Brinkman, and I. Honigberg, An HPLC Fluorescent Detection system for Amines, *J. High Res. Chromatogr. Chrom. Commun.*, **1**, 11 (1979).

Matsui, F., J. R. Watson and W. N. French, Use of the Acid Dye Technique for Quantitation of Pharmaceutical Amines Eluted from Thin-Layer Chromatograms, *J. Chromatogr.*, **44**, 109-115 (1969).

Reiss, T. J., Note on Gas Chromatographic Determination of Antihistamines Employing a Dual Column Direct Injection System, *J. Assoc. Off. Anal. Chem.*, **53**, 609–611 (1970).

Tompsett, S. L., The Detection of Basic and Other Drugs in Urine by Distillation and UV Spectrophotometry, *Clin. Chem.*, **15**, 591–599 (1969).

DOXORUBICIN

(8S-*cis*)-10-[(3-Amino-2,3,6-trideoxy-α-L-lyxohexopuranosyl)oxy]-7,8,9,10-tetrahydro-6,8,11-trihydroxy-8-(hydroxyacetyl)-1-methoxy-5,12-naphthacene-dione.
Adriamycin, 14-hydroxydaunamycin. Antineoplastic agent.

$C_{27}H_{29}NO_{11}$; mol. wt. 543.54.

a. Therapeutic Concentration Range. Following intravenous doses of 20 to 100 mg doxorubicin/m^2, plasma levels of the drug rapidly decline during the first several hours and then persist at a low nanograms per milliliter concentration range over several days (Bachur et al., 1977; Creasey et al., 1976).

b. Metabolism. A significant portion of doxorubicin is metabolized prior to excretion into the bile (Riggs et al., 1977) and the urine (Creasey et al., 1976). Metabolic pathways in human beings include carbonyl reduction to doxorubicinol as the major product, hydrolytic and reductive cleavage of the glycosidic bond, and O-demethylation followed by O-sulfation and glycuronidation (Takanashi and Bachur, 1976). The aglycone metabolites, deoxydoxorubicinol and desmethyldeoxydoxorubicinol, account for some of the flourescent adriamycin constituents in plasma (Benjamin et al., 1977). Redox cycling of the quinone moiety of adriamycin in microsomal preparations to form radicals may participate in the mechanism of adriamycin tissue toxicity, in addition to DNA intercalation (Goodman and Hochstein, 1977). Furthermore, adriamycin-induced superoxide radical production and lipid peroxidation may be associated with the cardiac toxicity of the drug (Myers et al., 1977). The adriamycin analog, daunorubicin, undergoes similar metabolic reactions (Loveless et al., 1978), most notably carbonyl reduction (Feldsted et al., 1974) and reductive glycosidic cleavage (Asbell et al., 1972). The carbonyl reduction products, doxorubicinol and daunorubicinol, may contribute to the pharmacological effects.

Doxorubicin disappears from plasma following biphasic or triphasic curves. While elimination in the alpha phase has a half-life of only 0.3 to 1.5 hr (Creasey et al., 1976; Watson and Chan, 1976), disappearance half-lives at later times after the dose range between 14 and 30 hr

(Benjamin et al., 1977; Creasey et al., 1976; Riggs et al., 1977; Watson and Chan, 1976). Initial plasma elimination of doxorubicin can be significantly increased by administration of DNA-complexed doxorubicin (Kummen et al., 1978). One of the major limitations of doxorubicin therapy is its cardiotoxicity, which is frequently observed after a total dose of more than 500 mg adriamycin/m², regardless of dosage schedule and individual metabolism, (e.g., Guthrie and Gibson, 1977). Long retention of DNA-intercalated doxorubicin in the body may contribute to this phenomenon.[Review: *Cancer Chemother Rep (Part 3)*, 6(2), (1975).]

c. Analogous Compounds. Anthracycline antibiotics; daunorubicin (daunomycin).

d. Analytical Methods

1. Thin-Layer Chromatography–Fluorescence—Watson and Chan (1976)

Discussion. The TLC-fluorescence scanning assay of doxorubicin plasma levels by Watson and Chan (1976) represents one of the few examples where this method has been successfully applied with high sensitivity and precision. Using daunorubicin as the internal standard, with identical fluorescence behavior but different R_f values on the TLC plates, the assay is sensitive to 2 ng doxorubicin/ml plasma (0.5 ml sample) with an average relative deviation of 3.9% at 50 ng/ml. Following organic solvent extraction of plasma samples, the extracts are chromatographed on silica gel plates and quantitatively detected on the plates by a thin-layer fluorescence scanner. Choice of the silica gel plates is critical, since some commercially available plates cause rapid decomposition of doxorubicin upon standing in the open air. Furthermore, all glassware has to be silanized to avoid irreversible absorption of the anthracyclines onto the glass walls.

The TLC separation also allows the simultaneous analysis of the equally fluorescent major metabolite, doxorubicinol, as well as some of the aglycone metabolites, which were not further identified and contributed only a small fraction to the total doxorubicin fluorescence activity in plasma.

Detailed Method. Plasma (0.5 ml) and a small aliquot of daunorubicin solution as the internal standard (to be prepared freshly every week) are extracted with 2 ml ice-cold chloroform-isopropanol (1:1). The lower organic layer is transferred to another test tube and evaporated to dryness under a stream of nitrogen. The tube walls are washed with 3 × 10 μgl portions of chloroform-methanol (1:1), and the solvent is again evaporated. The residue is dissolved in 3 to 4 μl chloroform-ethanol (1:1), and the entire sample spotted on the silica gel plate, using a 2 μl Lang-Levy pipet along with appropriate standards.

Silica gel plates (20 × 20 cm, 0.25 mm thick; E. M. Merck, Elmsford, N.Y.) are divided into sixteen 1 cm wide lanes. The plates are first developed to 12 cm from the origin in a paper-linked tank containing chloroform-methanol-acetic acid (93:5:2). The first chromatography serves mainly to separate lipophilic material from the plate origin resulting in improved separation of doxorubicin and daunorubicin. After removal and drying of the plates, a second chromatography is performed in chloroform-methanol-acetic acid (76:20:4), again to a height of 12 cm. Fluorescence scanning is performed after a 10 min drying period on a spectrophotofluorimeter equipped with a thin-film scanner (Aminco Bowman, Silver Spring, Md.) and a filtering device (Spectrum Scientific, Newark, Del.). The excitation and emission wavelengths are set at 475 and 580 nm, respectively. Doxorubicin-daunorubicin peak height ratios are used to construct standard curves. The R_f values are 0.2 and 0.27, respectively, for doxorubicin and daunorubicin.

2. High Performance Liquid Chromatography—Barth and Conner (1977), Baurain et al.. (1979), Eksborg (1978), Hulhoven and Desager (1976), Langone and van Vunakis (1975), Whatman Report (1976)

Discussion. Hulhoven and Desager (1976) perform an HPLC assay of daunorubicin, using doxorubicin as the internal standard. The two compounds are extracted from slightly alkaline plasma into dichloromethane-isopropanol (92:8) and separated on a 5μ silica microsphere column with a quaternary solvent mixture as the eluent. Ultraviolet detection at 490 nm affords a sensitivity of about 10 ng daunomycin (or doxorubicin)/ml plasma. A similar HPLC procedure is used by Barth and Conner (1977) for the determination of doxorubicin and minor impurities in pharmaceutical dosage forms.

Urinary doxorubicin, doxorubicinol, and aglycone metabolites can be separated on a diphenyl corasil column with a linear solvent gradient of acetonitrile, water, and pH 4.0 ammonium formate buffer (Langone and van Vunakis, 1975). Resolution of the column is not sufficient to separate the individual aglycone metabolites. Eluting compounds are collected and detected by radioimmunoassay (see Section d4). The fact that aglycone is four times less reactive in the RIA than are doxorubicin and doxorubicinol, has to be considered for quantitative interpretations.

Harris and Gudauskas (Whatman Report, 1976) suggest the use of a fluorescence detector which improves HPLC assay sensitivity by at least one order of magnitude over UV detection. In addition, they contend that a microparticle cyano-type bonded phase (Partisil™-10PAc, Whatman) is a more suitable stationary phase than either reverse phase or silica gel columns. Approximately 10 metabolites of doxorubicin are detectable in a rabbit bile sample.

The chromatographic behavior of doxorubicin, daunorubicin, and their metabolites, doxorubicinol and daunorubicinol, has been studied by Eksborg (1978), using a silica support phase chemically bonded to hydrocarbon chains of varying length (li Chrosorb RP-2, RP-8, RP-18). Good separation is achieved with acetonitrile/water eluents with relatively low acetonitrile content.

A rapid and highly sensitive (1 ng/ml plasma) HPLC-fluorescence assay of doxorubicin and its metabolites has been presented by Baurain et al. (1979). The chromatographic system consists of a 7 μm silica gel column and a chloroform-methanol-glacial acetic acid mixture (720:210:40, v/v) containing a 0.3 mM $MgCl_2$ solution in water (30 parts by volume) for better resolution. This assay appears to be quite suitable for the analysis of biological samples.

3. Fluorescence—Bachur et al. (1977), Chan and Harris (1973), Schwartz (1973)

Discussion. Doxorubicin and many of its metabolites exhibit strong fluorescence with an excitation maximum at 490 nm and an emission maximum at 560 nm. Backur et al. (1977) measure the fluorescence of plasma samples after protein precipitation without further extractions. Results are equivalent to those obtained with a RIA for doxorubicin over 4 hr after a dose of doxorubicin; however, after 4 hr the fluorescence assay yields considerably higher plasma values, indicating the presence of fluorescent metabolites that are serologically inactive. The assay of Schwartz (1973) includes extraction of doxorubicin and tissue homogenates with isoamyl alcohol in the presence of $AgNO_3$, which precipitates proteins, flavines, nucleotides and so on, and considerably lowers the fluorescence background. Fluorescence is read in the organic phase with a sensitivity of 0.05μg doxorubicin/ml plasma. Doxorubicinol and other metabolites are also measured by this assay. Chan and Harris (1973) hydrolyze plasma doxorubicin in HCL to the aglycone, doxorubicinone, which is then extracted into benzene and separated from interfering substances on glass fiber thin-layer plates. Fluorescence is read after elution from the plates with a sensitivity of 0.02 ng doxorubicin/ml plasma. The metabolite doxorubicinone contributes only 1% to the doxorubicin levels, which can be determined by omitting the HCl hydrolysis.

4. Radioimmunoassay.—Bachur et al. (1977), Broughton and Strong (1976) (review), Langone and van Vunakis (1975), van Vunakis et al. (1974)

Discussion. Doxorubicin antibodies were raised in goats against doxorubicin conjugated with human serum albumin by the carbodiimide method, which covalently binds the amino group of doxorubicin to a carboxy function of the protein (van Vunakis et al., 1974). The radioactive tracer is prepared by coupling doxorubicin to p-hydroxyphenylacetic acid and labeling the derivative with ^{125}I by the lactoperoxidase method. Free and antibody-bound tracer are separated by using either immunoprecipitation or nitrocellulose membrane filtration. The assay is sensitive to 1 ng doxorubicin/ml plasma and does not require any extractions. Doxorubicin and daunorubicin are equally reactive with the antibody, while doxorubicinone is four times less reactive (Bachur et al., 1977). The major metabolite, doxorubicinol, possesses an antibody affinity equal to that of the parent drug (Langone and van Vunakis, 1975). Separation of metabolites by HPLC, followed by RIA of column eluent fractions (Langone and van Vunakis, 1975), can be an extremely sensitive and specific method of detection.

5. Mass Spectrometry.—Chan and Watson (1978), Maurer et al. (1976)

Discussion. Underivatized doxorubicin yields abundant quasi-molecular ions (MH$^+$) in the field desorption ionization mode (Maurer et al., 1976). With selected ion monitoring and daunorubicin as the internal standard, doxorubicin can be quantitated in amounts of 100 pg and more. This procedure may be applicable to biological samples.

On the other hand, GC-MS in the electron impact mode can be utilized to study doxorubicin and daunorubicin aglycone analogs, following petrimethylsilylation and methoxime formation (Chan and Watson, 1978). Four aglycone metabolites were identified by this method in the acid hydrolysate of a 1-butanol extract of bile from a doxorubicin-treated rabbit.

6. Comments. The major analytical approaches to doxorubicin exploit its native fluorescence or use immunoassay methods in order to obtain the sensitivity required for biological applications. High performance liquid chromatography with fluorescence detection appears to be an obvious choice that has not yet been sufficiently explored.

e. References

Asbell, M. A., E. Schwartzbach, F. J. Bullock, and D. W. Yesair, Daunomycin and Adriamycin Metabolism via Reductive Glycosidic Cleavage, *J. Pharmacol. Exp. Ther.*, **182**, 63–69 (1972).

Bachur, N. R. et al., Plasma Adriamycin and Daunorubicin Levels by Fluorescence and Radioimmunoassay, *Clin. Pharmacol. Ther.*, **21**, 70–77 (1977).

Barth, H. G. and A. Z. Conner, Determination of Doxorubicin Hydrochloride in Pharmaceutical Preparations Using High-Pressure Liquid Chromatography, *J. Chromatogr.*, **131**, 375–381 (1977).

Baurain, B., D. Deprez-De Campeneere, and A. Trovet, Rapid Determination of Doxorubicin and Its Fluorescent Metabolites by High-Pressure Liquid Chromatography, *Anal. Biochem.*, **94**, 112–116 (1979).

Benjamin, R. S., C. E. Riggs, Jr. and N. R. Bachur, Plasma Pharmacokinetics of Adriamycin and Its Metabolites in Humans with Normal Hepatic and Renal Function, *Cancer Res.*, **37**, 1416–1420 (1977).

Broughton, A. and J. E. Strong, Radioimmunoassay of Antibiotics and Chemotherapeutic Agents, *Clin. Chem.*, **22**, 726–732 (1976).

Chan, K. K. and P. A. Harris, A Fluorimetric Determination of Adriamycin and Its Metabolites in Biological Tissues, *Res. Commun. Chem. Pathol. Pharmacol.*, **6**, 447–463 (1973).

Chan, K. K. and E. Watson, GLC-Mass Spectrometry of Several Important Anticancer Drugs.II: Doxorubicin and Daunorubicin Aglycone Analogs, *J. Pharm. Sci.*, **67**, 1748–1752 (1978).

Creasey, W. A. et al., Clinical Effects and Pharmacokinetics of Different Dosage Schedules of Adriamycin, *Cancer Res.*, **36**, 216–221 (1976).

Eksborg, S., Reversed-Phase Liquid Chromatography of Adriamycin and Daunorubicin and Their Hydroxyl Metabolites Adriamycinol and Daunorubicinol, *J. Chromatogr.*, **149**, 225–232 (1978).

Felsted, R. L., M. Gee, and N. R. Bachur, Rat Liver Daunorubicin Reductase: An Aldo-Keto Reductase, *J. Biol. Chem.*, **249**, 3672–3679 (1974).

Goodman, J. and P. Hochstein, Generation of Free Radicals and Lipid Peroxidation by Redox Cycling of Adriamycin and Daunomycin, *Biochem. Biophys. Res. Commun.*, **77**, 797–803 (1977).

Guthrie, D. and A. L. Gibson, Doxorubicin Cardiotoxicity: Possible Role of Digoxin in Its Prevention, *Br. Med. J.*, **2**, 1447–1449 (1977).

236 *Drug Monographs*

Hulhoven, R. and J. P. Desager, Quantitative Determination of Low Levels of Daunomycin and Daunomycinol in Plasma by High-Performance Liquid Chromatography, *J. Chromatogr.*, **125**, 369–374 (1976).

Kummen, M., K. K. Lie, and S. O. Lie, A Pharmacokinetic Evaluation of Free and DNA-Complexed Adriamycin: A Preliminary Study in Children with Malignant Disease, *Acta Pharmacol. Toxicol.*, **42**, 212–218 (1978).

Langone, J. J. and H. van Vunakis, Adriamycin Metabolites: Separation by High-Pressure Liquid Chromatography and Quantitation by Radioimmunoassay, *Biochem. Med.*, **12**, 283–289 (1975).

Lovless, H., E. Arena, R. L. Feldsted, and N. R. Bachur, Comparative Mammalian Metabolism of Adriamycin and Daunorubicin, *Cancer Res.*, **38**, 593–598 (1978).

Maurer, K. H., U. Rapp, K. K. Chan,* and W. Sadee,* Possible Applications of Quantitative Field Desorption Mass Fragmentography to Pharmacokinetic Studies of Antineoplastic Drugs, in *Advances in Mass Spectrometry in Biochemistry and Medicine*, Vol. I, A. Frigerio and N. Castagnoli, Eds., Spectrum, Publ., New York, 1976, pp 541–551.

Myers, C. E. et al., Adriamycin: The Role of Lipid Peroxidation in Cardiac Toxicity and Tumor Response, *Science*, **197**, 165–167 (1977).

Riggs, C. E., Jr., R. S. Benjamin, A. A. Serpick, and N. R. Bachur, Biliary Disposition of Adriamycin, *Clin. Pharmacol. Ther.*, **22**, 234–241 (1977).

Schwartz, H. S., A Fluorometric Assay for Daunomycin and Adriamycin in Animal Tissues, *Biochem. Med.*, **7**, 396–404 (1973).

Takanashi, S. and N. R. Bachur, Adriamycin Metabolism in Man: Evidence from Urinary Metabolites, *Drug Metab. Dispos.*, **4**, 79–87 (1976).

Van Vunakis, H., J. J. Langone, L. J. Riceberg, and L. Levine, Radioimmunoassay for Adriamycin and Daunomycin, *Cancer Res.*, **34**, 2546–2552 (1974).

Watson, E. and K. K. Chan, Rapid Analytic Method for Adriamycin and Metabolites in Human Plasma by a Thin-Film Fluorescence Scanner, *Cancer Treat. Rep.*, **60**, 1611–1618 (1976).

Whatman Liquid Chromatography Applications Report (based on data from P. A. Harris and G. Gudauskas), Pharmaceuticals, *Cancer*—Therapy, Antibiotics and Metabolites, 76-5A (1976).

ETHAMBUTOL—by D. Smith

(*R*)-2,2'-(1,2-Ethanediyldiimino)bis-1-butanol.
Myambutol. Antibacterial (tuberculostatic) agent.

$C_{10}H_{24}N_2O_2$; mol. wt. 204.31; pK_a 6.6 and 9.5.

a. Therapeutic Concentration Range. From 3 to 5 μg/ml (Strauss and Erhardt, 1970).

b. Metabolism. Ethambutol is metabolized to an intermediary aldehyde and further to a butyric acid derivative neither of which has any antimicrobial activity (Peets et al., 1964). The renal route is the primary mode of elimination of the unchanged drug (80%) (Strauss and Erhardt, 1970), and the half-life for patients with normal renal function ranges from 4 to 5 hr (Lee et al., 1977).

c. Analogous Compounds. Aliphatic secondary amines.

d. Analytical Methods

1. Gas Chromatography—Lee and Benet (1976)

Discussion. A GLC method has been developed for ethambutol, using solvent extraction, derivatization with trifluoroacetic anhydride, and an electron capture detector for quantitation. Dextro-2,2'-(ethylene-diimino)-di-1-propanol was selected as the internal standard because of its similar chemical structure. Sensitivity is 0.1 μg ethambutol/ml plasma, using a 0.2 ml sample volume.

Detailed Method. Samples of 0.2 ml plasma or urine, to which 5 μg internal standard has been added, are extracted with 8 ml chloroform for 10 min under alkaline conditions. Portions of the chloroform layer are transferred to another tube and evaporated to dryness under nitrogen. Methylene chloride (0.5 ml) is added and evaporated to dryness to ensure the azeotropic removal of water. Residues are dissolved in 1 ml of a benzene-pyridine mixture. Derivatization with trifluoro-acetic anhydride is complete within 2 hr. Excessive derivatizing agent is removed by washing with 0.01 M HCL. An appropriate aliquot of the benzene layer is injected into the gas chromatograph.

A Varian Aerograph Model 2700 equipped with a scandium tritide electron capture detector is used under the following conditions: glass column, 6 ft x ⅛ in., 3% OV-17 on Gas Chrom Q, 100/120 mesh; carrier gas (nitrogen) flow rate, 20/ml/min; injector temperature, 210°C; oven temperature, 157°; detector temperature, 230°C.

2. Other GC Methods—Calo et al. (1968) Lee and Benet (1978), Richard et al. (1974)

Discussion. The assays of Calo et al. (1968) and Richard et al. (1974) employ the less sensitive flame ionization detection after silylation

(Richard et al., 1974), involving a temperature program (Calo et al., 1968). Five other antitubercular drugs (ethionamide, pyrazinamide, cycloserine, isoniazid, pyridoxine) did not interfer with the GC analysis of ethambutol. Lee and Benet (1978) describe a dual-column, dual-detector GC assay for the simultaneous micro (plasma and dialyzate) and macro (urine) determination of ethambutol. Trifluoroacetylation is utilized together with GC-EC, while silylation is followed by GC-FID in the macro assay.

3. Gas Chromatography-Mass Spectrometry.—Blair et al. (1976)

Discussion. Ethambutol and tetradeuterated ethambutol (labeled at the ethylene bridge) as the internal standard are extracted into chloroform and derivatized with trimethylsilylimidazole after evaporation of the chloroform layer. Addition of 3 drops of 1 M HCL to the chloroform layer prevents evaporation of the ethambutol free base. Silylation serves to eliminate interferences from plasma samples observed after GC-MS analysis of the underivatized drug. Chemical ionization (methane) MS is used for detection in the selected ion monitoring mode (m/e 349 and 353 for MH$^+$ of di-TMS of ethambutol and d_4-ethambutol, respectively). The assay sensitivity is about 10 ng ethambutol/ml plasma or urine.

4. Microbiological Assay—Gangadharam and Candler (1977), Place and Thomas (1963)

Discussion. Serum and urine samples are assayed for ethambutol, using *Mycobacterium smegmatis* ATCC 607 as the assay organism. Place and Thomas (1963) use an agar diffusion technique, while Gangadharam and Candler (1977) propose a paper disk and cup plate method. Disadvantages of these methods include reduced sensitivity, specificity, accuracy, and precision, as well as being more time consuming. Interference with the microbiological ethambutol assay is observed with streptomycin, kanamycin, erythromycin, lincomycin, gentamicin, tetracycline, chloramphenicol, and colistin.

5. Colorimetric Assay—Strauss and Erhardt (1970)

Discussion. Analysis of serum and urine is based on the formation and extraction of an ethambutol-bromthymol blue complex of a yellowish coloration, which is measured quantitatively on a spectrophotometer at 405 nm. Weaknesses of this method include low sensitivity (0.5 μg/ml) and specificity.

6. Comments. The GC-EC assay appears to be the most versatile method for biomedical applications. The high sensitivity of the GC-MS method (Blair et al., 1976) relative to the therapeutic range warrants application of this assay only in special research projects using very small sample volumes.

e. References

Blair, A. D., A. W. Forrey, T. G. Christopher, B. Maxwell, and R. E. Cutler, Determination of Ethambutol in Plasma and Urine by Chemical Ionization GC-MS Using a Deuterated Internal Standard, *Method. Dev. Biochem.*, **5**, 231–234 (1976).

Calo, A., C. Cardini, and V. Quercia, Gas Chromotographic Separation of Some Antitubercular Drugs, *J. Chromatogr.*, **37**, 194–196 (1968).

Gangadharam, P. R. J. and E. R. Candler, Microbiological Assay of Ethambutol, *J. Antimicrob. Chemother.*, **3**, 57–63 (1977).

Lee, C. S. and L. Z. Benet, Gas-Liquid Chromatographic Determination of Ethambutol in Plasma and Urine of Man and Monkey, *J. Chromatogr.*, **128**, 188–192 (1976).

Lee, C. S. and L. Z. Benet, Micro and Macro GLC Determination of Ethambutol in Biological Fluids, *J. Pharm. Sci.*, **67**, 470-473 (1978).

Lee, C. S., J. G. Gambertoglio, D. C.. Brater, and L. Z. Benet, Kinetics of Oral Ethambutol in the Normal Subject, *Clin. Pharmacol. Ther.*, **22**, 615–621 (1977).

Peets, E. A., W. M. Sweeney, V. A. Place, and D. A. Buyske. The Absorption, Excretion, and Metabolic Fate of Ethambutol in Man, *Am. Rev. Respir. Dis.*, **91**, 51–58 (1964).

Place, V. A. and J. P. Thomas, Clinical Pharmacology of Ethambutol, Am. Rev. Respir. Dis., **87**, 901–904 (1963).

Richard, B. M., J. E. Manno, and B. R. Manno, Gas Chromatographic Determination of Ethambutol, *J. Chromatogr.*, **89**, 80–83 (1974).

Strauss, I. and F. Erhardt, Ethambutol Absorption, Excretion and Dosage in Patients with Renal Tuberculosis, *Chemotherapy*, **15**, 148–157 (1970).

ETHOSUXIMIDE—by B. Stafford

2-Ethyl-2-methylsuccinimide
Zarontine, and so one. Anticonvulsant, antiepileptic (petit mal).

$C_7H_{11}NO_2$; mol. wt. 141.17; pK_a 9.0

a. Therapeutic Concentration Range. From 40 to 100 $\mu g/ml$ serum (Leal and Troupin, 1977; Sherwin et al., 1973).

b. Metabolism. The major metabolites of ethosuximide are the inactive 2-(1-hydroxyethyl)-2-methylsuccinimide, 2-acetyl-2-methyl-succinimide (Chang et al., 1972), 2-carboxymethyl-2-methylsuc-cinimide, and diastereomeric 2-ethyl-2-methyl-3-hydroxysuccinimide (Pettersen, 1978; Preste et al., 1974). Glucuronidation of the 1-hydroxy-ethyl metabolite has also been reported (Chang et al., 1972). The ethosuximide plasma elimination half-life in human beings is between 24 and 72 hr (Leal and Troupin, 1977). (Further references are listed in the anticonvulsants monograph.)

c. Analogous Compounds. Succinimides: methosuximide (Cel-ontin) and its major metabolite, N-desmethylmethsuximide, phensuxi-mide (Milontin); anticonvulsants.

d. Analytical Methods

Discussion. The major assay methods for ethosuximide in serum —HPLC, GC, enzyme immunoassay, and others—are discussed in detail in the anticonvulsants monograph. Additional assays suitable only for the detection of ethosuximide and related succinimides are listed below.

1. Gas Chromatography—Bonitati 1976, Least et al. (1975), Solow and Green (1971), van der Kleijn et al. (1973), Wallace et al. (1979)

Discussion. Succinimides can be analyzed by GC-FID either following alkylation (methylation, butylation: (Least et al., 1975) or without chemical derivatization (Solow and Green, 1971). The latter alternative is preferable for measurement of methsuximide, since methylation does not allow separate analysis of the parent drug and the predominant active metabolite, N-desmethylmethsuximide, in plasma. Therefore Bonitati (1976) proposes a GC micromethod without derivatization based on the assay of van ber Kleijn et al. (1973). The assay requires 10 to 100 μl serum, which is diluted with 0.1 ml saturated solution of KH_2PO_4 and extracted with 0.8 ml chloroform containing 1 to 5 μg fluorene as the internal standard. After extraction the chloroform layer is concentrated to 5 to 20 μl, and 1 μl is analyzed by GC FID, using a temperature program from 180 to 230°C (10° min). The stationary phase is 5% OV-17. Ethosuximide, methsuximide, *N*-desmethyl-methsuximide, and phensuximide can be simultaneously measured with a CG analysis time of 6 min and a sensitivity of 2 μg drug/ml serum.

Derivatization of ethosuximide and desmethylmethsuximide with pentafluorobenzoyl chloride produces derivatives of high sensitivity to the electron capture detector. Wallace et al. (1979) describe a sensitive GC-EC assay based on this reaction, using methyl-α-propylsuccinimide as the internal standard.

2. Comments. Ethosuximide and other succinimide anticonvulsants can be measured in serum for therapeutic monitoring by GC, HPLC, or EMIT techniques (see anticonvulsants monograph).

References

Bonitati, J., Gas Chromatographic Analysis for Succinimide Anticonvulsants in Serum: Macro- and Micro-Scale Methods, *Clin. Chem.*, **22**, 341–345 (1976).

Chang, T., A. R. Burkett, A. J. Glazko, Biotransformation, in *Antiepileptic Drugs*, D. M. Woodbury, J. K. Penry, and R. P. Schmidt, Eds., New York Press, 1972, pp. 425–429.

Leal, K. W. and A. S. Troupin, Clinical Pharmacology of Antiepileptic Drugs: A Summary of Current Information, *Clin. Chem.*, **23**, 1964–1968 (1977).

Least, C. J., Jr., G. F. Johnson, and H. M. Solomon, A Quantitative Gas Chromatographic Determination of Ethosuximide Based on *N*-Butylation, *Clin. Chim. Acta*, **60**, 285–292 (1975).

Pettersen, J. E., Urinary Metabolites of 2-Ethyl-2-methylsuccinimide (Ethosuximide) Studied by Combined Gas Chromatography Mass Spectrometry, *Biomed. Mass Spectrom.*, **5**, 601–603 (1978).

Preste, P. G., C. E. Westerman, N. P. Das, B. J. Wilder, and J. H. Duncan, Identification of 2-Ethyl-2-methyl-3-hydroxysuccinimide as a Major Metabolite of Ethosuximide in Humans, *J. Pharm. Sci.*, **63**, 467–469 (1974).

Sherwin, A. L., J. P. Robby, and M. Lechter, Improved Control of Epilepsy by Monitoring Plasma Ethosuximide, *Arch. Neurol.*, **28**, 178–181 (1973).

Solow, E. B. and J. B. Green, The Determination of Ethosuximide in Serum by Gas Chromatography: Preliminary Results of Clinical Applications, *Clin. Chim. Acta*, **33**, 87–90 (1971).

Van der Kleijn, E., P. Collste, B. Norlander, S. Agurell, and F. Sjoqvist, Gas Chromatographic Determination of Ethosuximide and Phensuximide in Plasma and Urine of Man, *J. Pharm. Pharmacol.*, **25**, 324–327 (1973).

Wallace, J. E., H. A. Schwertner, H. E. Hamilton, and E. L. Shimek, Jr., Electron-Capture Gas-Liquid Chromatographic Determination of Ethosuximide and Desmethylmethsuximide in Plasma or Serum, *Clin. Chem.*, **25**, 252–255 (1979).

ETHYNYLESTRADIOL—by K. Maloney

17α-Ethynyl-1,3,5(10)-estratriene-3,17β-diol.
Novestron, Progynon, and so on. Estrogen.

$C_{20}H_{24}O_2$; mol. wt. 296.39

a. Therapeutic Concentration Range. Daily doses of 0.05 mg result in ethynylestradiol levels of 60 to 500 pg/ml plasma (Pasqualini et al., 1977; Verma et al., 1975).

b. Metabolism. Ethynylestradiol is extensively metabolized to many products that are conjugated mainly by glucuronidation and excreted into the urine (Williams et al., 1975). Hydroxylation occurs in C_2, C_6, and C_{16} with 2-methoxyethynylestradiol as a major product. The fact that deethynylation to the natural estrogens estrone, estradiol, estriol, and 2-methoxyestradiol also occurs in human beings may contribute to the effects of ethynylestradiol (Williams et al., 1975). The plasma elimination half-life of ethynylestradiol is 6.5 hr (Pasqualini et al., 1977).

c. Analogous Compounds. Estrogenic substances (e.g., dienestrol, diethylstilbestrol, estradiol, estradiol valerate, estriol, estrone, mestranol); progestogens (e.g., ethynodiol, norethindrone, norethynodrel).

d. Analytical Methods

1. Radio Receptor-Ligand Assay—Verma et al. (1975)

Discussion. This competitive protein binding assay employs the target estrogenic receptor from a rabbit uterine cytosol preparation and therefore measures any potentially active ethynylestradiol metabolites and endogenous estrogens, if performed without chromatographic separation. Negligible blank plasma levels (~15 pg/ml) in patients without ethynylestradiol treatment are obtained with a preseparation on Sephadex LH-20 columns, which has also been employed by

Williams et al. (1975) to separate ethynylestradiol from its metabolites. This chromatographic system separates estrone and estriol, but not estradiol, from ethynylestradiol. Blank plasma readings are probably caused by estradiol which has been carried through the ether extraction and column chromatography. The sensitivity is 25 pg ethynylestradiol/ml plasma with an interassay C.V. of 10.7%.

Detailed Method. Uterine cytosol solution is prepared from the uteri of 6 day pregnant rabbits by homogenization in tris(hydroxymethyl)aminomethane, 10 mM, and ethylenediaminetetraacetic acid, 1 mM. After centrifugation the supernatant is adjusted to pH 8.0 with HCl. To 1 ml of plasma, 1000 dpm 6,7-³H-ethynylestradiol (>10 Ci/mM) in 50 μl methanol is added as the internal standard for extraction losses; the solution is thoroughly mixed for 1 min and incubated for 30 min at room temperature. The plasma is then extracted twice with 5 ml diethyl ether, and the ether phase is evaporated to dryness under an air jet, redissolved in 1 ml benzene-methanol (9:1, v/v), and chromatographed on a Sephadex LH-20 column in the same solvent system. The fraction corresponding to ethynylestradiol is evaporated to dryness; the residue is dissolved in 1 ml methanol, and 0.1 ml transferred to a counting vial for estimation of recovery by liquid scintillation counting. The remainder is pipetted into a 10 mm x 75 mm disposable culture tube, and after drying 1 ml aqueous solution of 6,7-³H-ethynylestradiol (10,000 dpm) is added. A volume of 0.04 ml of the uterine cytosol preparation is added; after mixing, the tubes are incubated overnight at 4°C. Following incubation, 0.2 ml of a dextran-coated charcoal suspension is added. The mixture is thoroughly mixed, incubated for 20 min, and centrifuged at 2000 *g* for 15 min. The supernatant is decanted into a counting vial containing 3 ml dioxane and 10 ml of a standard scintillation fluid. The tritium content of the vials is then determined in a liquid scintillation counter.

2. Other Radio Receptor-Ligand Assay—Warren and Fotherby (1973)

Discussion. The method of Warren and Fotherby (1973) is equivalent to that used by Verma et al. (1975). It includes either alumina TLC or Sephadex LH-20 column chromatography with a slightly modified eluent (benzene-methanol, 85:15) for the separation of ethynylestradiol. However, the authors report that neither chromatographic system separates ethynylestradiol from estradiol, which has to be subtracted as determined in blank plasma samples. Mestranol, which is the 3-

methoxy prodrug of ethynylestradiol, is separated from ethynylestradiol and also does not interact with the cytosol receptor.

3. Radioimmunoassay—Pasqualini et al. (1977), Rao et al. (1974)

Discussion. Rao et al. (1974) have raised antibodies in rabbits directed against 6-(*O*-carboxymethyl)oxime–bovine serum albumin conjugates prepared from 6-oxoethynylestradiol and 6-oxomestranol, respectively. The affinity of these antibodies to ethynylestradiol, mestranol, and their metabolites is compared to that of equivalent antibodies produced by conjugation in the C-7 position in other laboratories. Specificity of these antibodies is good for ethynylestradiol in the presence of its metabolites. The assay is performed in unextracted plasma, using ^{3}H-ethynylestradiol as tracer and dextran-coated charcoal for separation of free and bound fractions. Sensitivity is sufficient for analysis of therapeutic levels of ethynylestradiol.

The assay of Pasqualini et al. (1977) includes several methylene chloride extractions of plasma prior to RIA with ethynylestradiol-specific antibodies. The method is also applicable to the measurement of norethindrone in plasma samples, using a different antibody with specific affinity for norethindrone. The procedure for production of those antibodies is not given.

4. Gas Chromatography-Mass Spectrometry—Wilson et al. (1977)

Discussion. This GC-MS assay describes the detection of estradiol in plasma by GC-MS in the selected ion monitoring mode after silylation. 2,4-Dideuterioestradiol- serves as the internal standard. The sensitivity is 40 pg estradiol/ml plasma, using a 10 ml plasma volume. This technique could be useful to differentiate between endogenous estradiol and stable isotopic estradiol generated by administration of stable isotopic ethynylestradiol.

5. High Performance Liquid Chromatography—Bagon and Hammond (1978)

Discussion. A reverse phase HPLC separation of ethynylestradiol from related steroids is presented and applied to the analysis of ethynylestradiol tablets, using a UV monitor set at 210 to 215 nm. The separation technique may be useful together with competitive protein binding assays of ethynylestradiol biological samples.

6. Comments. The radio receptor-ligand binding assay, in conjunction with appropriate chromatographic techniques, is more generally applicable than the equivalent radioimmnoassays, since only active estrogenic substances are measured and the available binding protein is rather uniform in specificity and affinity. However, antibodies might be preferable to cytosol receptors as the binding protein if they are specific for ethynylestradiol in the presence of estradiol, since the two are not separated by the chromatographic purification steps. Most physicochemical methods are not applicable because of the very low plasma levels of ethynylestradiol. Perhaps GC-MS methods such as the estradiol assay by Wilson et al. (1977) might also be applicable to the measurement of ethynylestradiol plasma levels.

e. References

Bagon, K. R. and E. W. Hammond, Determination of Ethinyloestradiol in Single Tablets and Its Separation from Other Steroids by High-Performance Liquid Chromatography, *Analyst*, **103**, 156–161 (1978).

Pasqualini, J. R., R. Castellet, M. C. Portois, J. L. Hill, and F. A. Kincl, Plasma Concentrations of Ethynyl Oestradiol and Norethindrone after Oral Administration to Women, *J. Reprod. Fert.*, **49**, 189–193 (1977).

Rao, P. N., A. de la Pena, and J. W. Goldzieher, Antisera for Radioimmunoassay of 17α-Ethynylestradiol and Mestranol, *Steroids*, **24**, 803-808 (1974).

Verma, P., C. Curry, C. Crocker, P. Titus-Dillon, and B. Ahluwalia, A Competitive Protein Binding Radioassay for 17α-Ethynylestradiol in Human Plasma, *Clin. Chim. Acta*, **63**, 363–368 (1975).

Warren, R. J. and K. Fotherby, Plasma Levels of Ethynyloestradiol after Administration of Ethynyloestradiol or Mestranol to Human Subjects, *J. Endocrinol.* **59**, 369–370 (1973).

Williams, M. C., E. D. Helton, and J. W. Goldzieher, The Urinary Metabolites of 17α-Ethynylestradiol-9α,11xi-^{3}H in Women: Chromatographic Profiling and Identification of Ethynyl and Non-ethynyl Compounds, *Steroids*, **25**, 229–246 (1975).

Wilson, D. W., B. M. John, G. V. Groom, C. G. Pierrepoint, and K. Griffiths, Evaluation of an Oestradiol Radioimmunoassay by High-Resolution Mass Fragmentography, *J. Endocrinol.*, **74**, 503–504 (1977).

5-FLUOROURACIL

5-Fluoro-2,4(1H,3H)-pyrimidinedione.
Efudex, Fluril, Fluracil, Fluroplex. Antineoplastic agent.

$C_4H_3FN_2O_2$; mol. wt. 130.08; pK_a 8.0 and 13.0

a. Therapeutic Concentration Range. Following intravenous injections of 15 mg fluorouracil/kg, peak serum concentrations are about 50 μg/ml. Steady-state levels of free 5-fluorouracil during intravenous infusions (30 mg/kg·day) range between 0.1 and 1.0 μg/ml (Sadée and Wong, 1977).

b. Metabolism. 5-Fluorouracil is rapidly cleared from serum with a $t_{1/2}$ of about 10 min (Finn and Sadée, 1975) and extensively metabolized by degradative pathways via dihydro-5-fluorouracil to CO_2, NH_3, and α-fluoro-β-alanine, as well as to active anabolites, including 5-fluorouridine triphosphate and 5-fluoro-2'-deoxyuridine monophosphate (Mukherjee et al., 1963). The active nucleotide metabolites are intracellularly localized, and the relationship between free 5-fluorouracil serum levels and metabolite concentrations is poorly understood (Sadée and Wong, 1977).

c. Analogous Compounds. Endogenous pyrimidine bases, nucleosides, and nucleotides; pyrimidine antimetabolites (e.g., 5-fluorodeoxyuridine, ftorafur, 5-fluorocytosine, cytosine arabinoside, 5-trifluoromethyl-2'-deoxyuridine, idoxuridine, 5-azacytidine, 6-azauridine); propylthiouracil.

d. Analytical Methods

1. High Performance Liquid Chromatography—Au et al. (1979), Wu et al. (1978)

Discussion. The detection limit of 5-fluorouracil (UV$_{max}$ 265 nm) is below 5 ng per injection, measured by UV absorbance at 254 or 280 nm coupled with HPLC. Reverse phase columns such as μ-Bondapak C_{18} are capable of separating 5-fluorouracil from all other pyrimidine analogs. The plasma assay procedure of Wu et al. (1978) involves an ethyl acetate extraction of 5-fluorouracil and, as modified by Au et al (1979), yields a sensitivity of about 20 ng 5-fluorouracil/ml plasma. This assay was also designed for the simultaneous determination of ftorafur, a 5-fluorouracil prodrug. Results with the HPLC assay agreed well with those obtained by GC-MS (Finn and Sadée, 1975).

Detailed Method. Plasma samples (0.1 to 1.0 ml) are adjusted with distilled water to a total volume of 1 ml, and 0.1 ml 0.5 M NaH_2PO_4 buffer and 8 ml ethyl acetate are added. After extraction and centrifugation the organic layer is removed and evaporated under N_2 at 50°. The residue is redissolved in 50 μl methanol, and 5 to 25 μl is injected onto the HPLC column with 0.01 M sodium acetate buffer (pH 4)-acetonitrile (95:5, v/v) at a flow rate of 2 ml/min as eluent. The retention time is 2.5 min, and overall extraction recovery 58%. Sensitivity limit of quantitative analysis in plasma is 0.02 μg/ml, using external standards. Standard curves for 5-fluorouracil obtained from peak height measurements are linear over a wide concentration range (0.02 to 200 μg/ml).

Analysis of 5-fluorouracil was performed on a high performance liquid chromatograph (Waters Assoc., Milford, Mass.) equipped with a Model 6000A solvent delivery system, a Model 440 absorbance detector at 280 nm, a Model UK-6 injector, and a μ-Bondapak C_{18} column (4 mm x 30 cm; Waters Assoc.).

2. Other HPLC Assays—Christophidis et al. (1979), Cohen and Brown (1978)

Discussion. 5-Fluorouracil can be assayed in plasma by strong anion-exchange HPLC with a sensitivity of 100 ng/ml (Cohen and Brown, 1978). A significant increase in sensitivity (25 ng 5-fluorouracil/ml plasma) is afforded by a double-extraction procedure and reverse phase HPLC with UV detection at 254 nm (Christophidis et al., 1979). This assay employs 5-fluorocytosine as the internal standard; results correlate well with those obtained by a GC method.

3. Gas Chromatography—Cohen and Brennan (1973), de Leenheer and Gelijkens (1978), Finch et al. (1978), Van den Berg et al. (1978), Windheuser et al. (1972)

Discussion. The earlier GC assays require either dialysis (Windheuser et al., 1972) or solvent extraction (Cohen and Brennan, 1973) and silylation prior to GC with flame ionization detection. The workable sensitivity limit is above 0.5 μg/ml, which is greatly inferior to the HPLC limit. Furthermore, care has to be taken to avoid cleavage of the glycosidic bond of 5-fluorouracil nucleosides and nucleotides derivatives. Ftorafur can yield as much as 30% 5-fluorouracil during silylation (Wu et al., 1976).

The use of nitrogen-senitive GC-FID after flash methylation improves the assay sensitivity to 0.1 μg 5-fluorouracil/ml plasma (Finch et al., 1978). Thymine as the internal standard may cause a systematic error because of potentially high endogenous thymine levels; however, we have found rather low thymine plasma levels (<0.1 μM) using a GC-MS assay (unpublished). A further increase in assay sensitivity has been achieved by van den Berg et al. (1978), using GC with electron capture detection of the chloromethyldimethylsilyl derivative. As little as 10 pg 5-fluorouracil is detectable with an assay sensitivity of 20 ng/ml plasma. Thymine is again used as the internal standard in amounts of 50 or 625 ng/sample, depending on the 5-fluorouracil concentration. Endogenous thymine levels may cause interferences at the lower level.

The metabolite 5-fluorouridine can be assayed by GC–N-FID with a sensitivity of 10 ng/ml urine (1 ml specimen), using 5-chlorouridine as the internal standard. Sample purification requires two column chromatography steps and the extraction of the permethylated (dimethyl sulfate) derivatives (De Leenheer and Gelijkens, 1978).

4. Gas Chromatography-Mass Spectrometry—Finn and Sadée (1975), Hillcoat et al. (1976), Wu et al. (1976)

Discussion. The GC-MS assay of Finn and Sadée (1975) is similar to the GC assay of Cohen and Brennan (1973), using electron impact selected ion monitoring MS with highly enriched ^{14}C-5-fluorouracil as the internal standard. Wu et al. (1976) introduced ^{15}N$_2$-5-fluorouracil as internal standard and added a LH-20 column chromatography pre-separation of 5-fluorouracil in order to avoid assay interference by 5-fluorouracil derivatives such as ftorafur. Sensitivity limit for 5-fluorouracil is 1 ng/ml plasma. Hillcoat et al. (1976) utilized flash methylation prior to GC-MS and thymine as the internal standard. Assay sensitivity was 10 ng 5-fluorouracil/ml plasma; high endogenous thymine levels might interfere.

5. Microbiological Methods—Clarkson et al. (1964)

Discussion. A number of microbiological assays are available, none of which has sufficient specificity to differentiate between 5-fluorouracil and its active metabolites.

6. Column   Chromatography   and ^{14}C–5-Fluorouracil— Mukherjee et al. (1963)

Discussion. Separation and detection of 5-fluorouracil from its metabolites has been achieved using Dowex 1 (formate) column chromatography, followed by liquid scintillation counting of ^{14}C.

7. Radioimmunoassay—Schreiber and Raso (1978)

Discussion. The RIA method for 5-fluoro-2′-deoxyuridine is highly sensitive (1 to 10 pmole per sample) and specific for 5-fluorouracil nucleosides and nucleotides. The antibody was raised against a bovine serum albumin conjugate with 5-fluorouridine, prepared by periodate oxidation of the vicinal OH groups of the riboside to aldehydes and stabilization of the conjugate of this product with BSA by sodium borohydride reduction. Free and bound ^{3}H—5-fluoro-2′-deoxyuridine are separated by filtration through nitrocellulose filters. The relative affinities of 5-fluoro-2′-deoxyuridine, 5-fluorouridine, and 5-fluoro-2′-deoxyuridine 5′-monophosphate (1:4:2) are similar because of the antigen conjugation procedure, while the free base, 5-fluorouracil, and and other pyrimidines are less reactive by several orders of magnitude. This RIA method is therefore suitable for the detection of 5-fluorouracil metabolites in the presence of high concentrations of the parent drug.

8. Comments. The HPLC method represents the assay of choice for the detection of free 5-fluorouracil. Higher sensitivity can be achieved by GC-MS procedures.

e. References

Au, J. L., A. T. Wu, M. A. Friedman, and W. Sadée, Metabolism and Disposition of *R,S*-1-(Tetrahydrofuran-2-yl)-5-fluorouracil (Ftorafur) in Humans, *Cancer Chemother. Rep.*, **63,** 343–350 (1979).

Christophidis, N., G. Mihaly, F. Vajda, and W. Louis, Comparison of Liquid-

and Gas-Liquid Chromatographic Assays of 5-Fluorouracil in Plasma, *Clin. Chem.*, **25,** 83–86 (1979).

Clarkson, B. et al., The Physiologic Disposition of 5-Fluorouracil and 5-Fluoro-2'-deoxyuridine in Man, *Clin. Pharmacol. Ther.*, **5,** 581–610 (1964).

Cohen, J. L. and P. B. Brennan, A Gas Chromatographic Determination of 5-Fluorouracil in Biological Fluids, *J. Pharm. Sci.*, **62,** 572–575 (1973).

Cohen, J. L. and R. E. Brown, High-Performance Liquid Chromatographic Analysis of 5-Fluorouracil in Plasma, *J. Chromatogr.*, **151,** 237–240 (1978).

De Leenheer, A. P. and C. F. Gelijkens, Quantification of 5-Fluorouridine in Human Urine by Capillary Gas-Liquid Chromatography with a Nitrogen-Selective Detector, *J. Chromatogr. Sci.*, **16,** 552–555 (1978).

Finch, R. E., M. R. Bending, and A. F. Lant, Use of a Nitrogen Detector for GLC Determination of Fluorouracil in Plasma During Single- and Combined-Agent Chemotherapy, *J. Pharm. Sci.*, **67,** 1489–1490 (1978).

Finn, C. and W. Sadée, Determination of 5-Fluorouracil (NSC-19893) Plasma Levels in Rats and Man by Isotope Dilution-Mass Fragmentography, *Cancer Chemother. Rep.*, **59,** 279–286 (1975).

Hillcoat, B. L., M. Kawai, P. B. McCulloch, J. Rosenfeld, and C. K. O. Williams, A Sensitive Assay of 5-Fluorouracil in Plasma by Gas Chromatography-Mass Spectrometry, *Br. J. Clin. Pharmacol.*, **3,** 135–143 (1976).

Mukherjee, K. L., J. Boohar, D. Wentland, F. J. Ansfield, and C. Heidelberger, Studies on Fluorinated Pyrimidines. XVI: Metabolism of 5-Fluorouracil-2-^{14}C and 5-Fluoro-2'-deoxyuridine in Cancer Patients, *Cancer Res.*, **23,** 49–66 (1963).

Sadée, W. and C. Wong, Pharmacokinetics of 5-Fluorouracil: Inter-relationship with Biochemical Kinetics in Monitoring Therapy, *Clin. Pharmacokin.*, **2,** 437–450 (1977).

Schreiber, R. and V. Raso, Radioimmunoassay for the Detection and Quantitation of 5-Fluorodeoxyuridine, *Cancer Res.*, **38,** 1889–1892 (1978).

Van den Berg, H. W., R. F. Murphy, R. Hunter, and D. T. Elsmore, An Improved Gas-Liquid Chromatographic Assay for 5-Fluorouracil in Plasma, *J. Chromatogr.*, **145,** 311–314 (1978).

Windheuser, J. J., J. L. Sutter, and E. Auen, 5-Fluorouracil and Derivatives in Cancer Chemotherapy: Determination of 5-Fluorouracil in Blood, *J. Pharm. Sci.*, **61,** 301–303 (1972).

Wu, A. T., H. J. Schwandt, C. Finn, and W. Sadée, Determination of Ftorafur and 5-Fluorouracil Levels in Plasma and Urine, *Res. Commun. Chem. Pathol. Pharmacol.*, **14,** 89–102 (1976).

Wu, A. T., J. L. Au, and W. Sadée, Hydroxylated Metabolites of *R,S*-1-(Tetrahydro-2-furanyl)-5-fluorouracil (Ftorafur) in Rats and Rabbits, *Cancer Res.*, **38,** 210–214 (1978).

GENTAMICIN

Mixture of gentamicin components C_1, C_{1a}, and C_2 with gentamicin A as a minor component.
Cidomycin, Garamycin, Gentalyn, Genticin, Refobacin. Antibiotic.

C_1, $R_1 = R_2 = CH_3$; C_{1a}, $R_1 = R_2 = H$; C_2 $R_1 = CH_3$; $R_2 = H$.

$C_{21}H_{43}N_5O_7$; mol. wt.: C_1, 477.59; C_{1a}, 449.54; C_2, 463.57.

a. Therapeutic Concentration Range. From 4 to 12 μg/ml (Dahlgren et al., 1975). Serum levels exceeding 12 μg/ml may cause ototoxicity or nephrotoxicity (for references see Peng et al., 1977). Opposing views exist on the predictability of serum gentamicin levels based on kidney function tests and other available clinical data (Barza et al., 1975; Hull and Sarubbi, 1976). Assessment of individual gentamicin pharmacokinetics obtained from serum level determinations permits reasonably accurate calculations of dosage regimen to achieve a desired target level (Sawchuck et al., 1977) and appears to be the optimum approach to monitor and adjust gentamicin therapy. A series of analogous aminoglycosides exists to which similar therapeutic considerations apply. The analytical problem is a particularly difficult one. Early recognition of the clinical value of gentamicin serum level determination has led to a number of different assays of general analytical interest. Therefore the analysis of gentamicin is described here in greater detail than that of other drugs.

b. Metabolism. Between 80 and 90% of an intravenous or intramuscular dose of gentamicin (100 mg) is recovered in unchanged form in the urine within 24 hr. Glomerular filtration seems to be the main pathway of elimination (Begamey et al., 1973). Consequently, the serum half-life of gentamicin increases from its normal value of about 2 to 4 hr to 40 to 120 hr in patients with chronic renal failure (T. W. Wilson et al., 1973).

c. Analogous Compounds. Aminoglycosides (e.g., amikacin, kanamycin, netilmicin, sisomicin, tobramicin).

d. Analytical Methods

1. High Performance Liquid Chromatography—Anhalt (1977), Anhalt and Brown (1978), Anhalt et al. (1978)

Discussion. The technique of Anhalt (1977) involves extraction of gentamicin from serum by using a CM-Sephadex column, separation by reverse phase ion-pair chromatography, and fluorescence detection by continuous-flow, postcolumn derivatization with o-phthaladehyde. The method separates the three gentamicin components and is applicable to other aminoglycosides. The HPLC and detection procedure was previously utilized for the detection of kanamycin components (Mays et al., 1976). Sensitivity for gentamycin is 1 μg/ml, using 0.2 ml serum samples, with good precision. Other drugs did not interfere with the assay. In a subsequent report, Anhalt and Brown (1978) describe the application of this assay to additional aminoglycosides, that is, amikacin and tobramycin. Moreover, they introduce an internal standard for the gentamicin assay, namely, 1-N-acetylgentamicin C_1.

Similar HPLC techniques have been applied by Anhalt et al. (1978) to analyze the major and minor components of gentamicin C.

Detailed Method. Gentamicin is separated from interfering compounds in serum by ion-exchange gel chromatography. A short (1.5 cm) column with a bed volume of 1 ml is prepared from CM-Sephadex (C-25), using 0.2 M Na_2SO_4 as the initial buffer. Serum (400 μl) is applied to this column and then eluted in succession with 1 ml and 4 ml of initial buffer. The eluting buffer is changed to 0.01 M NaOH in 0.2 M Na_2SO_4, and 600 μl of this buffer is added. After the column has drained completely, a second volume of alkaline buffer (400 μl) is added, the eluate collected, and 15 μl of this eluate injected into the HPLC.

A Tracor Model 990 pump (Tracor Instruments) is used to deliver mobile phase. A Schoeffel Model 970 fluorometer (Schoeffel Instrument Corp.) is used to detect fluorescent products formed by continuous-flow, postcolumn derivatization with o-phthalaldehyde. Fluorescence excitation is at 340 nm, and a KV418 filter is used for emission. Solutions of o-phthalaldehyde (60 mg) are prepared in 1 ml methanol and 0.2 ml mercaptoethanol and diluted with 100 ml 0.4 M potassium borate buffer (pH 10.4). It is supplied from a pressurized glass vessel to a mixing tee and a delay coil between the tee and the detector.

Analysis is performed by using a μ-Bondapak C_{18} column, 30 cm x 3.9 mm, (i.d.) (Waters Assoc.), with a precolumn, 4.3 cm x 4.2 mm (i.d.), packed with Micropart C_{18} phase-bonded silica gel (Applied Science Lab.). The mobile phase used for analysis contains 0.2 M

Na$_2$SO$_4$, 0.02 *M* sodium pentanesulfonate, and 0.1% (v/v) acetic acid in a water-methanol (97:3) mixture. Column flow rate is 2 ml/min at 184 atm. *o*-Phthalaldehyde solution flow rate is about 0.5 ml/min. Concentrations of gentamicin are determined by peak height measurements, using external standard curves (1 to 10 μg/ml).

2. Other HPLC Methods—Maitra et al. (1977), Peng et al. (1977)

Discussion. The assay of Maitra et al. (1977) differs from that of Anhalt (1977) mainly in the use of precolumn derivatization with *o*-phthalaldehyde, followed by reverse phase separation of the reaction product of the three gentamicin components. Both pre- and postcolumn derivatization techniques require careful development of the assay conditions. The method of Anhalt (1977) was more reliable in our hands. Peng et al. (1977) employ precolumn derivatization with dansyl chloride. However, gentamicin C$_{1a}$ and C$_2$ are not separated by reverse phase HPLC. We have encountered difficulties in setting up this procedure in our laboratory.

3. Radioimmunoassay—Berk et al. (1974), Broughton and Strong (1976), Griffiths et al. (1977), Lewis et al. (1972), Longmore et al. (1976), Mahon et al. (1973), Watson et al. (1976b)

Discussion. The haptene gentamicin lends itself to ready linkage to antigenic proteins, and a series of slightly differing assay methods has been developed. These include the use of ^{3}H and ^{125}I tracers, incubation with the antibody, and separation by double-antibody techniques or dextran-coated charcoal. All of the RIA procedures are rapid and sensitive. The specificities of the assays in the presence of other aminoglycosides vary; some assays have a remarkable lack of cross-reactivity even with close analogs of gentamicin. Commercial RIA kits are available; however, each aminoglycoside usually requires a separate assay kit because of the specificity of the antibody, a circumstance that increases the cost per sample, particularly with small sample numbers. If large numbers of gentamicin samples are to be analyzed, RIA is the method of choice.

4. Fluoroimmunoassay—Shaw et al. (1977, 1979) Watson et al. (1976a)

Discussion. These methods utilize fluorescein-labeled gentamicin, which is incubated with the antibody and the gentamicin serum sample. Watson et al. (1976a) employ fluorescence polarization induction by binding of the label to the antibody, requiring only 1.25 μl serum with good precision (C.V.'s at 3.0 and 13.0 μg/ml are 11 and 5%, respectively). The method of Shaw et al. (1977) involves measurement of the fluorescence quenching upon binding with comparable precision and sensitivity using 10 μl serum, thus obviating the need for a specialized polarization fluorimeter. Insufficient data preclude comparison with RIA methods. The method has been successfully adapted to a continuous-flow automated system by Shaw et al. (1979).

5. Enzyme Immunoassays—Burd et al. (1977), Standefer and Saunders (1978)

Discussion. Standefer and Saunders (1978) describe a gentamicin assay in which a peroxidase-gentamicin conjugate competes with gentamicin for binding to a gentamicin antibody adsorbed to a polystyrene solid phase. After the competition reaction is completed, the extent of conjugate binding is observed by adding a chromogenic substrate for the peroxidase [2,2'-azino-di(3-ethylbenzthiazolin)-6'-sulfonate and hydrogen peroxide]. The assay can be completed in 30 min and requires 50 μl of diluted serum; the reagents are stable for several months. The precision and accuracy are equivalent to those of an established RIA when assaying serum samples of patients receiving gentamicin therapy. This solid phase enzyme immunoassay therefore appears to be suitable for clinical gentamicin analysis.

A homogeneous fluoroescence enzyme immunoassay has been developed by Burd et al. (1977). The nonfluorescent sisomicin-umbelliferyl-β-galactoside conjugate serves as the substrate for bacterial β-galactosidase, yielding a fluorescent product. In the presence of antigentamicin antibody the gentamicin conjugate is bound and rendered inactive. Gentamicin in the serum sample competes with this conjugate, thereby reversing the fluorescence inhibition. This homogeneous enzyme immunoassay is unique, since complete inhibition of the enzymatic reaction is achieved by the antibody binding reaction. The additional use of a fluorescence label makes the procedure highly sensitive, using only 1 μl serum per assay. The gentamicin antibody has comparable affinity for other aminoglycosides, for example, sisomicin and netilmicin. The interassay C.V. is about 10%. More general clinical acceptance of such a procedure is likely in the future.

6. Enzymatic Assay: Acetyltransferase—Broughall and Reeves (1975), Case and Mezel (1978), Holmes and Sanford (1974), Shannon and Phillips (1977), Williams et al. (1975)

Discussion. An *Escherichia coli*-derived aminoglycoside *N*-acetyltransferase is used for the assay. Detection is achieved either by colorimetry of the generated coenzyme A using sulfhydryl reagents (Williams et al., 1975) or (more commonly) with radioactive aminoglycoside product (e.g., Case and Mezel, 1978). This assay could be clinically useful, given the availability of high quality enzyme sources. It was judged superior in reliability to the adenyltransferase assay (Broughall and Reeves, 1975). The sensitivity is 0.2 μg gentamicin complex/ml plasma with a ^{14}C assay (Case and Mezel, 1978); sisomicin is also measured by this assay, while tobramycin and netilmicin are only slightly reactive. No interferences were found by other aminoglycosides, antibiotics, or endogenous serum substances.

7. Enzymatic Assay: Adenyltransferase—Harber and Asscher (1977), Krooden and Darrell (1974), Smith and Smith (1974), Smith et al. (1972)

Discussion. One of the R factors conveying bacterial resistance to aminoglycosides is an adenyltransferase that utilizes ATP to form an AMP-aminoglycoside conjugate and pyrophosphate. With the use of either ^{14}C- or α-^{32}P-ATP, this reaction can be readily monitored following specific absorption of the radioactive AMP-gentamicin conjugate to phosphocellulose paper. It is necessary to heat-inactivate ATP utilizing enzymes in the serum. The precision is within 10% (Smith and Smith, 1974), using 10 μl serum. Other aminoglycosides can be measured by the same procedure.

8. Enzymatic Assay: Urease—Bourne et al. (1974), Lode and Kemmerich (1976).

Discussion. The urease assay is based on the pH deviation of a medium containing urea by the hydrolytic activity of *Proteus mirabilis* bacteria, which are sensitive to aminoglycosides. It is therefore a microbiological assay in principle.

9. Microbiological Assays—Adeniyi-Jones et al. (1976), Deacon (1976a, 1976b), Lund et al. (1973), Peromet et al. (1974), Reeves and Bywater, (1975), Renshaw and Cornere (1974), Stessman et al. (1976, 1977), Tilton and Lieberman (1974).

Discussion. Only a small selection of published papers is presented here. Most assays are based on microdilution or plate diffusion (Tilton and Lieberman, 1974; Deacon, 1976). Specificity of the assay for aminoglycosides in the presence of other antibiotics can be improved by selecting a multiple-antibiotic-resistant strain of *Klebsiella pneumoniae* (Lund et al., 1973) or by adding β-lactamase to eliminate penicillin interference (Peromet et al., 1974). However, many problems remain (Stessman et al., 1976), including errors or interferences in uremic sera (Stessman et al., 1977) and in cerebrospinal fluid (Deacon, 1976), and interference by a heat-labile serum factor (Adeniyi-Jones et al., 1965), and by bilirubin (Renshaw and Cornere, 1974). A national survey revealed in 1975 (Reeves and Bywater) that 50% of the laboratories reported "highly misleading results" on unknown control standards. The current trend is to use assays that are more specific than microbiological methods.

10. Microbiological Assay-Firefly Bioluminescence—Harber and Asscher (1977)

Discussion. This bioassay of gentamicin utilizes a penicillin- and cephalosporin-resistant strain of *Klebsiella edwardsii* for the determination of serum gentamicin levels. The aminoglycoside activity is determined by measuring the bacterial ATP as a sensitive indicator of bacterial growth. Bacterial ATP can be readily detected by means of the firefly bioluminescence system, in which light emission is measured during the oxidation of ATP-activated firefly luciferin. Variability of assay results is less than 20%; however, specificity is low. Use of multiple-antibiotic-resistant bacterial may pose an environmental hazard.

11. Chromatography—Betina (1975), Kantor and Selzer (1968), Kartseva et al. (1975), Wagman et al. (1972), W. L. Wilson et al. (1973)

Discussion. These techniques include thin-layer, paper, and column chromatography with chemical (ninhydrin) or bioassay detection.

12. Mass Spectrometry—Games et al. (1976)

Discussion. Parent ions of the gentamicins are observed using field desorption ionization with little fragmentation; chemical ionization and electron impact ionization yield progressively more fragmentation. This technique is not readily adaptable to serum samples for quantitation.

13. Assay Comparison—(Andrews et al. (1974), Phillips et al. (1974), Stevens et al. (1975), Waterworth (1977)

14. Gentamicin Stability in Solution—Jones et al. (1976), Josephsen et al. (1979), McLaughlin and Reeves (1971), Weiner et al. (1976)

Discussion. Gentamicin was found to be rather stable in serum when stored in either glass or plastic tubes at −20, 4, and 25°C (Jones et al., 1976). However, Josephsen et al. (1979) found that dilution of sera containing gentamicin and tobramycin in glass containers results in substantial absorption of the antibiotic to the surface. Gentamicin in injection solutions was more stable in glass than in plastic syringes (Weiner et al., 1976). The presence of carbenicillin in the solution causes chemical degradation of gentamicin (McLaughlin and Reeves, 1971).

15. Comments. It is evident that gentamicin has attracted the interest of many analysts. Its peculiar chemical structure and lack of native UV absorbance resulted in the development of derivatization techniques, which are now the methods of choice when coupled with HPLC separation. However, the absence of major metabolic pathways has allowed clinical use of a large number of bioassays and immunoassays with less specificity. Knowledge of the relative concentrations of gentamicin components may not be of any advantage in clinical drug level monitoring, since they appear to exhibit nearly identical pharmacokinetic behavior. Whatever the potential clinical benefit may be, a recent survey has shown that only 20% of the gentamicin assays ordered for individual therapy were correctly drawn and appropriately utilized in making therapeutic decisions (Anderson et al., 1976). Not only reproducible assay methods, but also their proper utilization is needed.

e. References

Adeniyi-Jones, C., D. Stevens, B. Page, and S. Barbardoro, Letter: Interference with Gentamicin Assay by Heat-Labile Serum Factor, *Lancet*, **2,** 581 (1976).

Anderson, A. C., G. R. Hodges, and W. G. Barnes, Determination of Serum Gentamicin Sulfate Levels: Ordering Patterns and Use as a Guide to Therapy, *Arch. Intern. Med.*, **136,** 785–787 (1976).

Andrews, J., P. Gillette, J. D. Williams, and M. Mitchard, Analysis of Gentamicin in Plasma: a Comparative Study of Four Methods, *Postgrad. Med. J.*, **50,** 17–23 (1974).

Anhalt, J. P., Assay of Gentamicin in Serum by High-Pressure Liquid Chromatography, *Antimicrob. Agents Chemother.*, **11,** 651–655 (1977).

Anhalt, J. P., and S. D. Brown, High-Performance Liquid-Chromatographic Assay of Aminoglycoside Antibiotics in Serum, *Clin. Chem.*, **24**, 1940–1947 (1978).

Anhalt, J. P., F. D. Sancilio, and T. McCorkle, Gentamicin C-Component Ratio Determination by High-Pressure Liquid-Chromatography, *J. Chromatogr.*, **153**, 489–493 (1978).

Barza, M, R. B. Brown, D. Shen, M. Gibaldi, and L. Weinstein, Predictability of Blood Levels of Gentamicin in Man, *J. Infect. Dis.*, **132**, 165–174 (1975).

Begamey, C. and R. C. Gordon, Comparative Pharmacokinetics of Tobramycin and Gentamycin, *Clin. Pharmacol. Ther.*, **14**, 396–403 (1973).

Berk, L. S., J. L. Lewis, and J. C. Nelson, One-hour Radioimmunoassay of Serum Drug Concentrations, as Exemplified by Digoxin and Gentamicin, *Clin. Chem.*, **20**, 1159–1164 (1974).

Betina, V., Paper Chromatography of Antibiotics, *Methods Enzymol.*, **43**, 100–172 (1975).

Bourne, P. R., I. Phillips, and S. E. Smith, Modification of the Urease Method for Gentamicin Assays, *J. Clin. Pathol.*, **27**, 168–169 (1974).

Broughall, J. M. and D. S. Reeves, The Acetyltransferase Enzyme Method for the Assay of Serum Gentamicin Concentrations and a Comparison with Other Methods, *J. Clin. Pathol.*, **28**, 140–145, (1975).

Broughton, A. and J. E. Strong, Radioimmunoassay of Iodinated Gentamicin, *Clin. Chim. Acta*, **66**, 125–129 (1976).

Burd, J. F., R. C. Wong, J. E. Feeney, R. J. Carrico, and R. C. Boguslaski, Homogeneous Reactant-Labeled Fluorescent Immunoassay for Therapeutic Drugs Exemplified by Gentamicin Determination in Human Serum, *Clin. Chem.*, **23**, 1402–1408 (1977).

Case, R. V. and L. M. Mezel, An Enzymatic Radioassay for Gentamicin, *Clin. Chem.*, **24**, 2145–2150 (1978).

Dahlgren, J. G., E. T. Anderson, and W. L. Hewitt, Gentamicin Blood Levels: A Guide to Nephrotoxicity, *Antimicrob. Agents Chemother.*, **8**, 58–62 (1975).

Deacon, S., Factors Affecting the Assay of Gentamicin by the Plate Diffusion Method, *J. Clin. Pathol.*, **29**, 54–57 (1976a).

Deacon, S., Assay of Gentamicin in Cerebrospinal Fluid, *J. Clin. Pathol.*, **29**, 749–751 (1976b).

Games, D. E., M. Rossiter, M. S. Rogers, A. Weston, and R. T. Parfitt, Chemical Ionization and Field Desorption Mass Spectrometry of the Gentamicins, *Biomed. Mass Spectrom.*, **3**, 232–234 (1976).

Griffiths, W. C., P. Dextraze, and I. Diamond, Analysis of Serum for Gentamicin by Radioimmunoassay, *Ann. Clin. Lab. Sci.*, **7**, 141–145 (1977).

Harber, M. J. and A. W. Asscher, A New Method for Antibiotic Assay Based on Measurement of Bacterial Adenosine Triphosphate Using the Firefly Bioluminescence System, *J. Antimicrob. Chemother.*, **3**, 35–41 (1977).

Holmes, R. K. and J. P. Sanford, Enzymatic Assay for Gentamicin and Related Aminoglycoside Antibiotics, *J. Infect. Dis.*, **129**, 519–527 (1974).

Hull, J. H. and F. A. Sarubbi, Jr., Gentamicin Serum Concentrations: Pharmacokinetic Predictions, *Ann. Intern. Med.*, **85**, 183–189 (1976).

Jones, S. M., D. J. Blazewig, and M. H. Balfour, Stability of Gentamicin in Serum, *Antimicrob. Agents Chemother.*, **10**, 866–867 (1976).

Josephsen, L., P. Hovle, and M. Haggerty, Stability of Dilute Solutions of Gentamycin and Tobramycin, *Clin. Chem.*, **25**, 298–300 (1979).

Kantor, N. and G. Selzer, Chromatographic Separation and Bioassay of the Gentamicin Complex, *J. Pharm. Sci.*, **57**, 2170–2171 (1968).

Kartseva, V. D., I. A. Portnovi, N. A. Vakulenko, N. G. Kruzhkova, and E. M. Savitskaia, Gentamicin Stability and a Method for Its Quantitative Determination, *Antibiotiki*, **20**, 514–517 (1975).

Krooden, E. ten, and J. H. Darrell, Rapid Gentamicin Assay by Enzymatic Adenylylation, *J. Clin. Pathol.*, **27**, 452–456 (1974).

Lewis, J. E., J. C. Neson, and H. A. Elder, Radioimmunoassay of an Antibiotic: Gentamicin, *Nature*, **239**, 214–216 (1972).

Lode, H. and B. Kemmerich, Enzymatic Quick Determination of Aminoglycoside Antibiotics in Serum, *Int. J. Clin. Pharmacol. Biopharm.*, **13**, 59–64 (1976).

Longmore, P., R. C. Atkins, D. Casley, and C. I. Johnston, Radioimmunoassay as an Improved Method for Measurement of Serum Levels of Gentamicin, *Med. J. Aust.*, **1**, 738–740 (1976).

Lund, M. E., D. J. Blazevic, and J. M. Matsen, Rapid Gentamicin Bioassay Using a Multiple-Antibiotic-Resistant Strain of *Klebsiella pneumoniae*, *Antimicrob. Agents Chemother.*, **4**, 569–573, (1973).

Mahon, W. A., J. Ezer, and T. W. Wilson, Radioimmunoassay for Measurement of Gentamicin in Blood, *Antimicrob. Agents Chemother.*, **3**, 585–589 (1973).

Maitra, S. K. et al., Serum Gentamicin Assay by High-Performance Liquid Chromatography, *Clin. Chem.*, **23**, 2275–2278 (1977).

Mays, D. L., R. J. van Apeldoorn, and R. G. Lauback, High-Performance Liquid Chromatographic Determination of Kanamycin, *J. Chromatogr.*, **120**, 93–102 (1976).

McLaughlin, J. E. and D. S. Reeves, Clinical and Laboratory Evidence for Inactivation of Gentamicin by Carbenicillin, *Lancet*, **1**, 261–264 (1971).

Peng, G. W., M. A. F. Gadalla, A. Peng, V. Smith, and W. L. Chiou, High-Pressure Liquid Chromatographic Method for Determination of Gentamicin in Plasma, *Clin. Chem.*, **23**, 1838–1844 (1977).

Peromet, M., E. Schoutens, M. P. Vanderlinden, and E. Yourassowsky, Specific Assay of Gentamicin in the Presence of Penicillins and Cephalosporins: Use of Commercially Manufactured Beta-Lactamases, *Chemotherapy*, **20**, 1–5 (1974).

Phillips, I., C. Warren, and S. E. Smith, Serum Gentamicin Assay: A Comparison and Assessment of Different Methods, *J. Clin. Pathol.*, **27**, 447-451 (1974).

Reeves, D. S. and M. J. Bywater, Quality Control of Serum Gentamicin Assays-Experience of National Surveys, *J. Antimicrob. Chemother.*, **1**, 103-116 (1975).

Renshaw, S. and B. Cornere, The Effect of Bilirubin on the Assay of Gentamicin, *J. Clin. Pathol.*, **27**, 445-446 (1974).

Sawchuk, R. J., D. E. Zaske, R. J. Cipolle, W. A. Wargin, and R. G. Strate, Kinetic Model for Gentamicin Dosing with the Use of Individual Patient Parameters, *Clin. Pharmacol. Ther.*, **21**, 362-369 (1977).

Shannon, K. P. and I. Phillips, The Use of Aminoglycoside 2'-N-Acetyltransferase for the Assay of Gentamicin in Serum, Plasma and Urine, *J. Antimicrob. Chemother.*, **3**, 25-33 (1977).

Shaw, E. J., R. A. A. Watson, J. Landon, and D. S. Smith, Estimation of Serum Gentamicin by Quenching Fluoroimmunoassay, *J. Clin. Pathol.*, **30**, 526-531 (1977).

Shaw, E. J., R. A. A. Watson. and D. S. Smith, Continuous-Flow Fluoroimmunoassay of Serum Gentamycin, with Automatic Sample Blank Correction, *Clin. Chem.*, **25**, 322-325 (1979).

Smith, A. L. and D. H. Smith, Gentamicin: Adenine Mononucleotide Transferase: Partial Purification, Characterization, and Use in the Clinical Quantitation of Gentamicin, *J. Infect. Dis.*, **129**, 391-401 (1974).

Smith, D. H., B. van Otto, and A. L. Smith, A Rapid Chemical Assay for Gentamicin, *N. Engl. J. Med.*, **286**, 583-586 (1972).

Standefer, J. C. and G. C. Saunders, Enzyme Immunoassay for Gentamicin, *Clin. Chem.*, **24**, 1903-1907 (1978).

Stessman, J., J. Michel, and T. Sacks, Letter: Potential Pitfalls in Bioassay of Serum-Gentamicin, *Lancet*, **11**, (1976).

Stessman, J., J. Michel, and T. Sacks, Error in Recovery Rate of Aminoglycosides from Uraemic Sera, *Chemotherapy*, **23**, 142-148 (1977).

Stevens, P., L. S. Young, and W. L. Hewitt, Radioimmunoassay, Acetylating Radio-Enzymatic Assay, and Microbioassay of Gentamicin: A Comparative Study, *J. Lab. Clin. Med.*, **86**, (349-359 (1975).

Tilton, R. C. and L. Lieberman, Microdilution Assay of Antibiotics in Body Fluids, *Ann. Clin. Lab. Sci.*, **4**, 178-183 (1974).

Wagman, G. H. et al., Chromatographic Separation of Some Minor Components of the Gentamicin Complex, *J. Chromatogr.*, **70**, 171-173 (1972).

Waterworth, P. M., Which Gentamicin Assay Method Is the Most Practicable? *J. Antimicrob. Chemother.*,**3**, 1-3 (1977).

Watson, R. A., J. Landon, E. J. Shaw, and D. S. Smith, Polarisation Fluoroimmunoassay of Gentamicin, *Clin. Chim. Acta*, **73**, 51-55 (1976a).

Watson, R. A., E. J. Shaw, and C. R. Edwards, A. ^{125}I-Based Radioimmunoassay for Serum Gentamicin, in *Chemotherapy*, Vol. 2, J. D. Williams and A. M. Geddes, Eds., Plenum Press, New York, 1976b, pp. 107–110.

Weiner, B., D. J. McNeely, R. M. Kluge, and R. B. Stewart, Stability of Gentamicin Sulfate Injection Following Unit Dose Repackaging, *Am. J. Hosp. Pharm.*, **33**, 1245–1249 (1976).

Williams, J. W., J. S. Langer, and D. B. Northrop, A Spectrophotometric Assay for Gentamicin, *J. Antibiot.* (Tokyo), **28**, 982–987 (1975).

Wilson, T. W., W. A. Mahon, T. Inaba, G. E. Johnson, and D. Kadar, Elimination of Tritiated Gentamicin in Normal Human Subjects and in Patients with Severely Impaired Renal Function, *Clin. Pharmacol. Ther.*, **14**, 815–822 (1973).

Wilson, W. L., G. Richard, and D. W. Hughes, Thin-Layer Chromatographic Identification of the Gentamicin Complex, *J. Chromatogr.*, **78**, 442–444 (1973).

HALOPERIDOL—by L. Lambert

4-[4-*p*-Chlorophenyl)-4-hydroxypiperidino]-4′-fluorobutyrophenone. Haldol. Major tranquilizer.

$$F-\langle\text{phenyl}\rangle-\overset{\displaystyle O}{\underset{\displaystyle C}{\|}}-CH_2CH_2CH_2N\langle\text{piperidine-OH}\rangle-\langle\text{phenyl}\rangle-Cl$$

$C_{21}H_{23}ClFNO_2$; mol. wt. 375.87; pK_a 8.3.

a. Therapeutic Concentration Range. Concentration ranges noted in the literature for therapeutic levels vary from 6 to 245 ng/ml serum in psychotic patients following daily oral doses of 20 to 200 mg haloperidol (Clark et al., 1977).

b. Metabolism. Haloperidol is metabolized by oxidative *N*-dealkylation to β-(*p*-fluorobenzoyl)propionic acid, which is further metabolized and conjugated with glycine in the rat (Braun et al., 1967; Soudijn et al., 1967). Metabolism in human beings requires further study. The plasma half-life of haloperidol is 12 to 39 hr after intravenous and oral administration (Forsman and Ohman, 1976; Forsman et al., 1974). The apparent volume of distribution is about 1300 l, and the bioavailability of the oral dosage form is approximately 60% (Forsman and Ohman, 1976).

   c. **Analogous Compounds.** Droperidol, trifluperidol.

   d. **Analytical Methods**

      1. **Gas Chromatography**—Forsman et al. (1974)

Discussion. This GC assay requires electron capture detection after alkaline heptane extraction and back-extraction into an aqueous acid. The chlorobenzoyl analog is used as the internal standard. Sensitivity is 0.5 ng/ml with an overall precision of ±8.3% S.E.M. (n = 37, serum concentration range 0.5 to 10 ng/ml). Thioridazine and dibenzepim are the only drugs studied that seriously interfere with the determination of haloperidol. Chlorprothixene, chlorpenthixol, and propiomazine may interfere with baseline or standard peak determinations. The relation between peak height ratio (haloperidol/internal standard) and serum concentration of haloperidol is linear.

Detailed Method. To 5 ml serum in a glass centrifuge tube is added 20 ng of the internal standard (the *p*-chlorobenzoyl analog of haloperidol), dissolved in 20 μl ethanol, and the mixture is shaken. Then 1 ml 2 N NaOH and 6 ml *n*-heptane containing 1.5% isoamyl alcohol (v/v) are added, and the tubes are again vigorously shaken and centrifuged. Five milliliters of the resulting organic phase is transferred to another test tube, and 2 ml 0.005 N H_2SO_4 added. The tube is shaken and left for 5 min to separate. Then 1.8 ml of the resulting lower H_2SO_4 phase is transferred into a small test tube, together with 0.3 ml 1 N NaOH and 100 μl *n*-hexane–isoamyl alcohol (1.5%, v/v), followed by shaking and centrifugation as above. Fifty microliters of the upper phase is pipetted into a pointed minute glass tube, and then evaporated to dryness in a stream of nitrogen. A solvent mixture containing 8 ml heptane with 1.5% (v/v) isoamyl alcohol, 2 ml toluene, and 100 μl diethylamine is prepared, and 3.5 μl of this mixture added to redissolve the residue by 10 sec of sonication of the tube. This final solution is injected into the GC apparatus.
   Chromatographic analysis is performed on a Perkin-Elmer 990 gas chromatograph with a ^{63}Ni radiation source electron capture detector. Glass columns, 2 m × 2 mm (i.d.), are packed with 0.3% Versamid and 0.6% OV-17 on Gaschrom Q (80/100 mesh). Columns are conditioned at 310°C with a nitrogen flow of 48 hr before use and resilanized daily by vaporizing 100 μl Silyl-8 (Pierce Chemicals) in the column. Temperatures are as follows: column, 230°C; injection block, 290°C; manifold, 310°C; detector block, 345°C. Total nitrogen flow is 85 ml/min; column

flow is 40 ml/min. Approximate retention times for haloperidol and the internal standard are 4 and 8 min, respectively. Concentrations of haloperidol are determined by comparing haloperidol/standard peak height ratios to a calibration curve.

2. Other GC Methods—Bianchetti and Morselli (1978), Marcucci et al. (1971), Wells et al. (1975), Zingales (1971)

Discussion. These GC procedures are similar to the method of Forsman et al. (1974) except for different internal standards and different extraction procedures. Wells et al. (1975) use GC with flame ionization; sensitivity is only about 1 μg/ml. Use of nitrogen-sensitive FID improves the assay sensitivity sufficiently for the detection of haloperidol plasma levels after therapeutic doses ranging from 1.2 to 200 mg/day (Bianchetti and Morselli, 1978).

Marcucci et al. (1971) and Zingales (1971) employ GC with electron capture detection for an assay sensitivity in the low nanograms per milliliter serum range. These procedures are similar in specificity to the method by Forsman et al. (1974).

3. Radioimmunoassay—Clark et al. (1977)

Discussion. Antibodies to haloperidol are raised by immunizing rabbits with a haloperidol hydrazone-bovine serum albumin conjugate. This antiserum allows the determination of haloperidol concentrations as low as 1 ng/ml in unextracted human serum. Lack of cross-reactivity with any of the known metabolites is demonstrated; but the metabolism of haloperidol has not been conclusively elucidated in human beings. Therefore the possibility cannot yet be excluded that unknown metabolites interfere with the assay. The assay variability averaged 7% for triplicate measurements on six samples with a concentration range of 6 to 245 ng/ml.

4. Polarographic Determination—Mikolajek et al. (1974)

Discussion. Butyrophenones can be polarographically measured in blood after cyclohexane extraction. However, the described method is rather insensitive (10 μg haloperidol/ml) and nonspecific.

5. Comments. The GC methods and RIA are adequate to study the physiologic disposition of haloperidol following therapeutic doses. For large numbers of samples RIA is the method of choice.

e. References

Bianchetti, G. and P. L. Morselli, Rapid and Sensitive Method for Determination of Haloperidol in Human Samples Using Nitrogen-Phosphorus Selective Detection, *J. Chromatogr.*, **153,** 203–209 (1978).

Braun, G. A., G. I. Poos, and W. Soudijn, Distribution, Excretion and Metabolism of Neuroleptics of the Butyrophenone Type: Part II, *Eur. J. Pharmacol.*, **1,** 58–62 (1967).

Clark, B. R., B. B. Tower, and R. T. Rubin, Radioimmunoassay of Haloperidol in Human Serum, *Life Sci.*, **20,** 319–326 (1977).

Forsman, A. and R. Ohman, Pharmacokinetic Studies on Haloperidol in Man, *Curr. Ther. Res.*, **20,** 319–330 (1976).

Forsman, A., E. Martensson, G. Nyberg, and R. Ohman, A Gas Chromatographic Method for Determining Haloperidol, *N.S. Arch. Pharmacol.*, **286,** 113–124 (1974).

Marcucci, F., L. Airolde, E. Mussini, and S. Garattini, A Method for the Gas Chromatographic Determination of Butyrophenones, *J. Chromatogr.*, **59,** 174–177 (1971).

Mikolajek, A., A. Krzyzanowska, and J. Fidelus, Polarographische Bestimmung von Butyrophenonderivaten in Blut, *Z. Anal. Chem.*, **272,** 39–42 (1974).

Soudijn, W., I. Van Wijngaarden, and F. Allenijn, Distribution, Excretion and Metabolism of Neuroleptics of the Butyrophenone Type: Part I, *Eur. J. Pharmacol.*, **1,** 47–57 (1967).

Wells, J., G. Cimbura, and E. Kowes, The Screening of Blood by Gas Chromatography for Basic and Neutral Drugs, *J. Forensic Sci.*, **20,** 382–390 (1975).

Zingales, A. A., Gas Chromatographic Method for the Determination of Haloperidol in Human Plasma, *J. Chromatogr.*, **54,** 15–24 (1971).

HYDRALAZINE—by C. A. Gloff

1-Hydrazinophthalazine.
Apresoline. Antihypertensive agent.

$C_8H_8N_4$; mol. wt. 160.18; pK_a 7.3.

a. Therapeutic Concentration Range.

Therapeutic drug concentrations are not well defined. Most investigators report submicrograms per milliliter plasma levels after therapeutic doses (Lesser et al., 1974; Reidenberg et al., 1973; Zak et al., 1977).

b. Metabolism. Hydralazine is nonenzymatically degraded to phtalazine (Reidenberg et al., 1973) and metabolized to a number of products, including phthalazinone, methyltriazolophthalazine, triazolophthalazine, hydroxyhydralazine, and glucuronide and sulfate conjugates (Lesser et al., 1974; Zimmer et al., 1973). Acetylation followed by spontaneous ring closure to methyltriazolophthalazine appears to be mediated by the bimodally distributed *N*-acetyltransferase (Reidenberg et al., 1973). Furthermore, hydralazine pyruvic acid hydrazone appears to be a major metabolite in plasma (Reece et al., 1978). In addition, acetone and α-ketoglutarate hydrazone conjugates have been identified by GC-MS (Haegele et al., 1978). The pyruvate and α-ketoglutarate hydrazones are as active as hydralazine in an isolated aortic strip preparation and may contribute to the pharmacological effects of hydralazine *in vivo* (Haegele et al., 1978). The half-life of hydralazine is approximately 1 to 2 hr (Zak et al., 1977).

c. Analogous Compounds. Hydrazines.

d. Analytical Methods

    1. Gas Chromatograpy—Jack et al. (1975), Reece et al. (1978)

Discussion. Hydralazine is treated with nitrous acid to form tetrazolophthalazine, which is analyzed by GC-EC. 1-Hydrazino-4-methylphthalazine is used as an internal standard. The assay is sensitive to 10 ng/ml plasma. No interference from its metabolites has been reported. However, the pyruvate hydrazone may have been partially reconverted to hydralazine by the acid treatment (Reece et al., 1978; Zak et al., 1977).

Detailed Method. To 1 ml plasma is added 50 ng of the internal standard in 0.1 ml 0.1 *N* HCl. One milliliter 2 *N* HCl and 0.1 ml 50% aqueous sodium nitrite solution are added and allowed to react at room temperature for 15 min. The pH is adjusted to 10 by adding 2 ml 1 *N* NaOH and 4 ml of a pH 10 (0.03 *M* borax, 0.04 *M* NaOH) buffer, and the mixture extracted with 3 ml benzene. The organic layer is separated and evaporated under a stream of dry nitrogen at 45 °C. Immediately after the benzene has evaporated, the vials are removed from the water bath and refrigerated until final analysis. At that time the residue is dissolved in 300 to 900 μl toluene, and 5 μl is injected into the gas chromatograph. Extraction and evaporation steps have to be performed quickly in order to avoid chemical decomposition during analysis.

    A Pye Unicam Model 74, Series 104 gas chromatograph with a pulsed (150 μsec) electron capture detector (^{63}Ni, 10 mCi) is used. The column is 5 ft x 2 mm (i.d.), made of borosilicate glass, and packed with 3% of OV-225 on Chromosorb W-HP (80/100 mesh). Temperature of the column oven is 220°C, of the detector 300°C, and of the injector 200°C. Nitrogen is the carrier gas, with a flow rate of 30 ml/min. The retention time for the derivative of hydralazine is 6.4 min, and that of the internal standard 9.1 min.

2. Other GC Methods—Reece et al. (1978), Zak et al. (1977)

Discussion. The assay of Zak et al. (1977) is similar to that published by Jack et al. (1975) except for the use of 0.5 *N* HCl, rather than 2 *N* HCl, in order to minimize hydrazone hydrolysis. However, both of these assays include acid treatment of the plasma sample, which has recently been shown to cause hydrolysis of the major metabolite, hydralazine pyruvic acid hydrazone (Reece et al., 1978). Therefore these assays measure an apparent hydralazine level which includes the keto conjugate. A specific hydralazine GC assay has been developed by Reece et al. (1978), involving selective extraction of the drug from plasma and its derivatization to 3-trifluoromethyl-5-triazolo[3,4-a]-phthalazine. This derivative can be measured by GC-EC with high sensitivity.

3. Gas Chromatography-Mass Spectrometry—Haegele et al. (1976), Zimmer et al. (1973)

Discussion. Haegele et al. (1976) utilize deuterium labeling in conjunction with GC-MS to quantitate hydralazine and identify metabolites in plasma. The presence of methyltriazolophthalazine as a major metabolite in human beings has been demonstrated using GC-MS analysis (Zimmer et al., 1973).

4. Colorimetry—Reidenberg et al. (1973), Zak et al. (1974)

Discussion. The colorimetric reaction utilizes either *p*-hydroxybenzaldehyde (Reidenberg et al., 1973) or *p*-methoxybenzaldehyde (Zak et al., 1974) to form the corresponding hydrazones. These techniques lack the specificity necessary to assay for hydralazine in the presence of its metabolites.

5. High Performance Liquid Chromatography—Honigberg et al. (1975)

Discussion. High performance liquid chromatography with UV detection has been used to separate various antihypertensive-diuretic mixtures that include hydralazine. This technique may be applicable to the determination of hydralazine in body fluids.

6. Comments. At the present time GC assays are the best tests available for hydralazine. They are sufficiently sensitive; however, only the assay of Reece et al. (1978) appears to be selective for the parent drug. It still requires a laborious work-up procedure that one might be able to circumvent by using appropriate HPLC techniques.

The instability of hydralazine and its metabolites in solution presents an analytical problem. Furthermore, the spontaneous reversible reaction of hydralazine with endogenous plasma components such as pyruvate is likely. This may make it difficult to determine the actual *in vivo* concentration of free hydralazine.

e. References

Haegele, K. D., H. B. Skrdlant, N. W. Robie, D. Lalka, and J. L. McNay, Determination of Hydralazine and Its Metabolites by Gas Chromatography-Mass Spectrometry, *J. Chromatogr.*, **126**, 517–534 (1976).

Haegele, K. D. et al., Identification of Hydralazine and Hydralazine Hydrazone Metabolites in Human Body Fluids and Quantitative *in vitro* Comparisons of Their Smooth Muscle Relaxant Activity, *Br. J. Clin. Pharmacol.*, **5**, 489–494 (1978).

Honigberg, I. L., J. T. Stewart, A. P. Smith, and D. W. Hester, Liquid Chromatography in Pharmaceutical Analysis. III: Separation of Diuretic-Antihypertensive Mixtures, *J. Pharm. Sci.*, **64**, 1201–1204 (1975).

Jack, D. B., S. Brechbuhler, P. H. Degen, P. Zbinden, and W. Riess, The Determination of Hydralazine in Plasma by Gas-Liquid Chromatography, *J. Chromatogr.*, **115**, 87–92 (1975).

Lesser, J. M., Z. H. Israili, D. C. Davis, and P. G. Dayton, Metabolism and Disposition of Hydralazine-^{14}C in Man and Dog, *Drug Metab. Dispos.*, **2**, 351–360 (1974).

Reece, P. A., P. E. Stanley, and R. Zacest, Interference in Assays for Hydralazine in Humans by a Major Plasma Metabolite, Hydralazine Pyruvic Acid Hydrazone, *J. Pharm. Sci.*, **67**, 1150–1153 (1978).

Reidenberg, M. M., D. Drayer, A. L. DeMarco, and C. T. Bello, Hydralazine Elimination in Man, *Clin. Pharmacol. Ther.*, **14**, 970–977 (1973).

Zak, S. B., M. F. Barlett, W. E. Wagner, T. G. Gillerman, and G. Lucas, Disposition of Hydralazine in Man and a Specific Method for Its Determination in Biological Fluids, *J. Pharm. Sci.*, **63**, 225–229 (1974).

Zak, S. B., G. Lukas, and T. G. Gillerman, Plasma Levels of Real and "Apparent" Hydralazine in Man and Rat, *Drug Metab. Dispos.*, **5**, 116–126 (1977).

Zimmer, H., J. Kokosa, and D. A. Garteiz, Identification of 3-Methyl-*s*-triazolo[3,4-a]phthalazine, A Human Hydralazine Metabolite, by Gas

Chromatography-Mass Spectrometry, *Arzneim.-Forsch.*, **23**, 1028–1035 (1973).

HYDROCHLOROTHIAZIDE—by B. Silber

6-Chloro-3,4-dihydro-2H-1,2,4-benzothiadiazine-7-sulfonamide-1,1-dioxide.
HydroDiuril, Esidrix. Diuretic, antihypertensive agent.

$C_7H_8ClN_3O_4S_2$; mol. wt. 297.72; pK_a 7.0 and 9.2.

a. Therapeutic Concentration Range. None defined. Oral doses of 50 mg hydrochlorothiazide result in peak plasma levels of about 450 ng/ml (Redalieu et al., 1978).

b. Metabolism. Metabolites of hydrochlorothiazide have thus far not been identified. The drug is incompletely (65%) absorbed from the gastrointestinal tract (Anderson et al., 1961). The elimination half-life of hydrochlorothiazide from plasma is 3 to 4 hr in the alpha phase and 7 to 10 hr in the beta phase (Williams and Benet, 1978).

c. Analogous Compounds. Chlorthalidone, chlorothiazide, diazoxide, furosemide.

d. Analytical Methods

1. High Performance Liquid Chromatography—Lin and Benet (1978)

Discussion. This reverse phase HPLC assay utilizes the sensitive detection of hydrochlorothiazide at its UV maximum at 271 nm. The bromo derivative of hydrochlorothiazide serves as the internal standard because of its structural similarity. The sensitivity is 40 ng hydrochlorothiazide/ml plasma with excellent assay reproducibility [r^2 = .9915 ± .0073 (mean ± S.D., n = 6)]. The method is also applicable to urine samples as well as to the measurement of chlorothiazide.

Detailed Method. The plasma sample (0.2 ml) is deproteinated with 0.4 ml acetonitrile which contains the bromo analog of hydrochlorothiazide (0.2 μg) as the internal standard. After the mixture is vigorously shaken and then centrifuged for 10 min, the supernatant is transferred and evaporated to approximately 0.15 ml under a slow stream of nitrogen. About 30 μl of the remaining solution is then injected into the column and analyzed by HPLC with UV detection. The concentration of hydrochlorothiazide is determined from the peak height ratio of the plasma sample versus the standard curve. Standard curves are linear in the range of 40 to 400 ng/ml.

A Perkin-Elmer Series 3 liquid chromatograph with LC-65 T variable-wavelength UV detector set at 271 nm is utilized. The reverse phase column, 30 cm x 3.9 mm (i.d.) (μ-Bondapak C_{18}, Waters Associates), is eluted with 15% CH_3CN in 0.01 M acetic acid (pH 3.7) at a flow rate of 2 ml/min. Under these conditions the retention times of hydrochlorothiazide and the internal standard are 6.8 and 7.5 min, respectively.

2. Other HPLC Methods—Christophersen et al. (1977), Cooper et al. (1976)

Discussion. The technique of Cooper et al. (1976) is also based on reverse phase HPLC and UV detection at 271 nm. However, an ethyl acetate extraction is required, and the assay sensitivity is inferior to that obtainable with the method of Lin and Benet (1978). Christophersen et al. (1977) use gel filtration of the plasma, followed by ethyl acetate extraction and adsorption HPLC with UV detection. The sensitivity is 50 ng hydrochlorothiazide/ml plasma.

3. Gas Chromatography—Lindstrom and Molander (1975), Redalieu et al. (1978), Vandenheuvel et al. (1975)

Discussion. There are several GC-EC assays with excellent sensitivity. The method of Vandenheuvel et al. (1975) includes on-column flash methylation with trimethylammonium hydroxide and is the least sensitive GC procedure (lower limit 50 ng/ml. In addition it requires extensive clean-up procedures. Lindstrom et al. (1975) propose extractive alkylation yielding a tetramethyl derivative of hydrochlorothiazide, followed by GC analysis (sensitivity 10 ng/ml). Redalieu et al. (1978) achieve a sensitivity of 5/ng hydrochlorothiazide/ml plasma, again using alkylation to improve the GC characteristics of the drug. All of the GC methods require chemical derivatization and more extensive prepurification steps than are needed for HPLC.

4. Spectroscopic Method—Sheppard et al. (1960)

Discussion. The method of Sheppard et al. (1960) is not sensitive and specific enough for measuring therapeutically active concentrations of hydrochlorothiazide.

5. Comments. The method of Lin and Benet (1978) can be considered the one of choice because of its speed, sensitivity, and accuracy. Potential interferences by other drugs will have to be carefully evaluated, however, if its routine clinical application is indicated.

e. References

Anderson, K. V., H. R. Brettell, and J. K. Aikawa, ¹⁴C-Labeled Hydrochlorothiazide in Human Beings, *Arch. Intern. Med.*, **107**, 168–174 (1961).

Christophersen, A. S., K. E. Rasmussen, and B. Salvesen, Determination of Hydrochlorothiazide in Serum by High Pressure Liquid Chromatography, *J. Chromatogr.*, **132**, 91–97 (1977).

Cooper, M. J., A. R. Sinaiko, M. W. Anders, and B. L. Mirkin, High Pressure Liquid Chromatographic Determination of Hydrochlorothiazide in Human Serum and Urine, *Anal. Chem.*, **48**, 1110–1111 (1976).

Lin, E. T. and L. Z. Benet, Paper presented at the 125th Annual Meeting of the American Pharmaceutical Association, Montreal, Canada, May, 1978.

Lindstrom, B. and M. Molander, Gas Chromatographic Determination of Hydrochlorothiazide in Plasma, Blood Corpuscles, and Urine Using an Extractive Alkylation Technique, *J. Chromatogr.*, **114**, 459–462 (1975).

Redalieu, E., V. V. Tipnis, and W. E. Wagner, J., Determination of Plasma Hydrochlorothiazide Levels in Humans, *J. Pharm. Sci.*, **67**, 726–728 (1978).

Sheppard, H., T. F. Mowles, and A. J. Plummer, Spectroscopic Determination of Hydrochlorothiazide, *J. Am. Pharm. Assoc.*, **49**, 722–725 (1960).

Vandenheuvel, W. J. A., V. F. Grober, R. W. Walker, and F. J. Wolf, GLC Analysis of Hydrochlorothiazide in Blood and Plasma, *J. Pharm. Sci.*, **64**, 1309–1312 (1975).

Williams, R. and L. Z. Benet, Hydrochlorothiazide Bioavailability and Pharmacokinetic Studies in Normal Subjects, in preparation (1978).

HYDROCORTISONE—by L. L. Tsai

11β,17α,21-Trihydroxy-4-pregnene-3,20-dione. Cortisol. Cortef, Cortril, Ficortil. Glucocorticosteroid.

$C_{21}H_{30}O_5$; mol. wt. 362.47.

a. Therapeutic Concentration Range. Cortisol is a physiological hormone with normal values ranging from 6.5 to 26.3 μg (average 14.2 μg)/100 ml plasma between 8 and 10 A.M. and 2 to 18 (average 8 μg)/100 ml plasma at 4 P.M. (Tietz, 1976). The apparent free cortisol concentrations are 0.9 ± 0.46 μg/100 ml plasma at 8 A.M. and 0.23 ± 0.16 μg/100 ml at 11 P.M. (Clerico et al., 1979).

b. Metabolism. Hydrocortisone is extensively metabolized. Major metabolites include cortisone, tetrahydrocortisol, tetrahydrocortisone, cortol and cortolone, 11β-hydroxyetiocholanolone, and 11-ketoetiocholanolone (Tietz, 1976). The plasma elimination half-life of hydrocortisone is about 60 min.

c. Analogous Compounds. Glucocorticosteroids: beclomethasone, betamethasone, dexamethasone, fludrocortisone, methylprednisolone, prednisolone, prednisone, triamcinolone.

d. Analytical Methods

1. Radioimmunoassay—Tilden (1977)

Discussion. This radioimmunoassay of serum hydrocortisone minimizes the problem of competition between hydrocortisone binding globulin and the immunoglobulin. Increasing incubation temperature tends to release cortisol from its carrier protein while favoring the immunoglobulin reaction. The intra-assay precision has a C.V. of 3.3%. The sensitivity is 3 μg hydrocortisone/l serum, using only a 5 μl serum sample. Prednisolone, diethylstilbestrol, and associated metabolites interfere with the assay.

Detailed Method. A Micromedic Model 25000 autodiluter is used to sample 5.0 μl serum and dispense it with 1 ml of a trisaminomethane buffer, containing the ^{125}I-labeled cortisol derivative, into test tubes coated with rabbit antihydrocortisone serum. Following a 60 min incubation at 45°C, the solvent is removed from the tubes by aspiration, and the tubes are counted in a gamma spectrometer.

2. Other Radioimmunoassays—Clerico et al. (1979)

Discussion. Among the many published RIA procedures, the assay of Clerico et al. (1979) is of particular interest, since it measures the apparent free cortisol concentration following dialysis of the serum.

3. Competitive Protein Binding Assay—Pena and Goldzieher (1974)

Discussion. Cortisol (hydrocortisone) binding globulin (CBG) serves as the binding protein. Hydrocortisone has to be extracted from plasma with dichloromethane prior to incubation. 11-Deoxycortisol, testosterone, 17α-hydroxyprogesterone, and progesterone can also bind to CBG and cause assay interferences. This method is therefore not suitable in the female luteal phase and during pregnancy, when progesterone levels are significantly increased.

4. Enzyme Immunoassay—Comoglio and Celada (1976), Ogihara et al. (1977)

Discussion. Alkaline phosphatase as the indicator enzyme is conjugated with hydrocortisone via a 21-hemisuccinate bridge, using the carbodiimide reaction (Ogihara et al., 1977). Antibody-bound and free hydrocortisone-enzyme conjugate are separated by double-antibody precipitation. Serum samples have to be heated at pH 3.3 prior to incubation in order to inactivate CBG and serum alkaline phosphatase. The method is suitable for routine clinical analysis of hydrocortisone.

Comoglio and Celada (1976) utilize an antiserum raised in rabbits against cortisol-21-hemisuccinate–bovine serum albumin conjugate and a cortisol–β-galactosidase conjugate, with O-nitrophenyl–β-D-galactopyranoside as the substrate. Separation of free and antibody-bound enzyme conjugate is based on a double-antibody (solid phase) method; production of o-nitrophenol from the substrate is measured spectrophotometrically. The assay is sensitive to 100 to 150 pg hdrocortisone per sample.

5. Fluorescence Immunoassay—Kobayashi et al. (1979)

Discussion. The fluorescence polarization immunoassay of hydro-cortisone is based on a commercial RIA assay kit; the radiolabeled tracer has been replaced by a fluorescence label. Hydrocortisone-21-amine can be readily coupled to fluorescein isothiocyanate to give a suitable fluorescent hydrocortisone tracer. Results of the fluorescence immunoassay and the equivalent RIA agree well with each other.

6. Colorimetry—Gold et al. (1960), Hadd (1975), Porter and Silber (1950)

Discussion. The colorimetric procedures of Porter and Silber (1950) for the analysis of C_{21} corticosteroids bearing the $17\alpha,21$-dihydroxy-20-keto side chain has been one of the principal means of determining the functional status of the adrenal cortex. A recent modification of the reagent increases the assay sensitivity (Hadd, 1975). 21-Desoxyhydrocortisone interferes with fluorimetric and competitive binding assays, but not with the colorimetric method (Gold et al., 1960). Thus potential overestimation of plasma hydrocortisone in patients with defects of the adrenal cortex 21-hydroxylase system can be avoided with the Porter-Silber method.

7. Fluorescence—Mattingly (1962)

Discussion. Hydrocortisone develops an intense fluorescence in concentrated sulfuric acid. This procedure consistently gives higher hydrocortisone plasma values than do radioimmunoassays (Carr et al., 1977), since acid-induced fluorescence is caused by many steroids, for example, corticosterone, 20-dihydrocortisol, and spironolactone.

8. High Performance Liquid Chromatography—Kabra et al. (1979), Reardon et al. (1979), van den Berg et al. (1977)

Discussion. Hydrocortisone is extracted from plasma with a recovery of 96% and analyzed by reverse phase HPLC with UV detection (van den Berg et al. 1977). Sensitivity of the method is 1 μg hydrocortisone/100 ml plasma with good specificity. Plasma sample volume is rather large when compared to other assay methods.

Reardon et al. (1979) describe a similar assay for hydrocortisone and, simultaneously, 11-desoxycortisol for use in evaluating pituitary functions after a challenge dose of metyrapone. After metyrapone administration, 11-desoxycortisol was increased, and the same values were obtained by HPLC and a RIA method, while hydrocortisone levels were decreased. However, the results obtained by HPLC were lower (50 to 90%) than those obtained by RIA, possibly owing to lack of specificity of the RIA under these clinical test conditions (Reardon et al., 1979).

Kabra et al. (1979) report a sensitivity of 2 ng hydrocortisone/ml plasma with a reverse phase HPLC assay at 50°C, using 1 ml plasma samples. However, prednisone and prednisolone may interfere under the HPLC assay conditions.

9. Gas Chromatography-Mass Spectrometry—Bjorkhem et al. (1974)

Discussion. Following extraction of plasma, derivatization of hydrocortisone with methoxylamine and trimethylsilylimidazol yields the dimethoxime-tritrimethylsilyl product, which is analyzed by GC-MS. Electron impact mass spectrometry is used in the selected ion monitoring mode with 4-^{14}C-hydrocortisone serving as isotopically labeled internal standard. This procedure is useful as a clinical reference method.

10. Comparison of Methods—Carr et al. (1977), Demers and Derck (1977), Narymberski (1971)

11. Comments.
None of the presently used clinical assays is entirely satisfactory with respect to both speed of analysis and specificity. Radioimmunoassays are the most widely accepted technique. Further improvement of HPLC assays might result in partial replacement of RIAs in the future. All of the assays are designed to measure total plasma concentrations, not free hydrocortisone, except for the assay of Clerico et al. (1979). Given the high and variable levels of CBG, total hydrocortisone levels may be only a poor indication of the systemic exposure to glucocorticoids. The fate of exogenously supplied hydrocortisone in the presence of endogenous levels has to be studied with isotopic tracers. The assay procedures presented here were selected from a large number of publications reporting equivalent techniques.

e. References

Bjorkhem, I., R. Blomstrand, O. Lantto, A. Lof, and L. Svensson, Plasma Cortisol Determination by Mass Fragmentography, *Clin. Chim. Acta,* **56,** 241–248 (1974).

Carr, P. J., R. P. Millar, and H. Crowley, A Simple Radioimmunoassay for

Plasma Cortisol: Comparison with Fluorimetric Method of Determination, *Ann. Clin. Biochem.*, **14**, 207–211 (1977).

Clerico, A., M. G. Del Chicca, S. Ghione, F. Materazzi, and G. C. Zucchelli, Radioimmunological Determination of Apparent Free Cortisol Concentrations: Some Physiological and Clinical Applications, *Clin. Chim. Acta*, **91**, 227–231 (1979).

Comoglio, S. and F. Celada, An Immuno-enzymatic Assay of Cortisol Using *E. coli* β-Galactisodase as Label, *J. Immun. Methods*, **10**, 161–170 (1976).

Demers, L. M. and D. D. Derck, Comparison of Competitive Protein Binding Analysis and Radioimmunoassay for the Determination of Cortisol in Serum and Urine, *Clin. Biochem.*, **10**, 104–108 (1977).

Gold, E. M., B. Serena, and J. Cook, The Combined Estimation of Cortisol and 11-Deoxycortisol in Plasma as Porter-Silber Chromogens, *J. Clin. Endocrinol. Metab.*, **20**, 315–326 (1960).

Hadd, H. E., Measurement of Blood Plasma Cortisol by *p*-Hydrazinobenzene Sulfonic Acid/H_3PO_4 Reagent; A Modified Porter-Silber Reagent, *Biochem. Med.*, **13**, 353–358 (1975).

Kabra, P. M., L. L. Tsai, and L. J. Marton, Improved Liquid Chromatographic Method for Determination of Serum Cortisol, *Clin. Chem.*, in press (1979).

Kobayashi, Y., K. Amitani, F. Watanabe, and K. Miyai, Fluorescence Polarization Immunoassays for Cortisol, *Clin. Chim. Acta*, **92**, 241–247 (1979).

Mattingly, D., A Simple Fluorometric Method for the Estimation of Free 11-Hydroxycorticoids in Human Plasma, *J. Clin. Pathol.*, **15**, 374–379 (1962).

Narymberski, J. K., Evaluation of GLC and Other Techniques for Estimation of Corticosteroids, *Clin. Chim. Acta*, **34**, 187–195 (1971).

Ogihara, T., K. Miyai, K. Nishi, K. Ishibashi, and Y. Kumahara, Enzyme-labeled Immunoassay for Plasma Cortisol, *J. Clin. Endocrinol. Metab.*, **44**, 91–95 (1977).

Pena, A. de la, and J. W. Goldzieher, Practical Determination of Total Plasma Cortisol by Use of Competitive Protein Binding, *Clin. Chem.*, **20**, 1376–1378 (1974).

Porter, C. C. and R. H. Silber, A Quantitative Color Reaction for Cortisone and Related 17,21-Dihydroxy-20-ketosteroids, *J. Biol. Chem.*, **185**, 201–207 (1950).

Reardon, G. E., A. M. Caldarella, and E. Canalis, Determination of Serum Cortisol and 11-Deoxycortisol by Liquid Chromatography, *Clin. Chem.*, **25**, 122–126 (1979).

Tietz, M., *Fundamentals of Clinical Chemistry*, 2nd edition, W. B. Saunders Co., 1976, p 721, 737.

Tilden, R. L., New, Advantageous Approach to the Direct Radioimmunoassay of Cortisol, *Clin. Chem.*, **23**, 211–215 (1977).

Van Den Berg, J. H. M., C.·H. R. Mol, R. S. Deelder, and J. H. H. Thijssen, A Quantitative Assay of Cortisol in Human Plasma by High Performance

Liquid Chromatography Using A Selective Chemically Bonded Stationary Phase, *Clin. Chim. Acta*, **78**, 165–172 (1977).

IMIPRAMINE—by K. Maloney

5-(3-Dimethylaminopropyl)-10,11-dihydro-5H-dibenz[b,f]azepine. Tofranil. Antidepressant.

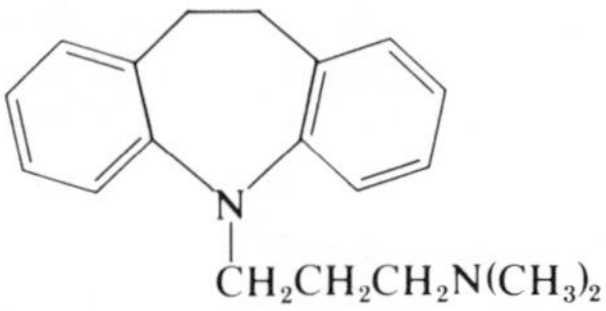

$$CH_2CH_2CH_2N(CH_3)_2$$

$C_{19}H_{24}N_2$; mol. wt. 280.40; pK_a 9.5.

a. Therapeutic Concentration Range. Steady-state plasma levels vary between 15 and 500 ng/ml (Gram et al, 1977; Vandemark et al., 1979). Therapeutic levels of the active metabolite desipramine vary between 30 and 400 ng/ml (Gram et al., 1977).

b. Metabolism. The major metabolite, desipramine, is also pharmacologically active and tends to accumulate to higher concentrations than does imipramine during maintenance therapy (Gram et al., 1977). The plasma elimination half-life of imipramine ranges from 4 to 20 hr (Alexanderson, 1972; Heck et al., 1978), and that of desipramine from 6 to 25 hr (Alexanderson, 1972).

c. Analogous Compounds. Tricyclic antidepressants: amitriptyline, clomipramine, cyproheptadine, doxepine, noripramine, nortriptyline, protriptyline.

d. Analytical Methods

1. High Performance Liquid Chromatography—Vandemark et al. (1978)

Discussion. The drug is extracted at pH 10.5 into hexane-isoamyl alcohol and chromatographed using silica gel adsorption and UV detection at 211 nm. The detection limit is 10 ng imipramine/ml plasma with a day-to-day C.V. of 7% at 100 ng/ml. Protriptyline is used as the internal standard. The method is also suitable for quantitation of desipramine, amitriptyline, and nortriptyline.

Detailed Method. A mixture of 2 ml plasma and 0.5 ml saturated Na_2CO_3 solution is extracted with 5 ml hexane-isoamyl alcohol (98:2, v/v) containing 500 ng protriptyline as the internal standard. After centrifugation the organic layer is separated and evaporated with dry air at 50°C. The residue is redissolved in the acetonitrile-NH_3 mobile phase and injected into the chromatograph. Standard curves are constructed, using peak area ratios, and are linear between 10 and 800 ng/ml plasma.

A Model 601 liquid chromatograph equipped with a Model LC-55 UV detector and a Rheodyne 7105 injection valve (Perkin-Elmer Corp.) is used. Analyses are performed on a 25 x 0.46 cm column packed with 5 μm (particle diameter) Silica B15 (Perkin-Elmer Corp.), using acetonitrile-concentrated ammonium hydroxide (99.3-0.7, v/v) at a flow rate of 1.5 ml/min at 65°C. The UV absorbance is measured at 211 nm, providing better sensitivity than the UV maximum of the tricyclic antidepressant at 245 nm.

2. Other HPLC Procedures—Watson and Stewart (1977a, 1977b)

Discussion. The plasma level assay for amitriptyline and nortriptyline utilizes noripramine as the internal standard and should also be suitable for the analysis of imipramine (Watson and Stewart, 1977a). A modification of this technique extends the applicability of the HPLC assay to the detection of tricyclic antidepressants in urine (Watson and Stewart, 1977b). Both assays are similar to the test of Vandemark et al. (1978).

3. Gas Chromatography—Bailey and Jatlow (1976), Dorrity et al. (1977)

Discussion. Bailey and Jatlow (1976) describe a sensitive plasma assay (5 ng/ml) of imipramine and desipramine, using GC with nitrogen-sensitive flame ionization detection. Desipramine is measured as its *N*-trifluoroacetyl derivative. Dorrity et al. (1977) present a similar assay suitable for therapeutic drug level monitoring of imipramine, amitriptyline, doxepine, and their corresponding *N*-desmethyl analogs. The secondary amines are analyzed without derivatization. The assay has a sensitivity of 10 ng drug/ml, using a 2 ml plasma sample.

4. Gas Chromatography-Mass Spectrometry—Belvedere et al. (1975), Biggs et al. (1976), Claeys et al. (1976), Dubois et al. (1976)

Discussion. These methods are similar to the N-FID–GC assays except for the use of SIM-MS as the detector system. Mass spectrometry is performed with chemical ionization (Claeys et al., 1976) or electron impact ionization (all others). The methods of Biggs et al. (1976) and Dubois et al. (1976) include several other tricyclic antidepressants in addition to imipramine. The sensitivity for these drugs ranges from 0.3 ng/ml plasma (Dubois et al., 1976) to about 10 ng/ml for the other GC-MS procedures, largely depending on the mass spectral conditions. All methods require acylation of the secondary amine antidepressants, including acetylation (Belvedere et al., 1975), trifluoroacetylation (Biggs et al., 1976; Claeys et al., 1976), and pentafluoropropylation (1976). Deuterated drugs serve as internal standards.

5. Field Ionization Mass Spectrometry—Heck et al. (1978)

Discussion. Imipramine is measured by direct insertion field ionization MS with hexadeuterated imipramine as internal standard. The field ionization MS method produces abundant molecular ions with little fragmentation, thus increasing the sensitivity and specificity of this technique in biological samples as compared to electron impact or chemical ionization MS. However, extensive prepurification of the plasma sample is needed, including a HPLC step, since field ionization MS at present cannot be utilized in the GC-MS mode. Sensitivity limit is about 1 ng imipramine/ml plasma, using a 4 ml sample. No advantage over GC-MS assays is apparent.

6. Thin-Layer Chromatography—Densitometry—Nagy and Treiber (1973)

Discussion. Following extraction from plasma and TLC separation, imipramine and desipramine are oxidized with a spray reagent and analyzed by densitometry. The method yields excellent sensitivity with a C.V. of 11 to 16% at 20 ng drug/ml plasma.

7. Radioimmunoassay—Hubbard et al. (1978), Maguire et al. (1978)

Discussion. The antiserum is directed against a nortriptyline-bovine serum albumin conjugate (Maguire et al., 1978). ^{3}H-Amitriptyline serves as the tracer. This RIA, although developed for the detection of nortriptylin, is also suitable for the other tricyclic antidepressants because of significant cross-reactivity of these drugs to the antibody. Thus specificity of the assay is poor.

The systematic development of drug haptens, including amitriptyline-nortriptyline and imipramine-desipramine, has been described by Hubbard et al. (1978). Antibodies directed against these haptens, conjugated to proteins, were found to be suitable for the sensitive RIA detection of these agents in serum. (See also chlorpromazine monograph.)

8. Comment. Although HPLC, GC, and GC-MS procedures are all useful for monitoring therapeutic levels of the tricyclic antidepressants, HPLC offers the most convenient approach at present. The tricyclic antidepressants also represent a good example for the application of GC-MS because of their low plasma levels and lack of sensitive GC-EC properties. The advances in HPLC and GC–N-FID, however, make quantitative MS obsolete because of its high cost, except for special applications in conjunction with stable isotope tracer techniques (Heck et al., 1978).

e. References

Alexanderson, B., Pharmacokinetics of Desmethylimipramine and Nortriptyline in Man after Single and Multiple Oral Doses—a Cross-Over Study, *Eur. J. Clin. Pharmacol.*, **5**, 1-10 (1972).

Bailey, N. and P. I. Jatlow, Gas-Chromatographic Analysis for Therapeutic Concentrations of Imipramine and Desipramine in Plasma, with Use of a Nitrogen Detector, *Clin. Chem.*, **22**, 1697–1701 (1976).

Belvedere, G., L. Burti, A. Frigerio, and C. Pantorotto, Gas Chromatographic-Mass Fragmentographic Determination of Steady-State Plasma Levels of Imipramine and Desipramine in Chronically Treated Patients, *J. Chromatogr.*, **111**, 313–321 (1975).

Biggs, J. T., W. H. Holland, S. Chang, P. P. Hipps, and W. R. Sherman, Electron Beam Ionization Mass Fragmentographic Analysis of Tricyclic Antidepressants in Human Plasma, *J. Pharm. Sci.*, **65**, 261–268 (1976).

Claeys, M., G. Muscettola, and S. P. Markey, Simultaneous Measurement of Imipramine and Desipramine by Selected Ion Recording with Deuterated Internal Standards, *Biomed. Mass Spectrom.*, **3**, 110–116 (1976).

Dorrity, Jr., F., M. Linnoila, and R. L. Habig, Therapeutic Monitoring of Tricyclic Antidepressants in Plasma by Gas Chromatography, *Clin. Chem.*, **23**, 1326–1328 (1977).

Dubois, J. P., W. Kund, W. Theobald, and B. Wirz, Measurement of Clomipramine, *N*-Desmethylclomipramine, Imipramine and Dihydroimipramine in Biological Fluids by Selective Ion Monitoring, and Pharmacokinetics of Clomipramine, *Clin. Chem.*, **22**, 892–897 (1976).

Gram, L. F. et al., Steady-State Kinetics of Imipramine in Patients, *Psychopharmacologia*, **54**, 255–261 (1977).

Heck, H. A., N. W. Flynn, S. E. Buttrill, R. L. Dyer, and M. Anbar, Determination of Imipramine in Plasma by High-Pressure Liquid Chromatography and Field Ionization Mass Spectrometry: Increased Sensitivity in Comparison with Gas Chromatography Mass Spectrometry, *Biomed. Mass. Spectrom.*, **5**, 250–259 (1978).

Hubbard, J. W., K. K. Midha, J. K. Cooper, and C. Charette, Radioimmunoassay for Psychotropic Drugs. II: Synthesis and Properties of Haptens for Tricyclic Antidepressants, *J. Pharm. Sci.*, **67**, 1571–1578 (1978).

Maguire, K. P., G. D. Burrows, T. R. Norman, and B. A. Scoggins, A Radioimmunoassay for Nortriptyline (and Other Tricyclic Depressants) in Plasma, *Clin. Chem.*, **24**, 549–554 (1978).

Nagy, A. and L. Treiber, Quantitative Determination of Imipramine and Desipramine in Human Blood Plasma by Direct Densitometry of Thin-Layer Chromatograms, *J. Pharm. Pharmacol.*, **25**, 599–603 (1973).

Vandemark, F. L., R. F. Adams, and G. J. Schmidt, Liquid-Chromatographic Procedure for Tricyclic Drugs and Their Metabolites in Plasma, *Clin. Chem.*, **24**, 87–91 (1978).

Watson, I. D. and M. J. Stewart, Quantitative Determination of Amitriptyline in Plasma by High-Performance Liquid Chromatography, *J. Chromatogr.*, **132**, 155–159 (1977a).

Watson, I. D. and M. J. Stewart, Assay of Tricyclic Structured Drugs and Their Metabolites in Urine by High-Performance Liquid Chromatography, *J. Chromatogr.*, **134**, 182–186 (1977b).

INDOMETHACIN—by D. Smith

1-(*p*-Chlorobenzoyl)-5-methoxy-2-methylindole-3-acetic acid.
Indocin. Anti-inflammatory, antipyretic, and analgesic.

$C_{19}H_{16}ClNO_4$; mol. wt. 357.81; pK_a 4.5.

a. Therapeutic Concentration Range. The plasma levels of indomethacin in patients receiving continuous treatment are between 0.5 and 3 μg/ml during the 4 to 5 hr after the last dose of 25 mg (Hvidberg et al., 1972).

b. Metabolism. The metabolites of indomethacin are *N*-deschlorobenzoylindomethacin, *O*-desmethylindomethacin, *O*-desmethyl-*N*-deschlorobenzoylindomethacin, and their respective glucuronides (Kwan et al., 1976; Duggan et al., 1972), all of which are devoid of anti-inflammatory activity (Duggan et al., 1972). The plasma elimination half-life in the beta phase ranges from 2.6 to 11.2 hr (Alvan et al., 1975).

c. Analogous Compounds. Aromatic carboxylic acid, indoles.

d. Analytical Methods

1. High Performance Liquid Chromatography—Lin and Benet (1977)

Discussion. The assay for indomethacin requires a reverse phase column with UV detection at 260 nm. Chlorpromazine is used as the internal standard, and plasma samples are deproteinated with acetonitrile prior to analysis. Standard curves run from 0.2 to 2 μg/ml give an r^2 (mean $\pm$ SD) of .9884 $\pm$ .0126 (n = 8), using peak height ratios.

Detailed Method. First, 0.2 ml plasma is deproteinated with 0.4 ml CH$_3$CN containing 1 μg chlorpromazine as the internal standard. The mixture is then vortexed for 1 min and centrifuged for 10 min at 4000 g. The resultant supernatant is evaporated to about 0.15 ml under nitrogen, and 20 to 30 μl injected into the column.

A Perkin-Elmer liquid chromatograph Series 3 with a LC-65T variable-wavelength detector at 260 nm is used, equipped with an analytical μ-Bondapak C$_{18}$ column (Waters Assoc.). The eluent is 70% methanol in 0.01 M sodium acetate (pH 3.2) with a flow rate of 2 ml/min. The retention times are 5.5 and 7 min for chlorpromazine and indomethacin, respectively.

2. Other HPLC Method—Skellern and Salole (1975)

Discussion. This method is similar to that described by Lin and Benet (1977) except for flufenamic acid as the internal standard and an ether extraction prior to HPLC analysis. The sensitivity is 0.1 μg indomethacin/ml plasma.

3. Gas Chromatography—Helleberg (1976), Moller Jensen (1978), Sibeon et al. (1978)

Discussion. Serum and urine samples are extracted at pH 5.0 with 1,2-dichloroethane and reacted with a solution of diazoethane in diethyl ether (Helleberg, 1976). Ethylation, in lieu of the more common methylation, avoids potential assay interference by the O-desmethyl metabolite. The ethyl ester derivative is then quantitated by GC with an electron capture detector. This method is sensitive (50 ng/ml), reproducible (SD 3%), and specific in the absence of griseofulvin and chlorcyclizin, which interfere with the analysis. The similar assay of Moller Jensen (1978) utilizes extractive alkylation to produce the ethyl ester of indomethacin.

Sibeon et al. (1978) describe a GC-EC assay of indomethacin following plasma extraction into dichloromethane and derivatization with pentafluorobenzylbromide. The sensitivity is 10 ng indomethacin/ml plasma, using 5-fluoroindomethacin as the internal standard.

4. Gas Chromatography-Mass Spectrometry—Palmer et al. (1974)

Discussion. Plasma samples containing the 5-fluoro analog of indomethacin as the internal standard are extracted into ethyl acetate and analyzed by electron impact GC-MS in the selected ion monitoring mode (m/e 139). This method is sensitive (25 ng/ml), accurate, and precise (S.D. for duplicate determinations, 7% mean). Specificity is questionable, however, since the derivatization for GC involves methylation with diazomethane, which yields the same product from indomethacin and its O-desmethyl metabolite.

5. Fluorimetry—Hucker et al. (1966), Hvidberg et al. (1972)

Discussion. Both methods involve a heptane extraction of plasma and measurement of the native fluorescence of indomethacin in the organic layer. Furosemide and aspirin interfere with the assay and have to be removed by additional extractions (Hucker et al., 1966) or chromatography (Hvidberg et al., 1971). Phenylbutazone, oxyphenbutazone, diazepam, chlordiazepoxide, and three common barbiturates do not interfere with the analysis (Hvidberg et al., 1971).

The fluorescence assays are nonspecific with respect to metabolites of indomethacin. Sensitivity is 0.1 μg/ml.

6. Radioimmunoassay—Hare et al. (1977)

Discussion. The antiserum is raised in rabbits against serum albumin and indomethacin, conjugated via its free carboxylic acid function. The conjugation reaction uses N-hydroxysuccinimide, forming the succinate amide bridge and coupling to protein with a carbodiimide reagent. The sensitivity is 50 ng indomethacin/ml plasma.

The O-desmethyl metabolite has a cross-affinity of 12% to the antibody, and the glucuronide of indomethacin is three times as reactive. The glucuronide interference necessitates an additional extraction step in urine samples in order to separate indomethacin from excess glucuronide.

7. Comments. It is clear that HPLC represents the method of choice in many applications, although GC using alkylation other than methylation is also a versatile analytical technique for monitoring indomethacin in plasma.

e. References

Alvan, G., M. Orme, L. Bertilsson, R. Ekstrand, and L. Palmer, Pharmacokinetics of Indomethacin, *Clin. Pharmacol. Ther.*, **18**, 364–373 (1975).

Duggan, D. E., A. F. Hogans, K. C. Kwan, and F. G. McMahon, The Metabolism of Indomethacin in Man, *J. Pharmacol. Exp. Ther.*, **181**, 563–575 (1972).

Hare, L. E., C. A. Ditzler, and D. E. Duggan, Radioimmunoassay of Indomethacin in Biological Fluids, *J. Pharm. Sci.*, **66**, 486–489 (1977).

Helleberg, L., Determination of Indomethacin in Serum and Urine by Electron-Capture Gas-Liquid Chromatography, *J. Chromatogr.*, **117**, 167–173 (1976).

Hucker, H. B., A. G. Zacchei, S. V. Cox, D. A. Brodie, and N. H. R. Cantwell, Studies on the Absorption, Distribution, and Excretion of Indomethacin in Various Species. *J. Pharmacol. Exp. Ther.*, **153**, 237–249 (1966).

Hvidberg, E., H. H. Lausen, and J. A. Jansen, Indomethacin: Plasma Concentrations and Protein Binding in Man, *Eur. J. Clin. Pharmacol.*, **4**, 119–124 (1972).

Kwan, K. C., G. O. Breault, E. R. Umbenhauer, F. G. McMahon, and D. E. Duggan, Kinetics of Indomethacin Absorption, Elimination, and Enterohepatic Circulation in Man, *J. Pharmacokin, Biopharm.*, **4**, 255–280 (1976).

Lin, E. and L. Z. Benet, HPLC Assay for Indomethacin, submitted for Protocol in Drug Studies Unit, School of Pharmacy, University of California—San Francisco (1977).

Moller Jensen, K., Determination of Indomethacin in Serum by an Extractive Alkylation Technique and Gas-Liquid Chromatography, *J. Chromatogr.*, **153**, 195–202 (1978).

Palmer, L. et al., Indomethacin: Quantitative Determination in Plasma by Mass Fragmentography Including Pilot Pharmacokinetics in Man, in *Prostaglandin Synthetase Inhibitors*, H. J. Robinson and J. R. Vane, (Eds.). Raven Press, New York, 1974, p. 91 ff.

Sibeon, R. G., J. D. Baty, N. Baber, K. Chan, and M. L'E. Orme, Quantitative Gas-Liquid Chromatographic Method for the Determination of Indomethacin in Biological Fluids, *J. Chromatogr.*, **153**, 189–194 (1978).

Skellern, G. G. and E. G. Salole, A High-Speed Liquid Chromatographic Analysis of Indomethacin in Plasma, *J. Chromatogr.*, **114**, 483–485 (1975).

ISONIAZID

Isonicotinic acid hydrazide.
INH, Isocid, and so on. Antitubercular agent.

$C_6H_7N_3O$; mol. wt. 137.15; pK_a 3.6.

a. Therapeutic Concentration Range. Intravenous administration of 10 mg isoniazid/kg results in isoniazid plasma levels of about 10 μg/ml 1 hr after the dose (Boxenbaum and Riegelman, 1974).

b. Metabolism. Isoniazide is metabolized to a host of derivatives that are mostly inactive as antitubercular agents. These include *N*-acetylisoniazid, spontaneously formed pyruvate and α-ketoglutarate hydrazones, isonicotinic acid, isonicotinylglycine, acetylhydrazine, and diacetylhydrazine (Boxenbaum and Riegelman, 1974; Russel, 1972; Timbrell et al., 1977a). Moreover, isoniazid may displace nicotinamide in the cofactor, nicotinamide adenine dinucleotide (Diaugustine, 1976). A liver isoniazid *N*-acetyltransferase is responsible for the major metabolic pathway to *N*-acetylisoniazid (Hearse and Weber, 1973). The activity of this enzyme is bimodally distributed, dividing the population into slow and fast isoniazid acetylators. The distribution within the general population of the United States is roughly equal, while either phenotype may dominate in specific ethnic groups (e.g., Eidus et al., 1971). The plasma elimination half-life of isoniazid is 45 to 80 min in fast acetylators and 140 to 200 min in slow acetylators (Boxenbaum and Riegelman, 1974). The polymorphic acetylation has several important clinical consequences for isoniazid and other basic drugs, such as procainamide, hydralazine, phenelzine, and salicylazosulfapyridine, which are metabolized by the same isoniazid N-acetyltransferase (Drayer and Reidenberg, 1977). All of these drugs induce a higher rate of side effects in slow acetylator individuals; side effects include systemic lupus erythematosus (isoniazid, procainamide, hydralazine), drowsiness and nausea (phenelzine), cyanosis and hemolysis (salicylazosulfapyridine), and polyneuropathy (isoniazid) (Drayer and Reidenberg, 1977). Furthermore, spontaneous lupus erythematosus occurs predominantly in slow acetylators of isoniazid (Larsson et al., 1977). Finally, phenytoin intoxication caused by concomitant administration of isoniazid, which inhibits phenytoin metabolism, is associated with the slow acetylator phenotype (Brennan et al., 1970).

However, the major metabolite, acetylisoniazid, is a metabolic precursor to the potent hepatotoxin acetylhydrazine (Timbrell et al., 1977a) and may be a lung carcinogen in mice (Toth and Shimizu, 1973). Mitchell et al. (1975a, 1975b) have demonstrated that 10 to 20% of patients ingesting isoniazid for preventive therapy develop liver injury as measured by routine liver function tests. In some cases the abnormal liver functions return to normal without known cause, whereas other patients, mostly rapid acetylators, develop clinically overt hepatitis. Urinary excretion of acetylhydrazine after an isoniazid dose is $1.8 \pm 0.4\%$ of the dose in slow acetylators and $2.5 \pm 0.5\%$ in rapid acetylators; the corresponding figures for diacetylhydrazine are $4.9 \pm 0.9\%$ and $23 \pm 2\%$, respectively (Timbrell et al., 1977b). Nelson et al. (1976) have proved that acetylhydrazine is activated by hepatic cytochrome P-450-catalyzed oxidations to potent acylating and alkylating species, which covalently bind to macromolecules as the presumed cause of hepatic necrosis.

Consequently, the analytical chemistry of *in vivo* isoniazid disposition has focused not only on the many metabolites, but also on the determination of isoniazid and N-acetylisoniazid as a test for the acetylator phenotype of individual patients. Knowledge of the individual phenotype may help in therapeutic decisions. For instance, the administration of isoniazid once weekly benefits slow acetylators but is of marginal value for rapid acetylators (Drayer and Reidenberg, 1977). The following sections review the assay techniques relevant to these biomedical problems.

c. **Analogous Compounds.** Hydrazides and hydrazines: iproniazid, procarbazine, and so on; nicotinamide.

d. **Analytical Methods**

1. **Phenotyping of Isoniazid Inactivators**—Eidus and Hodgkin (1973)

Discussion. Slow and rapid isoniazid acetylators can be differentiated by the relative amounts of isoniazid and *N*-acetylisoniazid excreted into the urine. Urine samples are obtained between 6 and 8 hr after an oral 10 mg/kg isoniazid dose in order to maximize the differences between the two groups. Inactivation indices (I.I.s) are obtained from the ratio of *N*-acetylisoniazid to isoniazid urine concentrations. Patients yielding I.I.s of 3.0 or less are considered slow inactivators; the I.I. values of fast acetylators are higher than 5.0. The assay is based on color formation of *N*-acetylisoniazid, but not isoniazid, with potassium cyanide, chloramine-T, and acetone. Isoniazid is converted in a separate urine sample to *N*-acetylisoniazid with acetic anhydride and also measured colorimetrically; color intensity, then, represents the sum of parent drug and metabolite. Urine samples are pretreated with acid in order to reconvert hydrazone metabolites to isoniazid. Without this hydrolysis step, results may be erratic. The color intensity can be measured either by spectrophotometer or by simple visual inspection. In slow acetylators the contrast between acetic anhydride treated and untreated samples is marked, whereas in fast acetylators this difference is insignificant because of much higher *N*-acetylisoniazid than isoniazid concentrations. The assay is simple and reliable in differentiating the two acetylator phenotypes.

Detailed Method. The patient receives an oral dose of 10 mg isoniazid/kg, and the urine is voided 6 hr after the dose. Another urine sample is collected between 6 and 8 hr for analysis. A 4 ml aliquot is acidified with 2 ml 0.5 *N* HCl and kept at room temperature for 15 min in order to hydrolyze isoniazid hydrazones. Of this solution, two aliquots of 1.5 ml each are transferred to separate test tubes, and one of the samples is thoroughly mixed with 1 drop of acetic anhydride for 1 min to quantitatively convert isoniazid to *N*-acetylisoniazid. This is followed by 1 drop of 1 *N* NaOH. To the other sample, 2 drops of distilled water are added, and each samples is then mixed with 0.5 *N* NaOH and diluted with 5 volumes of water. Optimal dilution for the color test may be determined by a spotting test on a white tile, using the color reagents described below.

An aliquot of 2 ml of each of these solutions is mixed with the following reagents in consecutive order: (a) 1 ml 0.5 M potassium phosphate (pH 6) buffer; (b) 1 ml freshly prepared 20% aqueous potassium cyanide solution; (c) 4 ml freshly prepared 12.5% chloramine-T solution (mix and wait for 1.5 min); (d) 5 ml acetone. Absorbance is measured at 550 nm, or the color intensity estimated visually using a Hellige Comparator without color disks and cells of 26 mm viewing depth.

2. Other Isoniazid Acetylator Phenotyping Tests—Eidus and Hodgkin (1977), Eidus et al. (1971), Raghupati et al. (1976), Russell and Eidus (1972), Vallon et al. (1976), Varughese et al. (1974)

Discussion. Although some investigators prefer other test substances, such as dapsone and sulfa drugs, which are also subject to acetylation by isoniazid N-acetyltransferase (Reidenberg and Martin, 1974; Schröder, 1972), most clinical tests are still performed with isoniazid and are based on the same principles as outlined in Section d1. Most assays employ color reactions (see Section d3) of isoniazid and N-acetylisoniazid and can be automated (Varughese et al., 1974). Vallon et al. (1976) utilize alternate current polarography, which permits the simultaneous detection of isoniazid and N-acetylisoniazid in diluted urine by superimposed sinusoidal tension polarography.

3. Colorimetry—Boxenbaum and Riegelman (1974), Devani et al. (1978), Dymond and Russell (1970), Eidus and Harnanansingh (1971), Ellard et al. (1972), Eswara Dutt and Mottola (1977), Peters et al. (1965), Russell (1971), Shah and Raje (1977)

Discussion. The chemical reactivity of isoniazid permits the application of many color reagents, such as vanillin, *p*-dimethylaminobenzaldehyde (Ehrlich's reagent), *trans*-cinnamaldehyde, 4-nitrobenzaldehyde, and pyridoxal to give hydrazones (Boxenbaum and Riegelman, 1974; Eidus and Harnanansingh, 1971; Shah and Raje, 1977), 2,4,6-trinitrobenzenesulfonic acid (Dumond and Russell, 1970), vanadium (v) (Eswara Dutt and Mottola, 1977), and 2,3-dichloro-1,4-naphthoquinone with ammonia (Devani et al., 1978). The pyridine ring can be assayed by a color reaction based on polymethine dye formation with cyanide, chloramine T, and C-H active compounds, for example, acetone (see Section d1) or barbituric acid (Boxenbaum and Riegelman, 1974). Employing a combination of these color reactions, together with acid hydrolysis of N-acetylisoniazid to give either hydrazine or isoniazid and with ion-exchange chromatography, it is possible to measure separately the plasma and urine concentrations of isoniazid and its metabolites,

N-acetylisoniazid, isonicotinic acid, isonicotinylglycine, acetylhydrazine, and diacetylhydrazine (Boxenbaum and Riegelman, 1974; Ellard et al., 1972; Peters et al., 1965).

4. Fluorescence—Miceli et al. (1975), Olson et al. (1977), Scott and Wright (1967), Wilson et al. (1973)

Discussion. Isoniazid fluorescence can be induced by reactions with pentanedione (Wilson et al., 1973) and salicylaldehyde, followed by reduction of the hydrazone with mercaptoethanol and quenching of salicylaldehyde fluorescence with bisulfite (Scott and Wright, 1967). The latter method has a sensitivity of 0.01 µg isoniazid/ml serum. Miceli et al. (1975) introduce ascorbic acid in lieu of mercaptoethanol as the reducing agent. The same technique can be applied to *N*-acetylisoniazid by acid hydrolysis to isoniazid (Olson et al. 1977). Free isoniazid is first inactivated by conversion to the nonfluorescent azide with sodium nitrate and acid. The method of Olson et al. (1977) has been scaled down for application in pediatric studies.

5. Gas Chromatography—Frater-Schroder and Zbinden (1975), Timbrell et al. (1977b)

Discussion. Timbrell et al. (1977b) propose a GC–N-FID assay of isoniazid and some of its metabolites based on the reaction of monoacylhydrazines with *p*-chlorobenzaldehyde. The resulting hydrazones are extracted into an organic solvent and subjected to GC analysis. The diacylhydrazines, *N*-acetylisoniazid and diacetylhydrazine, are first converted by acid hydrolysis to isoniazid and monoacetylhydrazine, respectively, and then derivatized and determined by GC.

Both isoniazid and *N*-acetylisoniazid react with *N,O*-bis(trimethylsilyl)trifluoroacetamide containing 1% trimethylchlorosilan to yield bis(trimethylsilyl) derivatives that can be analyzed by GC-FID, using OV-17 as the stationary phase. Urine samples are extracted with isoamyl alcohol-dichloromethane in the presence of ammonium sulfate and NaOH. With eicosan as the internal standard, sensitivity is on the order of 20 µg/ml plasma or urine (Frater-Schroder and Zbinden, 1975).

6. High Performance Liquid Chromatography—Bailey and Abdou (1977), Saxena et al. (1977), Stewart et al. (1976)

Discussion. Application of HPLC to the analysis of isoniazid in dosage forms has been reported, using a pellicular silica gel column with UV detection at 254 nm (Bailey and Abdou, 1977) and a reverse phase octadecyl column with UV detection at 293 nm (Stewart et al., 1976). Detection at 293 nm provides significantly greater sensitivity. Both methods are suitable for the separation of isoniazid from degradation products and condensation products with tablet ingredients such as lactose, which may impair its bioavailability. The reverse phase method, employing ion-pair chromatography with dioctyl sodium sulfosuccinate at pH 2.5 (column stability questionable), is directly applicable to the measurement of isoniazid and *N*-acetylisoniazid in plasma and urine samples, giving a precision of 5% for isoniazid and 7.5% for *N*-acetylisoniazid (Saxena et al., 1977).

7. Radioimmunoassay—Schwenk et al. (1975)

Discussion. The antibodies are raised against isoniazid coupled to human serum albumin. *N*-Acetylisoniazid, isonicotinic acid, and niacinamide have antibody cross-affinities of 2, 1, and below 1%, respectively. The assay procedure employs ^{3}H-isoniazid as the tracer and ammonium sulfate precipitation of antibody-bound ^{3}H activity; it is sensitive to 50 ng isoniazid/ml plasma.

8. Microbiological Assays—Bilyk and Dudchik (1977), Le Lirzin et al. (1971), Reiss et al. (1967)

Discussion. These assays are capable of measuring the active concentration of isoniazid in serum samples. Reiss et al. (1967) compare their microbiological assay to a fluorescence assay. Bilyk and Dudchik (1977) measure the protein-bound and non-protein-bound active isoniazid fraction in blood. Microbiological assays are subject to interferences by other antimicrobials active against the test organism.

9. Drug Stability in Plasma Samples—Huffman and Dujovne (1976), Olson et al. (1977)

Discussion. Huffman and Dujovne (1976) note that isoniazid and *N*-acetylisoniazid are unstable in plasma even when kept frozen. Ethylenediamine tetraacetatic acid should not be used as the anticoagulant (Olson et al. 1977). Storage of plasma samples at 4°C preserves intact isoniazid over a period of 1 month, but *N*-acetylisoniazid deteriorates appreciably after 24 hr unless the samples are deproteinized (Olson et al., 1977).

10. Comments.
Colorimetric assays prevail in most laboratories because of their simplicity. However, sophisticated studies of isoniazid

disposition require more specific and sensitive assays. High perform-ance liquid chromatography with electrochemical detection represents a yet unexplored approach to the sensitive analysis of underivatized isoniazid and its metabolites.

e. References

Bailey, L. C. and H. Abdou, High-Performance Liquid Chromatographic Analysis of Isoniazid and Its Dosage Forms, *J. Pharm. Sci.*, **66**, 564–567 (1977).

Bilyk, M. A. and G. K. H. Dudchik, Microbiologic Method of Separate Determination of Protein-Bound and Non-Protein-Bound Active INH Frac-tion in the Blood of Tuberculosis Patients During Treatment with Isoniazid, *Lab. Delo*, 11, 669–671 (1977).

Boxenbaum, H. G. and S. Riegelman, Determination of Isoniazid and Metabolites in Biological Fluids, *J. Pharm. Sci.*, **63**, 1191–1197 (1974).

Brennan, R. W., H. Dehejia, H. Kutt, K. Verebely, and F. McDowell, Diphenylhydantoin Intoxication Attendant to Slow Inactivation of Isoniazid, *Neurology*, **20**, 687–693 (1970).

Devani, M. B., C. J. Shishoo, M. A. Patel, and D. D. Bhalara, Spectro-photometric Determination of Isoniazid in Presence of Its Hydrazones, *J. Pharm. Sci.*, **67**, 661–663 (1978).

Diaugustine, R. P., Formation *in vitro* and *in vivo* of the Isonicotinic Acid Hydrazide Analogue of Nicotinamide Adenine Dinucleotide by Lung Nico-tinamide Adenine Dinucleotide Glycohydrolase, *Mol. Pharmacol.*, **12**, 291–298 (1976).

Drayer, D. E. and M. M. Reidenberg, Clinical Consequences of Polymorphic Acetylation of Basic Drugs, *Clin. Pharmacol. Ther.*, **22**, 251–258 (1977).

Dymond, L. C. and D. W. Russell, Rapid Determination of Isonicotinic Acid Hydrazide in Whole Blood with 2,4,6-Trinitrobenzenesulphonic Acid, *Clin. Chim. Acta*, **27**, 513–520 (1970).

Eidus, L. and A. M. Harnanansingh, A More Sensitive Spectrophotometric Method for Determination of Isoniazid in Serum or Plasma, *Clin. Chem.*, **17**, 492–494 (1971).

Eidus, L. and M. M. Hodgkin, Simplified Screening Test for Phenotyping of Isoniazid Inactivators, *Int. J. Clin. Pharmacol.*, **7**, 82–86 (1973).

Eidus, L. and M. M. Hodgkin, Letter: Acetylation of Isoniazid, *J. Antimicrob. Chemother.*, **3**, 626–627 (1977).

Eidus, L., A. M. Harnanansingh, and A. G. Jessamine, Urine Test for Phenotyping Isoniazid Inactivators, *Am. Rev. Respir. Dis.*, **104**, 587–591 (1971).

Ellard, G. A., P. T. Gammon, and S. M. Wallace, The Determination of Isoniazid and Its Metabolites Acetylisoniazid, Monoacetylhydrazine,

Diacetylhydrazine, Isonicotinic Acid and Isonicotinylglycine in Serum and Urine, *Biochem. J.*, **126**, 449–458 (1972).

Eswara Dutt, V. V., and H. A. Mottola, Repetitive Determination of Isonicotinic Acid Hydrazide in Flow-Through Systems by Series Reactions, *Anal. Chem.*, 776–779 (1977).

Frater-Schroder, M. and G. Zbinden, A Specific Rapid Gas Chromatographic Assay for the Determination of Isoniazid *N*-Acetylation: Observation in Rats with Induced Constant Urine Flow, *Biochem. Med.*, **14**, 274–284 (1975).

Hearse, D. J. and W. W. Weber, Multiple *N*-Acetyltransferases and Drug Metabolism: Tissue Distribution, Characterization and Significance of Mammalian *N*-Acetyltransferase, *Biochem. J.*, **132**, 519–526 (1973).

Huffman, D. H. and C. A. Dujovne, The Instability of Isoniazid in Frozen Plasma, *Res. Commun. Chem. Pathol. Pharmacol.*, **15**, 203–204 (1976).

Larsson, A., E. Karlsson, and L. Molin, Spontaneous Systemic Lupus Erythematosus and Acylator Phenotype, *Acta Med. Scand.*, **201**, 223–226 (1977).

Le Lirzin, M., J. N. Vivien, A. Lepeuple, R. Thibier, and C. Pretet, Rapid Microbiological Determination of Serum Isoniazid, *Rev. Tuberculol. Pneumol.*, **35**, 350–356 (1971).

Miceli, N. J., W. A. Olson, and W. W. Weber, An Improved Micro Spectrofluorometric Assay for Determining Isoniazid in Serum, *Biochem. Med.*, **12**, 348–355 (1975).

Mitchell, J. R., M. W. Long, U. P. Thorgeirsson, and D. J. Jollow, Acetylation Rates and Monthly Liver Function Tests During One Year of Isoniazid Preventive Therapy, *Chest*, **68**, 181–190 (1975a).

Mitchell, J. R. et al., Increased Incidence of Isoniazid Hepatitis in Rapid Acetylators: Possible Relation to Hydrazine Metabolites, *Clin. Pharmacol. Ther.*, **18**, 70–79 (1975b).

Nelson, S. D., J. R. Mitchell, J. A. Timbrell, W. R. Snodgrass, and G. B. Corcoran, Isoniazid and Iproniazid: Activation of Metabolites to Toxic Intermediates in Man and Rat, *Science*, **193**, 901–903 (1976).

Olson, W. A., P. G. Dayton, Z. H. Israili, and A. W. Pruitt, Spectrophotofluorometric Assay for Isoniazid and *N*-Acetylisoniazid in Plasma Adapted to Pediatric Studies, *Clin. Chem.*, **23**, 745–748 (1977).

Peters, J. H., K. S. Miller, and P. Brown, The Determination of Isoniazid and Its Metabolites in Human Urine, *Anal. Biochem.*, **12**, 379–394 (1965).

Raghupati, S. G. et al., Classification of Subjects as Slow or Rapid Inactivators of Isoniazid Based on the Ratio of Acetylisoniazid in Urine Determined by a Simple Colorimetric Method, *Indian J. Med. Res.*, **64**, 1456–1461 (1976).

Reidenberg, M. M. and J. H. Martin, The Acetylator Phenotype of Patients with Systemic Lupus Erythematosus, *Drug Metab. Dispos.*, **2**, 71–73 (1974).

Reiss, O. K., W. C. Morse, and R. W. Putsch, A Chemical Determination of Isoniazid in Serum, *Am. Rev. Respir. Dis.*, **96,** 111–114 (1967).

Russell, D. W., Determination of Isonicotinic Acid Hydrazide in Urine, *Clin. Chim. Acta*, **31,** 367–373 (1971).

Russell, D. W., Low Circulating Levels of Acid-Labile Hydrazones after Oral Administration of Isonicotinic Acid Hydrazide, *Clin. Chim. Acta*, **41,** 163–168 (1972).

Russell, D. W. and L. Eidus, Simplified Isoniazid Acetylator Phenotyping, *Can. Med. Assoc. J.*, **106,** 1155–1156 (1972).

Saxena, S. J., J. T. Stewart, I. L. Honigberg, J. G. Washington, and G. R. Keene, Liquid Chromatography in Pharmaceutical Analysis. VIII: Determination of Isoniazid and Acetyl Derivative in Plasma and Urine Samples, *J. Pharm. Sci.*, **66,** 813–816 (1977).

Schröder, H., Simplified Method for Determining Acetylator Phenotype, *Br. Med. J.*, **3,** 506–507 (1972).

Schwenk, R., K. Kelly, K. S. Tse, and A. H. Sehon, A Radioimmunoassay for Isoniazid, *Clin. Chem.*, **21,** 1059–1062 (1975).

Scott, E. M. and R. C. Wright, Fluorometric Determination of Isonicotinic Acid Hydrazide in Serum, *J. Lab. Clin. Med.*, **70,** 355–360 (1967).

Shah, P. R. and R. R. Raje, Hydrazones of Isoniazid for Colorimetric Analysis, *J. Pharm. Sci.*, **66,** 291–292 (1977).

Stewart, J. T. et al., Liquid Chromatography in Pharmaceutical Analysis V: Determination of Isoniazid-Pyridoxine Hydrochloride Mixture, *J. Pharm. Sci.*, **65,** 1536–1539 (1976).

Timbrell, J. A., J. M. Wright, and T. A. Baillie, Monoacetylhydrazine as a Metabolite of Isoniazid in Man, *Clin. Pharmacol. Ther.*, **22,** 602–608 (1977a).

Timbrell, J. A., J. M. Wright, and C. M. Smith, Determination of Hydrazine Metabolites of Isoniazid in Human Urine by Gas Chromatography, *J. Chromatogr.*, **138,** 165–172 (1977b).

Toth, B. and H. Shimizu, Lung Carcinogenesis with 1-Acetyl-2-isonicotinoylhydrazine, the Major Metabolite of Isoniazid, *Eur. J. Cancer*, **9,** 285–289 (1973).

Vallon, J. J., A. Badinand, and C. Bichon, Determination of Acetylator Phenotype of Isoniazid by Superimposed Sinusoidal Tension Polarography, *Eur. J. Toxicol. Environ. Hyg.*, **9,** 155–164 (1976).

Varughese, P., E. J. Hamilton, and L. Eidus, Mass Phenotyping of Isoniazid Inactivators by Automated Determination of Acetylisoniazid in Urine, *Clin. Chem.*, **20,** 639–641 (1974).

Wilson, D. M., M. Lever, and C. W. Small, An Evaluation of the Fluorometric Determination of Serum Isoniazid with Pentanedione, *Am. J. Med. Technol.*, **39,** 451–453 (1973).

ISOSORBIDE DINITRATE

1,4:3,6-Dianhydro-D-glucitol dinitrate.
Carvanil, Isoket, Isordil, and so on. Coronary vasodilator.

$C_6H_8N_2O_8$; mol. wt. 236.14.

a. Therapeutic Concentration Range. Sublingual administration of 5 mg isosorbide dinitrate yields peak plasma levels of 9 ± 3 (S.D.) ng parent drug/ml after 15 min, while the same peroral dose produces peak levels of 2 to 5 ng/ml (Assinder et al., 1977; Chasseaud et al., 1975).

b. Metabolism. The principal metabolic products are the two mononitrate derivatives, 2-mononitrate and 5-mononitrate isosorbide (Assinder et al. 1977; Chasseaud et al. 1975). Both metabolites accumulate to plasma levels substantially higher than the concentration of the parent drug and are eliminated rather slowly (Chasseaud et al., 1975; Chin et al., 1977). The mononitrates may therefore contribute to the pharmacologic effects of isosorbide dinitrate. The plasma elimination half-life of isosorbide dinitrate ranges between 30 and 50 min. Reviews on the metabolism of organic nitrates have been published by Needleman (1976) and Johnson et al. (1972).

c. Analogous Compounds. Organic Nitrates: glyceryl trinitrate, isomannide trinitrate, pentaerythritol tetranitrate.

d. Analytical Methods

1. Gas Chromatography—Chin et al. (1977)

Discussion. The GC-EC assay of Chin et al. (1977) overcomes the problem of irreversible absorption of small amounts of the nitrate compounds by use of an unusually high (30%) liquid phase loading of the solid support in order to mask active binding sites. Isosorbide dinitrate and its two mononitro metabolites can be measured with equal sensitivity (2 to 5 ng/ml plasma), using external standard curves for quantitation. The sensitivity is limited by small interfering background peaks (equivalent to < 2 ng isosorbide dinitrate/ml). No statistical data are given on the assay performance.

Detailed Method. Plasma (3 to 4 ml), 2 ml 1 *N* NaOH, and 30 ml ether are shaken for 20 min and centrifuged, and the aqueous layer is discarded. Anhydrous magnesium sulfate (0.5 g) is added to the ether phase; the mixture is shaken and centrifuged. A 25 ml aliquot of the ether layer is transfered to another tube and evaporated to dryness under a gentle stream of nitrogen at 30°C. The residue is redissolved in 1.0 ml ethyl acetate, and 1 to 3 μl of the solution is injected onto the column.

A gas chromatograph (Microtek MT-220 or Varian Model 2100-20) equipped with a [63]Ni electron capture detector is used. Carrier gas flow rate is 60 ml/min on a silanized glass column, 2 mm (i.d.) x 1.83 m, packed with 30% SE-30 on 60/80 mesh Gas Chrom Q (Applied Science Labs). Injection port, column, and detector temperatures are maintained at 190, 165, and 250°C, respectively.

2. Other GC Methods—Chin et al. (1977), Malbica et al. (1977), Rosseel and Bogaert (1973), Sherber et al. (1970)

Discussion. Earlier GC methods for plasma isosorbide dinitrate and its mononitrate metabolites utilize GC with flame ionization detection, with limited sensitivity (Sherber et al., 1970), or electron capture detection, with interference of the mononitrates by endogenous substances (Malbica et al., 1977; Rosseel and Bogaert, 1973). The assay of Rosseel and Bogaert (1973) is also suitable for the measurement of glyceryl trinitrate plasma levels. Malbica et al. (1977) emphasize the need to optimize the electron capture detector temperature (~175°) in order to achieve maximum sensitivity. With isoiodide dinitrate as the internal standard, their assay is sensitive to 0.5 ng isosorbide dinitrate/ml plasma. Irreversible absorption of the nitrates at low concentrations is overcome by repetitive prior injections of 10 ng priming doses of isosorbide dinitrate onto the GC column. The C.V. is less than 20% above 10 ng/ml. Irreversible drug absortion can also be minimized by the use of a high liquid phase leading of 30% SE-30 (Chin et al., 1977).

3. Polarography—Silvestri (1975), Turner and Lenkiewicz (1976)

Discussion. These studies demonstrate the ability of the organic nitrates to undergo facile electrochemical redox reactions. Mono- and oligonitrate derivatives are additive and cannot be differentiated without preseparation. Polarography has been applied only to the assay of isosorbide dinitrate in pharmaceutical dosage forms. Combination of an electrochemical detector with HPLC might prove suitable for organic nitrate analysis in biological fluids.

4. Comments. Gas chromatographic-electron capture assays are of high sensitivity and specificity for the detection of isosorbide dinitrate, 2-mononitrate, and 5-mononitrate in biological samples. Other organic nitrates may require derivatization prior to GC analysis. If HPLC with electrochemical detection can be successfully adapted to biological assays, it may prove to be a more widely applicable tool.

e. References

Assinder, D. F., L. F. Chasseaud, and T. Taylor, Plasma Isosorbide Dinitrate Concentrations in Human Subjects after Administration of Standard and Sustained-Release Formulations, *J. Pharm. Sci.*, **66**, 775–778 (1977).

Chasseaud, L. F., W. H. Down, and R. K. Grundy, Concentrations of the Vasodilator Isosorbide Dinitrate and Its Metabolites in the Blood of Human Subjects, *Eur. J. Clin. Pharmacol.*, **8**, 157–160 (1975).

Chin, D. A., D. G. Prue, J. Mitchelucci, B. T. Kho, and C. R. Warner, Quantitative Determination of Isosorbide Dinitrate and Two Metabolites in Plasma, *J. Pharm. Sci.*, **66**, 1143–1145 (1977).

Johnson, E. M., Jr., A. B. Harkey, D. J. Blehm, and P. Needleman, Clearance and Metabolism of Organic Nitrates, *J. Pharmacol. Exp. Ther.*, **182**, 56–62 (1972).

Malbica, J. O., K. Monson, K. Neilson, and R. Sprissler, Electron-Capture GLC Determination of Nanogram to Picogram Amounts of Isosorbide Dinitrate, *J. Pharm. Sci.*, **66**, 384–386 (1977).

Needleman, P., Organic Nitrate Metabolism, *Ann. Rev. Pharmacol. Toxicol.*, **16**, 81–93 (1976).

Rosseel, M. T. and M. G. Bogaert, GLC Determination of Nitroglycerin and Isosorbide Dinitrate in Human Plasma, *J. Pharm. Sci.*, **62**, 754–758 (1973).

Sherber, D. A., M. Marcus, and S. Kleinberg, Rapid Clearance of Isosorbide Dinitrate from Rabbit Blood: Determination by Gas Chromatography, *Biochem. Pharmacol.*, **19**, 607–612 (1970).

Silvestri, S., Quantitative and Automatic Polarographic Determination of Isosorbide Dinitrate in Pharmaceutical Preparations, *Pharm. Acta Helv.*, **50**, 304–307 (1975).

Turner, W. R. and R. S. Lenkiewicz, Polarographic Determination of Isosorbide Dinitrate, *J. Pharm. Sci.*, **65**, 118–121 (1976).

LIDOCAINE

2-(Diethylamino)-N-(2,6-dimethylphenyl)acetamide;
diethylglycinexylidide.
Xylocaine, Xylotox, and so on. Local anesthetic, antiarrhythmic agent.

$C_{14}H_{22}N_2O$; mol. wt. 234.33; pK_a 7.85.

a. Therapeutic Concentration Range. From 1.5 to 7 μg lidocaine/ml serum, with toxicity prevalent above 9 μg/ml (Burney et al., 1974; Collinsworth et al., 1974; Rowland et al., 1971).

b. Metabolism. Lidocaine is extensively metabolized in the liver by N-deethylation to monoethylglycinexylidide and glycinexylidide, followed by hydrolysis to 2,6-xylidine (review: Collinsworth et al., 1974). Other pathways include 4-hydroxylation to 4-hydroxy-2,6-xylidine (Thomas and Meffin, 1972), N-hydroxylation, and cyclization of monoethylglycinexylidide with an alcohol metabolite to imidazolidinones (Nelson et al., 1973). Monoethylglycinexylidide possesses significant antiarrhythmic potency (Burney et al., 1974; Smith and Duce, 1971) (83% of lidocaine potency: Narang et al., 1978) and may contribute to the lidocaine effects *in vivo*, where metabolite levels higher than those of the parent drug have been observed (Halkin et al., 1975; Narang et al., 1978; Strong et al., 1973). Both monoethylglycinexylidide and glycinexylidide possess convulsant activity and may contribute to the toxic central nervous system effects of lidocaine (Blumer et al., 1973; Halkin et al., 1975). The rapid and slow phases of lidocaine plasma elimination show average half-lives of 17 and 100 min, respectively (Collinsworth et al., 1974) in normal human subjects. Following 24 hr infusions in patients with uncomplicated myocardial infarction, however, the mean half-life of the slow elimination phase was found to be 3.2 hr (Lelorier et al., 1977).

c. Analogous Compounds. Benzocaine, bupivicaine, mepivicaine, prilocaine, procainamide.

d. Analytical Methods.

1. Gas Chromatography—Benowitz and Rowland (1973); modified procedure used in the Clinical Pharmacokinetics Laboratory, UCSF

Discussion. Lidocaine is measured in serum samples after a double back-extraction by GC with flame ionization detection. *N,N*-Diisobutylglycinexylidide serves as the internal standard. The extraction volumes are scaled down in order to permit direct GC injection of the final organic extraction phase (CS_2) without evaporation. The sensitivity is 0.3 μg lidocaine/ml serum, using a 0.2 ml sample volume. The C.V. is ±5% at 5 μg/ml, and GC analysis requires 6 min.

Detailed Method. To 0.2 ml serum are added 1 μg internal standard in 50 μl aqueous solution (*N,N*-diisobutylglycinexylidide, Astra Pharm. Prod., Worcester, Mass.), 0.1 ml 2 *N* NaOH saturated with NaCl crystals, and 4 ml anhydrous ether. The solution is vigorously shaken for 30 sec and centrifuged for 10 min at 1000 *g*. The aqueous layer is then frozen in an acetone-dry ice bath, and the top ether layer decanted into a nipple-bottomed (100 μl) centrifuge tube containing 0.2 ml 0.5 *N* HCl, saturated with NaCl crystals. Again, vigorously shake for 30 sec and centrifuge for 10 min at 1000 *g*. The aqueous layer is frozen in acetone-dry ice, the ether layer decanted and discarded, and the aqueous layer kept at 50 to 55°C in a water bath to evaporate the remaining ether. Then 0.1 ml 2 *N* NaOH saturated with NaCl crystals and 40 μl carbon disulfide are added. After mixing vigorously for 30 sec and then centrifuging for 10 min at 1000 *g*, 2 to 4 μl of the CS_2 layer is injected into the GC column.

The GC analysis is performed on a Varian Aerograph Series 2700 equipped with a 2 m x 2 mm (i.d.) glass column containing 3% OV-17 on Gas Chrom Q, 100/120 mesh. Column, detector, and injector temperatures are 185, 270, and 270°C, respectively. Gas flow rates are 250 ml/min for O_2, 30 ml/min for N_2, and 30 ml/min for H_2. Retention times of lidocaine and its internal standard are 3.5 and 5 min, respectively. Standard curves are constructed using peak height ratio measurements.

2. Other GC Methods—Adjepon-Yamoah and Prescott (1974), Caille et al. (1977), Hucker and Stauffer (1976), Irgens et al. (1976), Kline and Martin (1978), Naito et al. (1977), Nation et al. (1976), Rosseel and Bogaert (1978), Spechtmeyer and Steinbach (1969)

Discussion. Adjepon-Yamoah and Prescott (1974) describe an assay for lidocaine, monoethylglycinexylidide, and glycinexylidide in plasma and urine. 4-Hydroxyxylidine as a major excretion product is analyzed after treatment of urine samples with glusulase in order to hydrolyze sulfate and glucuronide conjugates. Plasma and urine samples are extracted into benzene, and the extraction residues treated with acetic anhydride in pyridine, yielding acetyl derivatives of the metabolites. The products are separated and detected by GC–N-FID, using a temperature program. Quantitation with aceto-*p*-toluidine as the internal standard is sensitive to 10 to 30 ng lidocaine and its metabolites/ml plasma (2 ml sample volume). A simplified GC–N-FID assay of lidocaine in plasma is reported by Kline and Martin (1978). The assay involves extraction of lidocaine and the internal standard, mepivacaine, from deproteinized alkaline plasma into a microvolume of CS_2 and direct injection of the CS_2 phase into the gas chromatograph. Assay sensitivity is 0.25 μg lidocaine/ml plasma (1 ml specimen).

The major plasma components—lidocaine, monoethylglycinexylidide, and glycinexylidide—can also be detected without temperature programming, thereby requiring less analysis time, by capillary GC (Rosseel and Bogaert, 1978). This GC assay differs from the procedure of Adjepon-Yamoah and Prescott (1974) in the use of trimecaine as the internal standard, rapid derivatization with trifluoroacetic acid, and glass capillary columns at constant temperatures. The GC analysis is completed within 6 min and is sensitive to 20 ng lidocaine and its metabolites/ml plasma.

Further assays of lidocaine plasma levels with GC–FID are reported by Hucker and Stauffer (1976) and Irgens et al. (1976). Naito et al. (1977) present a GC assay of lidocaine in body tissue samples after extraction with liquid nitrogen. The GC assay of Spechtmeyer and Steinbach (1969) is applicable to the analysis of several local anesthetics in biological samples. Further variations of the lidocaine assay procedure have been published by Nation et al. (1976) and Caille et al. (1977).

3. Gas Chromatography-Mass Spectrometry—Hignite et al. (1978), Strong and Atkinson (1972)

Discussion. The assay of Strong and Atkinson (1972) is essentially equivalent to current GC assays except for the detection of GC column eluents by electron impact MS in the selected ion monitoring mode. Monoethylglycinexylidide is chromatographed without derivatization, and trimecaine serves as the internal standard. The ion peaks at *m/e* 58, 68, and 120 are monitored by quadropole mass selection. These ions correspond to breakage of the C-C bond of the glycine moiety (m/e 68 for lidocaine and m/e 58 for its metabolite) and of the amide bond (m/e 120 for both). Peak height ratios of two ions of the same compound over the internal standard are monitored to ensure specificity in the presence of biological material. The sensitivity is 0.5 and 0.3 μg/ml for lidocaine and monoethylglycinexylidide, respectively.

Lidocaine, monoethylglycinexylidide, and glycinexylidide can be measured with a sensitivity of 50 ng/ml plasma by the GC-MS assay of Hignite et al. (1978). Propyl derivatives are formed from the desethylated metabolites by reaction with propionaldehyde and sodium cyanoborohydride in order to differentiate parent drug and metabolites. With mepivacaine as the internal standard, quantitation is achieved with high sensitivity in the selected ion monitoring mode.

4. Direct Insertion Mass Spectrometry—Garland et al. (1974), Nelson et al. (1977)

Discussion. Lidocaine and its metabolites are analyzed in urine and plasma after organic solvent extraction by direct insertion, chemical ionization MS (Garland et al., 1974). Since the mass spectra of these compounds consist primarily of the MH$^+$ parent ions in the chemical ionization mode, simultaneous detection of several metabolites in biological samples is possible without GC preseparation. Deuterated lidocaine and metabolites serve as the internal standards for quantitative analysis. Using this procedure, Nelson et al. (1977) could determine the amounts of seven metabolites in the urine, including monoethylglycinexylidide, glycinexylidide, xylidine, and the corresponding p-hydroxy derivatives of the metabolites and of the parent drug.

5. High Performance Liquid Chromatography—Adams et al. (1976), Narang et al. (1978)

Discussion. Lack of appreciable UV absorbance of lidocaine above 220 nm has prevented widespread application of HPLC techniques to the analysis of plasma samples. Adams et al. (1976) overcame the problem of insufficient UV detector sensitivity with a variable-wavelength UV detector set at 205 nm. The assay procedure involves ether extraction of lidocaine and its internal standard, procaine, at pH 11, followed by reverse phase HPLC analysis. With a plasma sample of only 50 μl, the method provides linear standard curves for lidocaine concentrations between 1 and 20 μg/ml with a C.V. of 8 to 10% at the 1 μg/ml level. Simultaneous administration of lidocaine and procainamide to patients may cause assay interference, since the *N*-acetylprocainamide metabolite cochromatographs with the internal standard, procaine. The internal standard has to be added to plasma immediately prior to analysis in order to avoid its decomposition by serum esterases. The authors suggest that this method may be suitable for routine clinical analysis of lidocaine; however, no reports have been published so far to confirm assay specificity under the stringent requirements of randomly selected patient samples. The same procedure is also applicable to the determination of procainamide, using a different internal standard.

This HPLC assay has been recently extended to the measurement of lidocaine and, simultaneously, two of its active metabolites, that is, monoethylglycinexylidide and glycinexylidide (Narang et al., 1978). Pharmacologically significant levels of monoethylglycinexylidide were found in all of the three patients studied.

6. Enzyme Immunoassay—Pape et al. (1978), Walburg (1978)

Discussion. The homogeneous enzyme immunoassay of lidocaine is based on bacterial glucose-6-phosphate dehydrogenase, which is coupled to lidocaine (EMIT). Binding to a lidocaine-specific antibody decreases the enzyme-catalyzed oxidation of glucose-6-phosphate in the presence of NAD^+. Binding and inactivation of the enzyme are prevented by free lidocaine in the serum sample. The assay requires only 50 μl serum and gives a sensitivity of below 1 μg/ml and a C.V. of ~5%. Pape et al. (1978) and Walburg (1978) both find excellent agreement of the lidocaine EMIT assay with a specific GC assay (r = .98 to .99).

7. Comments. The published GC methods are well suited for determining lidocaine disposition in biological samples. Choice of a specific assay method depends on whether levels of lidocaine alone or of lidocaine and its metabolites are to be measured.

e. References

Adams, R. F., F. L. Vandemark, and G. Schmidt, The Simultaneous Determination of Lidocaine and Procainamide in Serum by Use of High Pressure Liquid Chromatography, *Clin. Chim. Acta*, **69**, 515–524 (1976).

Adjepon-Yamoah, K. K. and L. F. Prescott, Gas-Liquid Chromatographic Estimation of Lignocaine, Ethylglycylxylidide, Glycylxylidide and 4-Hydroxyxylidine in Plasma and Urine, *J. Pharm. Pharmacol.*, **26**, 889–893 (1974).

Benowitz, N. and M. Rowland, Determination of Lidocaine in Blood and Tissues, *Anesthesiology*, **39**, 639–641 (1973).

Blumer, J., J. M. Strong, and A. J. Atkinson, Jr., The Convulsant Potency of Lidocaine and Its *N*-Dealkylated Metabolites, *J. Pharmacol. Exp. Ther.*, **186**, 31–36 (1973).

Burney, R. G., C. A. Difazio, M. J. Peach, K. A. Petrie, and M. J. Silvester, Anti-arrhythmic Effects of Lidocaine Metabolites, *Am. Heart J.*, **88**, 765–769 (1974).

Caille, G., J. Lelorier, Y. Latour, and J. G. Besner, GLC Determination of Lidocaine in Human Plasma, *J. Pharm. Sci.*, **66**, 1383–1385 (1977).

Collinsworth, K. A., S. M. Kalman, and D. C. Harrison, The Clinical Pharmacology of Lidocaine as an Antiarrhythmic Drug, *Circulation*, **50**, 1217–1230 (1974).

Garland, W. A., W. F. Trager, and S. D. Nelson, Direct (Non-chromatographic) Quantification of Drugs and Their Metabolites from Human Plasma Utilizing Chemical Ionization Mass Spectrometry and Stable Isotope Labeling: Quinidine and Lidocaine, *Biomed. Mass. Spectrom.*, **1**, 124–129 (1974).

Halkin, H., P. Meffin, K. L. Melmon, and M. Rowland, Influence of Congestive Heart Failure on Blood Levels of Lidocaine and Its Active Monoethylated Metabolite, *Clin. Pharmacol. Therap.*, **17**, 669–676 (1975).

Hignite, C. E., C. Tschanz, J. Steiner, D. H. Huffman, and D. L. Azarnoff, Quantitation of Lidocaine and Its Deethylated Metabolites in Plasma and Urine by Gas Chromatography-Mass Fragmentography, *J. Chromatogr.*, **161**, 243–249 (1978).

Hucker, H. B., and S. C. Stauffer, GLC Analysis of Lidocaine in Plasma Using a Novel Nitrogen Sensitive Detector, *J. Pharm. Sci.*, **65**, 926–927 (1976).

Irgens, T. R., W. M. Henderson, and W. H. Shelver, GLC Analysis of Lidocaine in Blood Using an Alkaline Flame Ionization Detector, *J. Pharm. Sci.*, **65**, 608–610 (1976).

Kline, B. J. and M. F. Martin, Simplified GLC Assay for Lidocaine in Plasma, *J. Pharm. Sci.*, **67**, 887–889 (1978).

Lelorier, J. et al., Pharmacokinetics of Lidocaine after Prolonged Intravenous Infusions in Uncomplicated Myocardial Infarction, *Ann. Intern. Med.*, **87**, 700–706 (1977).

Naito, E., M. Matsuki, and K. Shimoji, A Simple Method for Gas Chromatographic Determination of Lidocaine in Tissues, *Anesthesiology*, **47**, 466–467 (1977).

Narang, P. K., W. G. Crouthamel, N. H. Carliner, and M. L. Fisher, Lidocaine and Its Active Metabolites, Clin. Pharmacol. Ther., **24**, 654–662 (1978).

Nation, R. L., E. J. Triggs, and M. Selig, Gas Chromatographic Method for the Quantitative Determination of Lidocaine and its Metabolite Monoethylglycinexylidide in Plasma, *J. Chromatogr.*, **116**, 188–193 (1976).

Nelson, S. D., G. D. Breck, and W. F. Trager, *In vivo* Metabolite Condensations: Formation of N-1-Ethyl-2-methyl-N3-(2,6-dimethylphenyl)-4-imidazolidinone from the Reaction of a Metabolite of Alcohol with a Metabolite of Lidocaine, *J. Med. Chem.*, **16**, 1106–1112 (1973).

Nelson, S. D., W. A. Garland, G. D. Breck, and W. F. Trager, Quantification of Lidocaine and Several Metabolites Utilizing Chemical-Ionization Mass Spectrometry and Stable Isotope Labeling, *J. Pharm. Sci.*, **66**, 1180–1190 (1977).

Pape, B. E., R. Whiting, K. M. Parker, and R. Mitra, Enzyme Immunoassay, and Gas-Liquid Chromatography Compared for Determination of Lidocaine in Serum, *Clin. Chem.*, **24**, 2020–2022 (1978).

Rosseel, M. T. and M. G. Bogaert, Determination of Lidocaine and Its Desethylated Metabolites in Plasma by Capillary Column Gas-Liquid Chromatography, *J. Chromatogr.*, **154**, 99–102 (1978).

Rowland, M., P. D. Thompson, A. Guichard, and K. L. Melmon, Disposition Kinetics of Lidocaine in Normal Subjects, *Ann. N.Y. Acad. Sci.*, **173**, 383–398 (1971).

Smith, E. R. and B. R. Duce, The Acute and Antiarrhythmic and Toxic Effects in Mice and Dogs of 2-Ethylamino-2',6'-acetoxylidine (L-86), a Metabolite of Lidocaine, *J. Pharmacol. Exp. Ther.*, **179**, 580–585 (1971).

Spechtmeyer, H. and H. Steinbach, Gas Chromatographic Determinations of Some Local Anesthetics and Their Alkylamino-Metabolites in Biological Material, *Arzneim.-Forsch.*, **19**, 1754–1756 (1969).

Strong, J. M. and A. J. Atkinson, Simultaneous Measurement of Plasma Concentrations of Lidocaine and Its Desethylated Metabolite by Mass Fragmentography, *Anal. Chem.*, **44**, 2287–2290 (1972).

Strong, J. M., M. Parker, and A. J. Atkinson, Jr., Identification of Glycinexylidide in Patients Treated with Intravenous Lidocaine, *Clin. Pharmacol. Ther.*, **14**, 67–72 (1973).

Thomas, J. and P. Meffin, Aromatic Hydroxylation of Lidocaine and Mepivacaine in Rats and Humans, *J. Med. Chem.*, **15**, 1046–1049 (1972).

C. B. Walberg, Lidocaine by Enzyme Immunoassay, *J. Analt. Toxicol.*, **2**, 121–122 (1978).

LITHIUM CARBONATE

Camcolit, Candamide. Treatment of manic psychosis.
Li_2CO_3; mol. wt. 73.89.

a. Therapeutic Concentration Range. From 0.6 to 1.2 mEq lithium/l plasma is the currently accepted "standardized 12 hr Li^+ serum concentration," obtained 12 hr after the dose in order to avoid fluctuation of the level over time (Amdison, 1977). Levels above 2 mEg/l lead to serious toxicity in most cases.

b. Metabolism. Lithium is almost entirely excreted into the urine with a normal urinary clearance of 20 ml/min (Thomsen and Shou, 1968). The terminal plasma elimination half-life of lithium ranges between 15 and 20 hr (Groth et al., 1974; Amdisen, 1977). Cooper et al. (1973, 1976) suggest that the Li^+ plasma level obtained at 24 hr after a dose of 600 mg Li_2CO_3 is a good prognosticator of individual dose requirements.

c. Analogous Compounds. Salts of metals of the alkali and alkaline earth groups.

d. Analytical Methods

1. Atomic Emission Spectrometry—Robertson et al. (1973)

Discussion. The emission wavelength of lithium at 670.8 permits its detection at much lower concentrations than are possible with atomic absorption spectrometry (3 pg lithium/ml versus 5 ng lithium/ml). Atomic emission offers the additional advantage of a linear standard curve over a much wider concentration range. Following appropriate dilution of plasma, blood, and urine samples, lithium levels are determined with a precision of 1.5% (S.D.). The fuel-oxidant combination for the burner may be critical to the intensity of the emission signal. Furthermore, the effect of urine dilution on emission intensity for each burner and fuel-oxidant combination should be carefully monitored. Urine samples should be acidified prior to analysis in order to prevent coprecipitation of lithium with calcium phosphate.

Detailed Method. Plasma (1 ml) is diluted with water to a volume of 100 ml and refrigerated until analysis. Lithium standard solutions in the range of 0.002 to 0.02 mEg/l in a 1 : 100 plasma dilution, containing 0.025% EDTA as preservative, are used.

A modular form Heath EV-703 flame photometer fitted with an air–acetylene-fueled Techtron 5 cm slot premix burner is used. The optics consist of a Czerny-Turner diffraction grating and an RCA 1P28A photomultiplier. Emission is read at 670.8 nm.

2. Atomic Absorption Spectrophotometry—Birch et al. (1978), Cooper et al. (1973, 1976)

Discussion. Serum lithium determinations are carried out utilizing a Perkin-Elmer Model 303 atomic absorption spectrophotometer fitted with a three-slot Boling burner head. Serum is diluted with deionized water, and the standard solutions are matched for viscosity and electrolyte content by dilution with a lithium-free serum. As an additional control, another set of standards is also diluted with 6% dextran in saline solution. The C.V. is 1.5%. Whereas the standard method (Cooper et al., 1973) consumes 0.5 ml serum, the micromethod suggested by Cooper et al. in 1976 requires only 50 μl of blood obtained by the finger prick method.

The ^{7}Li isotope of the normotopic ^{6}Li can be utilized to measure the lithium kinetics of single doses during maintenace therapy (Birch et al., 1978). The serum levels of ^{6}Li and ^{7}Li can be simultaneously determined by atomic absorption spectroscopy. The results obtained by Birch et al. (1978) suggest that the rate of lithium appearance in the blood is unaffected by the previous state of lithium loading.

3. Comments. Accurate determinations of lithium in plasma can be useful in monitoring lithium therapy of patients with manic depression. Both atomic emission and atomic absorption spectrophotometry can yield satisfactory data if the laboratory techniques are well controlled.

e. References

Amdisen, A., Serum Level Monitoring and Clinical Pharmacokinetics of Lithium, *Clin. Pharmacokin.*, **2**, 73–92 (1977).

Birch, N. J., D. Robinson, R. A. Inie, and R. P. Hullin, Lithium-6 Stable Isotope Determination by Atomic Absorption Spectroscopy and Its Application to Pharmacokinetic Studies in Man, *J. Pharm. Pharmacol.*, **30**, 683–685 (1978).

Cooper, T. B. and G. M. Simpson, The 24 Hour Lithium Level as a Prognosticator of Dosage Requirements: A 2-Year Follow-Up Study, *Am. J. Psychiatr.*, **133**, 440–443 (1976).

Cooper, T. B., P.-E. E. Bergner, and G. M. Simpson, The 24 Hour Serum Lithium Level as a Prognosticator of Dosage Requirements, *Am. J. Psychiatr.*, **130**, 601–603 (1973).

Groth, V., W. Prellwitz, and E. Jähnchen, Estimation of Pharmacokinetic Parameters of Lithium from Saliva and Urine, *Clin. Pharmacol. Ther.*, **16**, 490–498 (1974).

Robertson, R., K. Fritze, and P. Grof, On the Determination of Lithium in Blood and Urine, *Clin. Chim. Acta*, **45,** 25–31 (1973).

Thomsen, K. and M. Shou, Renal Lithium Excretion in Man, *Am. J. Physiol.*, **215,** 823–827 (1968).

LYSERGIDE (LSD)—by K. Maloney

N,N-Diethyl-D-lysergamide. Psychotomimetic drug.

$C_{20}H_{25}N_3O$; mol. wt. 323,42; pK_a 7 to 8

a. Therapeutic Concentration Range. An oral dose of 160 μg LSD results in plasma levels of 1 to 5 ng/ml (Upshall and Wailling, 1972).

b. Metabolism. Lysergide is extensively metabolized with only 1% excreted into the urine as the parent drug (Ratcliffe et al., 1977). A major metabolite is 2-oxy-LSD (Axelrod et al., 1957), which is pharmacologically inactive. Further metabolites include 13- and 14-hydroxy-LSD (Ratcliffe et al., 1977). The plasma elimination half-life of LSD is about 3.5 hr (Upshall and Wailling, 1972).

c. Analogous Compounds. Ergot alkaloids (e.g., dihydroergotamine, ergonovine, ergotamine, methylergonovine, methylergotamine, methysergide).

d. Analytical Methods

1. Radioimmunoassay—Ratcliffe et al. (1977)

Discussion. Antibodies are raised against two different bovine serum albumin-LSD conjugates. Rabbit anti-LSD serum is obtained by immunizing with a conjugate linked via the indole N—H to BSA in a

Mannich condensation with formaldehyde. Sheep anti-LSD serum is directed against a LSD-BSA conjugate with a 5-carboxyamyl spacer bridge in place of one of the ethylamide groups of LDS. The assay procedures with these two antibodies utilize ^{3}H-LSD as tracer and double-antibody precipitation for the separation of bound and free ^{3}H activity. The spectrum of specificity is quite different for the two antibodies with respect to similar indole alkaloids that might interfere with the unambiguous detection of LSD. Therefore application of both procedures to the same sample provides a high degree of confidence in the detection of LSD in plasma and urine. The antibody with the amyl spacer bridge gives lower values for urine samples than does the Mannich condensation antibody, presumably because of a higher cross-affinity of the latter to LSD metabolites with altered indole moiety. Both assays give similar values in plasma, suggesting that the results represent mostly unchanged LSD. The sensitivity is 10 pg LSD in 25 μl unextracted urine or plasma.

Detailed Method. A standard curve is prepared along with the samples, and all dilutions of reagents are made with a sodium diethylbarbital buffer (pH 8.6, 50 mM) containing bovine serum albumin (1 g/l). First, 100 μl diluent buffer is added, then 25 μl serum or urine, 100 μl tracer solution [20 pg 2(n)-^{3}H-LSD], 100 μl LSD antiserum (20,000-fold final dilution), and a mixture of rabbit antigoat serum and nonimmune sheep serum as the precipitating agent for the sheep anti-LSD serum. The rabbit anti-LSD serum is precipitated with a donkey antirabbit serum. The assay tubes are incubated for 3 hr at room temperature and then centrifuged at 2000 g for 30 min to remove the antibody-bound fraction of LSD. The supernatants are decanted into counting vials, and 10 ml scintillation fluid is added.

The vials are counted in a beta counter for a sufficient amount of time to accumulate 10,000 counts in the vial containing the total tracer amount added to the assay (about 5 min). Concentrations are determined from a standard curve obtained by plotting percent ^{3}H activity bound against known LSD concentrations.

2. Other RIA Methods—Loeffler and Pierce (1973), Taunton-Rigby et al. (1973)

Discussion. Loeffler and Pierce (1973) use a lysergic acid-human albumin conjugate as the antigen to produce LSD-directed sheep antiserum. The step to separate free and bound ^{3}H-LSD employs dextran-coated charcoal. Taunton-Rigby et al. (1973) raise rabbit antibodies against a conjugate prepared by Mannich condensation of LSD, formaldehyde, and human serum albumin. Both methods are highly sensitive, but lack the high degree of specificity achieved by use of the 4-carbon spacer in the LSD conjugate of Ratcliffe et al. (1977).

3. High Performance Liquid Chromatography—Christy et al. (1976), Twitchett et al. (1978)

Discussion. The HPLC assay of Christy et al. (1976) utilizes a silica gel (6 μm average particle size) adsorption column and fluorescence detection of LSD. A combination of HPLC and TLC is satisfactory for the isolation and identification of LSD in biological samples. No sensitivity limits are given.

Twitchett et al. (1978) describe a similar HPLC-fluorescence assay to be used in conjunction with RIA before and after HPLC separation of LSD. Levels as low as 0.5 ng LSD/ml body fluid can be determined for clinical and forensic purposes.

4. Fluorescence—Axelrod et al. (1957), Upshall and Wailling (1972)

Discussion. In the method of Axelrod et al. (1957), LSD is isolated from NaCl-saturated plasma at an alkaline pH by extraction into hexane and back-extraction into HCl. Fluorimetric analysis of LSD in the HCl phase yields a sensitivity of 1 ng LSD/ml plasma, assuming that plasma blank fluorescence is nonvariable. The fluorescence method of Upshall and Wailling (1972) exploits the UV-light-catalyzed hydration of LSD to the non fluorescent lumiderivative. The difference in fluorescence of plasma hexane extracts before and after intense UV irradiation is a measure of LSD concentration. The method is sufficiently sensitive to follow LSD plasma levels after oral doses of 2 μg/kg. Specificity with respect to LSD metabolites is questionable.

5. Comments.
The RIA procedures are presently the methods of choice for LSD plasma determinations. In conjunction with separation techniques (HPLC), RIA methods can be extremely sensitive and specific. High performance liquid chromatography with fluorescence detection could be further improved to allow its ready application in biological samples.

e. References

Axelrod, J., R. O. Brady, B. Witkop, and E. V. Evarts, The Distribution and Metabolism of Lysergic Acid Diethylamide, *Ann. N. Y. Acad. Sci.*, **66**, 435–444 (1957).

Christy, J., M. W. White, and J. M. Wiles, A Chromatographic Method for the Detection of LSD in Biological Liquids, *J. Chromatogr.*, **120,** 496–501 (1976).

Loeffler, L. J. and J. V. Pierce, Radioimmunoassay for Lysergide (LSD) in Illicit Drugs and Biological Fluids, *J. Pharm. Sci.*, **62,** 1817–1820 (1973).

Ratcliffe, W. A. et al., Radioimmunoassay of Lysergic Acid (LSD) in Serum and Urine by Using Antisera of Different Specificities, *Clin. Chem.*, **23,** 169–174 (1977).

Taunton-Rigby, A., S. E. Sher, and P. R. Kelley, Lysergic Acid Diethylamide: Radioimmunoassay, *Science*, **181,** 165–166 (1973).

Twitchett, R. J., S. M. Fletcher, A. T. Sullivan, and A. C. Moffat, Analysis of LSD in Human Body Fluids by High-Performance Liquid Chromatography, Fluorescence Spectroscopy and Radioimmunoassay, *J. Chromatogr.*, **150,** 73–84 (1978).

Upshall, D. G. and D. G. Wailling, The Determination of LSD in Human Plasma Following Oral Administration, *Clin. Chim. Acta*, **36,** 67–73 (1972).

MELPHALAN—by M. Cohen

4-[Bis(2-chloroethyl)amino]-L-phenylalanine.
Alkeran. Antineoplastic alkylating drug.

$C_{13}H_{18}Cl_2N_2O_2$; mol. wt. 305.20.

a. Therapeutic Concentration Range. None defined. An intravenous dose of 6 mg melphalan/kg in rats results in plasma levels of about 1 μg/ml at 1 hr after the dose.

b. Metabolism. The *in vitro* degradation of melphalan occurs by hydrolysis of the chloroethyl groups of the molecule (Chang et al., 1978b). Both mono- and dihydroxy metabolites have also been detected *in vivo* in the plasma of several species (Chang et al., 1978a; Furner et al., 1976). The monohydroxy metabolite retains the alkylating potential of the parent drug, whereas the dihydroxy metabolite is considered to be inactive. In dogs the plasma elimination half-lives of melphalan and its monohydroxy metabolite are about 30 min (Furner et al., 1976).

c. Analogous Compounds. Nitrogen mustard derivatives, amino acids; chlorambucil; cyclophosphamide.

d. Analytical Methods

1. High Performance Liquid Chromatography—Chang et al. (1978a)

Discussion. This assay requires reverse phase HPLC with UV detection at 254 nm. Plasma proteins are precipitated with methanol prior to analysis. Dansylproline is used as the internal standard. The sensitivity of the method is 50 ng melphalan/ml serum. No important interfering peaks at the position of either melphalan or the internal standard are observed.

Detailed Method. To 1 ml plasma, 5 µg dansylproline and 2 ml cold methanol (0°C) are added and mixed vigorously on a vortex mixer for 20 sec. The sample is then cooled to −60° (acetone and dry ice) for 3 min. The plasma-methanol mixture is centrifuged for 3 min, and the clear methanolic solution injected directly onto the column.

The HPLC analysis is performed on a liquid chromatograph (Waters Assoc., Model 6000) equipped with a Waters 440 UV detector (254 nm) and Waters µ-Bondapak C_{18} reverse phase column. An isocratic solvent system of water and methanol (1:1) with 1% acetic acid is delivered at a rate of 2 ml/min. Total analysis time per sample is approximately 20 min. External standard curves are used for quantitation of melphalan by peak height measurements.

2. Other HPLC Method—Furner et al. (1976)

Discussion. Melphalan and its mono- and dihydroxy metabolites are separated on a reverse phase column, using an eluent gradient program, and detected by UV absorption. Quantitation is achieved using peak area determinations with a sensitivity of 10 ng melphalan/ml plasma. The monohydroxy metabolite is detectable in dog, but not mouse, plasma.

3. Fluorescence—Chirigos and Mead (1964)

Discussion. Sensitivity of the assay is 50 ng melphalan/ml. The mono- and dihydroxy metabolites of melphalan cannot be differentiated from melphalan, and they give an increased fluorescence yield. Endogenous compounds also fluoresce under the assay conditions and may contribute to variable plasma blanks. Therefore this assay is not directly applicable to the study of melphalan pharmacokinetics, although the native fluorescence of melphalan at 268 → 365 nm might provide a suitable detection method after chromatographic separation of the parent drug and hydroxy metabolites.

4. Colorimetry—Epstein et al. (1955), Klatt et al. (1960)

Discussion. The reagent γ-(4-nitrobenzyl)pyridine was introduced by Epstein et al. (1955) to measure the alkylation potentials of a large number of chemicals, including alkylating antineoplastic agents. Upon alkylation of the pyridine nitrogen, a blue color is generated by the addition of strong bases. This technique has been modified by Klatt et al. (1960) to measure melphalan and its alkylating metabolites in plasma. The achieved sensitivity of 1 μg alkylating melphalan equivalent is not sufficient to study melphalan pharmacokinetics.

5. Comments.
High performance liquid chromatographic methods are suitable for the measurement of melphalan and its hydrolyzed metabolites after the administration of therapeutic doses of the drug to human beings. Plasma samples containing melphalan must be stored at low temperature in order to minimize *in vitro* hydrolysis and alkylation of plasma proteins (Chang et al., 1978b).

e. References

Chang, S. Y., D. S. Alberts, L. R. Melnick, P. D. Walson, and S. E. Salmon, High Pressure Liquid Chromatographic Analysis of Melphalan in Plasma, *J. Pharm. Sci.*, **5**, 679–682 (1978a).

Chang, S. Y. et al., Hydrolysis and Protein Binding of Melphalan, *J. Pharm. Sci.*, **5**, 682–684 (1978b).

Chirigos, M. A., and J. A. R. Mead, Experiments on Determination of Melphalan by Fluorescence: Interaction with Protein in Various Solutions, *Anal. Biochem.*, **7**, 259–268 (1964).

Epstein, J., R. W. Rosenthal, and R. J. Ess, Use of γ-(4-Nitrobenzyl)pyridine as Analytical Reagent for Ethylenimines and Alkylating Agents, *Analyt. Chem.*, **27**, 1435–1439 (1955).

Furner, R. L., L. B. Mellet, R. K. Brown, and G. Duncan, A Method for the Measurement of L-Phenylalanine Mustard in the Mouse and Dog by High Pressure Liquid Chromatography, *Drug Metab. Dispos.*, **4**, 577–583 (1976).

Klatt, O., A. C. Griffin, and J. S. Stehlin, Jr., Method for Determination of Phenylalanine Mustard and Related Alkylating Agents in Blood, *Soc. Exp. Biol. Med.*, **104**, 629–631 (1960).

MEPERIDINE—by D. R. Doose

N-Methyl-4-phenyl-4-carbethoxypiperidine.
Demerol, Dolantin. Narcotic analgesic.

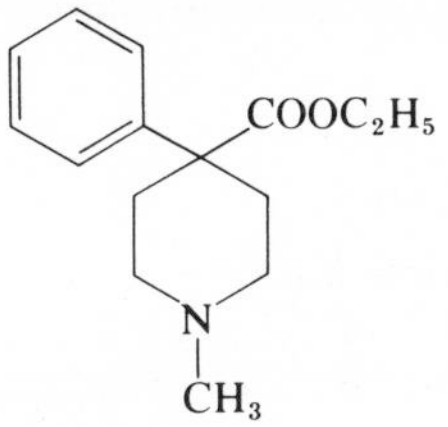

$C_{15}H_{21}NO_2$; mol. wt. 247.34; pK_a 8.7.

a. Therapeutic Concentration Range. Therapeutic, toxic, and lethal concentrations are about 0.6, 5, and 30 μg/ml blood, respectively (Winek, 1976).

b. Metabolism. The major metabolic pathways involve hydrolysis to form meperidinic acid and *N*-demethylation to normeperidine, which may also be hydrolyzed to normeperidinic acid (Plotnikoff et al., 1952). A minor pathway giving a N-oxide metabolite has been shown (Mitchard et al., 1972). Normeperidine is the only metabolite with significant pharmacological activity, being half as potent as meperidine as an analgesic and twice as potent as a convulsant (MacDonald et al., 1946). Normeperidine may be present in higher serum concentrations than the parent drug (Miller and Anderson, 1954). The plasma elimination half-life of meperidine is 3.2 to 3.7 hr in normal human subjects (Mather et al., 1975).

c. Analogous Compounds. Narcotic analgesics (e.g., α-prodine, methylphenidate).

d. Analytical Methods

1. Gas-Liquid Chromatography—Szeto and Inturrisi (1976)

Discussion. The method includes a double back-extraction prior to GC with flame ionization detection. Normeperidine is analyzed as the heptafluorobutyryl derivative to decrease on-column adsorption. Sensitivity is 0.02 μg/ml serum for meperidine and normeperidine, a finding that has been difficult to reproduce in our laboratory. Precision and accuracy are not specifically reported. No interferences are present in blank human plasma extracts. Using a very similar extraction and detection method, Chan et al. (1974) have also not observed any interferences in plasma samples from patients treated with morphine, pentazocine, and methadone.

Detailed Method. To plasma (0.5 to 2.0 ml) in a siliconized glass tube with a PTFE-lined screw cap are added 0.15 ml of the internal standard solution (aqueous solution of 2-ethyl-5-methyl-3,3-diphenylpyrrolline, 4 μg/ml), 0.25 ml of 2.5 N NaOH, and 2 drops of octyl alcohol. The sample is extracted with 5 ml anhydrous ethyl ether by shaking for 5 min and is then centrifuged for 5 min at 500 g. The ethyl ether layer (upper) is carefully removed and saved. The extraction is repeated again with 5 ml ethyl ether. The ether layers are combined and extracted with 5 ml 0.2 N HCl by shaking for 10 min and centrifuging. The ether phase is discarded. The acid phase is then washed with 5 ml hexane by shaking for 5 min and centrifuging. The washed acid phase is made alkaline by the addition of 3 drops of 50% sodium hydroxide (pH 11), and extracted by shaking with 7 ml ether for 10 min. After centrifugation the ether layer is carefully transferred to a siliconized glass tube and evaporated to dryness in a water bath at 42°. The residue is reacted with a 3% solution (40 μl) of heptafluorobutyrylimidazole in ethyl acetate at room temperature in the dark for 30 min. The mixture is then dried by a gentle stream of nitrogen. The sample is reconstituted with 30 μl cyclohexane by warming at 40°, and 1 to 2 μl is injected onto the gas chromatograph.

The GC analysis is performed on a Varian Aerograph Model 2700 equipped with a flame ionization detector. The glass column is 6 ft × 2 mm (i.d.). The packing material consists of 3% OV-17 on 80/100 mesh Gas Chrom Q. The temperatures of the injection port, column, and detector are 250, 175, and 275°C, respectively. Helium carrier gas, hydrogen, and air flow are 34, 30, and 300 ml/min, respectively. The concentrations of meperidine and normeperidine are assessed by comparing the resulting peak height ratio of the two compounds to the internal standard with a standard calibration curve.

2. Other GLC Methods—Chan et al. (1974), Goehl and Davison (1973), Klotz et al. (1974)

Discussion. The various GLC assays are similar to the one described by Szeto and Inturrisi (1976) except for the choice of column systems, derivatization, and internal standards and the sensitivity reported. Some of these assay methods do not measure the active metabolite normeperidine.

3. Colorimetry—Way et al. (1949)

Discussion. The colorimetric method is based on the spectrophotometric measurement of an extractable methyl orange complex at 540 nm. This technique is generally applicable to lipophilic organic bases and therefore nonspecific. Nicotine levels in smokers may interfere with the meperidine assay (Beckett et al., 1965).

4. Fluorescence—Dal Cortivo et al. (1970)

Discussion. This assay method exploits the observation that a characteristic fluorophore is produced when meperidine is incubated at elevated temperatures in a mixture of formaldehyde and concentrated sulfuric acid. The reported sensitivity of 0.3 μg/ml plasma is sufficient for the detection of meperidine; however, the procedure does not measure the active metabolite and suffers from possible interferences.

5. Comments.
Plasma level determinations of meperidine and N-normeperidine have to be sensitive to at least 20 to 50 ng/ml for pharmacokinetic studies. The sensitivity of reported GC assays is marginal in this range and, in our hands, cannot readily be improved by use of a nitrogen-sensitive FID. There remains a need for techniques with better sensitivity.

e. References

Beckett, A. H., M. Rowland, and E. J. Triggs, Significance of Smoking in Investigations of Urinary Excretion Rates of Amines in Man, *Nature*, **207**, 200–201 (1965).

Chan, K., M. J. Kendall, and M. Mitchard, A Quantitative Gas-Liquid Chromatographic Method for the Determination of Pethidine and Its Metabolites, Norpethidine and Pethidine N-oxide, in Human Biological Fluids, *J. Chromatogr.*, **89**, 169–176 (1974).

Dal Cortivo, L. A., M. M. DeMayo, and S. B. Weinberg, Fluorometric Determination of Microgram Amounts of Meperidine, *Anal. Chem.*, **42**, 941–942 (1970).

Goehl, T. J. and C. Davison, GLC Determination of Meperidine in Blood Plasma, *J. Pharm. Sci.*, **62**, 907–909 (1973).

Klotz, U., T. S. McHorse, G. R. Wilkinson, and S. Schenker, The Effect of Cirrhosis on the Disposition and Elimination of Meperidine in Man, *Clin. Pharmacol. Ther.*, **16**, 667–675 (1974).

MacDonald, A. D., G. Woolfe, F. Bergel, A. L. Morrison, and H. Rinderknecht, Analgesic Action of Pethidine Derivatives and Related Compounds, *Brt. J. Pharmacol.*, **1**, 4–14 (1946).

Mather, L. E., G. T. Tucker, A. E. Pflug, M. J. Lindop, and C. Wilkerson, Meperidine Kinetics in Man, *Clin. Pharmacol. Ther.*, **17**, 21–30 (1975).

Miller, J. W. and H. H. Anderson, The Effect of *N*-Demethylation on Certain Pharmacologic Actions of Morphine, Codeine, and Meperidine in the Mouse, *J. Pharmacol. Exp. Ther.*, **112**, 191–196 (1954).

Mitchard, M., M. J. Kendall, and K. Chan, Pethidine N-oxide: A Metabolite in Human Urine, *J. Pharm. Pharmacol.*, **24**, 915 (1972).

Plotnikoff, N. P., H. W. Elliot, and E. L. Way, The Metabolism of N-C^{14}H$_3$ Labeled Meperidine, *J. Pharmacol. Exp. Ther.*, **104**, 377–386 (1952).

Szeto, H. H. and C. E. Inturrisi, Simultaneous Determination of Meperidine and Normeperidine in Biofluids, *J. Chromatogr.*, **125**, 503–510 (1976).

Way, E. L. et al., The Absorption, Distribution and Excretion of Isonipecaine (Demerol), *J. Pharmacol. Exp. Ther.*, **96**, 477–484 (1949).

Winek, C. L., Tabulation of Therapeutic, Toxic, and Lethal Concentrations of Drugs and Chemicals in Blood, *Clin. Chem.*, **22**, 832–836 (1976).

6-MERCAPTOPURINE—by T. L. Ding

Purinethol. Antineoplastic agent.

$C_5H_4N_4S$; mol. wt. 152.19; pK_a 2.5, 7.8, and 10.8.

a. Therapeutic Concentration Range. Undefined.

b. Metabolism. Mercaptopurine is activated to nucleotides of 6-mercaptopurine (e.g., thioinosinic acid), 6-methylmercaptopurine, and 6-thioguanine (Allan et al., 1966; Zimmerman et al., 1974). It is inactivated by xanthine oxidase to thiouric acid, which is excreted into the urine (Elion, 1967). Inorganic sulfate resulting from cleavage of the sulfur from the purine ring can also be detected in the urine.

c. **Analogous Compounds.** Purine analogs (e.g., thioguanine, aza-thioprine); pyrimidines (see also 5-fluorouracil monograph).

d. **Analytical Methods**

1. **High Performance Liquid Chromatography**—Ding and Benet (1979)

Discussion. The assay involves ethyl acetate extraction of 6-mercaptopurine from plasma together with the standard 9-methyl-mercaptopurine. Reverse phase HPLC and UV detection at 325 nm provide a sensitivity of 5 ng/ml for 6-mercaptopurine with a C.V. of 4.6% ($n = 8$ at 50 ng/ml). The compound is stable in frozen plasma for at least 7 weeks. Average recovery of 6-mercaptopurine from spiked plasma is $41 \pm 4.8\%$. Irreversible loss of 6-mercaptopurine during the extraction step through binding to plasma proteins is prevented with dithioerythritol. Azathioprine, a 6-mercaptopurine prodrug, is removed prior to the extraction by washing the acidified plasma sample with ethyl acetate. This step prevents interference of the 6-mercaptopurine assay by azathioprine, which cleaves to 6-mercaptopurine in the presence of dithioerythritol.

Detailed Method. To 1 ml plasma, 120 ng 9-methylmercaptopurine as an internal standard for 6-mercaptopurine, 200 μl $2N$ HCl, and 5 ml ethyl acetate are added. The mixture is shaken for 10 min and cent-rifuged. The organic layer is discarded. Another 5 ml ethyl acetate is added to the plasma, and the process is repeated. To the remaining aqueous layer, 10 μl of a 1% solution of dithioerythritol in distilled water, 1 ml 1 M sodium acetate buffer (pH 5.1), and 10 ml ethyl acetate are added. The sample is shaken for 10 min and centrifuged. The organic layer is transferred to another tube and evaporated to dryness under nitrogen. The sample is reconstituted with a mixture of 50 μl HPLC buffer eluent, 50 μl 0.2 N sulfuric acid, and 100 μl ethyl acetate. The tube is vortexed for 1 min and centrifuged, and 90 μl of the aqueous phase is injected onto the HPLC column. 6-Mercaptopurine concentra-tions are determined by comparing the peak height ratio of 6-mercaptopurine to its internal standard with a standard curve.

The HPLC analysis is performed using a Perkin-Elmer liquid chromatograph Series 2, equipped with a LC-55 Perkin-Elmer spec-trophotometer for detection at 325 nm and a LiChrosorb RP-18 column, 10 μ particle size, 4.6 mm (i.d.) $\times$ 25 cm (E. Merck, Darmstat, West

Germany). The eluent, containing 1% methanol, 0.5% acetonitrile, and 60 mg/l dithioerythritol in 0.005 M potassium phosphate buffer (pH 4.0), is used at a flow rate of 2 ml/min.

2. Other HPLC Assays—Breter (1977), Breter and Zahn (1976, 1977), Day et al. (1978), Nelson et al. (1973), Tidd and Dedhar (1978), Zimmerman et al. (1974)

Discussion. The earlier HPLC techniques are mainly employed to separate various thiopurines in tissue extracts. Day et al. (1978) assayed for mercaptopurine in plasma using paired ion HPLC. A sensitivity of 0.2 μg of mercaptopurine/ml is reported. The method of Tidd and Dedhar (1978) employs HPLC with fluorescence detection for the analysis of 6-thioguanine nucleosides and nucleotides as active metabolites of 6-mercaptopurine. 6-Thioguanine (and 6-thioxanthine) derivatives are oxidized to fluorescent products by treatment with alkaline permanganate prior to analysis. The method is highly selective and sensitive.

3. Fluorescence—Finkel (1967)

Discussion. Fluorescence is measured after 6-mercaptopurine is oxidized to purine-6-sulfonate by potassium permanganate. The method is subject to potential interference by metabolites and is not sensitive enough for adequate study of the disposition kinetics of 6-mercaptopurine.

4. Colorimetry—Loo et al. (1968)

Discussion. 6-Mercaptopurine is reduced, diazotized, coupled with Bratton-Marshall reagent, and determined colorimetrically. The method lacks specificity in the presence of 6-mercaptopurine metabolites.

5. Spectrophotometry—Chalmers (1975).

Discussion. Urinary 6-mercaptopurine and thiouric acid levels are measured with a UV spectrophotometer after column chromatography.

6. Gas Chromatography—Bailey et al. (1975)

Discussion. 6-Mercaptopurine is extracted from plasma, together with the internal standard theophyllin, derivatized by flash methylation with trimethylphenyl ammonium hydroxide, and analyzed by GC-FID. The sensitivity is 0.2 μg/ml plasma.

7. Gas Chromatography-Mass Spectrometry—Rosenfeld et al. (1977)

Discussion. Electron impact GC-MS is used in the selected ion monitoring mode to sensitively detect 6-mercaptopurine. The drug is removed from plasma by alkylative extraction with methyl iodide into an organic solvent. The alkylation yields two methylated products with the methyl groups in the C_6—S position and in the N—9 or N—7 position. Both derivatives give suitable ion peaks for quantitative analysis and are formed in a constant ratio. S-Methyl-N-9-trideuteromethyl-6-mercaptopurine serves as the internal standard. Analytical recovery from plasma is 18%, and sensitivity is 20 ng/ml plasma.

8. Comments. The advantages of HPLC methods include high sensitivity, specificity, precision, and simplicity. Thiouric acid, the chief metabolite, is very poorly extracted from plasma and is readily separated from 6-mercaptopurine. Azathioprine can decompose to 6-mercaptopurine and interfere with the assay, unless separated prior to analysis (Ding and Benet, 1978). The relationship between plasma levels of 6-mercaptopurine and its intracellular active nucleotide metabolites remains unknown. Nucleotide metabolites are best separated by ion-exchange HPLC, although ion-pair–reverse phase HPLC can also be applicable.

e. References

Allan, P. W., H. P. Schnebli, and L. L. Benett, Conversion of 6-Mercaptopurine and 6-Mercaptopurine Ribonucleoside to 6-Methylmercaptopurine Ribonucleotide in Human Epidermoid Carcinoma No. 2 Cells in Culture, *Biochim. Acta*, **114**, 647–650 (1966).

Bailey, D. G., T. W. Wilson, and G. E. Johnson, A Gas Chromatographic Method for Measuring 6-Mercaptopurine in Serum, *J. Chromatogr.*, **111**, 305–311 (1975).

Breter, H. J., The Quantitative Determination of Metabolites of 6-Mercaptopurine in Biological Materials. I: A Separation Method for Purine and 6-Thiopurine Bases and Nucleosides Using High-Pressure Liquid Chromatography, *Anal. Biochem.*, **80**, 9–19 (1977).

Breter, H. J. and R. K. Zahn, Determination of Radioactively Labeled Derivatives of (8-^{14}C)-6-mercaptopurine in Homogenates of L5178Y Mouse Lymphoma Cells Using High Pressure Liquid Cation Exchange Chromatography, *Z. Anal. Chem.*, **279**, 151–152 (1976).

Breter, H. J. and R. K. Zahn, The Quantitative Determination of Metabolites of 6-Mercaptopurine in Biological Materials. II: Advantages of a Variable Wavelength HPLC spectrophotometric detector for the Determination of 6-Thiopurines, *J. Chromatogr.*, **137**, 61–68 (1977).

Chalmers, A. H., A Spectrophotometric Method for the Estimation of Urinary Azathioprine, 6-Mercaptopurine, and 6-Thiouric Acid, *Biochem. Med.*, **12**, 234–241 (1975).

Day, J. L., L. Tterlikkis, R. Niemann, A. Mobley, and C. Spikes, Assay of Mercaptopurine in Plasma Using Paired-Ion High-Performance Liquid Chromatography, *J. Pharm. Sci.*, **67**, 1027–1028 (1978).

Ding, T. L. and L. Z. Benet, Determination of 6-Mercaptopurine and Azathioprine in Plasma by High-Pressure Liquid Chromatography, *J. Chromatogr. Biomed. Applic.*, **163**, 281–288 (1979).

Elion, G. B., Biochemistry and Pharmacology of Purine Analogs, *Fed. Proc.*, **26**, 898–904 (1967).

Finkel, J. M., A Fluorometric Method for the Estimation of 6-Mercaptopurine in Serum, *Anal. Biochem.*, **21**, 362–371 (1967).

Loo, T. L., J. K. Luce, M. P. Sullivan, and E. Frei, Clinical Pharmacologic Observations on 6-Mercaptopurine and 6-Methyl-thiopurine Ribonucleoside, *Clin. Pharmacol. Ther.*, **9**, 180–194 (1968).

Nelson, D. J., C. J. L. Bugge, H. C. Krasny, and T. P. Zimmerman, Separation of 6-Thiopurine Derivatives on DEA-Sephadex Column and the High Pressure Liquid Chromatograph, *J. Chromatogr.*, **77**, 181–190 (1973).

Rosenfeld, J. M., V. Y. Taguchi, B. L. Hillcoat, and M. Kawal, Determination of 6-Mercaptopurine in Plasma by Mass Spectrometry, *Anal. Chem.*, **49**, 725–727 (1977).

Tidd, D. M., and S. Dedhar, Specific and Sensitive Combined High-Performance Liquid Chromatographic-Flow Fluorometric Assay for Extracellular 6-Thioguanine Nucleotide Metabolites of 6-Mercaptopurine and 6-Thioguanine, *J. Chromatogr.*, **145**, 237–246 (1978).

Zimmerman, T. P. and al., Identification of 6-Methyl-mercaptopurine Ribonucleoside 5′-Diphosphate and 5′-Triphosphate as Metabolites of 6-Mercaptopurine in Man, *Cancer Res.*, **34**, 221–224 (1974).

METHADONE—by M. Raeder-Schikorr

6-(Dimethylamino)-4,4-diphenyl-3-heptanone
dl-form: Algidon, l-form: Levadone, Polamidon, Narcotic analgesic.

$$CH_2CH_3$$
$$C=O \quad CH_3$$

$$-C-CH_2CH-N(CH_3)_2$$

$C_{21}H_{27}NO$; mol. wt. 309.46; pK_a 8.25.

a. Therapeutic Concentration Range. The average dose of 1 mg methadone/kg·day during a maintenance program for heroin addicts yields methadone plasma levels between 100 and 400 ng/ml (Hachey et al., 1977).

b. Metabolism. Methadone is mainly metabolized in the liver by *N*-demethylation, followed by cyclization to 2-ethyl-1,5-dimethyl-3,3-diphenyl-1-pyrroline as the major metabolite of methadone in the urine (Beckett et al., 1968). The plasma elimination half-life is 28.8 ± 4.8 hr (S.D.) (Hachey et al., 1977).

c. Analogous Compounds. 1-Acetylmethadol, diphenoxylate, propoxyphene, SKF 525-A.

d. Analytical Methods

1. Gas Chromatography—Lynn et al. (1977)

Discussion. Methadone is analyzed in whole-blood samples, saliva, and gastric juice with GC, using flame ionization detection. Quantities as low as 5 ng/ml can be reproducibly determined; however, sample volume requirements are rather large (15 ml) in order to achieve this sensitivity. The internal standard, 2-dimethylamino-4,4-diphenyl-5-nonanone, is a chemical analog of methadone. The sample preparation consists of three liquid-liquid extraction steps without evaporation.

Detailed Method. Blood samples, as well as saliva or gastric juice, can be extracted as follows: 15 ml sample fluid is added to 30 ml 1-chlorobutane, 3 ml 0.4 *M* sodium carbonate, and about 1 μg of the internal standard. The tubes are shaken for 20 min and centrifuged at 1500 *g* for 10 min. The organic phase (25 ml) is equilibrated with 2 ml 0.5 *M* sulfuric acid and again centrifuged at 1500 *g* for 10 min. The aqueous layer is alkalinized by addition of 1 ml 13.5% ammonium hydroxide and extracted with chloroform. After centrifugation, 1 to 5 μl of the chloroform layer is injected onto the GC column.

Analysis is performed on a gas chromatograph (Hewlett-Packard Model 5830), using flame ionization detection. The column is 6 ft x 2 mm (i.d.) glass, packed with 1.5% OV-101 on Gas Chrom Q, 100/120 mesh. The oven temperature is programmed from 170 to 250° at 10°/min. The injector and detector temperatures are 250 and 275°C, respectively. The hydrogen, air, and nitrogen flow rates are 30, 250, and 30 ml/min, respectively. Standard curves are prepared by plotting peak area ratios of methadone to the internal standard versus methadone levels.

2. Other GC Methods—Beckett et al. (1971), Inturrisi and Verebely (1972)

Discussion. The GC method of Inturrisi and Verebely (1972) employs the chemical methadone analog, SKF 525-A, as the internal standard and SE-30 as the stationary phase. The sensitivity is 15 ng methadone/ml plasma. Two pyrrolidine metabolites, other than the major quaternary metabolite, have also been detected in human urine following organic solvent extraction. Beckett et al. (1971) describe a GC method for the detection of methadone and a series of its metabolites, including pyrroline derivatives. Methadone N-oxide can also be measured following chemical reduction to the parent tertiary amine.

3. Gas Chromatography-Mass Spectrometry—Hachey et al. (1977), Sullivan et al. (1975)

Discussion. Hachey et al. (1977) describe the synthesis of penta-deuterated *d*-, *l*-, and *dl*-methadone labeled in one phenyl ring in order to study the pharmacokinetics of the stereoisomers of methadone. The assay of methadone in plasma involves a double back-extraction and GC-chemical ionization (isobutane)-MS in the selected ion monitoring mode [MH$^+$ (d_0) m/e 310 and (d_5) m/e 315]. The sensitivity is 3 ng methadone/ml plasma. Sullivan et al. (1975) use a single extraction method followed by GC-30 eV electron impact MS in the selected ion monitoring mode (M$^+$ − 15 at m/e 294) with dimethylamino-4,4-diphenyl-5-octanone (m/e 308) as the internal standard. 7-Trideuteriomethadone (m/e 297) is administered to human subjects on a methadone maintenance program in order to determine the pharmacokinetic parameters of methadone without interrupting the therapy. The sensitivity is about 5 ng methadone/ml plasma.

4. Ultraviolet Spectrophotometry—Hamilton et al. (1974)

Discussion. The detection of methadone in urine and tissue is difficult because of its low UV absorption ($\epsilon_{292\ nm} = 554$). Oxidation of methadone to benzophenone with ceric sulfate increases the molar absorptivity approximately 34 times (ϵ_{247} nm $= 18,713$). Amounts of methadone as low as 5 μg can be detected in biological specimens. The procedure involves an alkaline extraction into n-hexane and subsequent back-extraction into a ceric sulfate-sulfuric acid solution. After refluxing the acid with n-heptane for 30 min, the generated benzophenone is measured in the n-heptane layer by UV spectrophotometry. Interference may occur because of oxidation of analogous compounds to benzophenone- or acetophenone-related chemical structures. A GC assay of methadone is also described in this report.

5. Thin-Layer Chromatography—Ho and Loh (1972)

Discussion. Methadone can be detected by TLC in concentrations as low as 100 ng/ml in unhydrolyzed urine samples. Extensive sample preparation is not necessary, as urinary salts and pigments do not influence R_f values or spot detection with iodoplatinate reagent. This method can also be utilized for combined detection of morphine, codeine, quinine, and meperidine. There is no interference by amphetamine or barbiturates. This method is mainly used as a qualitative tool for urinary methadone detection in clinical laboratories.

6. Radioimmunoassay—Digregorio et al. (1977)

Discussion. This study employs a RIA for the detection of methadone in parotid saliva and blood of rats. A commercial RIA kit (Roche Diagnostics Lab.) is used. Details of the assay conditions are not given.

7. Enzyme Immunoassay—Broughton and Ross (1975)

Discussion. The methadone screening assay is based on the EMIT procedure, utilizing lysozyme enzyme and bacterial lysis. Assay sensitivity is about 0.5 μg/ml urine.

8. Comments. Gas chromatography and GC-MS are the most versatile tools to measure methadone in biological fluids.

e. References

Beckett, A. H., J. F. Taylor, A. F. Casey, and M. M. A. Hassan, The Biotransformation of Methadone in Man: Synthesis and Identification of a Major Metabolite, *J. Pharm. Pharmacol.* **20,** 754–762 (1968).

Beckett, A. H., M. Mitchard, and A. A. Shihab, Identification and Quantitative Determination of Some Metabolites of Methadone, Isomethadone and Normethadone, *J. Pharm. Pharmacol.*, **23,** 347–352 (1971).

Broughton, A., and D. L. Ross, Drug Screening by Enzymatic Immunoassay with the Centrifugal Analyzer, *Clin. Chem.*, **21,** 186–189 (1975).

Digregorio, G. J., A. J. Piraino, and E. K. Ruch, Radioimmunoassay of Methadone in Rat Parotid Saliva, *Drug Alcohol Depend.*, **2,** 295–298 (1977).

Hachey, D. L., M. J. Kreek, and D. H. Matson, Quantitative Analysis of Methadone Using Deuterium-Labeled Methadone and GLC-Chemical Ionization Mass Spectrometry, *J. Pharm. Sci.*, **66,** 1579–1582 (1977).

Hamilton, H. E., J. E. Wallace, and K. Blum, Improved Methods for Quantitative Determination of Methadone, *J. Pharm. Sci.*, **63,** 741–745 (1974).

Ho, I. K. and H. H. Loh, Mini Thin-Layer Chromatography in the Detection of Narcotics in Urine from Subjects on a Methadone Maintenance Program, *J. Chromatogr.*, **65,** 577–579 (1972).

Inturrisi, C. E. and K. Verebely, A Gas Liquid Chromatographic Method for the Quantitative Determination of Methadone in Human Plasma and Urine, *J. Chromatogr.*, **65,** 361–369 (1972).

Lynn, R. K., R. M. Leger, W. P. Gordon, G. D. Olsen, and N. Gerber, New Gas Chromatographic Assay for the Quantitation of Methadone, *J. Chromatogr.*, **131,** 329–340 (1977).

Sullivan, H. R. et al., Mass Fragmentographic Determination of Unlabeled and Deuterium Labeled Methadone in Human Plasma: Possibilities for Measurement of Steady State Pharmacokinetics, Biomed. Mass Spectrom., **2,** 197–200 (1975).

METHAQUALONE

2-Methyl-3-*o*-tolyl-4(3H)-quinazolinone.
Mandrax, Quaalude, Revonal, and so on. Sedative, hypnotic.

$C_{16}H_{14}N_2O$; mol. wt. 250.29; pK_a 2.54.

a. Therapeutic Concentration Range. Therapeutic oral doses of 4 mg methaqualone/kg yield peak plasma levels of 2 to 3 μg/ml (Alvan et al., 1973; Morris et al., 1972). Toxic levels resulting from methaqualone abuse and overdose range from 5 to 30 μg/ml, and unconsciousness usually occurs above 8 μg/ml (Bailey and Jatlow, 1973).

b. Metabolism. Methaqualone is extensively metabolized by hydroxylation, followed by glucuronidation and excretion into the urine (Kazyak et al., 1977). While urinary excretion of unchanged methaqualone is insignificant ($<1\%$, Heck et al., 1978), approximately one third of the dose can be accounted for in the urine by conjugated hydroxy metabolites (Ericsson and Danielsson, 1977). 4'-Hydroxy-and 2'-hydroxymethylmethaqualone represent the major oxidative metabolites (Kazyak et al., 1977; Ericsson and Danielsson, 1977), with high levels of the 4'-OH metabolite occurring in the bile (Bonnichsen et al., 1975). Other metabolites include 2-hydroxymethyl-, 6-hydroxy-, 3'-hydroxy-, and 5'-hydroxymethaqualone (Bonnichsen et al., 1975; Ericsson and Danielsson, 1977; Heck et al., 1978; Kazyak et al., 1977). In addition, dihydroxylated metabolites (Bonnichsen et al., 1974; Preuss and Hassler, 1970), a dihydrodiol (Kazyak et al., 1977; Stillwell et al., 1975), 4'-hydroxy- 5'-methoxymethaqualone, and methaqualone N-oxide (Ericsson and Danielsson, 1977) have been isolated from urine specimens.

Methaqualone possesses a high tissue affinity due to its lipophilicity and therefore exhibits multiphasic elimination kinetics. The plasma elimination half-life is initially 2.6 hr (Morris et al., 1972). However, the $t_{1/2}$ is on the order of 37 hr when methaqualone plasma levels are measured over several days with a more sensitive assay (Alvan et al., 1974). The terminal elimination $t_{1/2}$ may be as long as 72 hr, as determined by a highly sensitive MS analysis of urinary methaqualone excretion over a 30 day period after a single dose (Heck et al., 1978). Substantial drug accumulation during repetitive dosing has to be expected with such slow elimination rates.

c. Analogous Compounds. Hypnotics, sedatives.

d. Analytical Methods

1. Gas Chromatography—Kazyak et al. (1977)

Discussion. Kazyak et al. (1977) determine methaqualone and five principal hydroxylated metabolites in the urine by high resolution capillary column GC with flame ionization detection. Packed columns containing either the nonpolar SE-30 or the moderately polar OV-17 and OV-21 liquid phases did not provide sufficient resolution of the isomeric hydroxylated metabolites. Analytical recoveries of the extraction procedure average 91% in the concentration range of 0.1 to 1 μg/ml urine for the hydroxylated metabolites, while the recovery of methaqualone is 95%. The hydroxylated metabolites are silanized and separated in SE-30-coated glass tubes under isothermal conditions. External standard curves of known amounts of methaqualone and four chemically prepared metabolites are utilized for quantitation. Urine samples are treated with a β-glucuronidase-aryl sulfatase prior to extraction in order to determine the urinary conjugates, which are present at higher concentrations than the respective hydroxylated metabolites.

Detailed Method. Urine (5 ml) is incubated with glusulase (Endo Labs., Garden City, NY), containing 6000 Fishman units of aryl sulfatase and 10,000 Fishman units of β-glucuronidase, in 3 ml sodium acetate buffer (pH 4.6) for 18 hr at 37°C. The pH is then adjusted to 8 $\pm$ 0.5 with a few drops of saturated potassium hydroxide, and the mixture extracted with 50 ml chloroform. The higher pH results in less interference during GC without decreasing the extraction yield of phenolic methaqualone metabolites. A 40 ml aliquot of the solvent extract is filtered through Whatman 41 filter paper to remove traces of water and evaporated in a stream of nitrogen at 50°C. The residue is dissolved in 25 μl of N,O-bis(trimethylsilyl)acetamide-dimethylformamide (1:1), and 1 μl injected onto the column.

The GC analysis is performed on a 50 m open tubular glass column (0.76 mm i.d.) coated with SE-30 (Applied Science Lab., State College, Pa.) for a liquid phase. The column is maintained at 2(5°C, and eluting compounds are measured with a flame ionization detector. Peak areas are measured by integration using a Hewlett-Packard (Avondale, Pa.) Model 3370B electronic integrator. A linear calibration curve is maintained by computer on-line, as determined from the first three standardized mixtures of methaqualone and synthetic metabolites in concentrations ranging from 0.1 to 1 μg/ml urine. Metabolite standards include 2'-hydroxymethyl-, 3'-hydroxy-, 4'-hydroxy-, and 6-hydroxymethaqualone.

2. Other GC Assays—Berry (1969), Morris et al. (1972)

Discussion. These GC assays have been designed for the detection of methaqualone in plasma during the first hours after a therapeutic dose.

Selective extraction of methaqualone is achieved at pH 10 into hexane, and the sensitivity is 0.5 μg/ml using GC-FID. Codeine (Morris et al., 1972) and butobarbitone (Berry, 1969) serve as the internal standards. These GC assays are reported to be specific for methaqualone.

3. Mass Spectrometry—Alvan et al. (1973), Bonnichsen et al. (1974), Ericsson and Danielsson (1977), Heck et al. (1978), McReynolds et al. (1975), Permisohn et al. (1976)

Discussion. The need for greater sensitivity than is obtainable by GC-FID spurred interest in the use of MS as the detection mode. Alvan et al. (1973) describe a GC-MS assay, using 50 eV electron impact ionization in the selected ion monitoring mode, with a sensitivity of 0.1 μg methaqualone/ml plasma. The GC liquid phase is 5% SE-52 at 200°C. Methaqualone and its internal standard, the 2-ethyl analog, are monitored at their respective parent ions, *m/e* 250 and 264. Permisohn et al. (1976) include hydroxylated methaqualone metabolites with the GC-MS (dodecapole) analysis of urine samples. Conjugates are liberated by hot acid hydrolysis; however, treatment with mineral acid causes partial decomposition of metabolites (Kazyak et al., 1977; Bonnichsen et al., 1974), and subsequent studies employ enzymatic deconjugations.

The oxidative urinary metabolites can be determined after glucuronidase treatment and acetylation, using electron impact GC-MS with 6% QF-1 on Gas Chrom Q, which provides excellent separation efficiency (Ericsson and Danielsson, 1977). Ion currents are monitored at *m/e* 308 and 266, the latter being common to all metabolites with unchanged quinazoline ring moiety (i.e., not 6-hydroxymethaqualone). 6'-Hydroxymethaqualone, which is not a metabolite, serves as the internal standard. The GC-MS method of Bonnichsen et al. (1974) is similar to that of Ericsson and Danielsson (1977), except for derivatization by silylation and repetitive scanning rather than selected ion monitoring. Ion currents of interest are then determined, and mass chromatograms are constructed with the aid of a computer. Although this procedure is less sensitive than selected ion monitoring, it provides greater flexibility for metabolite screening. Screening for and quantitation of methaqualone metabolites in urine extracts is performed with mass chromatograms, utilizing 20 different *m/e* units that are potentially characteristic for methaqualone congeners.

The most sensitive MS assay for methaqualone and its 6-hydroxy metabolite presently available employs direct insertion, field ionization MS, which suppresses fragmentations, thereby providing greater selectivity and high relative ion abundance at the quasi-molecular ions

(Heck et al., 1978; McReynolds et al., 1975). Heptadeuterated methaqualone and 6-hydroxymethaqualone serve as the internal standard-analytical carrier, providing a sensitivity limit of 1 ng/ml urine for both compounds. In spite of the high selectivity of the detection mode, extensive sample prepurification by column chromatography, organic solvent extraction, and two-dimensional TLC is necessary to achieve high sensitivity in biological samples.

4. Spectrophotometry—Bailey and Jatlow (1973)

Discussion. Spectrophotometric assays of methaqualone in biological samples usually yield higher values than GC assays because of the presence of hydroxylated metabolites with similar UV spectra. Using a series of extraction steps, including partitioning into hexane, Bailey and Jatlow (1973) are able to specifically measure the lipophilic methaqualone at 235 and 255 nm in the presence of more polar oxidative metabolites, thus obtaining results comparable to GC data. Because of lack of sensitivity, this method is useful only as a screening procedure for methaqualone abuse and overdose.

5. Comments.
The increasing abuse of methaqualone has made it one of the major street drugs and has fostered many studies on the detection of this compound and its metabolites in urine and plasma. One of the outstanding pharmacokinetic features of methaqualone is its very long terminal elimination half-life, which can be measured with highly sensitive assays, such as GC-MS. Detection of methaqualone ingestion remains possible several weeks after the dose was taken.

e. References

Alvan, G., J.-E. Lindgren, C. Bogentoft, and O. Ericsson, Plasma Kinetics of Methaqualone in Man after Single Oral Doses, *Eur. J. Clin. Pharmacol.*, **6,** 187–190 (1973).

Alvan, G., O. Ericsson, S. Levander, and J.-E. Lindgren, Plasma Concentrations and Effects of Methaqualone after Single and Multiple Oral Doses in Man, *Eur. J. Clin. Pharmacol.*, **7,** 449–454 (1974).

Bailey, D. N. and P. I. Jatlow, Methaqualone Overdose: Analytical Methodology, and the Significance of Serum Drug Concentrations, *Clin. Chem.*, **19,** 615–620 (1973).

Berry, D. J., Gas Chromatographic Determination of Methaqualone, 2-Methyl-3-*o*-tolyl-4(3H)-quinazolinone, at Therapeutic Levels in Human Plasma, *J. Chromatogr.*, **42,** 39–47 (1969).

Bonnichsen, R., Y. Marde, and R. Ryhage, Identification of Free and

Conjugated Metabolites of Methaqualone by Gas Chromatography-Mass Spectrometry, *Clin. Chem.*, **20**, 230–235 (1974).

Bonnichsen, R., R. Dimberg, Y. Marde, and R. Ryhage, Variations in Human Metabolism of Methaqualone Given in Therapeutic Doses and in Overdose Cases Studied by Gas Chromatography-Mass Spectrometry, *Clin. Chim. Acta*, **60**, 67–75 (1975).

Ericsson, O. and B. Danielsson, Urinary Excretion Pattern of Methaqualone Metabolites in Man, *Drug Metab. Dispos.*, **5**, 497–502 (1977).

Heck, H. d'A., K. Maloney, and M. Anbar, Long-Term Urinary Excretion of Methaqualone in a Human Subject, *J. Pharmacokin. Biopharm.*, **6**, 111–122 (1978).

Kazyak, L. et al., Methaqualone Metabolites in Human Urine after Therapeutic Doses, *Clin. Chem.*, **23**, 2001–2006 (1977).

McReynolds, J. H., H. d'A. Heck, and M. Anbar, Determination of Picomole Quantities of Methaqualone and 6-Hydroxymethaqualone in Urine, *Biomed. Mass Spectrom.*, **2**, 299–303 (1975).

Morris, R. N., G. A. Gunderson, S. W. Babcock, and J. F. Zarolinski, Plasma Levels and Absorption of Methaqualone after Oral Administration to Man, *Clin. Pharmacol. Ther.*, **13**, 719–723 (1972).

Permisohn, R. C., L. R. Hilpert, and L. Kazyak, Determination of Metaqualone in Urine by Metabolite Detection via Gas Chromatography, *J. Forensic Sci.*, **21**, 98–107 (1976).

Preuss, R. and H. Hassler., Zur Biotransformation des 2-Methyl-3-o-tolyl-4(3H)-chinazolinon (Methaqualon), *Arzneim.-Forsch.*, **20**, 1920–1922 (1970).

Stillwell, W. G., P. A. Gregory, and M. G. Horning, Metabolism of Methaqualone by the Epoxide-diol Pathway in Man and the Rat, *Drug Metab. Dispos.*, **3**, 287–294 (1975).

METHOTREXATE

N-{*p*-{[(2,4-Diamino-6-pteridinyl)methyl]methylamino}benzoyl}glutamic acid.
Amethopterin. Dihydrofolate reductase inhibitor; antineoplastic agent.

$C_{20}H_{22}N_8O_5$; mol. wt. 454.46; pK_a 3.76, 4.83, and 5.60.

a. Therapeutic Concentration Range. Following high dose (1 to 9 g/m^2) methotrexate therapy with citrovorum factor rescue, methotrexate serum levels persisting above 4.5×10^{-6} M at 48 hr (Wang et al., 1976b) and above 2×10^{-7} M at 96 hr (Frei et al., 1975) are likely to cause bone marrow depression. This toxicity may be partially reversible by prolonged and higher rescue factor treatment (Frei et al., 1975).

b. Metabolism. 7-Hydroxymethotrexate appears to be a major metabolite following high doses of methotrexate (Jacobs et al., 1976). This metabolite may contribute to renal toxicity. The elimination half-life of methotrexate (high dose) ranges from 4 to 24 hr (Frei et al., 1975). The predominant mode of elimination is via the kidneys. High dose methotrexate therapy is therefore contraindicated in patients with renal failure.

c. Analogous Compounds. Folic acid cofactors, folinic acid (citrovorum factor); dichloromethotrexate, methopterine, pyrimethamine, triampterene, trimethoprine.

d. Analytical Methods

1. High Performance Liquid Chromatography—C. Canfell and W. Sadée, (1979)

Discussion. Separation and high sensitivity of detection of methotrexate and 7-hydroxymethotrexate are achieved on a reverse phase alkylphenyl column with UV detection at 303 nm. The selected alkylphenyl column gives better separation and chromatographic behavior of methotrexate and its 7-OH metabolite than do C_{18} chemically bonded reverse phase columns. Serum samples are deproteinized and concentrated in volume by an n-butanol extraction step, which removes interfering compounds while leaving methotrexate and its metabolite in the aqueous phase. Sensitivity is related to the volume of the serum sample because of negligible blank interference and is about 5 ng/ml ($\sim 10^{-8}$ M) for both compounds, using a 2 ml sample. The external standard method is used for quantitation with a precision of about ±5% (S.D.) at 50 ng/ml (1 ml serum samples). Standard curves are linear over a broad concentration range. The required volume for therapeutic drug level monitoring is 0.5 to 1.0 ml serum. 7-Hydroxymethotrexate appears to be the predominant species in serum 24 to 48 hr after the dose (W. Sadée, unpublished data; Watson et al., 1978). Folic acid and folinic acid do not interfere with the assay.

Detailed Method. To 1 ml serum, 1 ml acetonitrile is added to precipitate the proteins, followed by vortexing for 30 sec and centrifuging at 3000 *g* for 5 min. The supernatant is transferred to a glass tube with a narrow stem (100 μl volume) and extracted with a mixture of 5.5 ml ether and 2.9 ml *n*-butanol by shaking for 2 min. After centrifugation the upper organic layer is removed, and the remaining aqueous phase (≥ 100 μl) is reextracted with 2.5 ml ether. After aspiration and evaporation of the ether under a stream of nitrogen, the entire remaining aqueous phase (≥ 100 μl) is centrifuged at 12,800 *g* in an Eppendorf microcentrifuge and injected onto the column.

Chromatography is performed on a Model 6000A liquid chromatograph equipped with a U6K injector and an alkylphenyl μ-Bondapak column (all Waters Assoc.). The variable-wavelength detector (Schoeffel Instruments) is set at 303 nm (UV absorbance maximum of methotrexate). The mobile phase is 0.15 *M* sodium acetate buffer (pH 4.6)-acetonitrile (89:11, v/v) at a flow rate of 2 ml/min. The retention times of methotrexate and its 7-OH metabolite are 6 and 9 min, respectively. Quantitation is achieved using peak height measurements and external standard curves (10 to 10,000 ng/ml serum).

2. Other HPLC Methods—Lankelma and Poppe (1978), Nelson et al. (1977), Watson et al. (1978), Wisnicki et al. (1978)

Discussion. Watson et al. (1978) also utilize HPLC with UV detection. The assay involves $HClO_4$ ion-pair extraction of methotrexate into organic solvents followed by anion-exchange chromatography. This is the first reported method that allows simultaneous determination of methotrexate and its 7-hydroxy metabolite; however, the sensitivity is one order of magnitude lower (10^{-7} *M*) for both compounds than that obtained by us (Canfell and Sadée, 1979).

The method of Wisnicki et al. (1978) employs paired ion (tetrabutylammonium hydroxide) reverse phase HPLC with a μ-Bondapak C_{18} column. The ion pair has suitable chromatographic properties on this column. The primary purpose of this method is to detect by UV monitoring impurities in the methotrexate dose, which might contribute to the effects of high dose methotrexate treatment. It is also applicable to the analysis of urine samples without further purification.

Lankelma and Poppe (1978) propose a dual-column HPLC-UV assay of methotrexate in plasma with a sensitivity of 2×10^{-8} *M*. A short concentration column packed with C_8-alkyl chemically bonded silica gel is placed into a loop between the injector and the analytical column (an anion exchanger chemically bonded to silica gel, i.e., Partisil SAX).

The concentration column is first flushed to remove polar material with the eluent diverted away from the analytical column and then eluted and analyzed with a pH 4.9 phosphate buffer-methanol (4:1). Plasma proteins are precipitated with trichloroacetic acid prior to HPLC analysis.

The HPLC method of Nelson et al. (1977) requires oxidation by potassium permanganate to 2,4-diaminopteridine-6-carboxylic acid, which is the same reaction product as obtained in previous fluorescence assays (e.g., Freeman, 1958). 2-Hydroxyfolic acid undergoes a similar reaction and is used as the internal standard. Analysis is performed on a C_{18} reverse phase column with fluorescence detection. Sensitivity is good (10 ng/ml = 2×10^{-8} M) with excellent specificity; however, the presence of 7-hydroxymethotrexate was not considered in this study.

3. Radioimmunoassay—e.g., Hendel et al. (1976), Loeffler et al. (1976), Paxton (1979), Raso and Schreiber (1975)

Discussion. The published RIA methods are generally more sensitive than HPLC detection of methotrexate and are specific in the presence of endogenous folates, as well as folinic acid. However, RIA specificity in the presence of 7-hydroxymethotrexate is questionable at present and needs further investigation. Commercial RIA kits are available. Paxton (1979) contends that 7-hydroxymethotrexate has insignificant cross-reactivity to the antibody in his studies; however, the potential metabolite 4-amino4-deoxy-N^{10}-methylpteroic acid, when added *in vitro*, does interfere with the methotrexate assay. Thus RIA specificity remains to be investigated.

4. Competitive Protein Binding Assays—Arons et al. (1975), Kamen et al. (1976), Myers et al. (1975)

Discussion. These assays are equivalent to the methotrexate RIA except for substitution of the antibody with dihydrofolate reductase. The binding of methotrexate to this enzyme is very tight in the presence of the cofactor NADPH and exceeds the binding affinities of endogenous folate and folinic acid by several orders of magnitude, thereby allowing specific and sensitive detection of methotrexate in serum. The dynamic range of the standard curve is rather small, necessitating serial dilutions for unknown samples. Interference by 7-hydroxymethotrexate has not yet been thoroughly studied. Watson et al. (1978) reported negligible interference by this metabolite with the assay of Myers et al. (1975).

5. Enzymatic Assays—Epstein et al. (1976), Falk et al. (1976), Finley and Williams (1977)

Discussion. Increasing methotrexate concentrations depress the conversion of NADPH to NADP$^+$ by dihydrofolate reductase in the presence of dihydrofolate. The reaction can be followed by UV absorption at 340 nm. This technique is equivalent to the competitive protein binding assay; however, it requires more time unless automated (Finley and Williams, 1977). Abnormal serum constituents such as high bilirubin, lipid, or hemoglobin may interfere (Finley and Williams, 1977).

6. Fluorescence—Chakrabarti and Bernstein (1969), Freeman (1958), Kinkade et al. (1974), Overdijk et al. (1975)

Discussion. The oxidation of methotrexate with permanganate to a fluorescent product has already been described (Section d2) in connection with the HPLC assay of Nelson et al. (1977). The major problem of all fluorescence assays without chromatography is interference by other pteridines (e.g., folinic acid) which undergo similar reactions. Therefore nonspecificity eliminates direct fluorescence assays from clinical monitoring of high dose methotrexate therapy with folinic acid rescue.

7. Microbiologic Assay—Noble et al. (1975), Wyatt et al. (1976)

Discussion. Several bioassays have been published that are based on the bacterocidal or static effects of methotrexate. Wyatt et al. (1976) propose an interesting variant involving measurement of laser light scattering from suspensions of drug-sensitive bacteria. Assay sensitivity is sufficient for pharmacokinetic studies; however, specificity is low in the presence of other toxic agents.

8. Mass Spectrometry—Hignite and Azarnoff (1978), Kirk et al. (1976)

Discussion. Field desorption MS yields abundant molecular ions. Applicability of this technique to quantitative serum analysis is doubtful. Methotrexate and folic acid derivatives can be identified after isolation by MS (Hignite and Azarnoff, 1978).

9. Comments. The emerging awareness of the presence of 7-hydroxymethotrexate as a potentially toxic or active agent makes HPLC with UV detection the method of choice for drug level monitoring. For the detection of methotrexate alone, HPLC-fluorescence, RIA, and enzyme protein binding methods are suitable. A particular problem arises from the clinical need to study very large concentration ranges of methotrexate (from 10 ng to 10 μg/ml serum), which requires several dilutions of plasma samples when using methods with a narrow analytical detection range (e.g., competitive protein binding assays). The HPLC procedures overcome this problem. Assay specificity has to be reevaluated for most of the published assays, since the 7-OH metabolite may exceed methotrexate levels by an order of magnitude at 24 to 48 hr after the dose. High dose methotrexate therapy has also fostered interest in minor chemical contaminants in methotrexate preparations (Wang et al., 1976a; Wisnicki et al., 1978).

e. References

Arons, E., S. P. Rothenberg, M. D. Costa, C. Fischer, and M. P. Iqbal, A Direct Ligand-Binding Radioassay for the Measurement of Methotrexate in Tissue and Biological Fluids, *Cancer Res.*, **35**, 2033–2038 (1975).

Canfell, C. and W. Sadée, Methotrexate and 7-Hydroxymethotrexate Serum Level Monitoring by High Performance Liquid Chromatography, *Cancer Treat. Rep.*, in press (1979).

Chakrabarti, S. G. and I. A. Bernstein, A Simplified Fluorometric Method for Determination of Plasma Methotrexate, *Clin. Chem.*, **15**, 1157–1161 (1969).

Epstein, E., E. S. Baginski, and B. Zac, Monitoring the Administration of Methotrexate in Antimetabolite Therapy, *Ann. Clin. Lab. Sci.*, **6**, 312–317 (1976).

Falk, L. C., D. R. Clark, S. M. Kalman, and T. F. Long, Enzymatic Assay for Methotrexate in Serum and Cerebrospinal Fluid, *Clin. Chem.*, **22**, 785–788 (1976).

Finley, P. R. and R. J. Williams, Methotrexate Assay by Enzymatic Inhibition with Use of the Centrifugal Analyzer, *Clin. Chem.*, **23**, 2139–2141 (1977).

Freeman, M. V., The Fluorometric Measurement of the Absorption, Distribution, and Excretion of Single Doses of 4-Amino-10-methylpteroylglutamic Acid in Man, *J. Pharmacol. Exp. Ther.*, **122**, 154–158 (1958).

Frei, E., N. Jafee, M. H. N. Tattersall, S. Pitman, and L. Parker, New Approaches to Cancer Chemotherapy with Methotrexate, *N. Engl. J. Med.*, **292**, 846–851 (1975).

Hendel, J., L. J. Sarek, and E. F. Hvidberg, Rapid Radioimmunoassay for Methotrexate in Biological Fluids, *Clin. Chem.*, **22**, 813–816 (1976).

Hignite, A. J. and A. C. Azarnoff, Identification of Methotrexate and Folic Acid Analogs by Mass Spectrometry, *Biomed. Mass Spectrom.*, **5**, 161–165 (1978).

Jacobs, S. A., R. G. Stoller, B. A. Chabner, and D. G. Johns, 7-Hydroxymethotrexate as a Urinary Metabolite in Human Subjects and Rhesus Monkeys Receiving High Dose Methotrexate, *J. Clin. Invest.*, **57**, 534–548 (1976).

Kamen, B. A., P. L. Takach, R. Vatev, and J. D. Caston, A Rapid, Radiochemical Ligand Binding Assay for Methotrexate, *Anal. Biochem.*, **70**, 54–63 (1976).

Kinkade, J. M., Jr., W. R. Vogler, and P. G. Dayton, Plasma Levels of Methotrexate in Cancer Patiens as Studied by an Improved Spectrophotofluorometric Method, *Biochem. Med.*, **10**, 337–350 (1974).

Kirk, M. C., W. C. Coburn, Jr., and J. R. Piper, Field Desorption Mass Spectra of Methotrexate and Folic Acid Analogs, *Biomed. Mass Spectrom.*, **3**, 245–247 (1976).

Lankelma, J. and H. Poppe, Determination of Methotrexate in Plasma by On-Column Concentration and Anion-Exchange Chromotography, *J. Chromatogr.*, **149**, 587–598 (1978).

Loeffler, L. J., M. R. Blum, and M. A. Nelsen, A Radioimmunoassay for Methotrexate and Its Comparison with Spectrofluorimetric Procedures, *Cancer Res.*, **36**, 3306–3311 (1976).

Myers, C. E., M. E. Lippman, H. M. Elliot, and B. A. Chabner, Competitive Protein Binding Assay for Methotrexate, *Proc. Natl. Acad. Sci.*, **72**, 3683–3686 (1975).

Nelson, J. R., B. A. Harris, W. J. Decker, and D. Farquhar, Analysis of Methotrexate in Human Plasma by High-Pressure Liquid Chromatography with Fluorescence Detection, *Cancer Res.*, **37**, 3970–3973 (1977).

Noble, W. C., M. R. Path, P. M. White, and H. Baker, Assay of Therapeutic Doses of Methotrexate in Body Fluids of Patients with Psoriasis, *J. Invest. Dermatol.*, **64**, 69–76 (1975).

Overdijk, B., W. M. J. van der Kroef, A. A. M. Visser, and G. J. M. Hooghwinkel, The Determination of Methotrexate in Serum and Urine, *Clin. Chim. Acta*, **59**, 177–182 (1975).

Paxton, J. W., Interference of 4-Amino-4-deoxy-N^{10}-methyl-pteroic Acid with the Radioimmunoassay of Methotrexate, *Clin. Chem.*, **25**, 491–492 (1979).

Raso, V. and R. Schreiber, A Rapid and Specific Radioimmunoassay for Methotrexate, *Cancer Res.*, **35**, 1407–1410 (1975).

Wang, Y., P. Kim, D. C. van Eys, and W. W. Sutow, Study of Contaminants and Metabolites During Therapy with High Doses of Methotrexate, *Clin. Chem.*, **22**, 1937 (1976a).

Wang, Y., E. Lantin, and W. W. Sutow, Methotrexate in Blood, Urine and Cerebrospinal Fluid of Children Receiving High Doses by Infusion, *Clin. Chem.*, **22**, 1053–1056 (1976b).

Watson, E., J. L. Cohen, and K. K. Chan, High-Pressure Liquid Chromatographic Determination of Methotrexate and Its Major Metabolite, 7-Hydroxymethotrexate, in Human Plasma, *Cancer Treat. Rep.*, **62**, 381–387 (1978).

Wisnicki, J. L., W. P. Tong, and D. B. Ludlum, Analysis of Methotrexate and 7-Hydroxymethotrexate by High-Pressure Liquid Chromatography, *Cancer Treat. Rep.*, **62**, 529–532 (1978).

Wyatt, P. J., R. F. Pittillo, L. S. Rice, C. Wooley, and L. B. Mellett, Laser Differential Light-Scattering Bioassay for Methotrexate (NSC-740), *Cancer Treat. Rep.*, **60**, 225–233 (1976).

METHYLDOPA—by C. O. Gloff

L-3-(3,4-Dihydroxyphenyl)-2-methylalanine; α-methyldopa.

Aldomet, Aldomin, Sembrina. Antihypertensive agent.
$C_{10}H_{13}NO_4$; mol. wt. 211.21; pK_a 2.2, 9.2, 10.6, and 12.0.

a. Therapeutic Concentration Range. None available. Plasma levels of methyldopa in human beings 60 min after an intravenous dose of 250 mg are about 6 $\mu g/ml$ (Saavedra et al., 1975).

b. Metabolism. Methyldopa is extensively metabolized to a large number of products, mainly by decarboxylation, leading to pharmacologically active metabolites, and by 3-O-methylation. Metabolites include 3-O-methyl-α-methyldopa, α-methyldopamine, 3-O-methyl-α-methyldopamine, α-methylnorepinephrin, and 3,4-dihydroxyphenylacetone (Au et al., 1972). α-Methyldopamine and α-methylnorepinephrin are pharmacologically active. Sulfate conjugation of α-methyldopa has also been reported to occur in human beings (Saavedra et al., 1975). The plasma elimination half-life of α-methyldopa in normal adult subjects is 106 $\pm$ 12 min (S.D.) (Kwan et al., 1976) and 3.6 hr in patients with advanced renal failure (Myhre et al., 1972). Following the elimination of most of the drug (~95%), however, a second elimination phase appears with a half-life of 7 to 16 hr (Myhre et al., 1972).

c. Analogous Compounds. α-Amino acids; methyldopate (ethyl ester of methyldopa); dopa (levodopa), tyrosine, and related biogenic amines (e.g., dopamine, norepinephrin).

d. Analytical Methods

1. High Performance Liquid Chromatography—Walson et al. (1975)

Discussion. This assay utilizes cation-exchange HPLC with UV detection at 280 nm. It is sensitive to 5 μg/ml plasma and requires only plasma protein precipitation before injection onto the column. High levels of l-dopa interfere with this assay, but other catecholamines do not. The concentration of methyldopa is determined by comparison with an external standard curve, using peak height measurements. Although this method is rapid, it has the major drawback of the rather low sensitivity.

Detailed Method. Heparinized plasma (1 ml) is mixed with 0.1 ml 4 N HClO$_4$ and centrifuged. Aliquots of the supernatant are analyzed by HPLC. Appropriate standard samples are treated in the same manner.

Analysis is performed on a Chromatronix liquid chromatograph equipped with a 500 μl loop injector and a 6 mm (i.d.) x 500 mm glass column packed with Aminex A-6 cation-exchange resin. The mobile phase is 0.5 M ammonium formate adjusted to pH 4.3 with concentrated formic acid. A flow rate of 1 ml/min is maintained at a column temperature of 50°C.

2. Other HPLC Assay—Mell and Gustafson (1978)

Discussion. Urine samples (30 ml) are adsorbed to alumina, and a biogenic amine fraction is eluted from the alumina and analyzed by reverse phase (C$_{18}$) HPLC with 280 nm UV detection. The eluent is a dilute pH 2.7 acetate buffer. Methyldopa can be detected without interference with a sensitivity of 25 ng/ml urine.

3. HPLC and Electrochemical Detection—Kissinger et al. (1974), Riggin et al. (1974)

Discussion. Kissinger et al. (1974) propose that an electrochemical detector be used in conjunction with HPLC. The detector is a working electrode flow cell with a fixed potential to oxidize (or reduce) electroactive molecules. Catecholamines can be detected in picogram amounts. The technique has been applied to the measurement of methyldopa in dosage forms (Riggin et al., 1974) and of endogenous catecholamines, but not yet for methyldopa in body fluids (Kissinger et al., 1974). It may become the method of choice because of the high selectivity and sensitivity of the detector system.

4. Gas Chromatography—Marshall and Castagnoli (1973)

Discussion. Derivatization of the two stereoisomers of α-methyl-dopamine with diazomethane and subsequently with the chiral reagent, (S)-(−)-N-pentafluorobenzoylprolyl-1-imidazolide, gives a diastereomeric product that can be separated by GC. Following isolation of α-methyldopamine from a human subject on methyldopa therapy, Marshall and Castagnoli (1973) showed the absolute configuration of this methyldopa metabolite to be the (S)-(+)-α-methyldopamine.

5. Mass Spectrometry—Freed et al. (1977)

Discussion. The MS method of Freed et al. (1977) measures simultaneously nanomole quantities of dopamine, α-methyldopamine, norepinephrin, and α-methylnorepinephrine in rat brain tissue samples. Deuterium-labeled analogs of each of the four catecholamines are used as internal standards for quantitation. Catecholamines are extracted, separated from amino acids by ion-exchange chromatography, and derivatized with HCl-ethanol and pentafluoropropionic anhydride. Sample residues are introduced directly into the mass spectrometer and analyzed using chemical ionization. The mass regions of the peaks corresponding to the labeled and unlabeled derivatives are repeatedly scanned during sample evaporation, and concentrations are obtained by measuring the m/e peak height ratios of the specific catecholamine to its deuterated internal standard. Standard curves are constructed in the range of 0.2 to 2 nmol of the catecholamines per sample.

6. Fluorimetry—Kim and Koda (1977), Saavedra et al. (1975)

Discussion. These methods are based on oxidation to the fluorophor, dihydroxyindole. Plasma samples are adsorbed onto alumina and eluted with 0.2 N HCl prior to the fluorescene reaction. The assay also detects α-methyldopa, α-methyldopate, and α-methyldopamine, which form the same fluorophor. Optimization of the assay conditions results in a sensitivity of 100 ng methyldopa/ml plasma (Kim and Koda, 1977).

7. Comments. Several assay techniques are presently useful in studying the complex pharmacology of methyldopa. These include HPLC, GC, and MS. Current HPLC assays can be further improved with respect to sensitivity and simultaneous detection of methyldopa metabolites. Electrochemical detection appears to be a prime candidate for combination with HPLC separation.

e. References

Au, W. Y. W., L. G. Dring, D. G. Grahame-Smith, P. Isaac, and R. T. Williams, The Metabolism of ^{14}C-Labeled α-Methyldopa in Normal and Hypertensive Human Subjects, *Biochem. J.*, **129**, 1–10 (1972).

Freed, C. R., R. J. Weinkam, K. L. Melmon, and N. Castagnoli, Chemical Ionization Mass Spectrometric Measurement of α-Methyldopa and *l*-Dopa-Metabolites in Rat Brain Regions, *Anal. Biochem.*, **78**, 319–332 (1977).

Kim, B. K. and R. T. Koda, Fluorometric Determination of Methyldopa in Biological Fluids, *J. Pharm. Sci.*, **66**, 1623–1634 (1977).

Kissinger, P. T., L. J. Felice, R. M. Riggin, L. A. Pachla, and D. C. Wenke, Electrochemical Detection of Selected Organic Components in the Eluate from High-Performance Liquid Chromatography, *Clin. Chem.*, **20**, 992–997 (1974).

Kwan, K. C., E. L. Foltz, G. O. Breault, J. E. Baer, and J. A. Totaro, Pharmacokinetics of Methyldopa in Man, *J. Pharmacol. Exp. Ther.*, **198**, 264–277 (1976).

Marshall, K. S. and N. Castagnoli, Absolute Configuration of α-Methyldopamine Formed Metabolically from α-Methyldopa in Man, *J. Med. Chem.*, **16**, 266–270 (1973).

Mell, L. D. and A. B. Gustafson, Urinary Free Methyldopa Determined by Reversed Phase High-Performance Liquid Chromatography, *Clin. Chem.*, **24**, 23–26 (1978).

Myhre, E., E. K. Brodwall, O. Stenback, and T. Hansen, Plasma Turnover of Methyldopa in Advanced Renal Failure, *Acta Med. Scand.*, **191**, 343–347 (1972).

Riggin, R. M., L. Rau, R. L. Alcorn, and P. T. Kissinger, Determination of Phenolic Sympathetic Stimulants in Pharmaceuticals by Liquid Chromatography with Electrochemical Detection, *Anal. Lett.*, **7**, 791–798 (1974).

Saavedra, J. A., J. L. Reid, W. Jordan, M. D. Rawlins, and C. T. Dollery, Plasma Concentration of α-Methyldopa and Sulphate Conjugate after Oral Administration of Methyldopa and Intravenous Administration of Methyldopa and Methyldopa Hydrochloride Ethyl Ester, *Eur. J. Clin. Pharmacol.*, **8**, 381–386 (1975).

Walson, P. D. et al., Metabolic Disposition and Cardiovascular Effects of Methyldopate in Unanesthetized Rhesus Monkeys, *J. Pharmacol. Exp. Ther.*, **195**, 151–158 (1975).

METRONIDAZOLE

2-Methyl-5-nitroimidazole-1-ethanol.
Clont, Flagyl, Trichazol, and so on. Trichomoacide, antimicrobial agent, radiosensitizer.

$C_6H_9N_3O_3$; mol. wt. 171.16; pK_a 2.5.

a. Therapeutic Concentration Range. Antimicrobial doses of metronidazole (e.g., 2 × 200 mg/day) yield plasma levels averaging 5 μg/ml (Taylor et al., 1969; Welling and Monro, 1972). Plasma concentrations of metronidazole needed for sensitization to tumor radiation treatment are in the range of 100 to 200 μg/ml after oral doses of 100 to 200 mg metronidazole/kg (Deutsch et al., 1975; Foster et al., 1975).

b. Metabolism. The major metabolic pathways of metronidazole are side chain oxidations and conjugation (Stambaugh et al., 1968). Metabolites recovered in the urine of patients include the major metabolite, 2-hydroxymethylmetronidazol (35 to 40% of the excreted drug), 1-acetic acid metronidazole (15 to 20%), 2-carboxylic acid metro-nidazole (8–12%), conjugates (10 to 14%), and unchanged metronidazole (25 to 30%) (Stambaugh et al., 1968). The mutagenicity of metronida-zole in bacterial systems (and therefore its potential carcinogenicity) has been repeatedly observed (Connor et al., 1977; Minnich et al., 1976; Rosenkranz et al., 1976). Mutagenic activity of urine samples after metronidazole treatment is higher than that expected from urinary metronidazole concentrations alone, suggesting the presence of more active metabolites (Connor et al., 1977). 2-Hydroxymethyl metronida-zole is indeed a 10 times more potent mutagen than metronidazole itself, and still more mutagenic metabolites are likely to be present (Connor et al., 1977).

The one-electon redox potential appears to be critical in determining nitroimidazole antimicrobial effects, radiosensitization, and selective toxicity against hypoxic mammalian cells (Adams et al., 1976a, 1976b; Varghese et al., 1976). The metabolic formation, in the absence of oxygen, or reduction products and their subsequent binding to macromolecules may cause the preferential toxicity of nitro compounds to hypoxic mammalian cells (Varghese et al., 1976). The 2-nitroimidazoles (e.g., misonidazole) appear to be more effective radiosensitizers than the 5-nitroimidazoles (e.g., metronidazole and trinidazole), based on their different reduction potentials (Adams et al., 1976b; Foster et al., 1975).

The plasma elimination half-life of metronidazole ranges from 6 hr (Welling and Monro, 1972) to 10 to 14 hr (Urtasun et al., 1975; Taylor et al., 1969).

c. Analogous Compounds. 5-Nitroimidazoles (e.g., tinidazole); 2-nitroimidazoles (e.g., misonidazole); nitrofurantoins (nitrofurantoin).

d. Analytical Methods

1. High Performance Liquid Chromatography—Marques et al. (1978)

Discussion. The reverse phase HPLC assay of metronidazole and its major metabolite, 2-hydroxymethylmetronidazole, in plasma and urine employs UV detection at 324 nm. Proteins are precipitated with ethanol prior to analysis. The sensitivity is 0.5 μg/ml, using only a 10 μl sample volume; it can be considerably increased since biological background noise is low and as little as 5 ng metronidazole can be detected per injection. The method is equally suitable for the measurement of misonidazole, which is not separated from metronidazole on the C_{18} reverse phase column, and the metabolite desmethylmisonidazole. The coefficient of variation is ~5% with external standard curves.

A patient serum sample taken 27 hr following a metronidazole dose of 6 g/m² gave a metronidazole level of 14 μg/ml by UV spectrophotometry (Urtasun et al., 1975) but only 2.8 μg/ml by this HPLC assay. The discrepancy between these results is accounted for by the 2-hydroxymethyl metabolite, which interferes with the UV assay.

Detailed Method. Serum and diluted (1:2 in water) urine samples are mixed with equal volumes of ethanol and vigorously shaken for 10 sec;

proteins are allowed to precipitate for 15 min at room temperature. The mixture is then centrifuged at 1700 g for 10 min, and 20 μl of the supernatant is injected onto the HPLC column. Peaks eluting from the HPLC are quantitated using peak heights and compared to external standard curves at three different drug concentrations.

A Waters Associates (Milford, Mass. 01757) high performance liquid chromatograph is used, equipped with a Model 6000A pump, UK6 injector, and reverse phase μ-Bondapak C_{18} column (30 cm x 4 mm, average particle size 10 μ). Eluents are monitored by UV absorbance at 324 nm, utilizing a spectroflow monitor Model SF770 from Schoeffel Instrument Co. (Westwood, N.J. 07675). The HPLC eluent is 8% acetonitrile in 10^{-5} M phosphate buffer (pH 4.0) at a flow rate of 2.0 ml/min.

2. Other HPLC Assays—Lanbeck and Lindström (1979), Little et al. (1978), Workman et al. (1978)

Discussion. Similar HPLC assays have been adopted for the metronidazole analog, misonidazole (Workman et al. 1978). Some of the commercially available C_{18} reverse phase columns are capable of separating metronidazole and misonidazole, which can then be used as respective internal standards (G. Gudauskus, unpublished data). This is particularly desirable when analyzing small tissue samples (e.g., tumor tissue). Little et al. (1978) suggest that a C_{22}-bonded silica phase is superior to the common C_{18} columns for misonidazole separations. Chemical modification of the bonded phase promises to enhance the utility and applicability of reverse phase HPLC in the future.

A highly sensitive (25 ng/ml plasma) HPLC-UV assay has recently been described for the detection of metronidazole and tinidazole following antimicrobial doses (Lanbeck and Lindström, 1979).

3. Thin-Layer Chromatography—Welling and Monro (1972)

Discussion. The procedure of Welling and Monro (1972) involves chloroform extraction of metronidazole from plasma and TLC separation on silica gel, containing a 254 nm fluorescence indicator, with $CHCl_3$-CH_3COOH (9:1) as the eluent. The quenching of the plate fluorescence by metronidazole is quantitatively measured with a thin-layer densitometer. The short elimination half-life of metronidazole observed in this study (6.2 hr) is attributed by the authors to the high degree of specificity of the TLC assay.

4. Gas Chromatography—be Silva et al. (1970)

Discussion. Nitroimidazoles can be detected by GC with electron capture at the picogram level after derivatization with silylating reagents (for metronidazole hexamethyldisilazane-trimethylchlorosilane, 2:1). The silylating reagent has to be evaporated before injecting the product in *n*-hexane onto the GC column in order to avoid detector contaminations. The assay includes an additional TLC purification step before GC analysis and provides a sensitivity of 0.01 μg metronidazole/ml plasma.

5. Spectrophotometry and Colorimetry—de Silva et al. (1970), Urtasun et al. (1975)

Discussion. Metronidazole can be measured in blood by UV spectrophotometry at 314 nm after protein precipitation with ethanol (blood:ethanol = 1:9) (Urtasun et al., 1975). However, simple UV analysis lacks specificity (see Section d1) and has limited sensitivity ($\sim$10 μg/ml). The specificity can be improved by introducing a TLC separation step (de Silva et al., 1970). In addition, de Silva et al. (1970) propose the colorimetric analysis of nitrite, which is released from metronidazole (and its metabolites) upon treatment with acid (HCl) or, more reproducibly, with base and UV irradiation. The generated nitrite is measured by the Bratton-Marshall method, using sulfanilamide for the diazotation reaction. The sensitivity (0.5 μg metronidazole/ml plasma) is considerably improved over that of the UV assay.

6. Polarography—Brooks et al. (1976) de Silva et al. (1970) Deutsch et al. (1975), Kane (1961)

Discussion. Polarographic methods take advantage of the redox potential of the nitroimidazoles in a region where little endogenous interference occurs (Kane, 1961). Deutsch et al. (1975) suggest the displacement of oxygen with pure nitrogen in plasma samples in order to reduce the biological background noise. The assay sensitivity is in the range of 0.1 μg metronidazole/ml plasma, using differential pulse polarography with a three-electrode cell, after ethyl acetate extraction of the drug at pH 9 (de Silva et al., 1970). Precipitation of plasma proteins by hot acid treatment further reduces background interferences (Brooks et al., 1976). De Silva et al. (1970) and Brooks et al. (1976) include a TLC separation step, since polarography does not differentiate between metabolites with the intact nitroimidazole moiety.

7. **Comments.** The HPLC method described in Section d1 offers advantages in specificity, sensitivity, precision, and ease of analysis over all previous methods. The new radiosensitizer, misonidazole, can also be measured by the assay methods listed. Since misonidazole metabolites with intact nitroimidazole moiety are insignificant in plasma (Marques et al., 1978), the parent drug can be measured by simple UV analysis without chromatography. The extensive metabolism of metronidazole, however, makes chromatography before analysis a necessity.

e. References

Adams, G. E. et al., Mammalian Cell Toxicity of Nitro Compounds: Dependence upon Reduction Potential, *Biochem. Biophys. Res. Commun.*, **72**, 824–829 (1976a).

Adams, G. E. et al., Electron Affinic Sensitization. VII: Correlation Between Structures, One Electron Reduction Potential, and Efficiences of Nitroimidazoles as Hypoxic Cell Radiosensitizers, *Radiat. Res.*, **6**, 9–20 (1976b).

Brooks, M. A., L. d'Arconte, and J. A. F. de Silva, Determination of Nitroimidazoles in Biological Fluids by Differential Pulse Polarography, *J. Pharm. Sci.*, **65**, 112–114 (1976).

Connor, T. H., M. Stoeckel, J. Evrard, and M. S. Legator, The Contribution of Metronidazole and Two Metabolites to the Mutagenic Activity Detected in Urine of Treated Humans and Mice, *Cancer Res.*, **37**, 629–633 (1977).

de Silva, J. A. F., N. Munno, and N. Strojny, Absorptiometric, Polarographic, and Gas Chromatographic Assays for the Determination of N-1-Substituted Nitroimidazoles in Blood and Urine, *J. Pharm. Sci.*, **59**, 201–210 (1970).

Deutsch, G., J. L. Foster, J. A. Mc Fadzean, and M. Parnell, Human Studies with "High Dose" Metronidazole: A Non-toxic Radiosensitizer of Hypoxic Cells, *Br. J. Cancer*, **31**, 75–80 (1975).

Foster, J. L. et al., Serum Concentration Measurements in Man of the Radiosensitizer Ro-07-0582: Some Preliminary Results, *Br. J. Cancer*, **31**, 679–683 (1975).

Kane, P. O., Polarographic Methods for the Determination of Two Anti-protozoal Nitroimidazole Derivatives in Materials for Biological and Non-biological Origin, *J. Polarogr. Soc.*, **7**, 58–65 (1961).

Lanbeck, K., and B. Lindström, Determination of Metronidazole and Tinidazole in Plasma and Feces by High-Performance Liquid Chromatography, *J. Chromatogr.*, **162**, 117–121 (1979).

Little, C. J., A. D. Dale, and M. B. Evans, "C_{22}"—A Superior Bonded Silica for Use in Reverse-Phase High-Performance Liquid Chromatography, *J. Chromatogr.*, **153**, 543–545 (1978).

Marques, R. A., B. Stafford, N. Flynn, and W. Sadée, Determination of Metronidazole and Misonidazole and Their Metabolites in Plasma and Urine by High-Performance Liquid Chromatography, *J. Chromatogr.*, **146**, 163–166 (1978).

Minnich, V., M. E. Smith, D. Thompson, and S. Kornfeld, Detection of Mutagenic Activity in Human Urine Using Mutant Strains of *Salmonella typhimurium*, *Cancer*, **38**, 1253–1258 (1976).

Rosenkranz, H. S., Jr., W. T. Speck, and J. E. Stambaugh, Mutagenicity of Metronidazole: Structure-Activity Relationships, *Mutat. Res.*, **38**, 203–206 (1976).

Stambaugh, J. E., L. G. Feo, and R. W. Manthei, The Isolation and Identification of the Urinary Oxidative Metabolites of Metronidazole in Man, *J. Pharmacol. Exp. Ther.*, **161**, 373–381 (1968).

Taylor, J. A., Jr., J. R. Migliardi, and M. Schack von Wittenan, Tinidazole and Metronidazole Pharmacokinetics in Man and Mouse, *Antimicrob. Agents Chemother.*, **9**, 267–270 (1969).

Urtasun, R. C. et al., Phase I Study of High-Dose Metronidazole: A Specific *in vivo* and *in vitro* Radiosensitizer of Hypoxic Cells, *Ther. Radiat.*, **117**, 129–133 (1975).

Varghese, A. J., S. Gulyas, and J. K. Mohindra, Hypoxia-dependent Reduction of 1-(2-Nitro-1-imidazolyl)-3-methoxy-2-propanol by Chinese Hamster Ovary Cells in KHI Tumor Cells *in vitro* and *in vivo*, *Cancer Res.*, **36**, 3761–3765 (1976).

Welling, P. G. and A. M. Monro, The Pharmacokinetics of Metronidazole and Tinidazole in Man, *Arzneim.-Forsch.*, **22**, 2128–2132 (1972).

Workman, P. et al., Estimation of the Hypoxic Cell-Sensitiser Misonidazole and Its *O*-Demethylated Metabolite in Biological Materials by Reversed-Phase High-Performance Liquid Chromatography, *J. Chromatogr. (Biomed. Applic.)*, **145**, 507–512 (1978).

MORPHINE

Narcotic, analgesic.

$C_{17}H_{19}NO_3$; mol. wt. 285.33; pK_a 8.02 and 9.76.

a. Therapeutic Concentration Range. A subcutaneous 10 mg morphine dose in a 70 kg adult produces peak plasma levels of approximately 0.07 μg/ml (Berkowitz et al., 1975; Brunk and Delle, 1974).

b. Metabolism. The major metabolic route of morphine is formation of the inactive 3-*O*-glucuronide, which is sustained in plasma at much higher levels than the parent drug and represents the major urinary excretion product (Berkowitz et al., 1975; Brunk and Delle, 1974). Morphine-6-*O*-glucuronide, which retains analgesic activity, represents only a minor metabolite (Yoshimura et al., 1973). Other metabolic pathways lead to normorphine and norcodeine (Boerner et al., 1974), dihydromorphinone (Klutch, 1973), morphine N-oxide, and morphine 2,3-dihydrodiol (tentative structure) (Misra et al., 1973), some of which are pharmacologically active. Furthermore, α- and β-dihydromorphine, β- or γ-isomorphine, and hydroxylated metabolites have been isolated (Yeh et al., 1979).

Immunochemical measurements suggest that morphine-related material may persist in human plasma for several days or even much longer after a single dose of morphine (Spector and Vesell, 1971; Takahashi et al., 1975). The chemical nature of this persistent material remains unknown (Mullis et al., 1979). The plasma elimination half-life of morphine in the interval between 10 and 240 min after a dose of 10 mg morphine in human beings is 2 hr, but the terminal log-linear elimination of morphine may be much longer (Berkowitz et al., 1975; Spector and Vesell, 1971).

c. Analogous Compounds. Opiate analgesic agonists and antagonists (e.g., the metabolic morphine progenitors heroin, codeine, and ethylmorphine).

d. Analytical Methods

1. Gas Chromatography-Mass Spectrometry—Finn et al. (1975), Hipps et al. (1976)

Discussion. The selected ion monitoring electron impact GC-MS assay of Hipps et al. (1976) has been designed for the sensitive detection of morphine ($\sim$5 ng/g) in rat brain and is readily applicable to other biological samples. The problem of nonlinear standard curves, observed with many morphine assays at very low levels, is partially overcome and compensated for by using N-CD$_3$-morphine (d_3-morphine) as analytical carrier in high amounts (1 µg/sample) and as internal standard. Following extraction and derivatization to ditrifluoroacetylmorphine, GC-MS analysis is performed at both the molecular ions, m/e 477/480 (d_0/d_3), and at the base peaks, m/e 364/367 (d_0/d_3). Identical results obtained with both ion pairs provide strong supporting evidence for the specificity of the method. Interpretation of d_0/d_3 ratios must consider the contribution of d_3-morphine to the d_0 peak, which is caused by electron-impact-induced loss of several hydrogens from the parent ion and, to a smaller extent, by less than complete deuterium substitution. Given an amount of 1 µg d_3-morphine/g brain as carrier and analytical standard and a d_0 contribution of 1%, the brain blank level will show 10 ng morphine/g, which has to be subtracted from the results obtained. The precision of the measurement under these conditions is directly related to the sensitivity of the assay. By use of $^{13}CD_3$-morphine to lower the d_0 contribution (0.5%), ion counting MS, and computerized curve fitting with the Savitzky-Golay smoothing algorithm to increase the signal/noise ratio, we have been able to increase the precision and thus the sensitivity of this assay by a factor of 10 (Finn et al., 1975).

Detailed Method. Frozen brain samples are weighed and homogenized in 4.0 ml 0.05 M Tris buffer (pH 8.6) containing about 1 µg d_3-morphine. An aliquot of the homogenate (0.5 ml), 0.5 ml of the Tris buffer, and 1 ml saturated $NaHCO_3$ are extracted with 5.5 ml ethyl acetate-isopropanol (10:1) in polypropylene tubes. After shaking and centrifuging for 10 min each, 5 ml of the organic layer is transferred into another polypropylene tube containing 1.2 ml 1 N HCl. The mixture is again shaken and centrifuged, and the organic layer discarded. Then 1 ml of the aqueous layer is evaporated to dryness at room temperatures under N_2. The residue is allowed to react with 0.5 ml trifluoroacetic anhydride for 30 min, and the reagent is again evaporated under nitrogen. The final residue is redissolved in 0.1 ml dry ethyl acetate, and an aliquot injected into the gas chromatograph-mass spectrometer.

Analysis is performed on a computerized (PDP-12) LKB 9000 gas chromatograph-mass spectrometer operating in the selected ion monitoring mode and fitted with a 50 cm x 6 mm (i.d.) glass column packed with 3% SE-30 on Gas Chrom Q. Columns are conditioned at 250°C overnight with a helium flow of 50 ml/min. The GC conditions are as follows: helium carrier flow rate, 30 ml/min; column temperature, 210°C; injector temperature, 245°C. The MS analysis is performed at 70 eV electron impact with separator and ion source at 250 and 275°C, respectively, The resolution is approximately 400 $m/\Delta m$ (10% valley). The following ions are monitored: m/e 364, 367, 477, and 480. Ratios of

d_0/d_3 are determined by GC peak height ratios of the selected ion monitoring records of d_0- and d_3-morphine.

2. Other GC-MS Assay—Ebbighausen et al. (1973a, 1973b).

Discussion. The procedures of Ebbighausen et al. (1973a, 1973b) are similar to the GC-MS assay of Hipps et al. (1976). The GC liquid phase is 1% OV-17. A sensitivity of 5 to 10 pg morphine per injection of pure sample is claimed by these authors, with a considerable increase of the minimum detectable amount to 500 pg per injection when biological samples are used. Heptafluorobutyric anhydride is proposed as an alternative to trifluoroacetic anhydride for the derivatization of morphine. The metabolite normorphine can also be measured by GC-MS (Ebbighausen et al., 1973a).

3. Gas Chromatography—Dahlström and Paalzow (1975), Nicolau et al. (1977), Rasmussen (1976), Wilkinson and Way (1969)

Discussion. The GC assays are similar to the reported GC-MS procedures except for the mode of detection. Flame ionization detection usually limits the sensitivity to the high nanograms and micrograms morphine per milliliter range and is therefore applicable only to the assay of urine samples. Most of the GC assay procedures also include an acid hydrolysis step in HCl in order to liberate morphine from the major urinary excretion product, the 3-O-glucuronide. Derivatization to produce suitable GC characteristics is achieved by silylation either in solution (Wilkinson and Way, 1968) or on-column (Rasmussen, 1976).

In contrast, electron capture detection affords high sensitivity if suitable derivatives are employed (Nicolau et al., 1977) Rasmussen (1976) achieves a sensitivity of 2 pg morphine per injection, or 0.1 ng morphine/ml urine, using the bisheptafluorobutyryl derivative. Codeine serves as the internal standard; 3% OV-17 is the liquid GC phase. This procedure and the equivalent method of Dahlström and Paalzow (1975) represent the most sensitive morphine assays available at present. However, great care has to be asserted when analyzing subnanogram levels of morphine without isotopic tracers.

4. High Performance Liquid Chromatography—Jane and Taylor (1975), Mullis et al. (1979), Wu and Wittick (1977)

Discussion. Reverse phase columns are well suited for the separation of opiates in gum opium (Wu and Wittick, 1977) or morphine metabolites

(Mullis et al., 1979). The HPLC assay of Mullis et al. (1979) is designed in combination with ³H-labeled morphine for the detection of very small amounts (≥0.1 ng/ml) in plasma and brain.

Jane and Taylor (1975) take advantage of the $K_3Fe(CN)_6$ oxidation of morphine to the fluorescent dimer pseudomorphine, which can be detected by silica gel HPLC with a fluorescence flow cell detector, providing a sensitivity of 0.01 μg morphine/ml urine. The internal standard, dihydromorphine, undergoes the same reaction to the dihydromorphine dimer; however, the mixed morphine-dihydromorphine dimer is also formed. All three products are separated by HPLC, and their ratios are a function of the morphine and the dihydromorphine concentrations.

5. Radioimmunoassay—Berkowitz et al. (1974), Catlin (1977), Miller et al. (1975), Spector and Parker (1970), Steiner and Spratt (1978), van Vunakis et al. (1972)

Discussion. The article of Catlin (1977) demonstrates the high sensitivity achievable with most of these RIA procedures, but also the difficulties in interpreting morphine plasma level data. Two different antibodies are raised with a 2-diazomorphine conjugate with bovine serum albumin (Catlin, 1977). The protein coupling mode exposes both the 3-OH and the 6-OH to antibody recognition sites, resulting in a lower antibody affinity of the 3-*O*-glucuronide when compared to morphine itself. However, the degree of cross-affinity of the 3-*O*-glucuronide is different for the two antibodies, that is, 90 and 14 times less than morphine, in spite of antibody production against the same antigenic morphine conjugate. Furthermore, significant differences are obtained with these two antibodies when assaying plasma samples from rabbits that received a dose of morphine. These results indicate lack of specificity, particularly at the low plasma concentrations present many hours after the dose. Berkowitz et al. (1974) find detectable morphine-like activity in rat plasma for up to 20 days after the dose with a RIA method, which correlates well with a fluorescence assay for unchanged morphine at micrograms per milliliter range. Using ³H-morphine and HPLC analysis, we have recently shown that the long retention of ³H activity in plasma is not accounted for by unchanged morphine, which falls below 0.1 ng/ml 48 hr after a subcutaneous dose of 10 mg/kg in the rat (Mullis et al., 1979). Therefore it is likely that the persistence of immunologically active material is related to yet unknown morphine metabolites of unknown pharmacological significance.

Several RIA methods are based on morphine antibodies raised against 3-*O*-carboxymethylmorphine conjugates with various proteins

(Miller et al., 1975; Spector and Parker, 1970; van Vunakis et al., 1972). These assays do not discriminate as well between unchanged morphine and the 3-*O*-glucuronide.

The elegant solid phase RIA of Steiner and Spratt (1978) employs a morphine antibody raised against a morphine-6-hemisuccinate conjugate with bovine serum albumin. The immunoglobulin is purified on a morphine agarose affinity gel for selection of high affinity antibodies. Because of the coupling in C-6 this antibody has a 55 times lower affinity to morphine 3-*O*-glucuronide. Assays are performed in polystyrene tubes coated with the antibody, which facilitates separation of free and bound ^{3}H-morphine tracers.

6. Enzyme Immunoassay—Rowley et al. (1975), Rubenstein et al. (1972)

Discussion. These homogeneous EMIT assays of morphine are based on lysozyme activity against bacteria (Rubenstein et al., 1972) or on malate dehydrogenase-oxaloacetate-NADH (Rowley et al., 1975). Morphine is conjugated to both enzymes via the C_3-OH function, and binding to morphine-specific antibodies lowers the enzyme activity by 98 and 86%, respectively. Assay sensitivity is in the range of 1 ng morphine/ml test solution; however, the assays are nonspecific in the presence of codeine and, probably, morphine 3-glucuronide because of the methods of protein conjugation.

7. Hemagglutination-Inhibition Immunoassay—Adler and Liu (1971)

Discussion. The antibody is raised against a 3-carboxymethyl morphine conjugate with bovine serum albumin and therefore cross-reacts with the 3-*O*-glucuronide. The assay method is based on the agglutination of formaldehyde-stabilized red blood cells by the morphine antibody. The red blood cells have been sensitized against the antibody by coating with a morphine-rabbit serum albumin conjugate. The multivalent antibody can thus bind several blood cells. The agglutination reaction is inhibited by free morphine in the test sample, which inactivates the antibody. The assay detects as little as 25 pg morphine per sample; however, reproducible results with hemagglutination-inhibition tests depend largely on proper immunological manipulations; results are semiquantitive.

8. Spin Immunoassay—Montgomery and Holtzman (1974)

Discussion. The assay principle is similar to that published by the same authors for the assay of phenytoin (see phenytoin monograph).

Spin-labeled morphine and the antibody (unspecified cross-affinities to morphine metabolites) have been obtained from Syva Co. The assay is sensitive to 10 ng morphine and allows the direct measurement of unbound and plasma protein-bound morphine, if the spin-labeled morphine is incubated with plasma in the absence of antibody. The spin label assay gives a value of 70% for the free morphine fraction in plasma, and conventional ultrafiltration yields 65%.

9. Fluorescence—Kupferberg et al. (1964)

Discussion. Following extraction of plasma at pH 8.5 to 9.0 into 10% butanol in chloroform and back-extraction into acid, fluorescence is produced by $K_3Fe(CN)_6$ oxidation of morphine to pseudomorphine. The assay lacks specificity if performed without chromatography.

10. Assay Comparison—Garrett and Gürkan (1978)

Discussion. Garrett and Gürkan (1978) compare three morphine assays: GC-EC of the pentafluoropropionyl derivative, an isotope derivatization procedure employing ^{3}H-dansyl chloride and TLC separation of the product, and ^{14}C-morphine analysis following solvent extraction. The three methods gave equivalent results when applied to dog plasma samples containing 1 to 10 ng morphine/ml.

11. Comments.
Very small amounts of morphine (<10 ng) are readily destroyed by solvent impurities and absorption to active sites in glassware; therefore the glassware has to be silanized. Solvents and reagents should be of the highest quality.

The various assay techniques, that is, RIA, GC, GC-MS, HPLC, and so on, all serve different purposes and have to be selected to suit the analytical problem. The specificity of RIA methods at very low morphine plasma levels remains uncertain.

e. References

Adler, F. L. and C. T. Liu, Detection of Morphine by Hemagglutination-Inhibition, *J. Immunol.*, **106**, 1684–1685 (1971).

Berkowitz, B. A., K. V. Cerreta, and S. Spector, The Influence of Physiologic and Pharmacologic Factors on the Disposition of Morphine as Determined by Radioimmunoassay, *J. Pharmacol. Exp. Ther.*, **191**, 527–534 (1974).

Berkowitz, B. A., S. H. Ngai, J. C. Yang, J. Hempstead, and S. Spector, The Disposition of Morphine in Surgical Patients, *Clin. Pharmacol. Ther.*, **17**, 629–635 (1975).

Boerner, U., R. L. Roe, and C. E. Becker, Detection, Isolation and Characterization of Normorphine and Norcodeine as Morphine Metabolites in Man, *J. Pharm. Pharmacol.*, **26**, 393–398 (1974).

Brunk, S. F. and M. Delle, Morphine Metabolism in Man, *Clin. Pharmacol. Ther.*, **16**, 51–57 (1974).

Catlin, D. H., Pharmacokinetics of Morphine by Radioimmunoassay: The Influence of Immunochemical Factors, *J. Pharmacol. Exp. Ther.*, **200**, 224–235 (1977).

Dahlström, B. and L. Paalzow, Quantitative Determination of Morphine by Gas-Liquid Chromatography and Electron Capture Detection, *J. Pharm. Pharmacol.*, **27**, 172–176 (1975).

Ebbighausen, W. O. R., J. Mowat, and P. Vestergaard, Mass Fragmentographic Detection of Normorphine in Urine of Man after Codeine Intake, *J. Pharm. Sci.*, **62**, 146–148 (1973a).

Ebbighausen, W. O. R., J. H. Mowat, P. Vestergaard, and N. S. Kline, Stable Isotope Method for the Assay of Codeine and Morphine by Gas Chromatography-Mass Spectrometry: A Feasibility Study, *Adv. Biochem. Psychopharmacol.*, **7**, 135–146 (1973b).

Finn, C., H.-J. Schwandt, and W. Sadée, Application of Ion Counting Selected Ion Monitoring-Mass Spectrometry in Pharmacokinetics, in *Stable Isotopes in Chemistry, Biology, and Medicine*, E. R. Klein and P. D. Klein, Eds., Argonne Natl. Lab., 1975, pp. 129–137.

Garrett, E. R. and T. Gürkan, Pharmacokinetics of Morphine and Its Surrogates. I: Comparisons of Sensitive Assays of Morphine in Biological Fluids and Application to Morphine Pharmacokinetics in the Dog, *J. Pharm. Sci.*, **67**, 1512–1517 (1978).

Hipps, P. P., M. R. Eveland, E. R. Meyer, W. R. Sherman, and T. J. Cicero, Mass Fragmentography of Morphine: Relationship Between Brain Levels and Analgesic Activity, *J. Pharmacol. Exp. Ther.*, **196**, 642–648 (1976).

Ikekawa, N. and K. Takayama, Determination of Morphine in Urine by Gas Chromatography, *Anal. Biochem.*, **28**, 156–163 (1969).

Jane, I. and J. F. Taylor, Characterisation and Quantitation of Morphine in Urine Using High-Pressure Liquid Chromatography with Fluorescence Detection, *J. Chromatogr.*, **109**, 37–42 (1975).

Klutch, A., A Chromatographic Investigation of Morphine Metabolism in Rats: Confirmation of *N*-Demethylation of Morphine and Isolation of a New Metabolite, *Drug Metab. Dispos.*, **2**, 23–30 (1974).

Kupferberg, H., A. Burkhalter, and E. L. Way, A Sensitive Fluorimetric Assay for Morphine in Plasma and Brain, *J. Pharmacol. Exp. Ther.*, **145**, 247–251 (1964).

Miller, C., J. Nakamura, C. Y. Leung, W. D. Winters, and E. Benjamini, Immunological Specificity of Antibodies to Morphine and Their Effect on the Electroencephalographic Activity of Morphine, *Neuropharmacology*, **14**, 385–396 (1975).

Misra, A. L., N. L. Vadlamani, R. B. Pontani, and S. J. Mule, Evidence for a New Metabolite of Morphine-*N*-Methyl-^{14}C in the Rat, *Biochem. Pharmacol.*, **22**, 2129–2139 (1973).

Montgomery, M. R. and J. L. Holtzman, Determination of Serum Morphine by the Spin-Label Antibody Technique, *Drug Metab. Dispos.*, **2**, 391–395 (1974).

Mullis, K. B., D. C. Perry, A. M. Finn, B. Stafford, and W. Sadée, Morphine Persistence in Rat Brain and Serum after Single Doses, *J. Pharmacol. Exp. Ther.*, **208**, 228–231 (1979).

Nicolau, G., G. van Lear., B. Kaul, and B. Davidow, Determination of Morphine by Electron Capture Gas-Liquid Chromatography, *Clin. Chem.*, **23**, 1640–1643 (1977).

Rasmussen, K. E., Quantitative Morphine Assay by Means of Gas-Liquid Chromatography and On-Column Silylation, *J. Chromatogr.*, **120**, 491–495 (1976).

Rowley, G. L., K. E. Rubenstein, J. Huisjen, and E. F. Ullman. Mechanism by Which Antibodies Inhibit Hapten-Malate Dehydrogenase Conjugates, an Enzyme Immunoassay for Morphine, *J. Biol. Chem.*, **250**, 3759–3766 (1975).

Rubenstein, K. E., R. S. Schneider, and E. F. Ullman, "Homogeneous" Enzyme Immunoassay: A New immunochemical Technique, *Biochem. Biophys. Res. Commun.*, **47**, 846–851 (1972).

Spector, S. and C. W. Parker, Morphine: Radioimmunoassay, *Science*, **168**, 1347–1348 (1970).

Spector, S. and E. S. Vesell, Disposition of Morphine in Man, *Science*, **174**, 421–422 (1971).

Steiner, M. and J. L. Spratt, Solid-Phase Radioimmunoassay for Morphine, with Use of an Affinity-Purified Morphine Antibody, *Clin. Chem.*, **24**, 339–342 (1978).

Takahashi, M., M. Koida, H. Kaneto, J. Goto, and S. Haseba, Urinary Excretion Patterns of Morphine in Humans Followed by an Radioimmunoassay Method, *Jap. J. Pharmacol.*, **25**, 348–349 (1975).

Van Vunakis, H., E. Wasserman, and L. Levine, Specificities of Antibodies to Morphine, *J. Pharmacol. Exp. Ther.*, **180**, 514–521 (1972).

Wilkinson, G. R. and E. L. Way, Sub-microgram Estimation of Morphine in Biological Fluids by Gas-Liquid Chromatography, *Biochem. Pharmacol.*, **18**, 1435–1439 (1969).

Wu, C. Y. and J. J. Wittick, Separation of Five Major Alkaloids in Gum Opium and Quantitation of Morphine, Codeine, and Thebaine by Isocratic Reverse Phase High Performance Liquid Chromatography, *Anal. Chem.*, **49**, 359–363 (1977).

Yeh, S. Y., H. A. Krebs, and C. W. Gorodetzky, Isolation and Identification of Morphine N-Oxide, α- and β-dihydromorphines, β- or γ-isomorphine, and

Hydroxylated Morphine as Morphine Metabolites in Several Mammalian Species, *J. Pharm. Sci.*, **68**, 133–140 (1979).

Yoshimura, H., S. Ida, K. Oguri, and H. Tsukamoto, Biochemical Basis for Analgesic Activity of Morphine-6-glucuronide. 1: Penetration of Morpine-6-glucuronide in the Brain of Rats, *Biochem. Pharmacol.*, **22**, 1423–1430 (1973).

NICOTINE

1-Methyl-2-(3-pyridyl)pyrrolidine.

$C_{10}H_{14}N_2$; mol. wt. 162.23; pK_a 3.1 and 8.0.

a. Concentration Range in Smokers. The midmorning plasma levels of nicotine range between 15 and 38 ng/ml (Russell et al., 1975).

b. Metabolism. The major metabolite of nicotine is the oxidation product cotinine, which is much more slowly excreted and reaches higher plasma levels than does nicotine (Dumas et al., 1975). Other metabolites include nicotine 1'-N-oxide (Beckett et al., 1971), cotinine 1'-N-oxide (Dagne and Castagnoli, 1972b), γ-(3-pyridyl)-γ-oxo-*N*-methylbutyramide (Langone et al., 1974), hydroxycotinine (Dagne and Castagnoli, 1972a), and nicotine-cotinine analogs of nicotinamide nucleotides (Shen et al., 1977). The stereo selective metabolism of (+)- and (−)-nicotine (Jenner et al., 1971) and the formation of diastereo-meric *cis*- and *trans*-nicotine 1'-N-oxide (Beckett et al., 1971; Booth and Boyland 1970) have been well documented. Beckett et al. (1972) point out that urinary pH may be important in the renal elimination of nicotine.

Cigarette nicotine content and hence nicotine plasma levels may be a determinant of smoking behavior with implications for the relative exposure of smokers to tar and nicotine in low-yield nicotine cigarettes (Goldfarb et al., 1970). It is conceivable that cotinine plasma levels may represent a better indication of smoking behavior than nicotine because of the long elimination half-life of cotinine. Consequently, many analytical techniques include the measurement of nicotine and cotinine in plasma. Urinary excretion data are less reliable indicators, since they are subject to physiological variations of urine flow and pH.

c. Analogous Compounds. Nicotine alkaloids; cotinine; nicotinamide.

d. Analytical Methods

1. Gas Chromatography—Hengen and Hengen (1978)

Discussion. The GC assay of Hengen and Hengen (1978) employs a nitrogen-sensitive flame ionization detector and is sensitive to 0.1 ng nicotine or cotinine/ml plasma, using a 1 ml sample volume. The very high assay sensitivity is achieved by carefully selecting proper extraction and GC conditions in order to maximize sensitivity and minimize irreversible losses of these two compounds at subnanogram levels. The differential extraction method of nicotine with ether, followed by dichloromethane extraction of cotinine, is based on the procedures of Beckett and Triggs (1966) and Feyerabend et al. (1975). Evaporation of the volatile nicotine base is prevented by the addition of HCl to form the nonvolatile nicotine hydrochloride. Modaline and lidocaine serve as the internal standards for nicotine and cotinine, respectively. The day-to-day C.V. is 14% for nicotine and 6% for cotinine at concentrations between 1 and 100 ng/ml plasma. The high sensitivity of the assay allows one to measure nicotine levels in nonsmokers, which may be in the low nanograms per milliliter range, as well as environmental nicotine contamination of reagents. All analyses should be performed in rigorously tobacco-smoker-free laboratories with purified reagents (e.g., distillation).

Detailed Method. Plasma (1 ml), 1 ml 5 N NaOH, and 0.1 µl aqueous solution of the internal standard, modaline, are extracted twice with 3 ml diethyl ether by shaking for 30 min and subsequent centrifugation. The combined ethereal layers, containing the nicotine, are thoroughly mixed with 0.1 ml 2 N HCl, which prevents evaporation of the free base, and evaporated under nitrogen. The aqueous residue is washed with 0.5 ml ether, alkalinized with 0.3 ml 5 N NaOH, and mixed with 50 µl heptane-ether (1:1). Five microliters of the organic layer is injected into the gas chromatograph.

Cotinine is then extracted from the remaining plasma sample. Residual ether is evaporated under nitrogen, and 0.1 ml of the internal aqueous standard solution (lidocaine) added. The mixture is extracted with 5 ml dichloromethane by shaking for 10 min, followed by centrifugation. The organic layer is transferred to another test tube and evaporated under nitrogen. The residue is again transferred to a small tube with 0.4 ml dichloromethane. After evaporation of the dichloro-methane, the final residue is dissolved in 10 μl acetone, of which 1.5 μl is injected onto the GC column.

Nicotine analysis is performed on a Model 5711 A gas chromatograph (Hewlett-Packard, Avondale, Pa.), equipped with a nitrogen flame ioni-zation detector (Model 18789A). The column is a 1.80 m x 2 mm (i.d.) glass tube packed with 3% silicone SP-2250-DB on 100/120 mesh Supelcoport (Supelco, Bellefonte, Pa.). Column, injector, and detector temperatures are 155, 200, and 300°C, respectively. Carrier gas (helium) flow rate is 30 ml/min, air 60 ml/min, and hydrogen 3 ml/min. [Cotinine assay conditions are slightly different, requiring higher column temperatures (190°C).] Peak area ratios are used for quantitation.

2. Other GC Assays—Beckett and Triggs (1966), Beckett et al., (1971), Dow and Hall (1978), Dumas et al. (1975), Feyerabend et al. (1975), McNiven et al (1965), Neelakantan and Kostenbauder (1974)

Discussion. Early reports on GC assays of nicotine (McNiven et al., 1965) and cotinine (Beckett and Triggs, 1966) are applicable only to the analysis of urine samples, since they utilize the less sensitive and specific regular flame ionization detector. Reduction of the nicotine 1'-N-oxide diastereomers to nicotine with $TiCl_3$-HCl permits the detection of this metabolite after selective extraction of nicotine from the urine (Beckett et al., 1971). Langone et al. (1973) contend that the N-oxide is thermolabile under GC conditions and may cause erroneous results; however, selective extraction of nicotine in the presence of its more polar 1'-N-oxide circumvents this interference problem.

Introduction of the nitrogen-sensitive flame ionization detector permits GC analysis of nicotine and cotinine in plasma. The assays of Dumas et al. (1975) and Feyerabend et al. (1975) are similar to the method of Hengen and Hengen (1978) (see Section d1) except for dif-ferent GC conditions. Dow and Hall (1978) utilize capillary GC–N-FID (as well as MS) and quinoline as the internal standard with excellent results. Neelakantan and Kostenbauder (1974) explore the production of nicotine derivatives that are amenable to electron capture detection. Catalytic hydrogenation of nicotine yields *N*-methyl-4-(3'-piperidyl)-*n*-butylamine, which is detectable as the dipentafluoropropionyl deriva-tive at the 0.03 pmol level by GC-EC. The authors claim a sensitivity of 200 pg nicotine/ml plasma without providing statistical data.

3. Gas Chromatography-Mass Spectrometry—Dow and Hall (1977), Horning et al. (1973), Petrakis et al. (1978)

Discussion. Dow and Hall (1977) and Petrakis et al. (1978) propose GC-MS under electron impact for the sensitive detection of nicotine. Monitoring of the major fragment ion at m/e 84 provides a sensitivity of 25 pg nicotine per injection (Petrakis et al., 1978). The internal standard, $5',5'$-^{2}H-nicotine, gives rise to a major ion peak at m/e 86 and can be utilized for quantitation by the isotope dilution method. Cotinine and $3,3$-^{2}H-cotinine are similarly determined. Petrakis et al. (1978) utilize this technique to establish significant concentrations of nicotine and continine in the breast fluids of nonlactating women, when obtainable sample volume is limited (50 μl).

Horning et al. (1973) take advantage of the extreme sensitivity (femtomole quantities) of atmospheric pressure MS with a ^{63}Ni foil as the source of electrons. Organic solvent urine extracts can be directly injected into the external ionization chamber, which is maintained at atmospheric pressure. Selective ionization methods (e.g., by proton transfer to amines) and selected mass/charge analysis yield a sensitive probe of nicotine in small biological samples. Horning et al. (1973) measure nicotine in the urine of smokers and nonsmokers and in the air; they conclude that nicotine contaminations are transferred by air, rather than water.

4. High Performance Liquid Chromatography—Watson (1977)

Discussion. This silica gel HPLC assay with UV absorbance detection at 260 nm has a sensitivity of 5 ng nicotine per injection. Urine samples (3 ml) are extracted at alkaline pH into 0.3 ml $CHCl_3$, 10 μl of which is injected onto the column. the HPLC eluent is ethyl acetate–propan-2-ol–NH_3 (80:3:0.4). The limit of quantitative analysis in urine is 50 ng nicotine/ml, using desmethylimipramine or quinoline as the internal standard.

5. Radioimmunoassay—Haines et al. (1974), Langone et al. (1973, 1974), Shen et al. (1977)

Discussion. Nicotine antibodies have been raised against protein conjugates of 6-aminonicotine, obtained with succinic anhydride and carbodiimide (Haines et al., 1974), and of *trans*-3'-succinylmethyl-nicotine, formed by the carbodiimide method (Langone et al., 1973). These antibodies are highly specific and sensitive to nicotine. Cross-affinity to nicotine metabolites and other nicotine alkaloids is negligible. A cotinine-specific antibody is similarly obtained by immunization with a *trans*-4'-carboxycotinine protein conjugate (Langone et al.,

1973). Separation of free and bound tracer, ^{3}H-nicotine or ^{3}H-cotinine, is achieved by immunoprecipitation (Langone et al., 1973).

Specific RIA methods are also available for the picomole analysis of the nicotine metabolite γ-(3-pyridyl)-γ-oxo-N-methylbutyramide (Langone et al., 1974) and nicotine and cotinine analogs of nicotinamide nucleotides (Shen et al., 1977).

6. Comments. Very low plasma levels of nicotine can be detected by GC–N-FID, GC-MS, and RIA. Similar analysis of cotinine may also be valuable in assessing tobacco smoking behavior. The ubiquitous presence of nicotine as an environmental pollutant poses special problems when highly sensitive assay techniques capable of detecting these background levels are employed.

e. References

Beckett, A. H. and E. J. Triggs, Determination of Nicotine and Its Metabolite, Cotinine, in Urine by Gas Chromatography, *Nature*, **211**, 1415–1417 (1966).

Beckett, A. H., J. W. Gorrod, and P. Jenner, The Analysis of Nicotine-1'-N-oxide in Urine, in the Presence of Nicotine and Cotinine, and Its Application to the Study of *in vivo* Nicotine Metabolism in Man, *J. Pharm. Pharmacol.*, **23**, 55S–61S (1971).

Beckett, A. H., J. W. Gorrod, and P. Jenner, A Possible Relation Between pK_a 1 and Lipid Solubility and the Amounts Excreted in Urine of Some Tobacco Alkaloids Given to Man, *J. Pharm. Pharmacol.*, **24**, 115–120 (1972).

Booth, J. and E. Boyland, The Metabolism of Nicotine into Two Optically Active Stereoisomers of Nicotine-1'-oxide by Animal Tissues *in vitro* and by Cigarette Smokers, *Biochem. Pharmacol.*, **19**, 733–742 (1970).

Dagne, E. and N. Castagnoli., Jr., Structure of Hydroxycotinine, a Nicotine Metabolite, *J. Med. Chem.*, **15**, 356–360 (1972a).

Dagne, E. and N. Castagnoli., Jr., Cotinine N-Oxide, A New Metabolite of Nicotine, *J. Med. Chem.*, **15**, 840–841 (1972b).

Dow, J. and K. Hall, Estimation of Plasma Nicotine by Combined Capillary Column Gas Chromatography-Mass Spectrometry, *Br. J. Pharmacol.*, **61**, 159P (1977).

Dow, J. and K. Hall, Capillary-Column Combined Gas Chromatography-Mass Spectrometry Method for the Estimation of Nicotine in Plasma by Selective Ion Monitoring, *J. Chromatogr.*, **153**, 521–525 (1978).

Dumas, C., E. R. Badr, A. Viala, J. P. Cano, and R. Guillerm, A Micromethod for Determination of Nicotine and Cotinine in Blood and Urine by Gas Chromatography: Results Obtained from Various Smokers, *Eur. J. Toxicol. Environ. Hyg.*, **8**, 280–286 (1975).

Feyerabend, C., T. Levitt, and M. A. Russell, A Rapid Gas-Liquid Chromatographic Estimation of Nicotine in Biological Fluids, *J. Pharm. Pharmacol.*, **27**, 433–436 (1975).

Goldfarb, T. L., M. E. Jarvik, and S. D. Glick, Cigarette Nicotine Content as a Determinant of Human Smoking Behavior, *Psychopharmacologia*, **17**, 89–93 (1970).

Haines, C. F., Jr., D. K. Mahajan, C. D. Miljkov, C. M. Miljkov, and E. S. Vesell, Radioimmunoassay of Plasma Nicotine in Habituated and Naive Smokers, *Clin. Pharmacol. Ther.*, **16**, 1083–1089 (1974).

Hengen, N. and M. Hengen, Gas Liquid Chromatographic Determination of Nicotine and Cotinine in Plasma, *Clin. Chem.*, **24**, 50–53 (1978).

Horning, E. C., M. G. Horning, D. I. Carroll, R. N. Stillwell, and I. Dzidic, Nicotine in Smokers, Non-smokers and Room Air, *Life Sci.*, **13**, 1331–1346 (1973).

Jenner, P., J. W. Gorrod, and A. H. Beckett, Comparative *C*- and *N*-Oxidation of (plus)- and (minus)-Nicotine by Various Species, *Xenobiotica*, **1**, 497–498 (1971).

Langone, J. J., H. B. Gjika, and H. Van Vunakis, Nicotine and Its Metabolites: Radioimmunoassays for Nicotine and Cotinine, *Biochemistry*, **12**, 5025–5030 (1973).

Langone, J. J., J. Franke, and H. Van Vunakis, Nicotine and Its Metabolites: Gamma-(3-pyridyl)-gamma-oxo-*N*-methylbutyramide, *Arch. Biochem. Biophys.*, **164**, 536–543 (1974).

McNiven, N. L., K. H. Raisinghani, S. Patashnik, and R. I. Dorfman, Determination of Nicotine in Smokers' Urine by Gas Chromatography, *Nature*, **208**, 788–789 (1965).

Neelakantan, L. and H. B. Kostenbauder, Electron Capture Derivative for Determination of Nicotine in Sub-picomole Quantities, *Anal. Chem.*, **46**, 452–454 (1974).

Petrakis, N. L., L. D. Gruenke, T. C. Beelen, N. Castagnoli, Jr., and J. C. Craig, Nicotine in Breast Fluid of Nonlactating Women, *Science*, **199**, 303–305 (1978).

Russell, M. A., C. Wilson, U. A. Patel, C. Feyerabend, and P. V. Cole, Plasma Nicotine Levels after Smoking Cigarettes with High, Medium, and Low Nicotine Yields, *Br. Med. J.*, **02**, 414–416 (1975).

Shen, W. C., K. M. Greene, and H. Van Vunakis, Detection by Radioimmunoassay of Nicotinamide Nucleotide Analogues in Tissues of Rabbits Injected with Nicotine and Cotinine, *Biochem. Pharmacol.*, **26**, 1841–1846 (1977).

Watson, I. D., Rapid Analysis of Nicotine and Cotinine in the Urine of Smokers by Isocratic High-Performance Liquid Chromatography, *J. Chromatogr.*, **143**, 203–206 (1977).

NITROFURANTOIN—by D. R. Doose

N-(5-Nitrofurfurylidene)-1-aminohydantoin.
Furadantin, Furantoin. Urinary antibacterial agent.

$C_8H_6N_4O_5$; mol. wt. 238.16; pK_a 7.2.

a. Therapeutic Concentration Range. Normal therapeutic concentration is ~2 μg/ml blood (Winek, 1976).

b. Metabolism. Nitrofurantoin undergoes reduction of the nitro group in rat microsomal preparations, followed by spontaneous rearrangement to 1-[(3-cyano-1-oxopropyl)methylene]amino-2,4-imidazolidinedione (Aufrere et al., 1978). The reductive pathway is biologically important, since it may lead to carcinogenic intermediates of some of the nitrofuran and nitroimidazole derivatives. Also, hydrolysis of the azomethine linkage may occur, giving 5-nitrofuraldehyde and 1-aminohydantoin. The 5-nitrofuraldehyde may be further oxidized to the carboxylic acid. All of the known metabolites are thought to be devoid of significant antibacterial activity. The plasma elimination half-life of nitrofurantoin is 0.3 to 1.0 hr with 40% of the drug excreted in unchanged form in the urine (review: Cadwallader and Jun, 1976).

c. Analogous Compounds. 5-Nitrofuran antibacterial agents (e.g., furazolidone, nifuroxime, nitrofurazone); nitroimidazole antibacterials and radiosensitizers (e.g., metronidazole, misonidazole).

d. Analytical Methods

     1. High Performance Liquid Chromatography—Aufrere et al. (1977)

Discussion. Plasma samples are deproteinized and directly analyzed by reverse phase HPLC with UV detection at 365 nm, which is close to the UV maximum of nitrofurantoin at 370 nm. Furazolidone serves as the internal standard. The sensitivity is 20 ng nitrofurantoin/ml plasma, using a 0.2 ml plasma volume. The C.V. is less than 2%. With no modification, this procedure can also be used for the detection of nitrofurazone.

Detailed Method. Plasma (200 μl), 10 μl of 50 mg/l furazolidone internal standard solution (DMSO-water, 3:100, v/v), and 300 μl methanol are thoroughly mixed and centrifuged for 15 min at 10,000 *g* to remove the proteins. Then 15 to 60 μl of the supernatant is injected onto the column.

A Model ALC/GPC 244 (Waters Assoc.) high pressure liquid chromatograph is used, equipped with a U6-K injector and a 365 nm fixed-wavelength UV detector operated at 0.01 AUFS. The instrument is fitted with a 30 cm x 3.9 mm (i.d.) μ-Bondapak C_{18} reverse phase column (Waters Assoc.). The mobile phase consists of 20:80 (v/v) methanol-0.01 *M* sodium acetate buffer (pH 5.0). The mobile phase is degassed prior to use by applying reduced pressure to the stirred solvent. The retention times for nitrofurantoin and furazolidone are 6 and 8 min, respectively. Standard curves are constructed by plotting nitrofurantoin/furazolidone peak height ratios versus nitrofurantoin concentrations. A straight line fit of the data by least-squares linear regression analysis is obtained using the PROPHET system (NIH).

2. Other HPLC Assay—Vree et al. (1979)

Discussion. This HPLC procedure is equivalent to that published by Aufrere et al. (1977), also providing a sensitivity of 20 ng nitrofurantoin/ml plasma or urine. The assay is also applicable to the determination of hydroxymethylfurantoin (Urfadyn®).

3. Gas Chromatography—Ryhan et al. (1975)

Discussion. Nitrofurans are commonly used as animal feed additives. A GC method has been developed for the detection of these residual nitrofurans in animal tissues with a sensitivity of 10 to 50 ng/g. This method requires solvent extraction procedures, followed by hydrolysis of the nitrofurans to 5-nitro-2-furaldehyde, which is then measured by GC with electron capture detection. This method, although specific for nitrofurans in general, does not differentiate between the various nitrofuran derivatives and their metabolites.

4. Colorimetry—Buzard et al. (1958), Hollifield and Conklin (1970), Mattock et al. (1970)

Discussion. The method of Buzard et al. (1958) is based on the hydrolytic formation of 5-nitro-2-furaldehyde and its colorimetric estimation as the phenylhydrazone. The method requires relatively large plasma samples (3 ml) and is unsuitable for the determination of two or more different 5-nitrofurans or their metabolites in the same biological sample.

The nitromethane-Hyamine method (Hollifield and Conklin, 1970) is based on extraction of nitrofurantoin from biological samples and formation of a colored methylbenzethonium (Hyamine 10X) complex, which is measured at 400 nm. The method is proposed to be specific for nitrofurantoin in the presence of its metabolites and other antibacterial agents except for phenazopyridine, which can be separated by column chromatography. Mattock et al. (1970) improved the sensitivity of this assay in whole blood to 0.2 μg nitrofurantoin/ml.

5. Microbiological Assay—Gang and Shaikh (1972)

Discussion. This turbidimetric method measures the antimicrobial activity of nitrofurantoin in biological fluids. Interferences may occur from any compound possessing antimicrobial activity toward the test organism.

6. Comments. The reverse phase HPLC method (Aufrere et al., 1977) is clearly the method of choice at present. Its sensitivity for the detection of pure nitrofurantoin injected on-column is between 2 and 4 ng, which can result in highly sensitive procedures for biological samples if further prepurification steps are utilized.

e. References

Aufrere, M. B., B. A. Hoener, and M. E. Vore, High-Performance Liquid-Chromatography Assay for Nitrofurantoin in Plasma and Urine, *Clin. Chem.*, **23**, 2207–2212 (1977).

Aufrere, M. B., B. A. Hoener, and M. E. Vore, Reductive Metabolism of Nitrofurantoin *in vitro* in the Rat, *Drug Metab. Dispos.*, in press (1978).

Buzard, J. A., D. M. Vrablic, and M. F. Paul, Colorimetric Determination of Nitrofurazone, Nitrofurantoin, and Furazolidine in Plasma, *Antibiot. Chemother.*, **6**, 702–707 (1958).

Cadwallader, D. E. and H. W. Jun, Nitrofurantoin, in *Analytical Profiles of Drug Substances*, Vol. 5, S. K. Florey, Ed., Academic Press, New York, 1976, pp. 345–373.

Gang, D. M. and K. Z. Shaikh, Turbibimetric Method for Assay of Nitrofuran Compounds, *J. Pharm. Sci.*, **61**, 462–465 (1972).

Hollifield, R. D. and J. D. Conklin, A Method for Determining Nitrofurantoin in Urine in the Presence of Phenazopyridine Hydrochloride and Its Metabolites, *Clin. Chem.*, **16**, 335–338 (1970).

Mattock, G. L., I. J. McGilveray, and C. Charette, Improved Nitromethane-Hyamine Method for the Chemical Determination of Nitrofurantoin in Whole Blood, *Clin. Chem.*, **16**, 820–823 (1970).

Ryhan, J. J., J. A. Lee, Y. C. Dupont, and C. F. A. Charbonneau, A Screening Method for Determining Nitrofuran Drug Residues in Animal Tissues, *J. Assoc. Off. Anal. Chem.*, **58**, 1227–1231 (1975).

Vree, T. B., Y. A. Hekster, A. M. Baars, J. E. Damsma, and E. van der Kleijn, Determination of Nitrofurantoin (Furadantine®) and Hydroxymethylnitrofurantoin (Urfadyn®) in Plasma and Urine of Man by Means of High-Performance Liquid Chromatography, *J. Chromatogr.*, **162**, 110–116 (1979).

Winek, C. L., Tabulation of Therapeutic, Toxic, and Lethal Concentrations of Drugs and Chemicals in Blood, *Clin. Chem.*, **22**, 832–836 (1976).

OXAZEPAM—by S. Abdella

7-Chloro-1,3-dihydro-3-hydroxy-5-phenyl-2H-1,4-benzodiazepin-2-one. Serax, Praxiten. Minor tranquilizer.

$C_{15}H_{11}ClN_2O_2$; mol. wt. 286.74; pK_a 1.7 and 11.6.

a. Therapeutic Concentration Range. From 1 to 2 µg oxazepam/ml plasma (Goth, 1974).

b. Metabolism. The major metabolic and excretion product of oxazepam is its 3-O-glucuronide, which may exceed the concentration of the parent drug in plasma. The plasma elimination half-life ranges from 5.9 to 25 hr (review: Sjöquist and Sundwall, 1977).

c. Analogous Compounds. Benzodiazepines: clonazepam, chlorazepate, chlordiazepoxide, diazepam, flurazepam, lorazepam, medazepam, nitrazepam, prazepam.

d. Analytical Methods

1. Gas Chromatography—Vessman et al. (1977)

Discussion. Oxazepam is converted by extractive alkylation with methyl iodide to its N_1,O_3-dimethyl derivative, which is stable under the GC conditions and can be measured by electron capture detection with a sensitivity of 1 ng/ml serum. Lorazepam serves as the internal standard. Oxazepam is detectable by this method following the administration of diazepam and chlorazepate. However, 3-hydroxy-diazepam, another known minor metabolite of diazepam, yields the same methylation product as oxazepam and may therefore interfere with the assay.

Detailed Method. To 1 ml serum are added 4 ml of a pH 7.4 phosphate buffer, the internal standard lorazepam, and 8 ml methylene chloride followed by shaking for 15 min. After centrifugation the organic layer is filtered through silanized glass wool in a Pasteur pipet into a centrifuge tube and then evaporated to dryness. The tubes are cooled, and 1.5 ml 4 M methyl iodide and 0.5 ml 0.01 M tetramethylammonium hydrogen sulfate are added. The mixture is shaken for 5 min. The organic layer is again filtered thorough silanized glass wool into a centrifuge tube, and the solvents are evaporated. Then 1 ml hot saturated silver sulfate solution is added to the residue, which is dissolved by vortexing for 10 sec; 0.1 ml toluene is added and vibrated for 10 sec. One to two microliters of the toluene layer is injected into the gas chromatograph.

A Varian Model 1400 gas chromatograph with a scandium-type electron capture detector is used, equipped with a 1.5 m x 1.8 mm glass column, filled with 3% OV-225 on Chromosorb G (100/120 mesh, acid washed and silanized). The temperature of the column is kept at 265°C after conditioning with a glass flow for 2 hr at 290°C. The detector temperature is 300°C. The nitrogen flow rate is 30 ml/min. The standard curve is prepared from the oxazepam/lorazepam peak area ratios.

2. Other GC Assays—Sadée and van der Kleijn (1971), Vessman et al. (1972)

Discussion. The GC analysis of underivatized benzodiazepines using electron capture detection provides a sensitive assay for biological fluids. However, oxazepam is quantitatively converted to the corresponding quinazoline carboxaldehyde (Sadée and van der Kleijn, 1971) with a readily attainable sensitivity of only about 50 ng oxazepam/ml plasma (Vessman et al., 1977). Excellent sensitivity (about 1 ng/ml plasma) can be achieved by GC-EC analysis of the acid hydrolysis product of oxazepam, that is, 2-amino-5-chlorobenzophenone (Vessman et al., 1972). However, the specificity of this assay is low, since several benzodiazepines and their metabolites yield the same product.

3. High Performance Liquid Chromatography—Kabra et al. (1978)

Discussion. The method of Kabra et al. (1978) requires 2 ml blood for the determination of oxazepam by isocratic reverse phase HPLC with UV detection at 240 nm and prazepam as the internal standard. The sensitivity is 30 ng/ml blood. No derivatization is required following organic solvent extraction of the biological sample. Diazepam and its major plasma metabolite, *N*-desmethyldiazepam, can be simultaneously measured by this technique.

4. Thin-Layer Chromatography—Schwandt et al. (1976)

Discussion. There are several quantitative TLC methods employing UV scanning, fluorescence scanning, and colorimetry using the Bratton-Marshall reaction of 2-amino-5-chlorobenzophenone. The UV scanning method at 320 nm can detect 0.1 μg oxazepam directly on the plate and is applicable to the analysis of biological samples. All of the known metabolites of diazepam and oxazepam are separated from oxazepam on the TLC plate (Schwandt et al., 1976).

5. Polarography—Brooks and de Silva (1975)

Discussion. Differential pulse polarography of benzodiazepines in plasma and urine is based on the electrochemical reduction of the 4,5-azomethine functional group common to all benzodiazepines. It is therefore nonspecific unless employed in conjunction with chromatographic techniques. The sensitivity for benzodiazepines is >100 ng/ml plasma or urine.

6. Assay Comparison and Review—Hailey (1974)

Discussion. This article summarizes the chromatography of 1,4-benzodiazepines.

7. Comments. Both GC with electron capture and HPLC with UV detection are suitable techniques to determine the physiologic disposition of oxazepam.

e. References

Brooks, M. A. and J. A. F. de Silva, Determination of the 1,4-Benzodiazepines in Biological Fluids by Differential Pulse Polarography, *Talanta*, **22**, 849–860 (1975).

Goth, A., *Medical Pharmacology*, 7th Edition, C. V. Mosby Co., St. Louis, Mo., 1974.

Hailey, D. M., Chromatography of the 1,4-Benzodiazepines, *J. Chromatogr.*, **98**, 527–568 (1974).

Kabra, P. M., G. L. Stevens, and J. Marton, High Pressure Liquid Chromatographic Analysis of Diazepam, Oxazepam, and N-Desmethyldiazepam in Human Blood, *J. Chromatogr.*, **150**, 355–360 (1978).

Sadée, W. and E. van der Kleijn, Thermolysis of 1,4-Benzodiazepines during Gas Chromatography and Mass Spectrometry, *J. Pharm. Sci.*, **60**, 135–137 (1971).

Schwandt, H.-J., W. Sadée, K.-H. Beyer, and A. G. Hildebrandt, Effects of Enzyme Induction and Inhibition on Microsomal Oxidation of 1,4-Benzodiazepines, *N.S. Arch. Pharmacol.*, **294**, 91–98 (1976).

Sjöqvist, F. and A. Sundwall, The Pharmacokinetic Profile of Oxazepam, *Acta Pharmacol. Toxicol.* (Suppl. I), **40**, 1–70 (1977).

Vessman, J., G. Freij, and S. Stromberg, Determination of Oxazepam in Serum and Urine by Electron Capture Gas Chromatography, *Acta Pharm. Suec.*, **9**, 447–456 (1972).

Vessman, J., M. Johansson, P. Magnusson, and S. Strömberg, Determination of Intact Oxazepam by Electron Capture Gas Chromatography after an Extractive Alkylation Reaction, *Anal. Chem.*, **49**, 1545–1549 (1977).

PENICILLINS

Antimicrobials.

AMOXICILLIN

6-[D-(−)-α-Amino-p-hydroxyphenylacetamido]penicillanic acid. Amoxil, Clamoxyl, and so on.

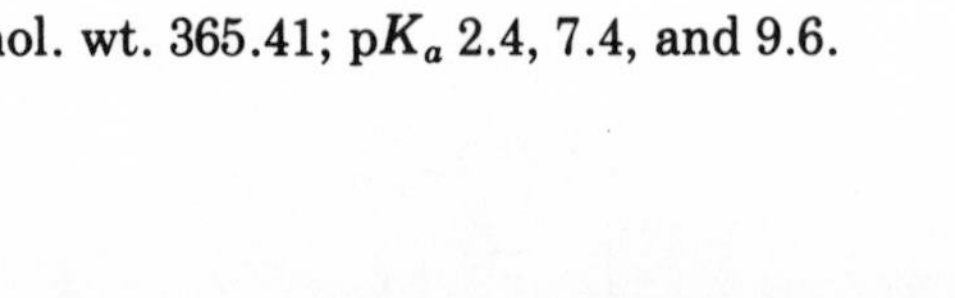

$C_{16}H_{19}N_3O_5S$; mol. wt. 365.41; pK_a 2.4, 7.4, and 9.6.

AMPICILLIN

6-[D-(−)-α-Aminophenylacetamido]penicillanic acid.
Ampicin, Binotal, and so on.

$C_{16}H_{19}N_3O_4S$; mol. wt. 349.42; pK_a 2.7 and 7.3.

CARBENICILLIN

6-(α-Carboxyphenylacetamido)penicillanic acid.

$C_{17}H_{18}N_2O_6S$; mol. wt. 378. 42.

CLOXACILLIN

[3-(o-Chlorophenyl)-5-methyl-4-isoxazolyl]penicillanic acid sodium salt.
Bactopen, and so on.

$C_{19}H_{17}ClN_3NaO_5S$; mol. wt. 457.89; pK_a 2.73.

DICLOXACILLIN

2,6-Dichlorophenyl analog of cloxacillin.
$C_{19}H_{16}Cl_2N_3NaO_5S$; mol. wt. 492.33; pK_a 2.8.

METHICILLIN

6-(2,6-Dimethoxybenzamido)penicillanic acid sodium salt.
Dimocillin, and so on.

$C_{17}H_{19}N_2Nao_6S$; mol. wt. 402.42; pK_a 2.76.

NAFCILLIN

6-(2-Ethoxy-1-naphthamido)penicillanic acid sodium salt.
Unipen.

$C_{21}H_2N_2NaO_5S$; mol. wt. 436.46; pK_a 2.7.

OXACILLIN

6-(5-Methyl-3-phenyl-2-isoxazoline-4-carboxamido)penicillanic acid
sodium salt. Micropenin, and so on.

$C_{19}H_{18}N_3NaO_5S.H_2O$; mol. wt. 441.45.

PENICILLIN G

6-(2-Phenylacetamido)penicillanic acid.

$C_{16}H_{18}N_2O_4S$; mol. wt. 334.39; pK_a 2.8.

PENICILLIN V

6-Phenoxyacetamidopenicillanic acid.
Oracillin, and so on.

$C_{16}H_{18}N_2O_5S$; mol. wt. 350.38; pK_a 2.73.

a. Therapeutic Concentration Range. Therapeutic concentrations are defined by the minimum inhibitory concentration specific for the infecting organism. Representative maximum plasma levels achieved after oral doses of 1150 mg ampicillin and 750 mg amoxicillin in adults are 11 and 14 μg/ml, respectively (Vree et al., 1978).

b. Metabolism. All penicillin antimicrobials are actively secreted into the renal tubules and excreted mostly unchanged into the urine, except for nafcillin, which exhibits high biliary excretion (Barza and Weinstein, 1976; Yaffee et al., 1977). Penicillin V and oxacillin are also extensively degraded to inactive products (Barza and Weinstein, 1976). Major metabolites are the respective penicilloic acids, generated by cleavage of the β-lactam ring, and α-amino substituted derivatives of ampicillin and related α-amino penicillins (Graber et al., 1976). Interindividual variability in the amount of excreted penicilloic acids suggest their enzymatic, rather than spontaneous, formation. Whereas practically all of an ampicillin dose can be accounted for in the urine, urinary recovery of amoxicillin and oxacillin amounts to only 50% of the dose (Graber et al., 1976).

Serum protein binding, half-life, and peak plasma levels are different for the individual penicillins and represent important determinants of antimicrobial efficacy (Barza and Weinstein, 1976; Egert et al., 1977). All of these parameters may vary considerably for the same penicillin among different patients. The need for penicillin plasma level monitoring has been emphasized in newborn infants, who show significantly longer elimination half-lives than adults (Colburn et al., 1976; Sarff et al., 1977; Schwartz et al., 1976).

Some of the penicillin plasma elimination half-lives in adults with normal renal function, as well as further pharmacokinetic data, are listed in the following table (references: Barza and Weinstein, 1976; Colburn et al., 1976; Vree et al., 1978).

Drug	Serum protein binding (%)	Peak serum level after oral dose of 500 mg (μg/ml)	$t_{1/2}$ (hr)[a]	Percentage of dose metabolized[b]
Amoxicillin	20	7–8	1.3	28 (po)
Ampicillin	20	2–6	1.3	21 (po)
Carbenicillin	50	—	1	2 (im)
Cloxacillin	93	7–14	0.5	22 (po)
Dicloxacillin	96	15–18	0.8	10 (po)
Methicillin	30–50	—	0.43	8 (im)
Nafcillin	90	—	0.55–1.0	
Oxacillin	90	5–6	0.4–0.7	49 (po)
Penicillin G	60	1.5–2.7	0.7	19 (im)
Penicillin V	80	3–5		56 (po)

[a] Normal renal function.
[b] po = peroral, im = intramuscular.

c. Analogous Compounds. Other penicillins; cephalosporins.

d. Analytical Methods

1. High Performance Liquid Chromatography—Vree et al. (1978)

Discussion. The reverse phase HPLC assay of Vree et al. (1978) utilizes UV absorbance detection at 225 nm for the measurement of amoxicillin and ampicillin in plasma and urine. Proteins are precipitated with perchloric acid, and the supernatants directly injected onto the column. With the use of external standards, quantitation is readily obtained above 1 μg/ml plasma. 6-Aminopenicillanic acid can also be measured by this technique; however, two unknown metabolites appeared in HPLC records of urine samples after administration of amoxicillin, while 6-aminopenicillanic acid was not detectable.

Detailed Method. Plasma (0.1 ml) is thoroughly mixed with 0.4 ml 0.33 N HClO$_4$ and centrifuged at 2600 g for 5 min. Then 100 μl of the clear supernatant is injected onto the column.

A Spectra Physics 3500B high performance liquid chromatograph is used, equipped with a 15 cm x 4.6 mm (i.d.) column packed with Li-Chrosorb RP-8 with a 5 μ particle size (Chromapack, Middleburg, The Netherlands). The UV absorbance detector is set at 225 nm. The eluent for amoxicillin is a 0.067 M, potassium dihydrogen phosphate buffer (pH 4.6) at a flow rate of 1.2 ml/min. The solvent for ampicillin is a mixture of the pH 4.6 KH$_2$PO$_4$ buffer and methanol (425:75) at a flow rate of 1.2 ml/min. The retention times are 6.5 and 8 min, respectively, for amoxicillin and ampicillin.

2. Other HPLC Assays—White et al. (1975), Yamaoka et al. (1979)

Discussion. The HPLC separation of penicillins is achieved by White et al. (1975) on a reverse phase porous packing material (C$_{18}$-Porasil B), while pellicular reverse phases (C$_{18}$) produce tailing peaks. These authors suggest the applicability of HPLC to biological samples; however, resolution of the proposed column is much lower than that reported for the RP-8 column by Vree et al. (1978). The reverse phase HPLC assay of Yamaoka et al. (1979) for urinary carbenicillin requires only a 10 μl sample, which is filtered through a Micropore membrane before injection. With this method, 91.5% of a dose of carbenicillin to a human subject was recovered in the urine.

3. Thin-Layer Chromatography—Hurwitz and Carney (1978)

Discussion. Quantitative and specific TLC analysis of ampicillin in plasma and urine is based on its conversion directly on a plate to a fluorescent product (no details given; see Section d5), followed by thin-layer fluorescence scanning. The amphoteric penicillins (e.g., ampicillin) can be extracted from plasma as an ion pair with tetra-heptylammonium chloride. The silica gel TLC eluent is ethyl acetate-acetone-water-formic acid (65:15::13:9). The limit of sensitivity is 0.5 μg ampicillin/ml plasma, using a 4 ml sample volume.

4. Polarography—Benner (1970)

Discussion. Oscillographic polarography with the dropping mercury electrode can measure penicillins and cephalosporins in plasma and urine in the range of 1 to 50 μg/ml. Good correlations are obtained with microbiological assays in spite of potential interferences by sulfhydryl compounds, tryptophane, histidine, and arginine. Plasma preparation includes an ultrafiltration step to remove interferences; therefore this procedure measures the unbound fraction of penicillins in the plasma.

5. Spectrophotometry and Fluorescence—Barbhaiya and Turner (1977), Barbhaiya et al. (1977), Davidson (1976), Dürr and Schatzmann (1975), Jusko (1971), Yu et al. (1977)

Discussion. The penicillins yield a yellow fluorescent product upon treatment with diluted strong bases or acids. The fluorescent species may be a diketopiperazine derivative (Jusko, 1971). The fluorescent yield is dependent on the reaction condition, and different optimal methods have been developed for the individual penicillins [ampicillin: Barbhaiya and Turner (1977), Dürr and Schatzmann (1975), Jusko (1971); amoxicillin: Barbhaiya et al. (1977), Davidson (1976); other penicillins: Yu et al. (1977)]. The fluorescence assays are usually sensitive to 0.05 μg/ml plasma; UV absorbance at 370 nm can also be utilized to determine the much higher urine levels of amoxicillin (Davidson, 1976). The corresponding penicilloic acids form the same reaction product and cannot be differentiated from unchanged penicillins unless separated (Jusko, 1971). The difference between concentrations obtained by biological assay, which does not detect the inactive penicilloic acids, and fluorescence assay should give an estimate of the level of penicilloic acid metabolites (Jusko, 1971). Cephalosporins also yield fluorescent products under similar conditions (Yu et al., 1977).

6. Radioimmunoassay—Wal (1976)

Discussion. The RIA technique is based on antibodies raised against the penicilloyl group generated from administration of penicillin G to cattle and human beings. It can be used for penicillin trace analysis in milk.

7. Microbiological Assays—Burnett and Sutherland (1970), Carlström et al. (1974), Ericsson and Malmborg (1973), Jalling et al.

(1972), Reeves and Holt (1975), Unertl et al. (1974), Whyatt et al. (1974)

Discussion. Microbiological assays have undoubtedly provided the most useful approach to the analysis of both bacterial susceptibility, in order to determine minimum inhibitory concentrations, and antimicrobial plasma and urine levels. The assays cited above represent only a fraction of the published articles on this subject. A problem is the lack of specificity, since several antimicrobials are often administered together. The use of multiple-resistant test organisms is questionable because of the environmental hazards. Careful selection of the organism with special consideration of the enzyme penicillinase appears to be of importance (Burnett and Sutherland, 1970). Specificity of the assay can be improved by including chromatographic separation steps, such as high voltage electrophoresis (Carlström et al., 1974; Reeves and Holt, 1975). Some assays (e.g., the microtube dilution method) can be semiautomated (Unertl et al., 1974). Assay reproducibility is usually on the order of 10 to 15%.

    8. Comments. Apart from the widely used biological assays, relatively little analytical effort has been expended on this important class of drugs. Penicillin drug level monitoring appears to be important in newborn infants and under special clinical conditions, in spite of the rather large therapeutic index of the penicillins. Furthermore, the mechanism of penicillin allergy is only poorly understood as yet. High performance liquid chromatography with UV detection at rather low wavelengths may become an important pharmacological tool in this area.

e. References

Barbhaiya, R. H. and P. Turner, Fluorimetric Determination of Ampicillin and Epicillin, *J. Antimicrob. Chemother.*, **3**, 423–427 (1977).

Barbhaiya, R. H., P. Turner, and E. Shaw, A Simple, Rapid Fluorimetric Assay of Amoxycillin in Plasma, *Clin. Chim. Acta*, **77**, 373–377 (1977).

Barza, M. and L. Weinstein, Pharmacokinetics of the Penicillins in Man, *Clin. Pharmacokin.*, **1**, 297–308 (1976).

Benner, E. J., Two-Hour Assay for Content of Penicillins and Cephalosporins in Serum, *Antimicrob. Agents Chemother.*, **10**, 201–204 (1970).

Burnett, J. and R. Sutherland, Procedures for the Assay of Carbenicillin in Body Fluids, *Appl. Microbiol.*, **19**, 264–267 (1970).

Carlström, A., K. Dornbusch, A. Hagelberg, and A. S. Malmborg, Microbiological Assay of Mixtures of Ampicillin and Cloxacillin in Plasma after

Separation by High Voltage Electrophoresis, *Acta Pathol. Microbiol. Scand.*, **82**, 67–74 (1974).

Colburn, W. A., M. Gilbaldi, H. Yoshioka, M. Takimoto, and H. D. Riley, Jr., Pharmacokinetic Model for Serum Concentrations of Ampicillin in the Newborn Infant, *J. Infect. Dis.*, **134**, 67–69 (1976).

Davidson, D. F., A Simple Chemical Method for the Assay of Amoxycillin in Serum and Urine, *Clin. Chim. Acta*, **69**, 67–71 (1976).

Dürr, A. and H. J. Schatzmann, A Simple Fluorimetric Assay for Ampicillin Serum, *Experientia*, **31**, 503–504 (1975).

Egert, J., J. Carrizosa, D. Kaye, and W. D. Kobasa, Comparison of Methicillin, Nafcillin, and Oxacillin in Therapy of *Staphylococcus aureus* Endocarditis in Rabbits, *J. Lab. Clin. Med.*, **89**, 1262–1268 (1977).

Ericsson, H. and A. S. Malmborg, A Micromethod for Determination of Antibiotic Concentrations in Body Fluids, *Acta Pathol. Microbiol Scand.* (B Suppl.), **241**, 107 (1973).

Graber, H., T. Perenyi, M. Arr, and E. Ludwig, On Human Biotransformation of Some Penicillins, *Int. J. Clin. Pharmacol. Biopharm.*, **14**, 284–289 (1976).

Holt, H. A., J. M. Broughall, M. McCarthy, and D. S. Reeves, Interactions Between Aminoglycoside Antibiotics and Carbenicillin or Ticarillin, *Infection*, **4**, 107–109 (1976).

Hurwitz, A. R. and C. F. Carney, Enhancement of Ampicillin Partition Behavior, *J. Pharm. Sci.*, **67**, 138–140 (1978).

Jalling, B., A. S. Malmborg, A. Lindman, and L. O. Boreus, Evaluation of a Micromethod for Determination of Antibiotic Concentrations in Plasma, *Eur. J. Clin. Pharmacol.*, **4**, 150–157 (1972).

Jusko, W. J., Fluorometric Analysis of Ampicillin in Biological Fluids, *J. Pharm. Sci.*, **60**, 728–732 (1971).

Reeves, D. S. and H. A. Holt, Resolution of Antibiotic Mixtures in Serum Samples by High-Voltage Electrophoresis, *J. Clin. Pathol.*, **28**, 435–442 (1975).

Sarff, L. D., G. H. McCracken, M. L. Thomas, L. J. Horton, and N. Threlkeld, Clinical Pharmacology of Methicillin in Neonates, *J. Pediatr.*, **90**, 1005–1008 (1977).

Schwartz, G. J., T. Hegyi, and A. Spitzer, Subtherapeutic Dicloxacillin Levels in a Neonate: Possible Mechanisms, *J. Pediatr.*, **89**, 310–312 (1976).

Unertl, K., F. D. Daschner, and W. Marget, A Semiautomatic Microtube Dilution Method for Determining Antibiotic Blood Concentrations, *Chemotherapy*, **20**, 331–338 (1974).

Vree, T. B., Y. A. Hekster, A. M. Baars, and E. van der Kleijn, Rapid Determination of Amoxycillin (Clamoxyl) and Ampicillin (Penbritin) in Body Fluids of Man by Means of High-Performance Liquid Chromatography, *J. Chromatogr.*, **145**, 496–501 (1978).

Wal, J. M., Radioimmunoassay for the Detection of Penicolloyl Groups in Biological Fluids after Therapy with Penicillin G, in *Drug Interference and Drug Measurement in Clinical Chemistry*, G. Siest and D. S. Young, Eds., Basel, Karger, 1976, pp. 99–102.

White, R. E., M. A. Carroll, J. E. Zaremba, and A. D. Bender, Reverse Phase High Speed Liquid Chromatorgraphy of Antibiotics, *J. Antibiot.*, **28**, 205–214 (1975).

Whyatt, P. L., R. E. Dann, G. W. Slywka, and M. C. Meyer, Rapid, Precise Turbidometric Assay for Low Levels of Ampicillin in Serum after Single-Dose Oral Administration, *Antimicrob. Agents Chemother.*, **6**, 811–814 (1974).

Yaffe, S. J. et al., Pharmacokinetics of Methicillin Patients with Cystic Fibrosis, *J. Infect. Dis.*, **135**, 828–831 (1977).

Yamaoka, K., S. Narita, T. Nakagawa, and T. Uno, High-Performance Liquid Chromatographic Analyses of Sulbenicillin and Carbenicillin in Human Urine, *J. Chromatogr.*, **168**, 187–193 (1979).

Young, L. S., G. Decker, and W. L. Hewitt, Inactivation of Gentamycin by Carbenicillin in the Urinary Tract, *Chemotherapy*, **20**, 212–220 (1974).

Yu, A. B., C. H. Nightingale, and D. R. Flanagan, Rapid, Sensitive Fluorometric Analysis of Cephalosporin Antibiotics, *J. Pharm. Sci.*, **66**, 213–216 (1977).

PHENCYCLIDINE—by L. Lambert

1-(1-Phenylcyclohexyl)piperidine.
Sernyl, Sernylan. Anesthetic (no longer used for human beings); PCP; drug of abuse.

$C_{17}H_{25}N$; mol. wt. 243.38; pK_a 8 to 9.

a. Therapeutic Concentration Range. None defined. Phencyclidine plasma levels between 6 and 240 ng/ml plasma have been reported after nonfatal overdoses (Pearce, 1976).

b. Metabolism. Three metabolites have been found in human urine: α1-phenylcyclohexylamine, 4-hydroxypiperidylphencyclidine, and 4-hydroxycyclohexylphencyclidine. The hydroxylated metabolites are further conjugated (Wong and Biemann, 1976). The plasma elimination half-life of phencyclidine in human beings has been reported to be about 11 hr (Marshman et al., 1976). This drug sequesters into the central nervous and adipose tissue because of its high lipophilicity, providing an explanation for the persistent pharmacological effects over several days in the absence of detectable blood levels (James and Schnoll, 1976).

c. Analogous Compounds. Anesthetics (e.g., ketamine), piperidine-type analgesics (e.g., meperidine), local anesthetics (e.g., mepivicaine).

d. Analytical Methods

1. Gas Chromatography-Mass Spectrometry—Lin et al. (1975)

Discussion. Following solvent extraction of phencyclidine, together with pentadeuterated phencyclidine as the internal standard, from plasma, analysis is carried out by GC-methane chemical ionization MS in the selected ion monitoring mode. Two ion pairs at m/e 159/164 and 243/248 (phencyclidine/d_5-phencyclidine) are used for quantitation in an isotope dilution assay. The sensitivity is 1 ng phencyclidine/ml plasma with a relative S. D. of below 3% at 100 ng/ml. Thermal decomposition of phencyclidine is noted to be proportional to injection port temperature and is minimal at injection port temperatures less than 200°C.

Detailed Method. One milliliter body fluid (whole blood, serum, urine, or cerebral spinal fluid) is pipetted into a glass tube, and 50 ng pentadeuterated phencyclidine hydrochloride in CH_3OH (1 ng/μl), 1 ml distilled water, and 1 ml pH 9.5 carbonate buffer are added. For phencyclidine concentrations greater than 1 μg/ml, 1 μg of the internal standard is used. After mixing the contents with a vortex mixer, 7 ml hexane is added and the tube gently rotated for 30 to 45 min. After centrifugation the organic layer is pipetted to another test tube and extracted with an approximately equal volume of 0.2 N H_2SO_4 by shaking for 5 min. The tubes are centrifuged, and the organic layer is discarded. The aqueous layer is adjusted to pH 9.5 by adding 3 N KOH, followed by 1 ml carbonate buffer, and extracted with an equal volume of hexane. After centrifuging, the organic layer is transferred to another tube and evaporated to near dryness with a nitrogen stream at a temperature below 50°C. The residue is dissolved in 10 μl hexane, and 1 to 3 μl injected onto the gas chromatograph-mass spectrometer.

The MS analysis is conducted on a Finnigan Model 3200 gas chromatograph-chemical ionization mass spectrometer controlled by a Computer System Industries 250 data system. The gas chromatograph is equipped with a 1.83 m x 2 mm (i.d.) glass column packed with 3% OV-1 on 100/120 mesh Gas Chrom (Applied Science Lab., State College, Pa.). The injection port temperature and the line between the gas chromatograph and the mass spectrometer are kept at 200°C, while the column is maintained at 180°C. Methane is the carrier gas at a flow of 20 ml/min, giving a pressure of 0.5 torr in the ion source. The concentration of phencyclidine in the body fluid is determined by measuring the peak heights of the ion current signals at the elution time of phencyclidine (approx. 2 min after injection). The peak height ratios, *m/e* 159/164 and *m/e* 243/248, are calculated and converted to phencyclidine concentrations, using a previously established standard curve.

2. Other GC-MS Methods—Domino and Wilson (1977), MacLeod et al. (1976), Pearce (1976), Wilson and Domino (1978)

Discussion. These methods are equivalent to the one of Lin et al. (1975), except for the use of electron impact ionization. The electron impact GC-MS-SIM assay of Wilson and Domino (1978) has a sensitivity comparable to that of the chemical ionization procedure of Lin et al. (1975). Again, d_5-phencyclidine is used as the internal standard, and the ion pairs at *m/e* 91/95 and 200/205 are monitored in a quadrupole mass analyzer. The method of McLeod et al. (1976) consists of an automated analysis of phencyclidine in urine by probability-based ion matching GC-MS, which is mainly used for its unequivocal qualitative identification.

3. Other MS Assay—Boerner et al. (1973)

Discussion. Biological sample extracts are introduced into the mass spectrometer via a Llewellyn three-stage membrane separation inlet system and analyzed by electron impact ionization. Automatic analysis is achieved by identification and quantitation of the characteristic fragmentation pattern of phencyclidine. The sensitivity is 5 μg phencyclidine/ml, with a total analysis time of 10 min.

4. Gas Chromatography—Froehlich and Ross (1977), Marshman et al. (1976), Reynolds (1976)

Discussion. These GC-FID procedures are specific for phencyclidine, but the levels of the drug in body fluids approach the useful limits of sensitivity with the flame ionization detector (Reynolds, 1976). Higher sensitivity may be obtainable with nitrogen-sensitive detectors. Marshman et al. (1976) note endogenous interference in samples extracted from plasma with chloroform. Froehlich and Ross (1977) describe three extraction methods, one of which includes chloroform and could therefore result in endogenous peak interference. The other two methods of extraction involve charcoal absorption and reversed phase XAD resin.

5. Comments. Because of the rather low levels of phencyclidine in biological fluids, GC-MS appears to be the method of choice at present. To avoid thermal decomposition of phencyclidine, GC temperatures should be kept below 200°C. Analysis of phencyclidine concentrations in the cerebrospinal fluid may be particularly useful to assess individual exposure to the drug.

3. References

Boerner, U. et al., Direct Mass Spectrometric Analysis of Body Fluids From Acutely Poisoned Patients, *Clin. Chim. Acta*, **49**, 445–454 (1973).

Domino, E. F. and A. E. Wilson, Effects of Urine Acidification on Plasma and Urine Phencyclidine Levels in Overdosage, *Clin. Pharmacol. Ther.*, **22**, 421–424 (1977).

Froehlich, P. M. and G. Ross, Separation and Detection of Phencyclidine in Urine by Gas Chromatography, *J. Chromatogr.*, **137**, 135–143 (1977).

James, S. H. and S. H. Schnoll, Phencyclidine: Tissue Distribution in the Rat, *Clin. Toxicol.*, **9**, 573–582 (1976).

Lin, D. C., A. F. Fentiman, R. L. Foltz, R. D. Forney, and I. Sunshine, Quantification of Phencyclidine in Body Fluids by Gas Chromatography Chemical Ionization Mass Spectrometry and Identification of Two Metabolites, *Biomed. Mass Spectrom.*, **2**, 206–214 (1975).

MacLeod, W. D., D. E. Green, and E. Seet, Automated Analysis of Phencyclidine in Urine by Probability Based Matching GC/MS, *Clin. Toxicol.*, **9**, 561–572 (1976).

Marshman, J. A., M. P. Ramsay, and E. M. Sellers, Quantitation of Phencyclidine in Biological Fluids and Application to Human Overdose, *Toxicol. Appl. Pharmacol.*, **35**, 129–136 (1976).

Pearce, J. S., Detection and Quantitation of Phencyclidine in Blood, Use of

[²H₅]Phencyclidine and Select Ion Monitoring Applied to Non-fatal Cases of Phencyclidine Intoxication, *Clin. Chem.*, **22**, 1623–1626 (1976).

Reynolds, P. C., Clinical and Forensic Experiences with Phencyclidine, *Clin. Toxicol.*, **9**, 547–552 (1976).

Wilson, A. E. and E. F. Domino, Plasma Phencyclidine Pharmacokinetics in Dog and Monkey, Using a Gas Chromatography Selected Ion Monitoring Assay, *Biomed. Mass Spectrom.*, **5**, 112–116 (1978).

Wong, L. K. and K. Biemann, Metabolites of Phencyclidine, *Clin. Toxicol.*, **9**, 583–591 (1976).

PHENYLBUTAZONE—by S. Abdellah

4-Butyl-1,2-diphenyl-3,5-pyrazolidinedione.
Artrizin, Butazolidin. Analgesic; anti-inflammatory and antiarthritic agent.

$C_{19}H_{20}N_2O_2$; mol. wt. 308.37; pK_a 9.1.

a. Therapeutic Concentration Range. Active phenylbutazone plasma levels are in the range of 100 μg/ml (Goth, 1976).

b. Metabolism. The major metabolites of phenylbutazone are the *p*-hydroxyphenyl derivative, oxyphenbutazone, and the γ-hydroxybutyl derivative, 3-hydroxyphenylbutazone (review: Disanto, 1976). Oxyphenbutazone exhibits a spectrum of pharmacological activity similar to that of phenylbutazone, while the 3-hydroxy metabolite has a marked uricosuric effect but little antirheumatic activity. Plasma protein binding of phenylbutazone and its metabolites is extensive. Phenylbutazone appears in the urine only in trace amounts, oxyphenbutazone to the extent of 3%, and 3-hydroxyphenylbutazone to 40% of the total dose. Further metabolism remains largely unknown. The average plasma elimination half-life of phenylbutazone in normal adults is 72 hr. Oxyphenbutazone is also slowly eliminated and may accumulate in the body in considerable amounts (Disanto, 1976).

c. Analogous Compounds. Aminopyrine, antipyrine, oxyphen-butazone, sulfinpyrazone.

d. Analytical Methods

   1. High Performance Liquid Chromatography—Pound and Sears (1975)

Discussion. Acidified plasma is partitioned with cyclohexane-ether containing an internal standard. The extract is chromatographed on an adsorption column and analyzed by UV absorbance at 254 nm (UV_{max} of phenylbutazone 265 nm). Phenylbutazone and oxyphenbutazone can be measured with a sensitivity of 0.25 μg/ml plasma. 3-Hydroxyphenyl-butazone, which is partially separated from oxyphenbutazone, is not detectable in plasma after single phenylbutazone doses.

Detailed Method. Plasma samples (1 ml), 1 ml 1 N HCl, and 6 ml cyclohexane-ether (1:1) containing 15 to 75 μg of the internal standard, 2,4-dinitrophenylhydrazone of 3,4-dimethoxybenzaldehyde, are shaken for 15 min and centrifuged. The organic layer is evaporated to dryness under nitrogen at 60°C. The residue is redissolved in 50 to 200 μl $CHCl_3$, and 10 μl of this solution is injected onto the column. Peak height ratios are used for quantitation.

A model 4000 Aerograph is used, equipped with a constant-temperature bath set at 35°C, a fixed-wavelength (254 nm) UV absorbance detector, and a Sil-X absorbent stainless steel column, 100 cm x 1.8 mm (i.d.) (Perkin-Elmer Corp.). The mobile phase, 0.002% acetic acid and 23% tetrahydrofurane in n-hexane, is used at a flow rate of 60 ml/hr.

   2. Gas Chromatography—Bogan (1977), Bruce et al. (1974), Perego et al. (1971)

Discussion. Perego et al. (1971) have developed a GC-FID assay for phenylbutazone in plasma and urine following organic solvent extraction. Derivatization of the anhydrous extracts with heptafluorobutyric anhydride (Bruce et al., 1974) extends the applicability of this method to the simultaneous determination of oxyphenbutazone. Bogan (1977) proposes oxidation of phenylbutazone with permanganate solution to azobenzene, which is detected by GC-FID. In spite of the reported specificity of this technique, analysis of a reaction product appears undesirable when assaying for unchanged phenylbutazone in biological samples.

3. Ultraviolet Spectrophotometry—Beckstead et al. (1968), Lukas et al. (1976), Wallace (1968)

Discussion. The method of Beckstead et al. (1968) combines organic-aqueous partitioning at various pH values and UV detection at 264 and 232 nm to obtain a specific assay of phenylbutazone in biological fluids. Application of an automated variation of this UV technique in plasma and urine is reported by Lukas et al. (1976). The permanganate oxidation of phenylbutazone (Wallace, 1968) produces azobenzene with a UV absorption maximum at 314 nm, which permits specific detection of phenylbutazone in biological fluids with sufficient sensitivity. The degree of UV assay specificity in biological samples, however, has not been clearly established.

4. Mass Spectrometry—Weinkam et al. (1977)

Discussion. Phenylbutazone is separated from plasma samples using a double back-extraction. The residue is analyzed by direct insertion isobutane chemical ionization MS with dideuterophenylbutazone as the internal standard. Specific quantitation of phenylbutazone is achieved by repetitive scanning over the MH^+ ion peaks and comparison to standard curves. Sensitivity is in the range of 1 $\mu g/ml$ plasma. Determinations of tolbutamide and two of its metabolites are also discussed in this paper.

5. Comments. The current methods of choice are HPLC and GC because of their specificity and general applicability.

e. References

Beckstead, H. D., K. K. Kaistha, and S. J. Smith. Determination and Thin-Layer Chromatography of Phenylbutazone in the Presence of Decomposition Products, *J. Pharm. Sci.*, **57**, 1952–1957 (1968).

Bogan, J. A., A Sensitive and Specific Method for the Determination of Phenylbutazone in Biological Samples, *J. Pharm. Pharmacol.*, **29**, 125–126 (1977).

Bruce, R. B., W. R. Maynard, and L. K. Dunning, Oxyphenbutazone and Phenylbutazone Determination in Plasma and Urine by GLC, *J. Pharm. Sci.*, **63**, 446–448 (1974).

Disanto, A. R., Phenylbutazone, *J. Am. Pharm. Assoc.*, **16**, 365–367 (1976).

Goth, A., *Medical Pharmacology*, 8th edition, C. V. Mosby Co., St. Louis, Mo., 1976, p. 335.

Lukas, G., C. B. Borman, and S. B. Zak, Fully Automated Analysis of Phenylbutazone in Plasma and Urine, *J. Pharm. Sci.*, **65**, 86–88 (1976).

Perego, R., E. Martinelli, and P. C. Vanoni, Gas Chromatographic Assay of Phenylbutazone in Biological Fluids, *J. Chromatogr.*, **54**, 280–281 (1971).

Pound, N. J. and R. W. Sears, Simultaneous Determination of Phenylbutazone and Oxyphenbutazone in Plasma by High-Speed Liquid Chromatography, *J. Pharm. Sci.*, **64**, 284–287 (1975).

Wallace, J. E., Spectrophotometric Determination of Phenylbutazone in Biological Specimens, *J. Pharm. Sci.*, **57**, 2053–2056 (1968).

Weinkam, R. J., M. Rowland, and P. J. Meffin, Determination of Phenylbutazone, Tolbutamide and Metabolites in Plasma and Urine Using Chemical Ionization Mass Spectrometry, *Biomed. Mass Spectrom.*, **4**, 42–47 (1977).

PHENYTOIN

5,5-Diphenyl-2,4-imidazolidinedione (diphenylhydantoin).
Dilantin. Anticonvulsant, antiarrhythmic agent.

$C_{15}H_{12}N_2O_2$; mol. wt. 252.26; pK_a 8.3

a. Therapeutic Concentration Range. From 10 to 20 μg phenytoin/ml plasma (Leal and Troupin, 1977).

b. Metabolism. Phenytoin is mainly metabolized to 5-(4-hydroxyphenyl)-5-phenylhydantoin(4'-OH-phenytoin), which is further glucuronidated and excreted into the urine (Hoppel et al., 1976, 1977). The metabolite 4'-OH-phenytoin is optically active (Glazko, 1972). Parahydroxylation of phenytoin is readily saturable at therapeutic doses; therefore a small increase in the phenytoin dose may cause unproportionally large increases of phenytoin plasma levels as a complicating factor in its clinical use (Atkinson and Shaw, 1973). The plasma elimination half-life of phenytoin shows large interindividual variability (24 ± 12 hr) (Leal and Troupin, 1977). Some patients have an unusually high metabolic clearance of phenytoin. Kutt et al. (1966) have pointed out three reasons for lack of therapeutic response to normally therapeutic doses of phenytoin: high metabolic clearance, poor gastrointestinal absorption (rare), and patient noncompliance. High metabolic clearance, on the one hand, and poor absorption or noncompliance, on the other hand, can be differentiated by measuring the urinary excretion of the 4'-hydroxyphenylglucuronide. This inactive urinary metabolite has therefore attracted the interest of clinical chemists, and many 4'-OH-phenytoin assay methods are now available.

Other phenytoin metabolites include a ring-*N*-glucuronide (Smith et al., 1977) and several aromatic hydroxylated derivatives (3'-OH: Atkinson et al., 1970; 3'4'-diOH: Borga et al., 1972; 3'-OCH$_3$-4'-OH: Hoppel et al., 1976; bis-3'-OH: Thompson et al., 1976; 3'4'-dihydrodiol: Hoppel et al., 1976). The dihydrodiol metabolite normally accounts for up to 20% of the urinary excretion products; all other hydroxylated metabolites are of minor importance. Species other than human beings, such as the dog, may produce 3'-OH-phenytoin as the major metabolite (Hoppel et al., 1977).

Phenytoin is highly plasma protein bound ($\sim$90%) with significant interindividual variability (Barth et al., 1976). However, all of the clinically used drug level assays determine only total concentration with the implicit assumption that the free, active fraction in plasma is similar in all patients.

c. Analogous Compounds. Hydantoins: ethotoin, mephenytoin; anticonvulsants.

d. Analytical Methods. The major assay methods for phenytoin in serum, for example, HPLC, GC, and EMIT, are discussed in detail in the anticonvulsants monograph. Additional assays suitable only for the detection of phenytoin are listed here.

1. High Performance Liquid Chromatography—Hermansson and Karlen (1977), Kabra and Marton (1976), Slonek et al. (1978)

Discussion. The assays of Hermansson and Karlen (1977) and Kabra and Marton (1976) are designed for the measurement of urinary 4'-OH-phenytoin glucuronide. Following acid hydrolysis of the glycosidic bond in HCl, analysis is performed on a μ-Bondapak C$_{18}$ reverse phase column with the 5-methyl analog as the internal standard. The UV absorbance is measured at 254 nm, which gives a sensitivity of only 300 ng 4'-OH-phenytoin per injection, since the UV absorbance maximum is at lower wavelengths (see anticonvulsants monograph, HPLC). However, this relatively low sensitivity is sufficient to measure the rather large amounts of glucuronide in the urine. The dihydrodiol metabolite of phenytoin is converted to equal amounts of 3'-OH- and 4'-OH-phenytoin and therefore can interfere with the assay of 4'-OH-phenytoin. Hermansson and Karlen (1977) eliminate this problem by preextracting urine samples with isoamyl alcohol, which selectively removes the dihydrodiol metabolite. In spite of the low UV absorptivity of phenytoin and 4'-OH-phenytoin at 254 nm, Slonek et al. (1978) claim an assay sensitivity of 1 μg/ml plasma for these two compounds. The procedures involve plasma protein precipitation with acetonitrile and reverse phase HPLC with 254 nm UV detection (sensitivity of the detector: 0.005 absorbance unit full-scale deflection), using only a 10 μl plasma sample volume.

2. Gas Chromatography—Dusci and Hackett (1976), Estas and Dumont (1973), Midha et al. (1976), Sampson et al. (1971)

Discussion. Sampson et al. (1971) utilize a GC-FID assay for phenytoin without derivatization; the other procedures include flash methylation with trimethylphenylammonium hydroxide and detection of phenytoin as well as 4'-OH-phenytoin. 5-(4-Methylphenyl)-5-phenylhydantoin is commonly used as the internal standard. The 4'-OH glucuronide is determined after either acid hydrolysis or enzymatic treatment with a β-glucuronidase. Dusci and Hackett (1976) point out a potential interference of the flash methylation GC assay by the plasticizer tributoxyethyl phosphate, which may have the same retention time as phenytoin. Care has to be taken when analyzing phenytoin plasma samples stored in Vacutainers. The assay interference can be eliminated by back-extraction of the organic phenytoin extracts into alkali solutions.

3. Gas Chromatography-Mass Spectrometry—Baty and Robinson (1977), Hoppel et al. (1976), Lehrer and Karmen (1975)

Discussion. These procedures employ derivatization of phenytoin and 4'-OH-phenytoin by silylation (Baty and Robinson, 1977), flash methylation (Lehrer and Karmen, 1975), and extractive methylation with tetrabutylammonium sulfate and methyl iodide in dichloromethane (Hoppel et al., 1976). Deuterium-labeled compounds usually serve as the internal standards. The GC-MS is performed in the electron impact mode at 70 or 20 eV. Additional hydroxylated metabolites are also measurable by these techniques, for example, 3'-OH-, 2',4'-diOH-, and 3'-OCH$_3$-4'-OH-phenytoin, none of which is present in significant amounts in human urine or plasma (Baty and Robinson, 1977; Hoppel et al., 1976). The 3',4'-dihydrodiol metabolite can be indirectly measured as the 3'-OH-phenytoin generated by acid treatment of urine samples (Hoppel et al., 1976). Sensitivity of the GC-MS assays is excellent (10 ng phenytoin/ml plasma, using a 100 μl sample) (Hoppel et al., 1976).

4. Radioimmunoassay—Paxton et al. (1977a, 1977b), Tigelaar et al. (1973)

Discussion. Phenytoin RIA assays are generally very sensitive and fairly specific. Paxton et al. (1977a) utilize a rabbit antiserum to the immunogen, phenytoin-3-ω-valerate–bovine serum albumin conjugate, and ^{14}C-, ^{3}H-, or ^{125}I-labeled phenytoin as the tracer. Plasma samples requirements are only 50 μl, and correlation with an established GC assay is good (r = .95). Cross-reactivity of the antibody with 4'-OH-phenytoin is 1.1% using the ^{14}C label and 7.5% using the ^{125}I label. This discrepancy is due to the different bindings of the ^{125}I-tyrosine derivative of phenytoin to the antibody. Other hydantoins (methotoin, ethotoin) and anticonvulsants possess minimal cross-reactivity.

5. Spin Immunoassay—Montgomery et al. (1975)

Discussion. The spin label is a nitroxyl radical substituent (pyrrolinidyl-1-oxyl) in the N-3 position of phenytoin. The unpaired electron of the spin-labeled compound, rotating freely in solution, gives rise to three sharp bands in the electron spin resonance (ESR) spectrum. However, the signals broaden and essentially disappear upon binding of the spin label to macromolecules. It is therefore possible to measure phenytoin-antibody-bound and free spin label in solution without separation. Spin label displacement curves by phenytoin in plasma samples can be constructed that are equivalent to RIA displacement curves. Using only 50 μl plasma, the assay is rapid and sensitive to 1 μg phenytoin/ml, with, however, significant cross-reactivity to 4'-OH-phenytoin. Since the 4'-OH metabolite is at low concentrations in plasma, good correlations are obtained with a specific phenytoin GC assay.

It is possible to determine drug plasma protein binding by a similar mechanism involving ESR peak broadening, if the incubation of spin label is carried out with undiluted plasma in the absence of antibody. Such an approach proved to be successful for an assay of morphine plasma protein binding. However, significantly lower phenytoin plasma protein binding was measured with the spin label technique (60%) as compared to ultrafiltration or equilibrium dialysis (90%). This discrepancy is caused by different plasma protein binding kinetics of the N-3 spin-labeled phenytoin and the unlabeled phenytoin. Attachement of the label to the 5-phenyl ring might prevent changes in plasma protein binding constants.

6. Spectrophotometry and Colorimetry—Dill et al. (1971), Fellenberg et al. (1975), Glazko (1972), Saito et al. (1973), Simon et al. (1971), Wallace (1969)

Discussion. Early UV spectrometric methods relied on the differential extraction of phenytoin in the presence of other UV absorbing substances, such as phenobarbital (see Glazko, 1972). Dill and coworkers (1971) introduced a colorimetric assay based on nitration of the phenyl substituents, followed by reduction to aromatic amines and formation of azo dyes with the Bratton-Marshall reaction. Phenytoin and 4'-OH-phenytoin can be measured separately by differential solvent extraction with a sensitivity of 0.2 µg/ml plasma in a 4 ml sample (Dill et al., 1971). Other anticonvulsants containing phenyl substituents interfere with the phenytoin assay. Simon et al. (1971) utilize a TLC step prior to colorimetry in order to improve assay specificity for phenytoin.

Wallace (1969) introduced a different assay principle based on permanganate oxidation of phenytoin to benzophenone (see Glazko, 1972, for review), which can be measured spectrophotometrically. 4'-OH-Phenytoin is oxidized to benzoic acid and does not interfere. Fellenberg et al. (1975) and Saitko et al. (1973) developed a microprocedure involving dichloromethane as the extraction solvent for blood and plasma, back-extraction into alkali, permanganate oxidation of phenytoin to benzophenone, and absorbance measurement of a final *n*-heptane extract. Using only 0.1 ml plasma and microcuvettes, these assays have a sensitivity of 1 µg phenytoin/ml. Phenobarbital does not interfere and the assay can be considered fairly specific.

7. Fluorescence—Dill et al. (1976)

Discussion. Back-extraction of the permanganate-generated benzophenone (see Section d6) from *n*-heptane into sulfuric acid induces a fluorescence that allows quantitative determination of phenytoin plasma levels above 1 µg/ml. Excellent correlation with a GC assay has been obtained for 154 clinical phenytoin plasma samples.

8. **Assay Comparison**—Chamberlain et al. (1977), Stauchansky et al. (1977)

9. **Comments.** Next to digoxin, determination of phenytoin plasma levels may be the most frequently requested clinical assay in the United States. Clinical laboratories use HPLC, GC, RIA, EMIT, and some of the spectrophotometric methods; all can produce reliable phenytoin plasma level data. The 4'-OH metabolite is best measured by GC or HPLC following acid hydrolysis of its glucuronide with usually minimal interference by the 3',4'-dihydrodiol metabolite.

e. References

Atkinson, A. J., Jr., and J. M. Shaw, Pharmacokinetic Study of a Patient with Diphenylhydantoin Toxicity, *Clin. Pharmacol. Ther.*, **14**, 521–528 (1973).

Atkinson, A. J., Jr., J. MacGee, J. Strong, D. Garteiz, and T. E. Gaffney, Identification of 5-*meta*-Hydroxyphenyl-5-phenylhydantoin as a Metabolite of Diphenylhydantoin, *Biochem. Pharmacol.*, **19**, 2483–2491 (1970).

Barth, N., G. Alvan, I. D. Borga, and F. Sjoqvist, Two-fold Interindividual Variation in Plasma Protein Binding of Phenytoin in Patients with Epilepsy, *Clin. Pharmacokin.*, **1**, 444–452 (1976).

Baty, J. D. and P. R. Robinson, Single and Multiple Ion Recording Techniques for the Analysis of Diphenylhydantoin and Its Major Metabolite in Plasma, *Biomed. Mass Spectrom.*, **4**, 36–41 (1977).

Borga, O., M. Garle, and M. Gutova, Identification of 5-(3,4-Dihydroxyphenyl)-5-phenylhydantoin as a Metabolite of 5,5-Diphenylhydantoin (Phenytoin) in Rats and Man, *Pharmacology*, **7**, 129–137 (1972).

Chamberlain, R. T., D. T. Stafford, A. G. Maijub, and B. C. MacNatt, High-Pressure Liquid Chromatographic and Enzyme Immunoassay Compared with Gas Chromatography for Determining Phenytoin, *Clin. Chem.*, **23**, 1764–1766 (1977).

Dill, W. A., J. Baukema, T. Chang, and A. J. Glazko, Colorimetric Assay of 5,5,-Diphenylhydantoin (Dilantin) and 5-(-p-Hydroxyphenyl)-5-phenylhydantoin, *Proc. Soc. Exp. Biol. Med.*, **137**, 674–679 (1971).

Dill, W. A., A. Leung, A. W. Kinkel, and A. J. Glazko, Simplified Fluorimetric Assay for Diphenylhydantoin in Plasma, *Clin. Chem.*, **22**, 908–911 (1976).

Dusci, L. J. and L. P. Hackett, Interference in Dilantin Assays, *Clin. Chem.*, **22**, 1236 (1976).

Estas, A. and P. A. Dumont, Simultaneous Determination of 5,5-Diphenylhydantoin and 5-(p-Hydroxyphenyl)-5-phenylhydantoin in Serum, Urine, and Tissues by Gas-Liquid Chromatography after Flash-Heater Methylation, *J. Chromatogr.*, **82**, 307–314 (1973).

Fellenberg, A. J., A. Magarey, and A. C. Pollard, An Improved Benzophenone Procedure for the Micro-Determination of 5,5-Diphenylhydantoin in Blood, *Clin. Chim. Acta*, **59**, 155–160 (1975).

Glazko, A. J., Diphenylhydantoin: Chemistry and Methods of Determination, in *Antiepileptic Drugs*, D. M. Woodbury, J. K. Penry, and R. P. Schmidt, Eds., Raven Press, New York, 1972, pp. 103–112.

Hermansson, J. and B. Karlen, Assay of the Major (4-Hydroxylated) Metabolite of Diphenylhydantoin in Human Urine by Reversed-Phase High-Performance Liquid Chromatography, *J. Chromatogr.*, **130**, 422–425 (1977).

Hoppel, C., M. Garle, and M. Elander, Mass Fragmentographic Determination of Diphenylhydantoin and Its Main Metabolite, 5-(4-Hydroxyphenyl)-5-phenylhydantoin, in Human Plasma, *J. Chromatogr.*, **116**, 53–61 (1976).

Hoppel, C., M. Garle, A. Rane, and F. Sjoqvist, Plasma Concentrations of 5-(4-Hydroxyphenyl)-5-phenylhydantoin in Phenytoin Treated Patients, *Clin. Pharmacol. Ther.*, **21**, 294–300 (1977).

Kabra, P. M. and L. J. Marton, High-Pressure Liquid Chromatographic Determination of 5-(4-Hydroxyphenyl)-5-phenylhydantoin in Human Urine, *Clin. Chem.*, **22**, 1672–1674 (1976).

Kutt, H., J. Haynes, and F. McDowell, Some Causes of Ineffectiveness of Diphenylhydantoin, *Arch. Neurol.*, **14**, 489–492 (1966).

Leal, K. W. and A. S. Troupin, Clinical Pharmacology of Anti-epileptic Drugs: A Summary of Current Information, *Clin. Chem.*, **23**, 1964–1968 (1977).

Lehrer, M. and A. Karmen, Quantitative Analysis of Diphenylhydantoin in Serum by Gas Chromatography-Mass Spectrometry, *Biochem. Med.*, **14**, 230–237 (1975).

Midha, K. K., I. J. McGilveray, and D. L. Wilson, Sensitive GLC Procedure for Simultaneous Determination of Phenytoin and Its Major Metabolite from Plasma Following Single Doses of Phenytoin, *J. Pharm. Sci.*, **65**, 1240–1243 (1976).

Montgomery, M. R., J. L. Holtzman, R. K. Leute, J. S. Dewees, and G. Bolz, Determination of Diphenylhydantoin in Human Serum by Spin Immunoassay, *Clin. Chem.*, **21**, 221–226 (1975).

Paxton, J. W., F. J. Rowell, and J. G. Ratcliff, The Evaluation of a Radioimmunoassay of Diphenylhydantoin Using an Iodinated Tracer, *Clin. Chim. Acta*, **79**, 81–92 (1977a).

Paxton, J. W., B. Whiting, and K. W. Stephen, Phenytoin Concentrations in Mixed, Parotid and Submandibular Saliva and Serum Measured by Radioimmunoassay, *Br. J. Clin. Pharmacol.*, **4**, 185–191 (1977b).

Saito, Y., K. Nishihara, F. Narkagawa, and T. Suzuki, Improved Micro-determination for Diphenylhydantoin in Blood by UV Spectrophotometry, *J. Pharm. Sci.*, **62**, 206–210 (1973).

Sampson, D., I. Harasymir, and W. J. Hensley, Gas Chromatographic Assay of

Underivatized 5,5-Diphenylhydantoin (Dilantin) in Plasma Extracts, *Clin. Chem.*, **17**, 382–385 (1971).

Simon, G. F., P. J. Jatlow, H. T. Seligson, and D. Seligson, Measurement of 5,5-Diphenylhydantoin in Blood Using Thin-Layer Chromatography, *Am. J. Clin. Pathol.*, **55**, 145–151 (1971).

Slonek, J. E., G. W. Peng, and W. L. Chiou, Rapid and Micro High-Pressure Liquid Chromatographic Determination of Plasma Phenytoin Levels, *J. Pharm. Sci.*, **67**, 1462–1464 (1978).

Smith, R. G., G. D. Daves, Jr., R. K. Lynn, and N. Gerber, Hydantoin Ring Glucuronidation: Characterization of a New Metabolite of 5,5-Diphenylhydantoin in Man and in the Rat, *Biomed. Mass Spectrom.*, **4**, 275–279 (1977).

Stauchansky, S., T. Ludden, J. P. Allen, and P. Wu, Correlation of the E.M.I.T. Diphenylhydantoin Assay in Blood Plasma with a G.L.C. and Spectrophotometric Method, *Anal. Chim. Acta*, **92**, 213–216 (1977).

Thompson, R. M., J. Beghin, W. K. Fife, and N. Gerber, 5,5-Bis(3-hydroxyphenyl)hydantoin, a Minor Metabolite of Diphenylhydantoin in the Rat and Human, *Drug Metab. Dispos.*, **4**, 349–356 (1976).

Tigelaar, R. E., R. L. Rapport, J. K. Inman, and H. J. Kupferberg, A Radioimmunoassay for Diphenylhydantoin, *Clin. Chim. Acta*, **43**, 231–241 (1973).

Wallace, J., Simultaneous Spectrophotometric Determination of Phenytoin and Phenobarbitone in Biological Specimens, *Clin. Chem.*, **15**, 323–330 (1969).

PRIMIDONE—by B. Stafford

5-Ethyldihydro-5-phenyl-4,6(1H,5H)-pyrimidinedione, 2-desoxyphenobarbital, primaclone.
Mysoline. Anticonvulsant, antiepileptic (grand mal).

$C_{12}H_{14}N_2O_2$; mol. wt. 218.25.

a. Therapeutic Concentration Range. The parent drug ranges between 5 and 15 μg/ml plasma, while the anticonvulsant activity is mostly due to therapeutic concentrations of its metabolite, phenobarbital (10 to 30 μg/ml) (Booker et al., 1970; Leal and Troupin, 1977).

b. Metabolism. Primidone is mainly converted to phenobarbital as the major active principle (Bogan and Smith, 1968) and to phenyl-ethylmalondiamide, which also possesses some anticonvulsant activity. The plasma elimination half-lives of primidone, phenobarbital, and phenylethylmalondiamide are 12 ± 6, 96 ± 12, and 25 to 30 hr, respectively (Leal and Troupin, 1977). The long half-life of phenobarbital is responsible for its accumulation during primidone treatment, making simultaneous therapeutic use of primidone and phenobarbital unnecessary.

c. Analogous Compounds. Barbiturates, anticonvulsants.

d. Analytical Methods. The major assay methods for the detection of primidone in serum, for example, HPLC, GC, and enzyme immunoassay, are discussed in detail in the anticonvulsants monograph. Additional assays suitable only for the detection of primidone and its metabolites are listed below.

1. Gas Chromatography—Couch et al. (1973), Gupta et al. (1977), Wallace et al. (1977)

Discussion. Most primidone GC procedures are similar to those discussed for general anticonvulsant assays. Primidone can be analyzed in the underivatized form or following derivatization, usually methylation (see Wallace et al., 1977). The precision of GC assays for primidone tends to show a larger C.V. than equivalent assays for the other anticonvulsants, because of the difficulty in achieving high yield extraction and the relatively high instability of primidone in basic medium. Wallace et al. (1977) propose a microassay using only 50 to 100 μl plasma. Ammonium sulfate-20% ethyl acetate in toluene as the extraction system gives a 92% yield of primidone, which is subsequently derivatized with pentafluorobenzoyl chloride. The resultant diamide derivative can be measured with high sensitivity using GC-EC.

2. Assay Comparison—Sun and Walwick (1976)

Discussion. Sun and Walwick (1976) compare a specific primidone GC assay with an enzyme immunoassay (EMIT). The correlation coefficient is .98, and least-square values of slope and intercept are, respectively, 0.97 and 0.51 μg/ml. Thus both assays give the same result and are readily interchangeable in a clinical laboratory.

3. Comments. Primidone is usually determined as one component of several anticonvulsants potentially present in plasma. Gas chromatography affords separation of primidone from its metabolites and other concomitantly administered anticonvulsants. The GC, HPLC, and EMIT assays are suitable for therapeutic drug level monitoring (see anticonvulsants monograph).

e. References

Bogan, J. and H. Smith, The Relation Between Primidone and Phenobarbitone Blood Levels, *J. Pharm. Pharmacol.*, **20**, 64–67 (1968).

Booker, H. E., K. Hosokowa, R. D. Burdette, and B. Darcey, A Clinical Study of Serum Primidone Levels, *Epilepsia*, **11**, 395–402 (1970).

Couch, M. W., M. Greev, and C. M. Williams, Determination of Primidone and Its Metabolites in Biological Fluids by Gas Chromatography, *J. Chromatogr.*, **87**, 559–561 (1973).

Gupta, R. N., K. Dobson, and P. M. Keane, Gas-Liquid Chromatographic Determination of Primidone in Plasma, *J. Chromatogr.*, **132**, 140–144 (1977).

Leal, K. W. and A. S. Troupin, Clinical Pharmacology of Anti-Epileptic Drugs: A Summary of Current Information, *Clin. Chem.*, **23**, 1964–1968 (1977).

Sun, L. and E. R. Walwick, Primidone Analyses: Correlation of Gas-Chromatographic Assay with Enzyme Immunoassay, *Clin. Chem.*, **22**, 901–902 (1976).

Wallace, J. E., H. E. Hamilton, E. L. Shimek, Jr., H. A. Schwertner, and K. Blum, Determination of Primidone by Electron Capture Gas Chromatography, *Anal. Chem.*, **49**, 903–906 (1977).

PROCAINAMIDE

p-Amino-*N*-(2-diethylaminoethyl)benzamide hydrochloride.
Novocamide, Procardyl, Pronestyl. Cardiac antiarrhythmic agent.

$$H_2N-\text{C}_6H_4-\overset{\displaystyle O}{\overset{\displaystyle \|}{C}}-NH-CH_2CH_2N(CH_2CH_3)_2$$

$C_{13}H_{21}N_3O$; mol. wt. 235.33; pK_a 9.23

a. Therapeutic Concentration Range. From 4 to 10 μg/ml serum. Therapeutic levels of the active metabolite *N*-acetylprocainamide have not yet been defined (Koch-Weser, 1977; Koch-Weser and Klein, 1971).

b. Metabolism. The major metabolite, N-acetylprocainamide, possesses similar pharmacologic activity (Lee et al., 1976) and may be present in higher serum concentrations than the parent drug (Reidenberg et al., 1975). The half-life of procainamide is 2.5 to 5 hr in normal subjects (Graffner et al., 1975). Procainamide N-acetylation is biomodally distributed (see isoniazid monograph).

c. Analogous Compounds. Antiarrhymthmics (e.g., lidocaine); aromatic amines.

d. Analytical Methods

1. High Performance Liquid Chromatography—Shukur et al. (1977)

Discussion. This assay requires reverse phase HPLC with UV detection at 280 nm following alkaline ether extraction. External standards are used for quantitation. Sensitivity is 0.3 μg/ml serum for procainamide and 0.6 μg/ml for N-acetylprocainamide with C.V.s of 8.4 and 10.5%, respectively (n = 20 at 7.5 μg/ml). No interferences are observed in blank patient sera or on addition of a large number of drugs, except for sulfathiazole. Comparison to a fluorescence assay (Matusik and Gibson, 1975) shows good correlation for procainamide and its metabolite N-acetylprocainamide.

Detailed Method. To 0.2 ml serum are added 0.3 ml water, 0.1 ml 1 N sodium carbonate buffer (pH 10.8), and 4 ml diethyl ether. The mixture is vigorously shaken for 30 sec and centrifuged. The aqueous layer is frozen in solid CO_2-acetone, and the organic layer decanted into another test tube and dried under a stream of nitrogen at room temperature. Higher temperatures ($>50°C$) should be avoided if results are not to be erratic. To the residue 100 μl of the acetonitrile-acetate HPLC eluent is added, and the mixture vigorously shaken. After being kept for 40 min at room temperature to assure complete dissolution of procainamide and N-acetylprocainamide, the tubes are again briefly agitated, and a volume of 10 to 25 μl is injected onto the HPLC column.

The HPLC analysis is performed with a liquid chromatograph (Waters Assoc., Model 6000A), equipped with a 30 cm x 3.9 mm (i.d.) μ-Bondapak C_{18} reverse phase column (Waters Assoc.) and a 280 nm UV detector. A 100 g/l acetonitrile solution in 0.75 N sodium acetate buffer (pH 3.4) serves as eluent with a 2 ml/min flow rate at 17.3 MPa (2500 psi) and room temperature. Concentrations of procainamide and

N-acetylprocainamide are assessed by comparing the resulting peak heights with those obtained from an external standard curve.

2. Other HPLC Methods—Adams et al. (1976), Butterfield et al. (1978), Dutcher and Strong (1977), Gadalla et al. (1978), Rocco et al. (1977)

Discussion. The HPLC assays of Adams et al. (1976) and Rocco et al. (1977) are similar to the one described by Shukur et al. (1977), except for internal standards (e.g., *N*-propionylprocainamide), which are required for quantitation. Dutcher and Strong (1977) propose a microparticulate silica column and *p*-nitro-*N*-(diethylaminoethyl)benzamide as the internal standard. This assay requires a sample volume of 1 ml plasma for sufficient sensitivity. A similar silica gel HPLC assay by Butterfield et al. (1978) includes a dichloromethane extraction step and is sensitive to 0.1 μg/ml for procainamide and *N*-acetylprocainamide (2 ml specimen), with pheniramine as the internal standard. The procedure of Gadalla et al. (1978) utilizes a cation-exchange column and UV absorption at 274 nm. A sample of 20 μl plasma is sufficient for the detection of therapeutic procainamide and *N*-acetylprocainamide concentrations. Plasma proteins are precipitated by adding acetonitrile; no liquid-liquid extraction is needed. Quantitation is achieved by the external standard method.

3. Gas Chromatography—Atkinson et al. (1972), Frislid et al. (1975), Simons and Levy (1975)

Discussion. Gas chromatographic-flame ionization detection procedures are specific for procainamide in the presence of *N*-acetylprocainamide. Measurement of the *N*-acetyl metabolite by GC is less sensitive, but has been successfully employed by Frislid et al. (1975) for the detection of therapeutic concentrations.

4. Gas Chromatography-Mass Spectrometry—Dutcher et al. (1975), Strong et al. (1975)

Discussion. Strong et al. (1975) measure *N*-acetylprocainamide after ethyl acetate extraction from plasma, using electron impact quadrupole MS in the selected ion monitoring mode. The dipropyl analog serves as the internal standard. The GC-MS assay overcomes the inherent insensitivity of conventional GC for *N*-acetylprocainamide. The same group of workers (Dutcher et al., 1975) also employed deuterated procainamide and *N*-acetylprocainamide in conjunction with GC-MS for pharmacokinetic studies.

5. Fluorescence—Koch-Weser and Klein (1971), Matusik and Gibson (1975), Sterling and Haney (1974)

Discussion. Sterling and Haney (1974) utilize fluorescamine-induced fluorescence with the disadvantage that other extractable aromatic amines interfer Koch-Weser and Klein (1971) and Matusik and Gibson (1975) exploit the native fluorescence of procainamide in alkaline medium and of *N*-acetylprocainamide in acidic medium. This procedure is adequate for clinical applications; however, it is inferior to HPLC because of lower specificity.

6. Colorimetry—Sitar et al. (1976)

Discussion. The primary amino group of procainamide is diazotized and coupled with the Bratton-Marshall reagent to form an azo dye. The Bratton-Marshall reaction is subject to potential interference by other extractable primary aromatic amines. *N*-Acetylprocainamide is measurable only after hydrolysis to procainamide.

7. Enzyme Immunoassay—Syva Corp., unpublished data

Discussion. The EMIT assay system by Syva Corp. is now commercially available for the rapid measurement of procainamide and *N*-acetylprocainamide. Two separate antibodies have to be utilized in two separate incubations in order to determine both compounds. Excellent correlation of results has been obtained in clinical serum samples with EMIT and with an established HPLC procedure (Shukur et al., 1977).

8. Assay Comparison—Frislid et al. (1975), Sterling et al. (1974)

9. Comments. In speed, accuracy, sensitivity, and precision, HPLC assays with UV detection are superior to other published techniques. The HPLC conditions should be further improved to allow direct injection serum samples without prior extraction by organic solvents. This has been recently achieved by Gadalla et al. (1978), utilizing a cation-exchange column. Further testing of assay specificity is needed to establish this technique in clinical laboratories. *N*-Acetylprocainamide readily hydrolyzes to procainamide at low pH. Some of the earlier methods were developed without considering this potential assay interference.

e. References

Adams, R. F., F. L. Vandemark, and G. Schmidt, The Simultaneous Determination of Lidocaine and Procainamide in Serum by Use of High Pressure Liquid Chromatography, *Clin. Chim. Acta*, **69**, 515–524 (1976).

Atkinson, A. J. Jr., M. Parker, and J. Strong, Rapid Gas Chromatographic Measurement of Plasma Procainamide Concentration, *Clin. Chem.*, **18**, 643–646 (1972).

Butterfield, A. G., J. K. Cooper, and K. K. Midha, Simultaneous Determination of Procainamide and *N*-Acetylprocainamide in Plasma by High-Performance Liquid Chromatography, *J. Pharm. Sci.*, **67**, 839–842 (1978).

Dutcher, J. S. and J. M. Strong, Determination of Plasma Procainamide and *N*-Acetylprocainamide Concentration by High-Pressure Liquid Chromatography, *Clin. Chem.*, **23**, 1318–1320 (1977).

Dutcher, J. S., J. M. Strong, W. K. Lee, and A. J. Atkinson, Stable Isotope Methods for Pharmacokinetic Studies in Man, in *Proceedings of the Second International Conference on Stable Isotopes*, E. R. Klein and P. D. Klein, Eds., Argonne Natl. Lab, Illinois, 1975; National Technical Information Service, Conf. 751027, pp. 186–197.

Frislid, K., J. E. Bredesen, and P. K. M. Lunde, Fluorimetric or Gas Chromatographic Determination of Procainamide, *Clin. Chem.*, **21**, 1180–1182 (1975).

Gadalla, M. A. F., C. W. Peng, and W. L. Chion, Rapid and Micro High-Pressure Liquid Chromatographic Method for Simultaneous Determination of Procainamide and *N*-Acetylprocainamide in Plasma, *J. Pharm. Sci.*, **67**, 869–871 (1978).

Graffner, C., G. Johnson, and S. Sjogren, Pharmacokinetics of Procainamide Intravenously and Orally as Conventional and Slow-Release Tablets, *Clin. Pharmacol. Ther.*, **17**, 414–423 (1975).

Koch-Weser, J., Serum Procainamide Levels in Therapeutic Guides, *Clin. Pharmacokin.*, **2**, 389–402 (1977).

Koch-Weser, J. and S. W. Klein, Procainamide Dosage Schedules, Plasma Concentrations, and Clinical Effects, J.A.M.A., **215**, 1454–1461 (1971).

Lee, W. K., J. Strong, R. F. Kehoe, J. S. Dutcher, and A. J. Atkinson, Jr., Antiarrhythmic Efficacy of *N*-Acetylprocainamide in Patients with Premature Ventricular Contractions, *Clin. Pharmacol. Ther.*, **19**, 508–514 (1976).

Matusik, E. and T. B. Gibson, Fluorimetric Assay for *N*-Acetylprocainamide, *Clin. Chem.*, **21**, 1899–1902 (1975).

Reidenberg, M. M., D. E. Dreyer, M. Levy, and H. Warner, Polymorphic Acetylation of Procainamide in Man, *Clin. Pharmacol. Ther.*, **17**, 722–730 (1975).

Rocco, R. M., D. C. Abbott, R. W. Giese, and B. L. Karger, Analysis for Procainamide and N-Acetylprocainamide in Plasma or Serum by High-Performance Liquid Chromatography, *Clin. Chem.*, **23,** 705–708 (1977).

Shukur, L. R., J. L. Powers, R. A. Marques, M. E. Winter, and W. Sadée, Measurement of Procainamide and N-Acetylprocainamide in Serum by High Performance Liquid Chromatography, *Clin. Chem.*, **23,** 636–638 (1977).

Simons, K. J. and R. H. Levy, GLC Determination of Procainamide in Biological Fluids, *J. Pharm. Sci.*, **64,** 1967–1970 (1975).

Sitar, D. S., D. N. Graham, R. E. Rango, L. R. Dusfresne, and R. I. Ogilvie, Modified Colorimetric Method for Procainamide in Plasma, *Clin. Chem.*, **22,** 379–380 (1976).

Sterling, J. M. and W. G. Haney, Spectrophotofluorimetric Analysis of Procainamide and Sulfadiazine in Presense of Primary Aliphatic Amines Based on Reaction with Fluorescamine, *J. Pharm. Sci.*, **63,** 1448–1450 (1974).

Sterling, J. M., S. Cox, and W. C. Haney, Comparison of Procainamide Analysis in Plasma by Spectrofluorometry, Colorimetry and GLC, *J. Pharm. Sci.*, **63,** 1744–1747 (1974).

Strong, J. M., J. S. Dutcher, W.-K. Lee, and A. J. Atkinson, Pharmacokinetics in Man of the N-Acetylated Metabolite of Procainamide, *J. Pharmacokin. Biopharm.*, **3,** 223–235 (1975).

PROCARBAZINE

N-Isopropyl-α-(2-methylhydrazino)-p-toluamide.
Matulan. Anticancer agent.

$$CH_3-NH-NH-CH_2-\underset{}{\overset{}{\bigcirc}}-\overset{\overset{O}{\|}}{C}-NH-\underset{CH_3}{\overset{CH_3}{CH}}$$

$C_{12}H_{19}N_3O$; mol. wt. 221.30; pK_a 6.8.

a. Therapeutic Concentration Range. None defined.

b. Metabolism. Procarbazine is rapidly degraded by enzymatic and nonenzymatic pathways with a plasma elimination half-life of less than 10 min in human beings (Prough et al., 1970; Schwartz et al., 1967; Weinkam and Shiba, 1978). Rapid appearance of the azo derivative in plasma, followed by the inactive isopropylterephthalamic acid in plasma and urine (Schwartz et al., 1967), indicates that oxidative cleavage *in vivo* may contribute to the activation mechanism of the inactive parent drug. Two microsomal enzyme systems participate in the oxidative hydrazine dealkylation to give either formaldehyde or methane (Prough et al., 1970). Weinkam and Shiba (1978) show that procarbazine is rapidly oxidized in solution in the presence of oxygen to the azo derivative, followed by liver microsomal oxidation to the azoxy metabolite and additional hydroxylation steps which result in the active alkylating agent. The azoxy derivatives of procarbazine are considerably more cell toxic than is the parent drug (Weinkam, unpublished data). Free radical intermediates in another potential activation pathway give methane and *N*-isopropyltoluamide (Weinkam and Shiba, 1978).

c. Analogous Compounds. Hydrazine derivatives.

d. Analytical Methods

    1. High Performance Liquid Chromatography—Weinkam and Shiba (1978)

Discussion. Plasma samples are extracted several-fold with ether, and the residue is chromatographed on a reverse phase column with UV detection at 254 nm. The sensitivity for procarbazine and its metabolites is on the order of 200 ng/ml. After 30 min of a dose of 60 mg procarbazine in rats, several metabolite peaks are detectable in plasma, representing the azo derivative, two azoxy isomers, *N*-isopropyl-terephthalic acid, and small amounts of *N*-isopropyltoluamide. Unchanged procarbazine cannot be measured by HPLC; however, with direct insertion chemical ionization MS, it can be detected in plasma samples (Weinkam, personal communication).

Detailed Method. Plasma (0.5 ml) is extracted with 10 × 2 ml ethyl ether, and the ether layer is evaporated. The residue is suspended in 0.5 ml methanol, and an aliquot subjected to HPLC analysis.

    A Chromatronix 3500 high performance liquid chromatograph is used, equipped with a 254 nm absorbance detector. Separation is achieved on a Waters 30 cm μ-Bondapak C_{18} column eluted with 20% methanol in water. The retention time of the latest eluted component, *N*-isopropyl-*p*-toluamide, is 90 min.

    2. Comments. In spite of the importance of the *in vivo* mechanism of activation of procarbazine, only one suitable chemical assay procedure is currently available for the detection of nonenzymatic and metabolic key intermediates in biological samples. This HPLC assay (Weinkam and Shiba, 1978) does not allow sensitive determination of the parent procarbazine in plasma because of the presence of interfering endogenous substrates. Additional use of MS circumvents such assay interference. Procarbazine rapidly (within 15 min) oxidizes in aqueous solution in the presence of molecular oxygen to the azo derivative and H_2O_2 (Weinkam and Shiba, 1978).

e. References

Prough, R. A., J. A. Wittkop, and D. J. Reed, Further Evidence on the Nature of Microsomal Metabolism of Procarbazine and Related Alkylhydrazines, *Arch. Biochem. Biophys.*, **140**, 450–458 (1970).

Schwartz, D. E., W. Bollag, and P. Obrecht, Distribution and Excretion Studies of Procarbazine in Animals and Man, *Arzneim.-Forsch.*, **17**, 1389–1393 (1967).

Weinkam, R. J. and D. A. Shiba, Metabolic Activation of Procarbazine, *Life Sci.*, **22**, 937–946 (1978).

PROPOXYPHENE—by D. R. Doose

4-Dimethylamino-3-methyl-1,2-diphenyl-2-butanol propionate
Darvon (dextropropoxyphene, α,d-isomer). Analgesic.

$C_{22}H_{29}NO_2$; mol. wt. 339.48; pK_a 10

a. Therapeutic Concentration Range. Normal therapeutic concentrations are between 0.05 and 0.2 $\mu g/ml$, toxic concentrations 5 and 20 $\mu g/ml$, and lethal concentrations above 50 $\mu g/ml$ blood (Winek, 1976).

b. Metabolism. The nonanalgesic N-norpropoxyphene represents the major metabolite in man (Due et al., 1976), which accumulates after multiple doses of propoxyphene to plasma levels several-fold higher than those of the parent drug (Verebely and Inturrissi, 1973). Aromatic hydroxylation, ester hydrolysis, further demethylation to dinorpropoxyphene, and acetylation of nor- and dinorpropoxyphene have also been identified in human beings (Due et al., 1976). The plasma elimination half-life of propoxyphene is approximately 3.5 hr.

c. Analogous Compounds. Methadone, SKF-525A.

d. Analytical Methods

1. Gas-Liquid Chromatography—Cleemann (1977)

Discussion. This assay requires GC with flame ionization detection following chemical derivatization of propoxyphene and its major metabolite norpropoxyphene to chromatographically stable compounds. The thermally labile norpropoxyphene is first rearranged by internal O to N migration of the propionyl function under basic conditions to give norpropoxyphene amide. After extraction, propoxyphene and norpropoxyphene amide are reduced by lithium aluminum hydride, which cleaves the propionyl ester linkage of propoxyphene and reduces norpropoxyphene amide to the corresponding propylamine derivative. Sensitivity is 2 ng/ml in plasma for both propoxyphene and norpropoxyphene, using promethazine as the internal standard. The method includes an extensive purification procedure, which is needed for the high sensitivity reported by Cleemann (1977). Other GC assay methods have been prone to on-column decomposition of propoxyphene and its major metabolite, leading to erroneous results.

Detailed Method. To 5.0 ml plasma are added 0.2 ml internal standard solution (promethazine-HCl, 25 μg/ml), 1 ml 1 M carbonate buffer (pH 9.8) and 5 ml n-butyl chloride. The mixture is shaken for 15 min and centrifuged for 5 min at 1700 g. After freezing ($-24°$), the upper phase (n-butyl chloride) is decanted into another tube containing 2 ml 0.2 N HCl. The frozen residue is washed with 1.0 ml n-butyl chloride, and the combined n-butyl chloride phases are shaken for 5 min with the acidic phase, followed by a 5 min centrifugation. The separated acidic aqueous phase is washed by shaking for 5 min with 5 ml diethyl ether and again with 1 ml n-butyl chloride. The aqueous phase is made alkaline by the addition of 0.3 ml 4 N NaOH in order to afford quantitative rearrangement of norpropoxyphene to norpropoxyphene amide, and 15 min later is extracted with 1 ml n-butyl chloride by shaking for 5 min. After a 5 min centrifugation the aqueous phase is frozen and the n-butyl chloride decanted. The organic phase is reacted with lithium aluminum hydride in 0.5 ml diethyl ether (100 mg LiAlH$_4$/10 ml) for 15 min at 34 to 40°C.

Excess lithium aluminum hydride is then destroyed by adding 0.2 ml water. The mixture is again centrifuged and frozen, and the organic layer decanted and evaporated to dryness under reduced pressure. The residue is treated with 3 drops of *n*-butyl chloride, and the solution is evaporated. This final residue is dissolved in 10 μl CCl$_4$, and 1 μl of the solution injected into the gas chromatograph. Standard curves are constructed by the peak height ratio method.

The gas chromatograph system consists of a Varian Aerograph Model 2100 chromatograph equipped with a flame ionization detector. The U-shaped glass column, 1.8 m x 2 mm (i.d.), is silanized. The packing material consists of 3% OV-17 on Chromosorb W AW DMCS, 80/100 mesh. The temperature of the oven is 180°C, of the injection port 200°C, and of the detector 250°C. The carrier gas is nitrogen at a flow rate of 40 ml/min. The hydrogen flow rate is 40 ml/min, and the oxygen flow rate 200 ml/min.

2. Other GC Methods—Serfontein et al. (1976), Sparacino et al. (1973), Verebely and Inturrisi (1973), Wolen and Gruber (1968)

Discussion. Early GLC assay procedures report quantification of propoxyphene in biological fluids without derivatization (Wolen and Gruber, 1968). However, during GC analysis propoxyphene readily undergoes thermolysis, which is difficult to prevent under routine conditions (Sparacino et al., 1973). Norpropoxyphene cannot be analyzed without derivatization. Serfontein et al. (1976) propose the reaction with nitrous acid to give the rather stable *N*-nitrosonorpropoxyphene, while Verebely and Inturrisi (1973) use the base-catalyzed internal rearrangement of norpropoxyphene to the stable norpropoxyphene amide. Sensitivity of the latter assay is 10 and 50 ng/ml plasma for propoxyphene and norpropoxyphene, respectively.

3. Gas Chromatography-Mass Spectrometry—McMahon and Sullivan (1976), Sillivan et al. (1974)

Discussion. Propoxyphene and norpropoxyphene are analyzed in separate procedures by electron impact GC-MS in the selected ion monitoring mode with heptadeuterated propoxyphene and trideuterated norpropoxyphene as the internal standards (Sullivan et al., 1974). Norpropoxyphene is rearranged to norpropoxyphene amide by reaction at pH 11 for 1 hr. Sensitivity is about 1 ng/ml plasma for both compounds. McMahon and Sullivan (1976) report the *in vivo* administration of dideuterated *d*- or *l*-propoxyphene, together with the respective unlabeled stereoisomer, in order to quantitate the metabolic fate of racemic mixtures of *dl*-propoxyphene by MS. The position of the benzylmethylene deuterium label does not produce a measurable metabolic isotope effect. The plasma elimination half-life of the *d*-isomer is longer than that of the *l*-isomer.

4. Fluorescence—Valentour et al. (1974)

Discussion. Propoxyphene is extracted at pH 5.3 into chloroform, which retains other potentially interfering amines in the aqueous phase. Reaction with 4-chloro-7-nitrobenzo-2,1,3-oxadiazole gives a highly fluorescent derivative that is normally observed only with secondary and primary amines, not with tertiary. The authors conclude that propoxyphene is *N*-demethylated under the assay conditions and then undergoes the fluorescence reaction. Norpropoxyphene is only partially extracted by the assay procedure. The assay is sufficiently sensitive for therapeutic propoxyphene levels.

5. Spectrophotometry—Wallace et al. (1965)

Discussion. The UV assay depends on hydrolysis of the propionyl ester, followed by conversion of the hydroxyl group to a tertiary chloride. Dehydrohalogenation of the chloride in weak alkali yields a steam-distillable product with high molar absorptivity. The method does not distinguish between propoxyphene and norpropoxyphene and lacks the sensitivity of current GC assays.

6. Comments. At present GC and GC-MS are the most versatile and specific tools for the analysis of propoxyphene and its metabolites in biological fluids.

e. References

Cleeman, M., Gas Chromatographic Determination of Propoxyphene and Norpropoxyphene in Plasma, *J. Chromatogr.*, **132,** 287–294 (1977).

Due, S. L., H. R. Sullivan, and R. E. McMahon, Propoxyphene: Pathways of Metabolism in Man and Laboratory Animals, *Biomed. Mass Spectrom.*, **3,** 217–225 (1976).

McMahon, R. E. and H. R. Sullivan, Simultaneous Measurement of Plasma Levels of *d*-Propoxyphene and *l*-Propoxyphene Using Stable Isotope Labels and Mass Fragmentography, *Res. Commun. Chem. Pathol. Pharmacol.*, **14,** 631–641 (1976).

Serfontein, W. J. and L. S. DeVilliers, A New GC Procedure, Based on Nitrosation, for the Simultaneous Determination of Propoxyphene and Norpropoxyphene in Biological Material, *J. Pharm. Pharmacol.*, **28**, 718–719 (1976).

Sparacino, C. M., E. D. Pellizzari, C. E. Cood, and M. W. Wall, A Re-examination of the Gas Chromatographic Determination of α-d-Propoxyphene, *J. Chromatogr.*, **77**, 413–418 (1973).

Sullivan, H. R., J. L. Emmerson, F. J. Marshall, P. G. Wood, and R. E. McMahon, Quantitation of Plasma Levels of Propoxyphene and Norpropoxyphene by Combined Use of Stable Isotope Labeling and Selected Ion Monitoring, *Drug Metab. Dispos.*, **2**, 526–532 (1974).

Valentour, J. C., J. R. Monforte, and I. Sunshine, Fluorometric Determination of Propoxyphene, *Clin. Chem.*, **20**, 275–277 (1974).

Verebely, K. and C. E. Inturrisi, The Simultaneous Determination of Propoxyphene and Norpropoxyphene in Human Biofluids Using Gas-Liquid Chromatography, *J. Chromatogr.*, **75**, 195–205 (1973).

Wallace, J. E., J. D. Biggs, and E. V. Dahl, A Rapid and Specific Spectrophotometric Method for Determining Propoxyphene, *J. Forensic Sci.*, **10**, 179–191 (1965).

Winek, C. L., Tabulation of Therapeutic, Toxic, and Lethal Concentrations of Drugs and Chemicals in Blood, *Clin. Chem.*, **22**, 832–836 (1976).

Wolen, R. L. and C. M. Gruber, Jr., Determination of Propoxyphene in Human Plasma by Gas Chromatography, *Anal. Chem.*, **40**, 1243–1246 (1968).

PROPRANOLOL

DL-1-(Isopropylamino)-3-(1-naphthyloxy)-2-propanol.
Dositon, Inderal, Adrenergic blocking agent.

$C_{16}H_{21}NO_2$; mol.wt. 259.34; pK_a 9.45.

a. Therapeutic Concentration Range. From 50 to 100 ng/ml serum (Nies and Shand, 1975). 4-Hydroxypropranolol is equally active and can contribute to the pharmacological effect. However, 4-hydroxypropranolol levels may be neglible at 6 hr after the dose (minimum levels during multiple dosing), obviating the need to measure it for therapeutic drug level monitoring.

b. Metabolism. Propranolol is rapidly converted to many metabolites (Walle and Gaffney, 1972), some of which retain beta-blocking activity (e.g., 4-hydroxypropranolol) or possess anticonvulsant activity (e.g. propranolol glycol) (Saelens et al., 1974) or sympathomimetic effects (e.g. isopropylamine) (Ishizaki et al., 1974). The plasma elimination half-life is 2 to 3 hr, with the active *l*-form being cleared more slowly than the inactive *d*-form (George et al., 1972). Walle et al. (1978) report that the interindividual variability in propranolol plasma levels is threefold at a dose level of 40 mg but only 1.3-fold at doses exceeding 600 mg/day. Nonlinear kinetics are observed at the low propranolol dosage level.

c. Analogous Compounds. Alprenolol, dichloroisoproterenol, isoproterenol (adrenergic agent), oxprenolol, practolol, pronethanol, sotalol, toliprolol.

d. Analytical Methods

    1. High Performance Liquid Chromatography—Nation et al. (1978)

Discussion. The native fluorescence of propranolol and its active metabolite, 4-hydroxypropranolol, provides a means of detecting picogram quantities by HPLC with a fluorescence flow cell detector. The chemical analog, 4-methylpropranolol, exhibits similar chemical and fluorescent properties and serves as a suitable internal standard for quantitation. Separation of propranolol, its metabolite, and internal standard analog is achieved on a reverse phase, chemically bonded alkylphenyl column with an acidic water-acetonitrile eluent. The rather low pH of the eluent (pH 2.6) may shorten the useful life span of the analytical column. Plasma samples are extracted at pH 9.5 into ethyl acetate in the presence of sodium hydrogen sulfite in order to prevent oxidation of 4-hydroxypropranolol. Using 1 ml plasma, the assay is sensitive to 1 ng propranolol/ml and 5 ng 4-hydroxypropranolol/ml with C.V.s at 10 ng/ml of 9.6 and 15.4%, respectively. No assay interferences have been observed with other drugs or in plasma samples from patients ingesting propranolol.

Detailed Method. Plasma (1 ml), 20 mg sodium hydrogen sulfite, 0.1 ml water containing 15 ng 4-methylpropranolol, and 1 ml of a 0.5 N sodium carbonate buffer (pH 9.5) are extracted with 3 ml ethyl acetate by vigorous shaking for 1 min. After centrifugation at 800 g for 2 to 3 min, the organic layer is transferred to another tube containing 0.1 ml of a dilute sulfuric acid solution (pH 2.2). The mixture is again vigorously shaken and centrifuged, and 50 μl of the lower aqueous layer is injected onto the column. Standard curves are prepared by the peak height ratio method.

The HPLC analysis is performed on a system consisting of a M-6000A solvent delivery pump, a UK6 injection loop, and a 30 cm μ-Bondapak alkylphenyl column (particle size 10 μ) (Waters Assoc., Milford, Mass.). The eluent, a mixture of 0.06% phosphoric aqueous acid solution and acetonitrile (73:27) is delivered at a rate of 2 ml/min. The fluorescence detector (FS970, Schoeffel, Instrument, Westwood, N.J.) is operated at an excitation wavelength of 205 nm using an emission filter (KV 340).

2. Other HPLC Assays—Mason et al. (1977), Nygard et al. (1979), Pritchard et al. (1979), Schmidt and Vandemark, (1977)

Discussion. These HPLC fluorescence assays are similar to the one described by Nation et al. (1978). Schmidt and Vandemark (1977) propose an assay for the propranolol plasma levels with protriptyline as the internal standard. The double-extraction procedure includes more steps than the extraction procedure of Nation et al. (1978). Fluorescence is measured with an excitation of 293 nm and an emission of 375 nm. Excitation at 293 nm results in a lower sensitivity than does excitation at 205 nm. The column is a C_{18} chemically bonded reverse phase column, again using a rather acidic mobile phase in order to achieve suitable chromatographic properties of the lipophilic propranolol. The single-extraction method of Nygard et al. (1979) utilizing HPLC-fluorescence is based on a reverse phase cyanopropylsilane column at pH 7, which may provide longer column life.

Mason et al. (1977) utilize a single-step solvent extraction. Both propranolol and 4-hydroxypropranolol are included with the assay. The quantitative determination of propranolol and six of its metabolites in human urine is described by Pritchard et al. (1979). The assay includes ether extraction at pH 9.8 of propranolol and neutral or basic metabolites, followed by extraction of acidic metabolites at low pH. Analysis is performed by reverse phase HPLC with UV detection.

3. Gas Chromatography—Caccia et al. (1978), DiSalle et al. (1973), Walle, (1974)

Discussion. Derivatization with trifluoroacetic anhydride and electron capture detection are used after solvent extraction of propanolol from serum (DiSalle et al., 1973). Other beta-blocking agents can serve as internal standards, for example, oxprenolol. The GC-EC method is specific and extremely sensitive. As little as 0.1 to 1.0 pg can be detected, and the sensitivity limit in 1 ml serum is 0.1 ng/ml (Walle, 1974). The method is applicable to propranolol metabolites, that is, the desisopropyl and glycol derivatives, as well as to other β-adrenergic blocking agents. Gas chromatographic separations of the *d*- and *l*- enantiomers of propranolol have been described by Caccia et al. (1978). Following derivatization of *dl*-propranolol with *N*-trifluoroacetyl-l-propyl chloride or N-heptafluorobutyryl-l-prolyl chloride and subsequent silylation, the generated diastereomers can be readily separated on packed and capillary glass columns. The method is also applicable to other β-adrenergic antagonists.

4. Gas Chromatography-Mass Spectrometry—(Ehrsson (1977), Walle et al. (1975)

Discussion. The GS-MS method of Walle et al. (1975) is capable of determining propranolol and its active 4-hydroxy metabolite. It utilizes electron impact ionization (18 eV) and selected ion monitoring with oxprenolol as the internal standard. Sodium bisulfite is added to the aqueous extraction medium to prevent oxidation of the labile 4-hydroxy metabolite. Sensitivity is 1 ng/ml and 5 ng/ml for propranolol and its metabolite, respectively. No specific advantage over GC-EC assays is apparent.

Ehrsson (1977) specifically labels the (+)-propranolol isomer by heating in deuterated sulfuric acid. Administration of d_2-(+)-propranolol and (−)-propranolol can be simultaneously followed by measuring the electron impact fragment ions at m/e 144 and 146 (d_2) (naphthol fragment) by GC-MS. The propranolol enantiomers were found to be eliminated from plasma at the same rate in this study.

5. Fluorimetry—Ambler et al. (1974), Kraml and Robinson (1978), Shand et al. (1970)

Discussion. Various organic extractions and back-extractions are combined with the detection of propranolol's native fluorescence. A number of problems are inherent in fluorescence assays. These include lack of sufficient sensitivity, high and variable blank readings, and lower specificity relative to HPLC and GC. Some of these problems may be overcome by using citric acid in ethanediol, which lowers the serum blank reading by a factor of 10 (Ambler et al., 1974), or by employing a scanning fluorimeter, which allows ready detection of interferences (Kraml and Robinson, 1978). Fluorescence techniques are still widely used in clinical laboratories.

6. Thin-Layer Chromatography—Hadzija and Mattocks (1978), Schaefer et al. (1977)

Discussion Propranolol and its metabolite *N*-desisopropylpropranolol are extracted from plasma into an organic solvent and separated by TLC (Schaefer et al., 1977). Direct fluorescence scanning of the TLC plates affords quantitative detection of the two compounds with sufficient sensitivity for therapeutic drug level monitoring. Preseparation by TLC increases the specificity of this assay over that of fluorescence assays with solvent extraction alone. However, the fact that the plates have to be sprayed with a citric acid solution and analyzed while still wet may contribute to assay variability.

Using UV densitometry at 288 nm in a similar TLC assay procedure, Hadzija and Mattocks (1978) report a sensitivity of 10 ng propranolol/ml plasma (1 ml specimen or less) with good reproducibility. The authors claim that their assay is suitable for clinical applications.

7. Radioimmunoassay—Kawashima et al. (1976)

Discussion. The RIA is sensitive to 1 ng propranolol/ml serum. Furthermore, it offers the advantage of stereoselectivity; thus active *l*-propranolol can be monitored selectively, a capability that may be important because of the different pharmacokinetics of the *d*- and *l*-forms. Propranolol metabolites do not interfere with the assay, except at much higher concentrations. Antibodies were prepared by injecting a propranolol-succinate-bovine serum albumin antigen into rabbits.

8. Comments.

At present, HPLC-fluorescence seems to offer the best method for a relatively small number of samples per day (<20). Stereoselectivity can be introduced by formation of a disastereomeric derivative, using optically active reagents (e.g., reactive amino acids). A problem with all methods may arise if artifacts are introduced prior to analysis. It has been observed that chemicals in the stopper of certain blood collection tubes may diffuse into the blood, thereby displacing plasma protein-bound propranolol (Cotham and Shand, 1976). This results in a shift of free propranolol into the red blood cells and an underestimation of propranolol serum levels.

e. References

Ambler, P. K., B. N. Singh, and M. Lever, A Simple and Rapid Fluorimetric Method for the Estimation of Propranolol in Blood, *Clin. Chim. Acta*, **54**, 373-375 (1974).

Caccia, S., C. Chiabrando, P. De Ponte, and R. Fanelli, Separation of Beta Adrenoceptor Antagonist Enantiomers by High Resolution Capillary Gas Chromatography, *J. Chromatogr. Sci.*, **16**, 543-546 (1978).

Cotham, R. H. and D. Shand, Spuriously Low Plasma Propranolol Concentrations Resulting from Blood Collection Methods, *Clin. Pharmacol. Ther.*, **18**, 535-538 (1976).

DiSalle, E. et al., A Sensitive Gas Chromatographic Method for the Determination of Propranolol in Human Plasma, *J. Chromatogr.*, **84**, 347-353 (1973).

Ehrsson, H., Simultaneous Determination of (−)- and (+)-Propranolol by Gas Chromatography Mass Spectrometry Using a Deuterium Labeling Technique, *J. Pharm. Pharmacol.*, **28**, 662-663 (1977).

George, C. F., F. Fenyvesi, M. E. Conolly, and C. T. Dollery, Pharmacokinetics of Dextro-, Laevo- and Racemic Propranolol in Man, *Eur. J. Clin. Pharmacol.*, **4**, 74-76 (1972).

Hadzija, B. W. and A. M. Mattocks, Quantitative TLC Determination of Propranolol in Human Plasma, *J. Pharm. Sci.*, **67**, 1307-1309 (1978).

Ishizaki, T., P. J. Privitera, T. Walle, and T. E. Gaffrey, Cardiovascular Actions of a New Metabolite of Propranolol, *J. Pharmacol. Exp. Ther.*, **189**, 626-632 (1974).

Kawashima, K., A. Levy, and S. Spector, Stereospecific Radioimmunoassay for Propranolol Isomers, *J. Pharmacol. Exp. Ther.*, **196**, 517-523 (1976).

Kraml, M. and W. T. Robinson, Fluorimetry of Propranolol and its Glucuronide: Applicability, Specificity, and Limitations, *Clin. Chem.*, **24**, 169-171 (1978).

Mason, W. D., E. N. Amick, and O. H. Weddle, Rapid Determination of Propranolol and 4-Hydroxypropranolol in Plasma by High Pressure Liquid Chromatography, *Anal. Lett.*, **10**, 515-521 (1977).

Nation, R. L., G. W. Peng, and W. L. Chiou, High Pressure Liquid Chromatographic Method for the Simultaneous Quantitative Analysis of Propranolol and 4-Hydroxypropranolol in Plasma, *J. Chromatogr.*, **145**, 429-436 (1978).

Nies, A. S. and D. G. Shand, Clinical Pharmacology of Propranolol, *Circulation*, **52**, 6-15 (1975).

Nygard, G., W. H. Shelver, and S. K. W. Khalil, Sensitive High-Pressure Liquid Chromatographic Determination of Propranolol in Plasma, *J. Pharm.. Sci.*, **68**, 379-381 (1979).

Pritchard, J. F., D. W. Schneck, and A. H. Hayes, Jr., Determination of Propranolol and Six Metabolites in Human Urine by High-Pressure Liquid Chromatography, *J. Chromatogr.*, **162**, 47-58 (1979).

Saelens, D. A., T. Walle, P. J. Privitera, D. R. Knapp, and T. Gaffney, Central Nervous System Effects and Metabolic Disposition of a Glycol Metabolite of Propranolol, *J. Pharmacol. Exp. Ther.*, **188**, 86–92 (1974).

Schaefer, M., H. E. Geissler, and E. Mutschler, Fluorimetrische Bestimmung von Propranolol und seines Metaboliten *N*-Desisopropylpropranolol in Plasma und Urin durch direkte Auswerkung von Dünnschichtchromatogrammen, *J. Chromatogr.* (*Biomed. Appl.*), **143**, 607–613 (1977).

Shand, D. G., E. M. Nuckolls, and J. A. Oates, Plasma Propranolol Levels in Adults, with Observations in Four Children, *Clin. Pharmacol. Ther.*, **11**, 112–120 (1970).

Schmidt, G. J. and F. L. Vandemark, The Determination of Propranolol in Serum Using Liquid Chromatography and Fluorescence Detection, *Chromatogr. Newslett*, **5**, 30–45 (1977).

Walle, T., GLC Determination of Propranolol, other β-Blocking Drugs, and Metabolites in Biological Fluids, *J. Pharm. Sci.*, **63**, 1885–1891 (1974).

Walle, T. and T. E. Gaffney, Propranolol Metabolism in Man and Dog: Mass Spectrometric Identification of Six New Metabolites, *J. Pharmacol. Exp. Ther.*, **182**, 83–92 (1972).

Walle, T., J. Morrison, K. Walle, and E. Conradi, Simultaneous Determination of Propranolol and 4-Hydroxypropranolol in Plasma by Mass Fragmentography, *J. Chromatogr.*, **114**, 351–359 (1975).

Walle, T., E. C. Conradi, U. K. Walle, T. C. Fagan, and T. E. Gaffney, The Predictable Relationship between Plasma Levels and Dose During Chronic Propranolol Therapy, *Clin. Pharmacol. Ther.*, **24**, 668–677 (1978).

QUINIDINE

by T. W. Guentert, Ph.D., School of Pharmacy, University of California, San Francisco, Cal. 94143

6-Methoxy-α-(5-vinyl-2-quinuclidinyl)-4-quinolinemethanol.
Quindex, Cardioquin, Cardiac Antiarrhythmic agent.

$C_{20}H_{24}N_2O_2$; mol.wt. 324.41; pK_a 4.0 and 8.6.

a. Therapeutic Concentration Range. Sokolow and co-workers, (Sokolow and Edgar, 1950; Sokolow and Ball, 1956) report a therapeutic range of 2 to 8 µg/ml with a mean peak blood level of 5.9 µg/ml for conversion of auricular fibrillation to normal sinus rhythm. Levels of 5 µg/ml or 80% of those needed for conversion were sufficient to prevent recurrent atrial arrhythmias. These results were obtained with a rather nonspecific single-extraction fluorescence method. With a specific HPLC assay, therapeutic quinidine concentrations are considerably lower with mean values of about 2 to 3 µg/ml (Drayer et al., 1978).

b. Metabolism. The mean plasma elimination half-life of quinidine is 6.3 to 7.3 hr (Ueda and Dzindzio, 1978; Greenblatt et al., 1977). The drug is extensively metabolized in human beings via oxidative pathways. The major metabolites in blood are 3-OH-quinidine (Carroll et al., 1974; Beermann et al., 1976), 2'-quinidinone (Brodie et al., 1951; Palmer et al., 1969), and a recently isolated N-oxide formed from quinidine (Guentert et al., 1979). O-Desmethylquinidine is a minor metabolite which accounts for 1 to 2% of the dose excreted into the urine (Drayer et al., 1978). The relative contributions of these metabolites to the pharmacological effects of quinidine remain unknown (Drayer et al., 1976, 1978; Huffmann et al., 1977; Conn and Luchi, 1964), although it has been suggested that 3-OH-quinidine may contribute significantly to the effects (S. Riegelman et al., unpublished).

c. Analogous Compounds. Cinchonine, cinchonidine, dihydroquinidine, quinine, synthetic antimalarials (e.g., chloroquine).

d. Analytical Methods

1. High-Performance Liquid Chromatography—Kates, et al. (1978)

Discussion. The method of Kates et al. (1978) uses HPLC on a cation-exchange column with UV detection at 230 nm. The assay includes extraction of the drug and the internal standard (cinchonine) into benzene and can be accomplished with an overall analysis time of 20 min. The sensitivity is 100 ng/ml. Over the range of 1 to 10 µg/ml the C.V. is 7 to 9% (n = 7). None of the known quinidine metabolites interferes with the assay. Dihydroquinidine, a quinidine contaminant (usually 5 to 8%, maximum 25% according to USP XVIII compendial standards), is separated from quinidine and can also be quantitatively measured. Dihydroquinidine has antiarrhythmic and pharmacokinetic properties similar to those of quinidine, but it may have greater toxicity (Lewis et al., 1922; Alexander et al., 1947; Ueda et al., 1976a). A series of potentially interfering drugs did not affect quinidine measurements. Comparison with a specific quantitative TLC assay (Ueda et al., 1976b) showed good correlation (slope: 0.994, r: .998).

Detailed Method. One milliliter of plasma is combined with 100 μl 5 *N* sodium hydroxide and 100 μl of a methanolic solutuion containing 2 μg cinchonine (internal standard)/ml and subsequently extracted with 3 ml benzene. After centrifugation the benzene layer is transferred into another test tube and evaporated to dryness at 55° under a stream of nitrogen. The residue is dissolved in 200 μl mobile phase, and an aliquot injected onto the column. Quantitation of quinidine and dihydro-quinidine is achieved using standard curves of drug internal standard peak height ratios plotted against drug concentrations.

The chromatographic system consists of a liquid chromatograph (Tracor, Model 995) connected with a 25 cm x 4.6 mm (i.d.) cation-exchange column (Partisil 10 SCX, Reeve Angel, Clifton N.J.) and a variable-wavelength UV detector (Tracor, Model 970) set at 230 nm. The solvent system, 10 mmol trimethylamine hydrochloride together with 10 mmol KOH in 1 l water (giving pH ~9)-methanol (1:4), is used at a flow rate of 2 ml/min. Retention times for quinidine, internal standard, and dihydroquinidine are 4,6, and 8 min, respectively.

2. Other HPLC Methods—Bonora et al. (1978), Conrad et al. (1977), Crouthamel et al. (1977), Drayer et al. (1977), Guentert and Riegelman (1979), Guentert et al.(1979), Kline et al. (1979), Powers and Sadée (1978), Sved et al. (1978), Weidner et al. (1979).

Discussion. Specificity of quinidine measurements has been unambiguously verified for only some of the HPLC procedures. Conrad et al. (1977) and Crouthamel et al. (1977) have obtained good correlations with established fluorescence methods and infer the specificity of their methods. The fluorimetric procedures, however, include metabolites with the quinidine detection (Guentert and Riegelman, 1979) and are therefore not useful as reference methods. The silica gel chromatographic procedure of Sved et al. (1978) cannot be recommended for clinical use because of interference of the 2'-quinidinone metabolite peak with quinidine. Although the concentration of this metabolite may be low in healthy human beings, nothing is known about its possible accumulation in various disease states.

The HPLC-UV assay described by Powers and Sadée (1978) uses direct sample injection after protein precipitation and centrifugation and has the advantage of speed over more time consuming procedures. With 0.75 M acetate buffer (pH 3.6)-acetonitrile (6:4) as the eluent and an alkylphenyl reverse phase column, the retention time of quinidine can be kept as low as 2 min. However, the recently isolated N-oxide metabolite (Guentert et al., 1979) is not separated from quinidine under these chromatographic conditions. Chromatographic conditions capable of separating quinidine from its N-oxide metabolite are presently under investigation. This assay can also be used with a fluorescence detector, which yields a greatly increased sensitivity at lower sample volumes.

Several HPLC methods include separate detection of quinidine metabolites. While all metabolites can be monitored by UV detection, 2'-quinidinone gives greatly reduced fluorescence yields. The procedure used by Drayer et al (1977) allows quantitation of 3-OH-quinidine, O-desmethylquinidine, and quinidine in the same sample by C_{18} reverse phase chromatography and fluorescence detection. Determination of 2'-quinidinone, however, requires a separate work-up procedure (Drayer et al., 1978). Bonora et al. (1978) have developed a reverse phase HPLC system that facilitates quantitation of quinidine and all of its metabolites in urine by UV detection in a single step. The method of Guentert et al. (1979) simultaneously measures quinidine, dihydroquinidine, and all of the known metabolites in plasma, using silica gel chromatography and UV detection. At present this is the only assay applicable to the specific detection of quinidine and all of its known metabolites in plasma. Because a normal phase system is used, the N-oxide metabolite is readily separated from quinidine. Under reverse phase conditions with pH <4, however, this metabolite may interfere with the quinidine peak (Bonora et al., 1978). A recent reverse phase HPLC assay of quinidine and 3-OH-quinidine employs either UV or flourescence detection (Weidner et al., 1979). Excellent correlations between this assay and fluorimetric non chromatographic methods indicate potential lack of specificity due to interfering quinidine metabolites in both assays. The reverse phase HPLC-fluorescence assay of Kline et al. (1979) separates quinidine from dihydroquinidine and is sensitive to 50 ng/ml plasma or 1 ng injected onto the column.

3. Thin-Layer Chromatography—Härtel and Korhonen (1968), Steyn and Hundt (1975), Ueda et al. (1976b), Wesley-Hadzija and Mattocks (1977)

Discussion. Because of its intensive fluorescence properties, quinidine is well suited for quantitation by TLC, using fluorescence scanning densitometry. These methods either include extraction of the biological fluids and subsequent spotting of the extracts on the TLC plates

(Härtel and Korhonen, 1968; Ueda et al., 1976b; Wesley-Hadzija and Mattocks, 1977) or use direct sample spotting after protein precipitation (Steyn and Hundt, 1975). Quantitation after development of the plate in a suitable solvent system is achieved either by direct scanning of the fluorescence or UV absorbance on the plate (Steyn and Hundt, 1975; Wesley-Hadzija and Mattocks, 1977) or by extraction of the silica gel and fluorometric examination of the extract (Härtel and Korhonen, 1968; Ueda et al., 1976b). Bias and precision of these procedures are comparable to those obtained by HPLC. Specificity is similar to that achieved by silica gel adsorption chromatography.

4. Fluorometry—Armand and Badinand (1972), Brodie and Udenfriend (1943), Cramer and Isaksson (1963), Hamfelt and Malers (1963), Kessler et al. (1974) Sokolow and Edgar (1950)

Discussion. Fluorescence methods with plasma protein precipitation (Brodie and Udenfriend, 1943; Hamfelt and Malers, 1963) or organic solvent extraction (Sokolow and Edgar, 1950; Cramer and Isaksson, 1963) are still the most widely used quinidine determination procedures in clinical laboratories. Cramer and Isaksson (1963) developed a double-extraction method (extraction from plasma into benzene and reextraction of basic compounds into H_2SO_4) that yields lower values than the protein precipitation method of Brodie and Udenfriend (1943) and the single-extraction procedure of Sokolow and Edgar (1950) (fluorometric determination of an ethylene dichloride extract after acidification with trichloracetic acid). However, a persisting significant interference of the Cramer Isakksson method by quinidine metabolites has been demonstrated with specific techniques such as TLC (Härtel and Korhonen, 1968; Wesley-Hadzija and Mattocks, 1977; Huynh-Ngoc and Sirois, 1977) and HPLC (Drayer et al., 1977; Guentert and Riegelman, 1979; Guentert et al., 1979). In contrast to these results, Huffmann and Hignite (1976) obtained an excellent correlation between their GC-MS method and the Cramer-Isaksson procedure. Several improvements of the Cramer-Isaksson double-extraction method have been proposed (e.g., Armand and Badinand, 1972; Kessler et al., 1974), none of which achieves the specificity obtained with TLC or HPLC (Guentert and Riegelman, 1979; Guentert et al., 1979; Bonora et al., 1978.

5. Gas Chromatography—Midha and Charette (1974), Moulin and Kinsun (1977), Valentine et al. (1976)

Discussion. Several GC procedures have been described for quantitation of quinidine in plasma or whole blood using flash methylation

(Midha and Charette, 1974; Valentine et al., 1976) or without a derivatization step (Moulin and Kinsun, 1977). The sensitivity of the assays depends on the GC detector and sample preparation prior to GC (0.5 μg/ml with nitrogen-sensitive detector; Moulin and Kinsun, 1977; 0.05 μg/ml with FID; Midha and Charette, 1974). Plasma quinidine levels obtained by GC assay are in good agreement with those obtained by the Cramer and Isaksson (1963) fluorescence assay (Midha and Charette, 1974) or are even higher (Valentine et al., 1976), suggesting lack of specificity of the GC methods employing flash methylation.

6. Gas Chromatography-Mass Spectrometry—Huffmann and Hignite (1976)

Discussion. Huffmann and Hignite (1976) use a modification of the Midha and Charette (1974) GC assay procedure and monitor the effluent by electron impact MS. The GC-MS assay monitors the ion fragment at *m/e* 136 and may therefore lack improved specificity over the GC-FID assay. This fragment is derived from the quinuclidine ring, which is formed not only from the parent drug but also from several of its metabolites.

7. Direct Insertion Mass Spectrometry—Garland et al. (1974)

Discussion. Chemical ionization (isobutane) MS has been used to quantitate quinidine in plasma after extraction of the drug and an internal standard (dideuterodihydroquinidine) into benzene. A limited mass range including the MH^+ peaks of quinidine and the internal standard is monitored by repetitive scanning after direct probe insertion.

8. Comments. The issue of assay specificity is of particular importance when monitoring therapeutic quinidine plasma levels because of the narrow therapeutic range of the drug and the formation of metabolites of similar structure and unknown activity in human beings. Recently, several new specific assay procedures for quantitation of quinidine in plasma or serum have been developed, which can be utilized in clinical laboratories.

Reverse phase HPLC (e.g., Powers and Sadée, 1978) is superior to other quantitative quinidine determination methods because of speed (direct sample injection) and potential specificity. Excellent sensitivity can be obtained by the use of fluorescence detection. However, in light of the isolation of new quinidine metabolites (Guentert and Riegelman, 1979; Guentert et al., 1979) HPLC assays have to be reevaluated for potential interference by these metabolites. Interference by even a minor metabolite cannot be accepted in clinically used assays, unless the absence of its accumulation has been demonstrated in various disease states under a variety of clinical situations. Furthermore, existing methods should be critically examined for interference by drugs that are likely to be coadministered in cardiac disorders requiring quinidine therapy (e.g., lidocaine, propranolol, furosemid, hydrochlorothiazide, spironolactone, diazepam). The assay of Kates et al. (1978) meets all of these requirements on the bases of our present knowledge of quinidine disposition in man. Modification of the method of Powers and Sadée (1978) in order to avoid the N-oxide interference would provide an even faster assay with comparable specificity.

e. References

Alexander, F., H. Gold, L. N. Katz, R. L. Levy, R. Scott and P. D. White, The Relative Value of Synthetic Quinidine, Dihydroquinidine, Commercial Quinidine and Quinine in the Control of Cardiac Arrhythmias, *J. Pharmacol. Exp. Ther.*, **90**, 191-201 (1947).

Armand, J. and A. Badinand, Dosage de la Quinidine (ou de la Quinine) dans les Milieux Biologiques, *Ann. Biol. Clin.*, **30**, 599-604 (1972).

Beermann, B., K. Leander, and B. Lindström, The Metabolism of Quinidine is Man: Structure of a Main Metabolite, *Acta Chem. Scand.*, **30**, 465 (1976).

Bonora, M. R., T. W. Guentert, R. A. Upton, and S. Riegelman, Determination of Quinidine and Metabolites in Urine by Reverse-Phase High-Pressure Liquid Chromatography, *Clin. Chim. Acta.*, **91**, 277-284 (1979).

Brodie, B. B. and S. Udenfriend, the Estimation of Quinine in Human Plasma with a Note on the Estimation of Quinidine, *J. Pharmacol. Exp. Ther.*, **78**, 154-158 (1943).

Brodie, B. B., J. E. Baer, and L. C. Craig, Metabolic Products of the Cinchona Alkaloids in Human Urine, *J. Biol. Chem.*, **188**, 567-581 (1951).

Carroll, F. I., D. Smith, and M. E. Wall, Carbon-13 Magnetic Resonance Study: Structure of the Metabolites of Orally Administered Quinidine in Humans, *J. Med. Chem.*, **17**, 985-987 (1974).

Conn, H. L. and R. J. Luchi, Some Cellular and Metabolic Considerations Relating to the Action of Quinidine as Prototype Antiarrhythmic Agent, *Am. J. Med.*, **37**, 685-699 (1964).

Conrad, K. A., B. L. Molk, and C. A. Chidsey, Pharmacokinetic Studies of Quinidine in Patients with Arrhythmias, *Circulation*, **55**, 1-7 (1977).

Cramer, G. and B. Isaksson, Quantitative Determination of Quinidine in Plasma, *Scand. J. Clin. Lab. Invest.*, **15**, 553-556 (1963).

Crouthamel, W. G., B. Kowarski, and P. K. Narang, Specific Serum Quinidine Assay by High-Performance Liquid Chromatography, *Clin. Chem.*, **23**, 2030-2032 (1977).

Drayer, D. E., C. E. Cook, and M. M. Reidenberg, Active Quinidine Metabolites, *Clin. Res.*, **24**. 623A (1976).

Drayer, D. E., K. Restivo, and M. M. Reidenberg, Specific Determination of Quinidine and (3S-)-3-hydroxyquinidine in Human Serum by High-Pressure Liquid Chromatography, *J. Clin. Lab. Med.*, **90**, 816–822 (1977).

Drayer, D. E., D. T. Lowenthal, K. M. Restivo, A. Schwartz, C. E. Cook, and M. M. Reidenberg, Steady-State Serum Levels of Quinidine and Active Metabolites in Cardiac Patients with Varying Degrees of Renal Function, *Clin. Pharmacol. Ther.*, **24**, 31–39 (1978).

Garland, W. A., W. F. Trager, and S. D. Nelson, Direct (Non-chromatographic) Quantification of Drugs and Their Metabolites from Human Plasma Utilizing Chemical Ionization Mass Spectrometry and Stable Isotope Labeling: Quinidine and Lidocaine, *Biomed. Mass Spectrom.*, **1**, 124–129 (1974).

Greenblatt, D. J., H. J. Pfeifer, H. R. Ochs, K. Franke, D. S. MacLaughlin, T. W. Smith, and J. Koch-Weser, Pharmacokinetics of Quinidine in Humans after Intravenous, Intramuscular and Oral Administration, *J. Pharmacol. Exp. Ther.*, **202**, 365–378 (1977).

Guentert, T. W. and S. Riegelman, Specificity of Quinidine Determination Methods, *Clin. Chem.*, in press (1979).

Guentert, T. W., P. E. Coates, R. A. Upton, D. L. Combs, and S. Riegelman, Determination of Quinidine and Its Major Metabolites by High-Performance Liquid Chromatography, *J. Chromatogr.*, **162**, 59–70 (1979).

Hamfelt, A. and E. Malers, Determination of Quinidine Concentration in Serum in the Control of Quinidine Therapy, *Acta Soc. Med. Ups.*, **68**, 181–191 (1963).

Härtel, G. and A. Korhonen, Thin-Layer Chromatography for the Quantitative Separation of Quinidine and Quinidine Metabolites from Biological Fluids and Tissues, *J. Chromatogr.*, **37**, 70–75 (1968).

Huffman, D. H. and C. E. Hignite, Serum Quinidine Concentrations: Comparison of Fluorescence, Gas-Chromatographic, and Gas-Chromatographic/Mass-Spectrometric Methods, *Clin. Chem.*, **22**, 810–812 (1976).

Huffmann, D. H., C. E. Hignite, and C. Tschanz, The Relationship between Quinidine, 3-OH-quinidine and the QTc Interval in Man, *Clin. Res.*, **25**, 553A (1977).

Huynh-Ngoc, T. and G. Sirois, Comparison of Two Spectrofluorometric Procedures for Quinidine Determination in Biological Fluids, *J. Pharm. Sci.*, **66**, 591–592 (1977).

Kates, R. E., D. W. McKennon, and T. J. Comstock, Rapid High-Pressure Liquid Chromatographic Determination of Quinidine and Dihydroquinidine in Plasma Samples, *J. Pharm. Sci.*, **67**, 169–270 (1978).

Kessler, K. M., D. T. Lowenthal, H. Warner, T. Gibson, W. Briggs, and M. M. Reidenberg, Quinidine Elimination in Patients with Congestive Heart Failure or Poor Renal Function, *N. Engl. J. Med.*, **290**, 706–709 (1974).

Kline, B. J., V. A. Turner, and W. H. Barr, Determination of Quinidine and Dihydroquinidine in Plasma by High Performance Liquid Chromatography, *Anal. Chem.*, **51**, 449–451 (1979).

Lewis, T., A. L. Drury, A. M. Wedd, and C. C. Iliescu, Oberservations upon the Activity of Certain Drugs upon Fibrillation of the Auricles, *Heart*, **9**, 207–267 (1922).

Midha, K. K. and C. Charette, GLC Determination of Quinidine from Plasma and Whole Blood, *J. Pharm. Sci.*, **63**, 1244–1247 (1974).

Midha, K. K., and I. J. McGilveray, C. Charette, and M. L. Rowe, Comparison of a Spectrofluorometric Method with a Gas-Liquid Chromatographic Procedure for the Determination of Quinidine Plasma Levels, *Can. J. Pharm. Sci.*, **12**, 41–44 (1977).

Moulin, M. A. and J. Kinsun, A Gas-Liquid Chromatographic Method for the Quantitative Determination of Quinidine in Blood, *Clin. Chim. Acta*, **75**, 491–495 (1977).

Palmer, K. H., B. Martin, B. Baggett, and M. E. Wall, The Metabolic Fate of Orally Administered Quinidine Gluconate in Humans, *Biochem. Pharmacol.*, **18**, 1845–1860 (1969).

Powers, J. L. and W. Sadée, Determination of Quinidine by High-Performance Liquid Chromatography, *Clin. Chem.*, **24**, 229–302 (1978).

Sokolow, M. and R. E. Ball, Factors Influencing Conversion of Chronic Atrial Fibrillation with Special Reference to Serum Quinidine Concentration, *Circulation*, **14**, 568–583 (1956).

Sokolow, M. and A. L. Edgar, Blood Quinidine Concentrations as a Guide in the Treatment of Cardiac Arrhythmias, *Circulation*, **1**, 576–592 (1950).

Steyn, J. M. and H. K. L. Hundt, A Thin-Layer Chromatographic Method for the Quantitative Determination of Quinidine in Human Serum, *J. Chromatogr.*, **111**, 463–465 (1975).

Sved, S., I. J. McGilveray and N. Beaudoin, The Estimation of Quinidine in Human Plasma by Ion-Pair Extraction and High-Performance Liquid Chromatography, *J. Chromatogr.*, **145**, 437–444 (1978).

Ueda, C. T. and B. S. Dzindzio, Quinidine Kinetics in Congestive Heart Failure, *Clin. Pharmacol. Ther.*, **23**, 158–164 (1978).

Ueda, C. T., B. J. Williamson, and B. S. Dzindzio, Disposition Kinetics of Dihydroquinidine Following Quinidine Administration, *Res. Commun. Chem. Pathol. Pharmacol.*, **14**, 215–225 (1976a).

Ueda, C. T., B. J. Williamson, and B. S. Dzindzio, Absolute Quinidine Bioavailability, *Clin. Pharmacol. Ther.*, **20**, 260–269 (1976b).

Valentine, J. L., P. Driscoll, E. L. Hamburg, and E. D. Thompson, GLC Determination of Quinidine in Human Plasma, *J. Pharm. Sci.*, **65**, 96–98 (1976).

Weidner, N., J. H. Ladenson, L. Larson, G. Kessler, and J. M. McDonald, A High-Pressure Liquid Chromatography Method for Serum Quinidine and (3S)-3-Hydroxyquinidine, *Clin. Chim. Acta*, **91**, 7–13 (1979).

Wesley-Hadzija, B. and A. M. Mattocks, Specific Thin-Layer Chromatographic Method for the Determination of Quinidine in Biological Fluids, *J. Chromatogr.*, **144**, 223–230 (1977).

RESERPINE—by M. Cohen

3,4,5-Trimethoxybenzoyl methyl reserpate.
Rau-Sed, Serpasil, Sandril. Antihypertensive agent, tranquilizer, sedative.

$C_{33}H_{40}N_2O_9$; mol. wt. 608.70; pK_a 6.6.

a. Therapeutic Concentration Range. Plasma levels are in the low nanograms per milliliter range after therapeutic doses.

b. Metabolism. The major metabolites of reserpine are formed by esterases leading to methyl reserpate, trimethoxybenzoic acid, and minor amounts of reserpate. An oxidative pathway, mediated by liver microsomal enzymes, involves demethylation of the 4-methoxy substituent of the trimethoxybenzoate moiety of reserpine. The product, syringoyl methyl reserpate, undergoes facilitated hydrolysis to syringic acid and methyl reserpate. None of these metabolites is active. Reserpine has a two-phase plasma elimination with a 4.5 hr $t_{1/2}$ in the first phase and a 46 hr $t_{1/2}$ in the second phase. Little reserpine is excreted unchanged (review: Stitzel, 1977).

c. Analogous Compounds. Other rauwolfia alkaloids, including deserpidine and rescinnamine, tetrabenazine.

d. Analytical Methods

1. Radioimmunoassay—Levy et al. (1976)

Discussion. Antisera against reserpine have been raised in rabbits with reserpine conjugated to bovine serum albumin, using either azo coupling of the diazotized p-aminobenzoic acid to the indole benzo ring followed by protein coupling in the mixed anhydride method or directly with formaldehyde through the indolic nitrogen in a Mannich reaction. Reserpate, methyl reserpate, and trimethoxybenzoic acid are 2000, 90, and 2000 times, respectively, less reactive with both antibodies than is unchanged reserpine. Therefore only methyl reserpate may interfere with the reserpine assay if present in high concentration relative to reserpine. ^{3}H-Reserpine serves as the tracer, and separation of free and bound reserpine is achieved by a polyethylene glycol (PEG) protein precipitation method. The sensitivity is 150 pg reserpine per sample or 15 ng reserpine/ml serum without an extraction procedure.

Detailed Method. Antibody solution (0.4 ml) is incubated at 4°C with 0.01 ml serum and 0.1 ml tritiated reserpine solution in 0.01 N HCL (2000 cpm, approximately 1 ng). Binding equilibrium is reached within a few minutes. First, 0.1 ml of 1% γ-globulin solution and then, after brief mixing, 0.7 ml of 25% PEG solution in water are added to the incubation mixture to precipitate the bound ^{3}H fraction. The tubes are immediately centrifuged at 2500 g for 30 min, and the supernatant is carefully aspirated. The remaining pellet is dissolved in 0.6 ml water and decanted into a counting vial containing 6 ml of a liquid scintillation fluid (Riafluor). The tubes are washed twice with 3 ml Riafluor, and the wash is added to the counting vial.

The samples are counted for ^{3}H content in a Beckman LS-250 liquid scintillation system. Standard curves with known concentrations of reserpine are constructed under identical condition.

2. Thin-Layer Chromatography–Fluorescence—Tripp et al. (1975)

Discussion. Reserpine and its derivatives readily undergo an oxidative reaction to highly fluorescent products when exposed to acid and/or light. After extraction from plasma and separation of reserpine from interfering compounds and metabolites by TLC, fluorescence is developed by exposing the TLC plates to acetic acid vapors for 24 hr. Quantitation of samples and known amounts of standards spotted on the same plates is achieved by fluorescence densitometry of the TLC plates. Sensitivity of the method is 20 ng/ml serum. This procedure should possess a rather high degree of specificity because of the TLC separation step.

3. Comments. Both RIA and TLC-fluorescence densitometry provide suitable analytical techniques to follow reserpine disposition

without the use of radioactive labels. Very high sensitivity and specificity, however, may be needed to study the slow phase of reserpine elimination from the body (Stitzel, 1977). A combination of the RIA procedure with chromatographic techniques should be readily applicable to the analysis of trace amounts of reserpine in biological samples. The native fluorescence of reserpine (300 → 375 nm) might be sufficient for the detection of reserpine in plasma in conjunction with HPLC separation (HPLC-UV in dosage forms: Honigberg et al., 1974).

e. References

Honigberg, I. L., J. T. Stewart, A. P. Smith, R. D. Plunkett, and D. W. Hester, Liquid Chromatography in Pharmaceutical Analysis. II: Determination of a Reserpine-Chlorothiazide Mixture, *J. Pharm. Sci.*, **63**, 1762–1764 (1974).

Levy, A., K. Kawashima, and S. Spector, Radioimmunoassay of Reserpine, *Life Sci.*, **19**, 1421–1430 (1976).

Stitzel, R., The Biological Fate of Reserpine, *Pharmacol. Rev.*, **28**, 179–208 (1977).

Tripp, S. L., E. Williams, W. E. Wagner, Jr., and G. Lukas, A Specific Assay for Subnanogram Concentrations of Reserpine in Human Plasma, *Life Sci.*, **16**, 1167–1178 (1975).

SALICYLIC ACID

o-Hydroxybenzoic acid. Analgesic and anti-inflammatory agent.

$C_7H_6O_3$; mol. wt. 138.12; pK_a 3.0 and 13.4.

a. Therapeutic Concentration Range. From 15 to 30 mg salicylate/100 ml plasma represent therapeutic anti-inflammatory concentrations in the treatment of rheumatoid arthritis (Mongan et al., 1972). Toxicity (e.g., tinnitus) is likely to occur above 30 mg/100 ml (Mongan et al., 1972).

b. Metabolism. The metabolites of salicylate include salicyluric acid (glycine conjugate), the acyl and phenolic glucuronides, gentisic acid (Gibson et al., 1975; Levy et al., 1972), and gentisuric acid (Wilson et al., 1978). The common salicylate prodrug, acetylsalicylic acid (aspirin), is completely converted by esterases to salicylate with a half-life of about 15 min (Rowland and Riegelman, 1968). The following amounts of metabolites of acetylsalicylic acid (400 mg total dose) are recovered in the urine: salicylate, 2 to 5%; salicyluric acid, 80%; salicylate glucuronides, 10 to 20%; and gentisic acid, 1% (Levy et al., 1972). The metabolic pathways leading to salicyluric acid by conjugation with glycine and to the salicyl phenolic glucuronide are readily saturable (Levy, 1965; Tsuchiya and Levy, 1972), resulting in a relative reduction of the urinary excretion of these metabolites at higher salicylate doses (Gibson et al., 1975; Levy et al., 1972). The plasma elimination half-life of salicylate increases from 2 to 4 hr after low salicylate doses to 15 to 30 hr after high doses (Done, 1960; Levy, 1965; Levy et al., 1972; Rowland and Riegelman, 1968). The average salicylate elimination half-life in children is 20 hr after toxic doses (Done, 1960). Ingestion of salicylate is the leading cause of drug poisoning in children, and Done (1960) has pointed out the importance of salicylate blood level measurements to assist in the clinical management of its toxic effects.

Acetyl salicylate and salicylate are equally potent as anti-inflammatory agents in the treatment of rheumatoid arthritis, which requires high salicylate plasma levels (average acetyl salicylate dose: 65 mg/kg/day). However, salicylate is less active than acetyl salicylate as an analgesic. Bioavailability studies of acetylsalicylic acid have to take into account the partial metabolism of acetyl salicylate to salicylate during the first passage through the gut wall and the liver (Leonards, 1962). In view of the widespread use of salicylates, it is not surprising that the analysis of these agents has been an area of interest for more than three decades.

c. Analogous Compounds. Other salicylates (e.g., salicylamide); phenols and benzoate derivatives.

d. Analytical Methods

1. High Performance Liquid Chromatography—Peng et al. (1978)

Discussion. Acetylsalicylic acid, salicylic acid, and salicyluric acid are quantitatively measured by HPLC-UV in small plasma samples with phthalic acid as the internal standard. Acidified plasma is extracted with an ethyl acetate-benzene mixture, and the extract is chromatographed on an octadecyl chemically bonded reverse phase column and measured at 237 nm. The assay is sensitive to 0.5 μg/ml for all three compounds using only 0.1 ml plasma samples. Linearity of response (peak areas) is obtained between 0.5 and 300 μg/ml. The C.V. is ~2% at 50 μg/ml. Small amounts of both acetylsalicylic acid and salicyluric acid can be detected in the presence of large amounts of salicylate. Salicyluric acid plasma levels were found to be lower than salicylate levels by two orders of magnitude.

Detailed Method. Plasma (0.1 ml), 1 ml phthalic acid solution in water (160 or 5 μg/ml), and 1 drop of 0.5% phosphoric acid are mixed and extracted with 0.5 ml ethyl acetate-benzene (1:1) by vigorously shaking for 30 sec. After centrifugation the organic layer is transferred to another tube and evaporated under a stream of nitrogen in an ice water bath in order to avoid drug evaporation. The residue is dissolved in 50 μl of the HPLC eluent, and 45 μl is injected onto the column.

Analysis is carried out on a Model 601 high performance liquid chromatograph (Perkin-Elmer Corp., Norwalk, Conn.), fitted with a valve injector (Glenco Scientific, Houston, Tex.) and a variable-wavelength spectrophotometer (Model LC-55, Perkin-Elmer Corp). A microparticulate reverse phase column (μ-Bondapak C$_{18}$, Waters Assoc., Milford, Mass.) is used with 30% (v/v) acetonitrile in diluted phosphoric acid (0.05%, pH 2.5; column stability?) as the eluent at a flow rate of 1 ml/min. The column is kept at 50°C, and the UV detector set at 237 nm. Standard curves are constructed using peak area ratios.

2. Other HPLC Assays—Ascione and Chrekian (1975), Stevenson and Burtis (1971)

Discussion. Liquid chromatography is performed with controlled pore glass support or solid glass cores with porous silica surfaces (Ascione and Chrekian, 1975), and with a pellicular anion-exchange resin (Stevenson and Burtis, 1971). These assays have not been applied to biological samples.

3. Gas Chromatography—Blakley (1966), Rance et al. (1975), Rowland and Riegelman (1967), Thomas et al. (1973), Walter et al. (1974)

Discussion. The report of Rowland and Riegelman (1967) demonstrated that silanized acetylsalicylic acid can be sensitively measured (0.4 μg/ml) by GC in the presence of large amounts of salicylate, which is separated by a TLC step. Subsequent GC assays for acetylsalicylic acid and, simultaneously, salicylic acid usually involve an organic solvent extraction from plasma at acidic pH and silylation by hexamethyldisilazane or bis(trimethylsilyl)trifluoroacetamide. Assay sensitivity ranges from 0.2 to 1 μg/ml plasma, using n-butylbenzoate (Walter et al., 1974), m-toluic acid (Rance et al., 1975), or p-toluic acid (Thomas et al., 1973) as the internal standard. Gentisic acid, a minor salicylate metabolite, can also be analyzed by GC following silylation (Blakley, 1966).

Walter et al. (1974) suggest the addition of fluoride in order to inhibit the esterase-catalyzed hydrolysis of acetylsalicylic acid in plasma and whole blood. However, even when the sample is stored frozen in the presence of fluoride, slow hydrolysis continues and should be accounted for.

4. Thin-Layer Chromatography—Morrison and Orr (1966)

Discussion. Quantitative TLC analysis has been applied to various pharmaceuticals, including salicylates. After silica gel plate development and spraying with ferric chloride, salicylates are quantitated by a TLC absorbance scanner. Application to biological samples has not been attempted.

5. Spectrophotometry and Colorimetry—Brodie et al. (1944), Levy and Procknal (1968), Routh et al. (1967), Smith et al. (1946), Trinder (1954)

Discussion. Acetylsalicylic acid and salicylic acid can be measured in the presence of each other by differential UV absorbance (Routh et al., 1967). Both drugs are extracted from acidified plasma into CH_2Cl_2 and back-extracted into 4% $NaHCO_3$ aqueous solution. The UV assay is based on the UV shifts specific for acetylsalicylic acid and salicylic acid, caused by a pH change. Differential absorption at low and high pH is measured at 300 nm for acetylsalicylic acid and at 319 nm for salicylic acid. Salicylic acid possesses an isosbestic point at 300 nm and does not interfere with the measurement of acetylsalicylic acid. The differential UV assay is rather specific in the presence of salicyluric acid, salicylamide, barbiturates, and other analgesics.

Levy et al. (1968) present a colorimetric assay for salicylate, salicyluric acid, and total salicylate content in plasma, based on earlier methods by Brodie et al. (1944) and Smith et al. (1946). Salicylic acid is selectively extracted from acidified plasma into CCl_4, followed by extraction of salicyluric acid into CH_2Cl_2. Total salicylate content is extracted into CCl_4 after acid hydrolysis of salicylate metabolites in

HCl. After back-extraction into aqueous medium the characteristically colored iron complex is formed with ferric nitrate. The salicyluric acid metabolite yields 82% of the color intensity of salicylate, while the salicyl glucuronide metabolites do not form colored iron complexes (Smith et al., 1946).

The purple color formation of salicylates with ferric salts in weakly acid solution is also applied by Trinder (1954) in a colorimetric assay that is still widely used in clinical laboratories. The method simply involves the dilution of a plasma sample (1 ml) with 5 volumes of an aqueous color reagent containing 4% ferric nitrate and 4% mercuric chloride in 0.12 N HCl to precipitate proteins. After centrifugation the absorbance is measured at 540 nm. The background absorbance is equivalent to 1 mg salicylate/100 ml, and the measured salicylate absorbance is linear between 10 and 50 mg/100 ml. The fact that the method can be scaled down to measure 0.2 ml blood samples, using 1 ml of the color reagent, is important in pediatric studies. This simple assay, although lacking specificity, is quite adequate for the measurement of salicylate plasma levels following high doses of acetylsalicylic acid or other therapeutic salicylates, since salicylate is the predominant species in plasma at high levels.

6. Fluorescence—Harris and Riegelman (1967), Rowland and Riegelman (1967)

Discussion. Salicylic acid is extracted into ether from acidified plasma or blood and back-extracted into a pH 7 phosphate buffer. The fluorescence of the phosphate buffer is measured at an excitation maximum of 315 nm and an emission maximum at 420 nm. Acetylsalicylic acid does not fluoresce, unless hydrolyzed to salicylic acid in hot HCl (Rowland and Riegelman, 1967).

7. Comments. For therapeutic (high anti-inflammatory salicylate doses) and toxic drug level monitoring, the simple colorimetric method of Trinder (1954) is still the method of choice, since potentially interfering metabolites are present at rather low concentrations in the plasma. Further development of the HPLC technique should increase the use of this method in the pharmacology and toxicology of salicylates.

e. References

Ascione, P. P. and G. P. Chrekian, Automated High-Pressure Liquid Chromatographic Analysis of Aspirin, Phenacetin and Caffeine, *J. Pharm. Sci.*, **64**, 1029–1033 (1975).

Blakley, E. R., Gas Chromatography of Phenolic Acids, *Anal. Biochem.*, **15**, 350–354 (1966).

Brodie, B. B., S. Udenfriend, and A. F. Coburn, The Determination of Salicylic Acid in Plasma, *J. Pharmacol. Exp. Ther.*, **80**, 114–117 (1944).

Done, A. K., Significance of Measurements of Salicylate in Blood in Cases of Acute Ingestion, *Pediatrics*, **26**, 800–807 (1960).

Gibson, T., G. Zaphiropoulos, J. Grove, B. Widdop, and D. Berry, Kinetics of Salicylate Metabolism, *Br. J. Clin. Pharmacol.*, **2**, 233–238 (1975).

Harris, P. A. and S. Riegelman, Acetyl Salicylic Acid in Human Blood and Plasma. I: Methodology and *in vitro* Studies, *J. Pharm. Sci.*, **56**, 713–716 (1967).

Leonards, J. R., The Influence of Solubility on the Rate of Gastrointestinal Absorption of Aspirin, *Clin. Pharmacol. Ther.*, **4**, 476–479 (1962).

Levy, G., Pharmacokinetics of Salicylate Elimination in Man, *J. Pharm. Sci.*, **54**, 959–967 (1965).

Levy, G. and J. A. Procknal, Drug Biotransformation Interactions in Man. I: Mutual Inhibition in Glucuronide Formation of Salicylic Acid and Salicylamide in Man, *J. Pharm. Sci.*, **57**, 1330–1335 (1968).

Levy, G., T. Tsuchiya, and L. P. Amsel, Limited Capacity for Salicylic Glucuronide Formation and Its Effects on the Kinetics of Salicylate Elimination in Man, *Clin. Pharmacol. Ther.*, **13**, 258–268 (1972).

Mongan, E. et al., Tinnitus as an Indication of Therapeutic Serum Salicylate Levels, *J.A.M.A.*, **226**, 142–145 (1972).

Morrison, J. C. and J. M. Orr, Analysis of Selected Pharmaceuticals by Quantitative Thin-Layer Chromatography, *J. Pharm. Sci.*, **55**, 936–941 (1966).

Peng, G. W., M. A. F. Gadalla, V. Smith, A. Peng, and W. L. Chiou, Simple and Rapid High-Pressure Liquid Chromatographic Simultaneous Determination of Aspirin, Salicylic Acid, and Salicyluric Acid in Plasma, *J. Pharm. Sci.*, **67**, 710–712 (1978).

Rance, M. J., B. J. Jordan, and J. D. Nichols, A Simultaneous Determination of Acetylsalicylic Acid, Salicylic Acid, and Salicylamide in Plasma by Gas Liquid Chromatography, *J. Pharm. Pharmacol.*, **27**, 425–429 (1975).

Routh, J. I., N. A. Shane, E. G. Arredondo, and W. D. Paul, Method for the Determination of Acetylsalicylic Acid in the Blood, *Clin. Chem.*, **13**, 734–743 (1967).

Rowland, M. and S. Riegelman, Determination of Acetylsalicylic Acid and Salicylic Acid in Plasma, *J. Pharm. Sci.*, **56**, 717–720 (1967).

Rowland, M. and S. Riegelman, Pharmacokinetics of Acetylsalicylic Acid and Salicylic Acid after Intravenous Administration in Man, *J. Pharm. Sci.*, **51**, 1313–1319 (1968).

Smith, P. K., H. L. Gleason, C. G. Stoll, and S. Ogorzalek, Studies on the Pharmacology of Salicylates, *J. Pharmacol. Exp. Ther.*, **87**, 237–255 (1946).

Stevenson, R. L. and C. A. Burtis, The Analysis of Aspirin and Related Compounds by Liquid Chromatography, *J. Chromatogr.*, **61**, 253–261 (1971).

Thomas, B. H., G. Solomraj, and B. B. Coldwell, The Estimation of Acetylsalicylic Acid and Salicylate in Biological Fluids by Gas Chromatography, *J. Pharm. Pharmacol.*, **25**, 201–204 (1973).

Trinder, P., Rapid Determination of Salicylate in Biological Fluids, *Biochem. J.*, **57**, 301–303 (1954).

Tsuchiya, T. and G. Levy, Biotransformation of Salicylic Acid to Its Acyl and Phenolic Glucuronides in Man, *J. Pharm. Sci.*, **61**, 800–801 (1972).

Walter, L. J., D. F. Biggs, and R. T. Coutts, Simultaneous GLC Estimation of Salicylic Acid and Aspirin in Plasma, *J. Pharm. Sci.*, **63**, 1754–1758 (1974).

Wilson, J. T. et al., Gentisuric Acid: Metabolic Formation in Animals and Identification as a Metabolite of Aspirin in Man, *Clin. Pharmacol. Ther.*, **23**, 635–643 (1978).

SPIRONOLACTONE

17-Hydroxy-7α-mercapto-3-oxo-17α-pregnen-4-ene-21-carboxylic acid γ-lactone,7-acetate.
Aldactone. Aldosterone antagonist.

$C_{24}H_{32}O_4S$; mol. wt. 416.59.

a. Therapeutic Concentration Range. Peak serum levels of canrenone, the major metabolite, are 415 ± 145 ng/ml (mean $\pm$ S.D.) following an oral dose of 200 mg spironolactone in alcoholic solution (Karim et al., 1976).

b. Metabolism. After rapid metabolic hydrolysis of spironolactone to the 7α-SH intermediate (Sadée et al., 1974a), further metabolism occurs either by elimination of H_2S to the 4,6-dienone, canrenone (Gochman and Gantt, 1962; Karim et al., 1976; Sadée et al., 1973), or by SH-methylation and S-oxidation, leading to a series of sulfur retaining metabolites (Abshagen and Rennekamp, 1976; Karim and Brown, 1972). Both sulfur retaining metabolites and canrenone are responsible for the antimineralocorticoid activity of spironolactone. Canrenone accounts for about 30% of the total activity following a single dose of 100 mg spironolactone (Ramsay et al., 1976). However, canrenone accumulates during multiple dosing (Sadée et al., 1974b) a cirumstance that may increase its share of the total pharmacological activity. The plasma elimination half-life of spironolactone is short (<10 min in rats: Sadée et al., 1974a), while the terminal log-linear elimination half-life of canrenone ranges from 13.5 to 24 hr (Karim et al., 1976; Sadée et al., 1973). Canrenone is in rapid enzymatic equilibrium with the γ-hydroxy-carboxylic acid, canrenoate (Sadée et al., 1973). The potassium salt of canrenoate is clinically used as Aldactone® *pro injectione*.

c. **Analogous Compounds.** Canrenoate potassium

d. **Analytical Methods**

1. **High Performance Liquid Chromatography**—Silber et al. (1979)

Discussion. Canrenone is extracted from serum with ether and analyzed by reverse phase HPLC with UV detection at 293 nm (canrenone UV maximum in the eluent), using peak height measurements and external standards. The sensitivity of the assay is 20 ng/ml, using 0.5 ml serum with a C.V. of 7% at 100 ng/ml. The known spironolactone metabolites are not likely to interfere with this assay, although definitive proof of canrenone assay specificity is missing.

Detailed Method. Serum (500 μl) is added to 4 ml diethyl ether. The solutions are mixed and then centrifuged at 1000 g for 5 min. After the aqueous layer is frozen in a solid CO_2-acetone bath, the supernatant is transferred and dried under nitrogen. To this residue, 200 μl acetonitrile-0.005 M phosphate buffer (pH 4.0) (1:1) is added. The mixture is allowed to stand for 30 min, and then 50 μl is injected onto the column.

The high performance liquid chromatograph (Waters Assoc., Milford, Mass.) is equipped with a U6K injector, Model 6000A pump, and a μ-Bondapak C_{18} column, 3.9 mm (i.d.) x 30 cm. The variable-wavelength UV detector (Schoeffel Instrument, Westwood, N.J.) is set at 293 nm. The eluent, acetonitrile-0.005 M phosphate buffer (pH 4.0)

(1:1), is used at a flow rate of 2 ml/min, giving a retention time of 4.5 min for canrenone.

2. Gas Chromatography—Chamberlain (1971)

Discussion. Plasma levels of canrenone are measured after organic solvent extraction by GC-EC with androst-4-ene-3,6,17-trione as the internal standard. The sensitivity of the assay is about 100 ng canrenone/ml. This assay is claimed to be specific by Chamberlain (1971). However, spironolactone as well as canrenoate thermolytically converts to canrenone at the GC temperatures (Sadée et al., 1972), as might also occur with other sulfur retaining metabolites.

3. Fluorescence—Gochman and Gantt (1962), Neubert and Koch (1977), Sadée et al., 1971, 1972)

Discussion. Gochman and Gantt (1962) introduced a sensitive fluorescence assay of canrenone based on dichloromethane extraction and back-extraction into 65% H_2SO_4. The fluorigenic product has been identified as a 4,6,8-(14)-trien-3-one (Sadée et al., 1971). This assay, however, also produces a partial fluorescence yield with spironolactone and probably with most of the sulfur retaining metabolites. Canrenone levels measured by HPLC after an oral dose of 100 mg spironolactone are approximately three times lower than those measured by fluorescence (U. Abshagen, personal communication). In spite of the lack of specificity of the fluorescence assay, it may be the most useful method for spironolactone bioavailability determinations, since both active canrenone and sulfur retaining metabolites are jointly detected. Sadée et al. (1972) propose a modified extraction procedure, which allows separation and determination of total concentrations of fluorigenic metabolites, canrenone and canrenoate, and eliminates canrenone assay interference by spironolactone and the 7α-SH metabolite. Neubert and Koch (1977) present an automated method for the simultaneous measurement of canrenone and canrenoate, using similar principles.

4. Comments.
The HPLC procedure is the method of choice for canrenone and may also be applicable to other metabolites if 240 nm UV detection (4-en-3-ones) is used instead of 293 nm (4,6-dien-3-ones). The fluorescence assays are rather nonspecific and determine both canrenone and sulfur retaining metabolites, a feature that may be advanta-

geous for bioavailability studies. Glucocorticosteroids usually do not fluoresce under the canrenone assay conditions, while canrenone does interfere with the fluorimetric hydrocortisone assay.

e. References

Abshagen, U. and H. Rennekamp, Isolation and Identification of a Sulfur-Containing Metabolite of Spironolactone from Human Urine, *Steroids*, **28**, 467–480 (1976).

Chamberlain, J., Gas Chromatographic Determination of Levels of Aldadiene in Human Plasma and Urine Following Therapeutic Doses of Spironolactone, *J. Chromatogr.*, **55**, 249–253 (1971).

Gochman, N. and C. L. Gantt, A Fluorimetric Method for the Determination of a Major Spironolactone (Aldactone) Metabolite in Human Plasma, *J. Pharmacol. Exp. Ther.*, **135**, 312–316 (1962).

Karim, A. and E. A. Brown, Isolation and Identification of Novel Sulfur-Containing Metabolites of Spironolactone (Aldactone), *Steroids*, **20**, 41–62 (1972).

Karim, A., J. Zagarella, J. Hribar, and M. Dooley, Spironolactone. I: Disposition and Metabolism, *Clin. Pharmacol. Ther.*, **19**, 158–169 (1976).

Neubert, P. and K. Koch, Simultaneous Automated Determination of Spironolactone Metabolites in Serum, *J. Pharm. Sci.*, **66**, 1131–1134 (1977).

Ramsey, L., J. Shelton, I. Harrison, M. Tidd, and M. Asbury, Spironolactone and Potassium Canrenoate in Normal Man, *Clin. Pharmacol. Ther.*, **20**, 167–177 (1976).

Sadée, W., S. Riegelman, and L. F. Johnson, On the Mechanism of Steroid Fluorescence in Sulfuric Acid. I: The Formation of Trienones, *Steroids*, **17**, 595–606 (1971).

Sadée, W., M. Dagcioglu, and S. Riegelman, Fluorimetric Microassay for Spironolactone and Its Metabolites in Biological Fluids, *J. Pharm. Sci.*, **61**, 1126–1129 (1972).

Sadée, W., M. Dagcioglu, and R. Schroeder, Pharmacokinetics of Spironolactone, Canrenone, and Canrenoate-K in Humans, *J. Pharmacol. Exp. Ther.*, **185**, 686–695 (1973).

Sadée, W., U. Abshagen, C. Finn, and N. Rietbrock, Conversion of Spironolactone to Canrenone and Disposition Kinetics of Spironolactone and Canrenoate-K in Rats, *N.S. Arch. Pharmacol.*, **283**, 303–318 (1974a)

Sadée, W., R. Schroeder, E. von Leitner, and M. Dagcioglu, Multiple Dose Kinetics of Spironolactone and Canrenoate-Potassium in Cardiac and Hepatic Failure, *Eur. J. Clin. Pharmacol.*, **7**, 195–200 (1974b).

Silber, B., L. B. Sheiner, J. L. Powers, M. E. Winter, and W. Sadée, Spironolactone-Associated Digoxin Radio Immuno Assay Interference, *Clin. Chem.*, **25**, 48–50 (1979).

SUCCINYLCHOLINE CHLORIDE

Bis(2-dimethylaminoethyl)succinate bis(methochloride). Suxamethonium chloride.
Anectine chloride, and so on. Neuromuscular blocking agent.

$$(CH_3)_3\overset{+}{N}{-}CH_2CH_2O{-}\underset{O}{\overset{\|}{C}}{-}CH_2CH_2{-}\underset{O}{\overset{\|}{C}}{-}O{-}CH_2CH_2N(CH_3)_3 \quad 2Cl^-$$

$C_{14}H_{30}Cl_2N_2O_4 \cdot 2H_2O$; mol. wt. 397.34.

a. Therapeutic Concentration Range. Not available (micrograms per milliliter range).

b. Metabolism. Succinylcholine is rapidly degraded by serum cholinesterase to the inactive succinylmonocholine, choline, and succinate (Agarwal and Goedde, 1976; Goedde et al., 1968). With the widespread use of succinylcholine during surgery, it soon became apparent that some patients (~1%) exhibit a very strong reaction to succinylcholine with paralysis lasting for many hours. This unusual sensitivity was traced to the existence of several phenotypes with genetically altered "atypical" serum cholinesterase, which usually retains the ability to metabolize acetylcholine, but not succinylcholine (La Du, 1972). Succinylcholine-induced apnoea in these patients can be treated with infusions of normal serum cholinesterase from human sources (Scholler et al., 1977). Measurements of serum cholinesterase activity provides, therefore, a means of predicting succinylcholine sensitivity, and such assays have been automated for clinical purposes (Baum, 1976). However, in patients with normal serum cholinesterase the determinants of the duration of activity may be, not variations of the enzyme activity, but rather blood flow to the target tissue, since little redistribution of succinylcholine occurs (Cook et al., 1976). The neuromuscular blocking effects can be readily measured using a nerve stimulator of the ulnar nerve, resulting in thumb adduction. Based on this pharmacodynamic measure, the plasma elimination half-life of succinylcholine is approximately 5 min (Cook et al., 1976). Wingard and Cook (1977) review the clinical pharmacokinetics of muscle relaxants.

c. Analogous Compounds. Quaternary ammonium bases (e.g., acetylcholine, betaine, benzoylcholine, carnitine, decamethonium, gallamine, hexafluorenium, pancuronium, D-tubocurarine).

d. Analytical Methods.

1. Thin-Layer Chromatography—Stevens and Moffat (1974)

Discussion. This assay includes an ion-pair extraction of succinylcholine from urine with bromthymol blue and subsequent TLC on either silica gel or cellulose plates. The quaternary ammonium bases are detected colorimetrically on the plates with an iodoplatinate spray reagent. Semiquantitative estimates of succinylcholine are obtained by visually comparing the characteristically reddish purple spots with known standards (1 μg). Alternatively, the TLC spots can be eluted from the plates, and the extracts treated with NaOH. The liberated succinic acid is derivatized by either methylation (diazomethane), propylation (boron trifluoride-propanol), or chloroethylation (boron trichloride–2-chloroethanol) and analyzed by GC (sensitivity-10 μg succinylcholine per sample). As a third means of identifying succinylcholine on the cellulose plates, half of the plate is sprayed with a serum cholinesterase solution, followed by spraying of the whole plate with the iodoplatinate reagents. Spots of succinylcholine pretreated with serum cholinesterase do not react with the color reagent.

All of these methods are semiquantitative; however, the GC analysis can be utilized for exact quantitation if an appropriate internal standard is used.

Detailed Method. Urine (20 ml or less) is adjusted to pH 7.5 with a saturated solution of trisodium orthophosphate in water and diluted with 10 ml of a 0.15% bromthymol blue solution in methanol-water (1:19). The resulting mixture is extracted with 30 ml dichloromethane by vigorously shaking for a few seconds. The organic layer is then filtered and evaporated to dryness in a boiling water bath. The orange residue containing the ion-pair complexes of amines extracted from the urine is redissolved in 1 ml dichloromethane, and aliquots are applied to the TLC plates in 1.5 to 2 cm bands.

Cellulose TLC plates [CEL 300-25 UV 254 plates with fluorescent indicator (Camlab, Cambridge, England)] are eluted with a mixture of 1% ammonium formate and 5% formic acid solution in water-tetrahydrofurane (90:210). Silica gel plates [SIL G-25 plates without gypsum (Camlab)] are eluted with methyl alcohol-0.2 *N* HCl (80:20). After chromatography the plates are dried in warm air and sprayed lightly with potassium iodoplatinum solution [5% chloroplatinic acid solution (5 ml), water (200 ml), and 30 g potassium iodide]. A light spray of 25% ammonium formate is then applied, and the plates are heated in an oven at 100°C for 2 to 3 min. The reddish purple color produced is fairly specific for succinylcholine and allows detection of 1 μg or more per sample. The R_f values of succinylcholine are 0.35 to 0.40 on the cellulose plates and 0.1 on the silica plates.

2. Other Thin-Layer and Paper Chromatography Methods—Agarwal and Goedde (1976), Fiori and Marigo (1967)

Discussion. Agarwal and Goedde (1976) use cellulose plates with isopropanol-methanol-water (5:10:2) as the eluent. Succinylcholine and its hydrolytic metabolites are detected as the ^{14}C-labeled tracers by direct scanning of the plates to locate and quantitate radioactive bands. Fiori and Marigo (1967) first isolate succinylcholine from urine on a cation-exchange column (Amberlite IRC 50), followed by precipitation of succinylcholine as reineckate and TLC or paper chromatography. The spray reagent is iodoplatinate. Isolated succinylcholine can also be assayed in a biological test using mice, which are very sensitive to the drug (LD_{50}-10 to 12 μg).

3. Gas Chromatography—Stevens and Moffat (1974), Szilagyi et al. (1972), Vidic et al. (1972)

Discussion. Pyrolysis GC of quaternary ammonium compounds is based on thermolytic N-demethylation to tertiary amines with suitable GC properties (Szilagyi et al., 1972; Vidic et al., 1972). For instance, acetylcholine can be detected by pyrolysis GC in tissue samples at low nanograms per gram levels (Szilagyi et al., 1972); however, succinylcholine results in three decomposition products, rendering this approach less suitable for quantitative analysis. Stevens and Moffat (1974) propose alkaline hydrolysis of succinylcholine and GC analysis of the resulting succinate (see Section d1).

4. Colorimetry—Tanaka et al. (1974)

Discussion. Quaternary ammonium bases can be complexed with a lipophilic strong acid anion, and the ion pair extracted into organic solvents. If the anion is colored and has an absorbance maximum distinct from that of the corresponding uncharged species, the amount of quaternary ammonium base in the organic phase can be quantitatively measured by colorimetry. Tanaka et al., (1974) discuss structural requirements of quaternary ammonium bases (e.g., lipophilicity) for efficient organic solvent extraction with bromphenol blue as the anion. The method lacks specificity.

5. Comments. Little effort has been expended on the chemical analysis of succinylcholine because of the availability of other techniques to readily monitor drug effects. In addition, the serum cholinesterase present in blood and sometimes in urine samples rapidly decomposes succinylcholine *in vitro*, making analysis for the drug very difficult. Such decomposition may be averted by the addition of fluoride (see cocaine monograph); however, some of the phenotypes possess a fluoride-resistant serum cholinesterase that is not inactivated by fluoride (La Du, 1972). Similar analytical and sample storage problems apply to all drugs that are subject to serum cholinesterase mediated hydrolysis (e.g., cocaine, procaine).

e. References

Agarwal, D. P. and H. W. Goedde, Thin-Layer Chromatographic Separation of ¹⁴C-Labelled Succinyldicholine, Succinylmonocholine and Choline, *J. Chromatogr.*, **121**, 170–172 (1976).

Baum, G., An Automated Kinetic Analysis of Cholinesterase Activity by a Substrate-Selective Ion-Exchange Electrode, in *Ion and Enzyme Electrodes in Biology and Medicine*, M. Kessler et al., Ed., University Park Press, Baltimore, 1976, pp. 193–197

Cook, D. R., L. B. Wingard, and F. H. Taylor, Pharmacokinetics of Succinylcholine in Infants, Children, and Adults, *Clin. Pharmacol. Ther.*, **20**, 493–498 (1976).

Fiori, A. and M. Marigo, A Method for the Detection of D-Tubocurarine, Gallamine, Decamethonium and Succinylcholine in Biological Materials: Modification and Development, *J. Chromatogr.*, **31**, 171–176 (1967).

Goedde, H. W., K. R. Held, and K. Altland, Hydrolysis of Succinyldicholine and Succinylmonocholine in Human Serum, *Mol. Pharmacol.*, **4**, 274–278 (1968).

La Du, B. N., Pharmacogenetics: Defective Enzymes in Relation to Reactions to Drugs, *Ann. Rev. Med.*, **23**, 453–468 (1972).

Scholler, K. L., H. W. Goedde, and H. G. Benkmann, The Use of Serum Cholinesterase in Succinylcholine Apnoea, *Can. Anaesth. Soc. J.*, **24**, 396–400 (1977).

Stevens, H. M. and A. C. Moffat, A Rapid Screening Procedure for Quaternary Ammonium Compounds in Fluids and Tissues with Special Reference to Suxamethonium (Succinylcholine), *J. Forensic Sci. Soc.*, **14**, 141–148 (1974).

Szilagyi, P. I., J. P. Green, O. M. Brown, and S. Margolis, The Measurement of

Nanogram Amounts of Acetylcholine in Tissues by Pyrolysis Gas Chromatography, *J. Neurochem.*, **19**, 2555–2566 (1972).

Tanaka, K., M. Hioki, and H. Shindo, Determination of Pancuronium Bromide and Its Metabolites in Human Urine by Dye-Extraction Method: Relation between the Extractability and Structure of Quaternary Ammonium Ions, *Chem. Pharm. Bull.* (Tokyo), **22**, 2599–2606 (1974).

Vidic, H. J., H. Dross, and H. Kewitz, A Gas Chromatographic Determination of Quaternary Ammonium Compounds, *Z. Klin. Chem. Klin. Biochem.*, **10**, 156–159 (1972).

Wingard, L. B. and D. R. Cook, Clinical Pharmacokinetics of Muscle Relaxants, *Clin. Pharmacokin.*, **2**, 330–343 (1977).

SULFONAMIDES—by D. C. Perry

Antibacterial agents.

SULFACETAMIDE

N^1-Acetylsulfanilamide; N-[(4-aminophenyl)sulfonyl]acetamide. Albucid, Beocid.

$$H_2N-\text{⟨⟩}-SO_2NH-\overset{\overset{\displaystyle O}{\|}}{C}-CH_3$$

$C_8H_{10}N_2O_3S$; mol. wt. 214.24; pK_a 5.4.

SULFANILAMIDE

p-Aminobenzenesulfonamide.
Deseptyl, Prontalbin, Streptocide.

$$H_2N-\text{⟨⟩}-SO_2NH_2$$

$C_6H_8N_2O_2S$; mol. wt. 172.21; pK_a 10.4.

SULFAPYRIDINE

N^1-2-Pyridylsulfanilamide; 4-amino-N-2-pyridinylbenzenesulfonamide.
Dagenan, Eubasin.

$$H_2N-\text{⟨⟩}-SO_2-NH-\text{⟨N⟩}$$

$C_{11}H_{11}N_3O_2S$; mol. wt. 249.29; pK_a 8.4.

SULFATHIAZOLE

N^1-2-Thiazolylsulfanilamide; 4-amino-N-2-thiazolylbenzenesulfonamide.
Cibazol, Eleudron.

$C_9H_9N_3O_2S$; mol. wt. 255.32; pK_a 7.2.

SULFISOMIDINE

(= Sulfadimetine or sulfaisodimidine.)
N^1-(2,6-Dimethyl-4-pyrimidinyl)sulfanilamide; 4-amino-N-(2,6-dimethyl-4-pyrimidinyl)benzenesulfonamide.
Aristamid, Elkosin.

$C_{12}H_{14}N_4O_2S$; mol. wt. 278.34; pK_a 7.4.

SULFISOXAZOLE

(= Sulfafurazole.)
N^1-(3,4-Dimethyl-5-isoxazolyl)sulfanilamide; 4-amino-N-(3,4-dimethyl-5-isoxazolyl)benzenesulfonamide.
Gantrisin.

$C_{11}H_{13}N_3O_3S$; mol. wt. 267.30; pK_a 4.9.

References for pK_a values: Su et al. (1976), Struller (1968a).

a. Therapeutic Concentration Range. Extrapolation from published data on peak levels, dosage schedules (Barker and Prescott,

1973), and pharmacokinetics (see below) gives an estimate of the therapeutic range for all of these six drugs of 30 to 150 μg/ml plasma. As an example, specific values are 100 to 150 μg/ml for sulfanilamide and 90 to 150 μg/ml for sulfisoxazole (Winek, 1976).

b. Metabolism. The listed sulfa drugs are rapidly absorbed with a relatively short elimination half-life (Struller, 1968a, 1968b.; Weinstein, 1975). Absorption after oral doses is largely completed within 30 min (Weinstein, 1975). One report on sulfisoxazole gives a volume of distribution of 16% of body weight for this drug (Kaplan et al., 1972). Plasma protein binding varies dramatically with the various sulfa drugs. Elimination half-lives and percentages of drug protein bound in normal serum are given in the following Table.

Drug	$t_{1/2}$ (hr)	Reference[a]	Plasma protein bound (%)	Reference[a]
Sulfanilamide	8.8	2	12	3
	11	3, 4		
Sulfacetamide	12.8	2	18	3
	12	3	15–17	1
	12–14	1		
Sulfapyridine	9.4	2	No data	
	12	3		
Sulfathiazole	3.5	2	77	3
	4	3		
Sulfisomidine	7	2, 3	86	3
	6–7	1	79–85	1
Sulfisoxazole	6.1	2	85	1
	6	3, 4	86	3
	6–10	1		

[a] 1: Barker and Prescott, 1973; 2: Krueger-Thiemer and Buender, 1969; 3: Struller, 1968a; 4: Struller, 1968b.

The major route of metabolism of sulfonamides is hepatic acetylation of the free amino group to give the N^4-acetylated compound. The extent of acetylation is different for each drug. Sulfadiazine and its methylated derivatives are acetylated from 10 to 40% (of total plasma sulfonamide) (Weinstein, 1975). Fries et al. (1971) found that 43.9% of a dose of sulfanilamide in human beings is excreted as the N^4-acetyl metabolite over 72 hr. Sulfisomidine is acetylated to a much lesser extent than the other sulfonamides (Barker and Prescott, 1973). Acetylation of some sulfa drugs (sulfapyridine, sulfamethazine, sulfadimidine) follows a genetic bimodal distribution similar to that of INH, while other sulfa congeners (sulfanilamide) show a unimodal distribution in the population which argues for the existence of several classes of N-acetyltransferases (Fries et al., 1971; Schroeder and Price Evans, 1972).

Although unchanged drug and the N^4-acetylated compound account for the bulk of the dose, several minor metabolic pathways have also been reported. These include for sulfanilamide the N-(4-sulfamoyl-phenyl)glycolamide, N^1,N^4-diacetylsulfanilamide, and 3-hydroxy sulfanilamide, all at less than 2% of a dose of ^{35}S-sulfanilamide in human beings (Fries et al., 1971). Other potential metabolites are the N^1-acetyl derivative, the N^4-hydroxylamino derivative (not yet shown in human beings), 5-hydroxymethylsulfamethoxazole (Rieder, 1973), N^4-sulfonates, and N^1,N^4-and heterocyclic ring-N-glucuronide conjugates (Rieder, 1972, 1973). None of the above metabolites has any known antibacterial effect (Rieder, 1972, 1973). None is present in major amounts except the N^1-glucuronide in urine (Rieder, 1973). The N^4-acetylated metabolites, while devoid of antibacterial effects, are at least equal in toxicity to the parent compound. Furthermore, they are generally less soluble than the parent drug and thus more likely to crystallize out in the urine, causing renal complications (Weinstein, 1975). For this reason, clinically useful sulfonamide assays should include measurements of both the parent drug and the acetylated metabolite.

c. Analogous Compounds. Sulfonamides: other antibacterial sulfonamides (some preparations include two or three different sulfonamides); carbonic anhydrase inhibitors (e.g., acetazolamide, dichlorphenamide, ethoxzolamide, methazolamide); diuretic benzothiadiazides (e.g., hydrochlorothiazide, furosemide); phenazopyridine (given with sulfisoxazole; might interfere with colorimetric determination); antihypertensives (e.g., diazoxide). Primary aryl amines (interfere with the Bratton-Marshall reaction): p-aminobenzoic acid, procainamide, and so on.

d. Analytical Methods

1. Colorimetry—Bratton and Marshall (1939), Rieder (1972)

Discussion. The method of Rieder (1972) is based on the Bratton-Marshall technique used for the last 40 years (Bratton and Marshall, 1939). The free aryl primary amine is converted to the diazonium salt, using nitrous acid (NaNO$_2$ and HCl); excess nitrous acid is removed from the reaction mixture with sulfamic acid, yielding molecular nitrogen. The diazonium salt is then coupled to a chromogen (1-naphthylethylenediamine dihydrochloride) to form the azo dye, which is measured spectrophotometrically at 545 nm. Earlier methods use similar procedures with and without acid hydrolysis to deacetylate N^4-acetyl metabolites. Values obtained with the additional hydrolysis step have been referred to as "total sulfonamide" concentration, as opposed to "free sulfonamide" without hydrolysis. Protein precipitation precedes these steps in most previous procedures, using trichloroacetic acid. Rieder (1971) notes that this step could lead to hydrolysis of N^4-glucuronides, which are then included with the "free" fraction. Other inactive metabolites, such as N^1- and ring-N-glucuronides, with free N^4-amino groups also interfere with the measurement of the "free" fraction. This may represent a significant problem in urine but not in blood, where glucuronide concentrations are insignificant. To avoid these problems, Rieder (1972) uses an ethyl acetate plasma extraction in place of TCA precipitation. No spurious hydrolysis occurs, and conjugated metabolites are too hydrophilic to partition significantly into the ethyl acetate. Total sulfonamide analyis is performed as reported in preceding assay procedures. It should be noted here that the term "free sulfonamide" levels refers to the concentrations of both plasma protein-bound and unbound sulfonamides, since the assay procedures cancel out protein binding effects.

The majority of pharmacokinetic data on sulfonamides is derived from the Bratton-Marshall method. Lack of specificity must be considered when interpreting assay results, since any primary aromatic amine may interfere. However, sulfonamide plasma levels are rather high, minimizing the chances for obtaining spurious data under controlled experimental conditions and, to a lesser extent, in the clinical use of this colorimetric technique. The method of Rieder (1972) has a sensitivity of 0.5 μg sulfonamide/ml plasma and 1.3 μg/ml urine (e.g., sulfamethoxazole). The C. V. is 4.48% at 30 μg sulfamethoxazole/ml plasma ($n = 20$). The ethyl acetate extraction efficiency is between 89 and 98% for five sulfa drugs with pK_a values ranging from 5.0 to 10.08.

Detailed Method. "Free Sulfonamides." Add 1 ml McIlvain buffer (pH 5) (made with 8.6 volumes 0.2 M citric acid. 11.4 volumes 0.4 M disodium phosphate), 0.2 ml plasma, and 5 ml ethyl acetate to a 15 ml stoppered tube, shake for 10 min, and centrifuge for 5 min. Transfer 3 ml supernatant to another tube, and add 0.5 ml acetonic 2 N HCl (prepared freshly by mixing 1 volume 8 N HCl with 3 volumes acetone) and 0.5 ml 0.1% sodium nitrite in acetone-water (3:1, v/v). After mixing, allow the solution to react for 6 min. Then add sulfamic acid in acetone-water (3:1, v/v), and tap the tube until no further N_2 bubbles can be detected. Wait for 3 min; then add 0.5 ml 0.1% 1-naphthylethylene-diamine·2HCl in acetone-water (3:1, v/v) and 0.5 ml methanol. Mix until the solution is homogeneous, stopper, and measure the absorbance spectrophotometrically between 20 and 60 min later.

"Total Sulfonamides." Dilute 0.2 ml plasma with 4 ml water in a 15 ml tube, and place for 4 min in a boiling water bath. Without allowing the solution to cool, add 1 ml 20% trichloroacetic acid, mix, and centrifuge as hot as possible for 10 min. Transfer 3 ml supernatant to another tube, add 0.5 ml 3 N HCl, mix, seal with aluminum foil, and place in a boiling water bath for 1 hr for hydrolysis. Cool in 20°C water, add 0.5 ml 0.1% aqueous sodium nitrite, mix, and wait for 6 min. Then add 0.5 ml 0.5% aqueous sulfamic acid, mix, and tap tube, until no more N_2 bubbles are detectable. Add 0.5 ml 0.1% aqueous 1-naphthylethylene-diamine·2HCl, mix, and wait 20 to 60 min to measure.

The absorption of the solution is measured in glass cuvettes with an optical path of 10 mm at a wavelength of 545 nm. Reaction products of seven sulfonamides with 1-naphthylethylenediamine produce absorption maxima within a narrow range (540 to 550 nm). Selection of specific maxima for the individual sulfa drugs is not required since the position of the maximum is primarily a function of the chromogene rather than of the sulfonamide substituents. Appropriate standards are also carried through the procedures.

2. Other Colorimetric Methods—Kraml and Boudreau (1971), Saris et al. (1969), Stewart et al. (1969)

Discussion. The Bratton-Marshall procedure is still the most widely reported assay in clinical and research publications on sulfonamides. The principles of the assay are outlined above. Kraml and Boudreau (1971) and Saris et al. (1969) offer automated versions of the Bratton-Marshall procedure. Stewart et al. (1969) suggest a different chromophor (9-chloroacridine).

3. Fluorimetry—Bridges et al. (1974), de Silva and Strojny (1975), Stewart and Wilkin (1972)

Discussion. Bridges et al. (1974) examine the fluorescent and luminescent properties of sulfonamides in detail, and investigate the direct detection of sulfonamides in serum after protein precipitation, using a spectrofluorometer. Although theoretical sensitivity is good (10 ng/ml), in practice significant background is seen at levels below 5 μg/ml. Also, the procedure does not detect N^4-substituted derivatives without prior hydrolysis. Only some of the sulfonamides are sufficiently fluorescent for analysis in biological samples. De Silva and Strojne (1975) examine the use of fluorescamine as a fluorescence derivatizing reagent and suggest its use as a TLC spray reagent for sulfonamides. The detection limit is 5 to 10 ng/ml with an upper limit of the linear analytical range of 1 μg per sample. Fluorescamine reacts with all primary amines, requiring chromatography in the presence of other interfering primary aromatic and aliphatic amines.

Stewart and Wilkin (1972) propose a fluorescence quenching method. Sulfonamides react with 9-chloroacridine and quench the fluorescence of the free 9-chloroacridine in solution. Biological fluids and metabolites were not examined.

4. Phosphorimetry—Hollifield and Winefordner (1966)

Discussion. Serum is extracted with ethanol and centrifuged, and the supernatant frozen to 77°K with liquid nitrogen to form a clear, rigid glass. Phosphorescence of the solution is measured in a suitable spectrofluorimeter after excitation with UV light. Detections of pure sulfonamide from 1 μg/ml to 0.1 ng/ml are reported; however, serum background noise limits the sensitivity to 10 μg/ml without further purification steps.

5. High Performance Liquid Chromatography—Allfred and Dumire (1978), Goehl et al. (1978), Jung and Oie (1978), Owerbach et al. (1978), Peng et al. (1977), Sharma et al. (1976), Vree et al. (1978)

Discussion. Several HPLC procedures have been recently reported for the analysis of sulfa drugs in biological fluids. Some of these will be discussed here.

Goehl et al. (1978) analyze three sulfa pyrimidines in human plasma (sulfadiazine, sulfamerazine, sulfamethazine) with sulfamethizole as the internal standard. Plasma proteins are precipitated with trichloroacetic acid, and the supernatant is injected onto a reverse phase HPLC column (μ-Bondapak C_{18}, Waters Assoc.) and analyzed by UV absorbance at 254 nm. Standard curves in plasma for the three drugs are linear in the range of 1 to 30 μg/ml. Plasma from a human subject given a dose of trisulfapyrimidines (oral suspension providing 0.33 g of each drug) is analyzed by this HPLC method and the Bratton-Marshall assay for free sulfonamides over a 48 hr period. Results are similar through 6 hr. Thereafter the colorimetric method produces significantly (10 to 100%) higher levels than the HPLC method, presumably because of interference of the Bratton-Marshall reaction by metabolites.

Although metabolite peaks are detectable in plasma samples, they have not been further identified.

Jung and Oie (1978) describe an HPLC assay of sulfisoxazole and its N^4-acetylated metabolite in plasma and urine with N^4-acetylsulfamethoxazole as the internal standard. Column and UV detection are the same as used by Goehl et al. (1978). Linear standard curves are obtained in the range of 0.05 to 200 μg/ml.

The HPLC procedure of Peng et al. (1977) measures sulfisoxazole in human plasma, which is treated with acetonitrile in order to precipitate proteins prior to analysis. A reverse phase column (Sil-X-1, Perkin-Elmer) is used with UV detection at 280 nm. Potential metabolite peaks were not further identified. The sensitivity is below 2 μg sulfisoxazole/ml plasma.

Sharma et al. (1976) separate and analyze three sulfa drugs (sulfamethazine, sulfamerazine, sulfathiazole), and their respective N^4-acetylated metabolites in cattle urine. Following methanol precipitation of urine samples, supernatants are directly analyzed on an Amino-Sil-X-1 (Perkin-Elmer) column with methanol as the eluent and 254 nm UV detection. Sensitivity is 2 μg/ml urine for all compounds. Analysis of plasma samples was not attempted.

The sensitive assay of sulfamethazine at low levels in nonmedicated swine feeds involves acetonitrile extraction, sample purification on a mixed reverse phase (C_{18})—anion-exchange minicolumn, and analysis by HPLC on a μ-Bondapak C_{18} reverse phase column with UV detection at 254 nm (Allfred and Dumire, 1978). The assay is sensitive to well below 1 μg/g specimen and uses external standard curves for quantitation.

Sulfapyridine and acetylsulfapyridine can be detected in plasma and saliva with a sensitivity of 0.25 μg/ml urine, using a cyano-bonded reverse phase HPLC column and UV detection at 254 nm (Owerbach et al., 1978). Only a 10 μl plasma specimen, after deproteinization with perchloric acid, is injected onto the column; sulfamethoxazole serves as the internal standard.

Vree et al. (1978) describe a reverse phase (Lichrosorb RP-8, 5 μ particle size) HPLC assay of sulfamethoxazole and its N^4-acetylated metabolite in body fluids. Plasma proteins are precipitated with $HClO_4$ prior to analysis. The sensitivity is 0.5 μg/ml plasma. Six other sulfa drugs are separated from sulfamethoxazole and its metabolite. Trimethoprim, which is frequently comedicated with sulfamethoxazole (Co-Trimoxazole), also does not interfere with the assay and can be simultaneously measured at a UV absorbance at 225 nm with a sensitivity of 0.75 μg/ml plasma.

6. Gas-Liquid Chromatography—Bye and Land (1977), Nose et al. (1976)

Discussion. Sulfonamides can be analyzed by GC following derivatization to form volatile substances. Electron capture detection provides a high degree of sensitivity to GC analysis. Nose et al. (1976) define chromatographic conditions for 14 sulfonamides following alkylation with dimethylformamide dialkylacetals. No biological applications, however, are reported.

Bye and Land (1977) present a specific assay for sulfadiazine and its N^4-acetyl metabolite in human plasma and urine. After extraction at acidic pH into chloroform and evaporation of the chloroform layer, the residue is methylated with diazomethane and analyzed on a column at 285°C with ^{63}Ni electron capture detection and 9-bromophenanthrene as the internal standard. As little as 1 ng sulfadiazine and its metabolite can be separated and detected by GC-EC. Exact sensitivity limits in biological samples are not stated, but values as low as 2 μg/ml are reported. Standard deviations of six determinations, each at six concentrations from 2 to 30 μg/ml, range from 4 to 7%.

7. Comments.

Colorimetry using the Bratton-Marshall reaction and, increasingly, HPLC represent the assay methods of choice for the detection of sulfonamide drugs in biological samples.

e. References

Allfred, M. C. and D. L. Dumire, High Performance Liquid Chromatography Determination of Sulfamethazine at Low Levels in Nonmedicated Swine Feeds, *J. Chromatogr. Sci.*, **16**, 533–537 (1978).

Barker, B. M. and F. Prescott, *Antimicrobial Agents in Medicine*, Blackwell Sci. Publ., Oxford, 1973.

Bratton, A. C. and E. K. Marshall, Jr., A New Coupling Component for Sulfanilamide Determination, *J. Biol. Chem.*, **128**, 537–550 (1939).

Bridges, J. W., L. A. Gifford, W. P. Hayes, J. N. Miller, and D. T. Burns, Luminescence Properties of Sulfonamide Drugs, *Anal. Chem.*, **46**, 1010–1017 (1974).

Bye, A. and G. Land, Gas-Liquid Chromatographic Determination of Sulfadiazine and Its Major Metabolite in Human Plasma and Urine, *J. Chromatogr.*, **139**, 181–185 (1977).

de Silva, J. A. and N. Strojny, Spectrofluorometric Determination of Pharmaceuticals Containing Aromatic or Aliphatic Primary Amino Groups as Their Fluorescamine (Fluram) Derivatives, *Anal. Chem.*, **47**, 714–717 (1975).

Fries, W., M. Kiese, and W. Lenk, Additional Route in the Metabolism of Sulphanilamide, *Xenobiotica*, **1**, 241–256 (1971).

Goehl, T. J. et al., Simple High-Pressure Liquid Chromatographic Determination of Trisulfapyrimidines in Human Serum, *J. Pharm. Sci.*, **67**, 404–406 (1978).

Hollifield, H. C. and J. D. Winefordner, A Phosphormimetric Investigation of Several Sulfonamide Drugs: A Rapid Direct Procedure for the Determination of Drug Levels in Pooled Human Serum with Specific Application to Sulfadiazine, Sulfamethazine, Sulfamerazine, and Sulfacetamide, *Anal. Chim. Acta*, **36**, 352–359 (1966).

Jung, D. and S. Oie, unpublished data (1978).

Kaplan, S. A., R. E. Weinfeld, C. W. Abruzzo, and M. Lewis, Pharmacokinetic Profile of Sulfisoxazole Following Intravenous, Intramuscular and Oral Administration to Man, *J. Pharm. Sci.*, **61**, 773–778 (1972).

Kraml, M. and A. Boudreau, An Automated Determination of Serum Sulfonamides, *Clin. Biochem.*, **4**, 123–127 (1971).

Krueger-Thiemer, V. E. and P. Buenger, Kumulation and Toxizität bei falscher Dosierung von Sulfanilamiden, *Arzneim.-Forsch.*, **11**, 867–874 (1961).

Nose, N., S. Kobayashi, A. Hirose, and A. Watanabe, Gas-Liquid Chromatographic Determination of Sulfonamides, *J. Chromatogr.*, **123**, 167–173 (1976).

Owerbach, J., N. F. Johnson, T. R. Bates, H. J. Pieniaszek, Jr., and W. J. Jusko, High-Performance Liquid Chromatographic Assay of Sulfapyridine and Acetylsulfapyridine in Biological Fluids, *J. Pharm. Sci.*, **67**, 1250–1253 (1978).

Peng, G. W., M. A. F. Gadalla, and W. L. Chiou, High Pressure Liquid Chromatographic Determination of Sulfisoxazole in Plasma, *Res. Commun. Chem. Pathol. Pharmacol.*, **18**, 233–245 (1977).

Rieder, J., Quantitative Determination of the Bacteriostatically Active Fraction of Sulfonamides and the Sum of Their Inactive Metabolites in the Body Fluids, *Chemotherapy*, **17**, 1–21 (1972).

Rieder, M., Metabolism and Techniques for Assay of Trimethoprim and Sulfamethoxazole, *J. Infect. Dis.*, **128**, (Suppl.), 567–573 (1973).

Saris, N-E., A. Sorto, and S.-L. Karonen, On the Assays of Free, Unconjugated and Total Sulfonamides, *Scand. J. Clin. Lab. Invest.*, **110**, (Suppl.), 28–31 (1969).

Schroeder, H. and D. A. Price Evans, The Polymorphic Acetylation of Sulfapyridine in Man, *J. Med. Genet.*, **9**, 168–171 (1972).

Sharma, J. P., E. G. Perkins, and R. F. Bevill, High-Pressure Liquid Chromatographic Separation, Identification, and Determination of Sulfa Drugs and Their Metabolites in Urine, *J. Pharm. Sci.*, **65**, 1606–1608 (1976).

Stewart, J. T. and R. E. Wilkin, Determination of Sulfonamides and Local

Anesthetics with 9-Chloroacridine by Quenching Fluorometry, *J. Pharm. Sci.*, **61**, 432–433 (1972).

Stewart, J. T., A. B. Ray, and W. B. Fackler, Colorimetric Determination of Some Sulfonamides with 9-Chloroacridine, *J. Pharm. Sci.*, **58**, 1261–1262 (1969).

Struller, T., Progress in Sulfonamide Research, *Prog. Drug Res.*, **12**, 389–452 (1968a).

Struller, T., Long-Acting and Short-Acting Sulfonamides: Recent Developments, *Antibiot. Chemother.*, **14**, 179–215 (1968b).

Su, S. C., A. V. Hartkopf, and B. L. Karger, High-Performance Ion-Pair Partition Chromatography of Sulfa Drugs: Study and Optimization of Chemical Parameters, *J. Chromatogr.*, **119**, 523–538 (1976).

Vree, T. B., Y. A. Hekster, A. M. Baars, J. E. Damsma, and E. van der Klein, Determination of Trimethoprim and Sulfamethoxazole (Co-Trimoxazole) in Body Fluids of Man by Means of High Performance Liquid Chromatography, *J. Chromatogr.*, in press (1979).

Weinstein, L. Sulfonamides and Trimethoprim-Sulfamethoxazole, in *The Pharmacological Basis of Therapuetics*, L. S. Goodman and A. Gilman, Eds. The Macmillan Co., New York, 1975, pp. 1113–1129.

Winek, C. L., Tabulation of Therapeutic, Toxic, and Lethal Concentrations of Drugs and Chemicals in Blood, *Clin. Chem.*, **22**, 832–836 (1976).

TETRACYCLINES

Antimicrobials.

DOXYCYCLINE

6-Deoxy-5-hydroxytetracycline.
Vibramycin.

$C_{22}H_{24}N_2O_8$; mol. wt. 444.43.

MINOCYCLINE

7-Dimethylamino-6-demethyl-6-deoxytetracycline.
Minocin, Klinomycin, Vectrin (hydrochlorides).

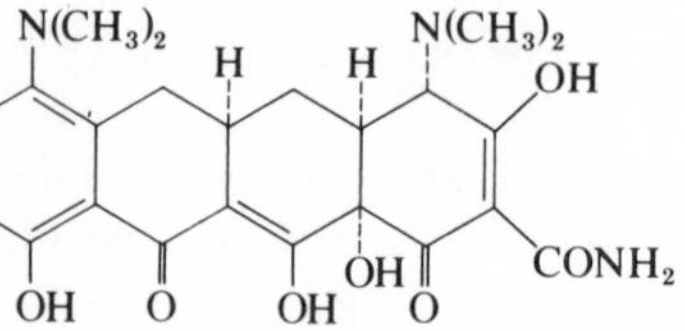

$C_{23}H_{27}N_3O_7$; mol. wt. 457.49; pK_a 2.8, 5.0, 7.8, and 9.3.

OXYTETRACYCLINE

5-Hydroxytetracycline.
Oxymycin, Terramycin, and so on.

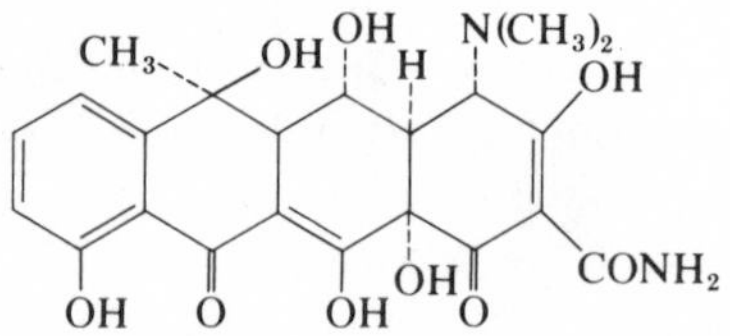

$C_{22}H_{24}N_2O_9 \cdot 2H_2O$; mol. wt. 496.46; pK_a 3.3, 7.3, and 9.1.

TETRACYCLINE

4-(Dimethylamino)-1,4,4a,5,5a,6,11,12a-octahydro-3,6,10,12-pentahy-
droxy-6-methyl-1,11-dioxo-2-naphthacenecarboxamide.
Achromycin, Cyclomycin, and so on.

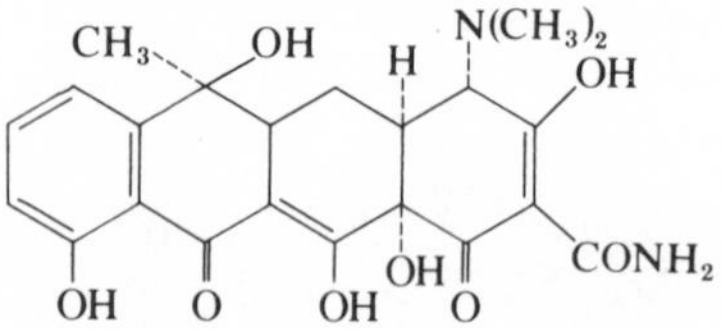

$C_{22}H_{24}N_2O_8$; mol. wt. 444.43; pKa 3.3, 7.7, and 9.5.

a. Therapeutic Concentration Range. The therapeutic levels of
the tetracycline antibiotics usually range from 0.5 to 5 μg/ml plasma
(Hall, 1976; Neuvonen et al., 1976).

b. Metabolism. The elimination of most tetracycline occurs by renal excretion, with the exception of doxycycline, which is partially cleared by extrarenal routes (Gavend et al., 1972; Neuvonen et al., 1976; Simon et al., 1975). Little is known of potential metabolic degradation of tetracyclines, and metabolites do not appear to play a significant pharmacological role. The plasma elimination half-lives of some tetracyclines are as follows: doxycycline, 12 to 15 hr (Neuvonen et al., 1976; Simon et al., 1975); oxytetracycline, 8 to 10 hr (Scales and Assinder, 1973); tetracycline, 8 to 9 hr (Neuvonen et al., 1976). Bioavailability of orally administered tetracyclines may vary considerably with the pharmaceutical dosage formulation (e.g., Brice and Hammer, 1969; Rhaguram and Krishnaswami, 1977).

c. Analogous Compounds. Other tetracyclines (e.g., β-cetotetrine, demethylchlortetracycline, chlortetracycline, rolitetracycline).

d. Analytical Methods

1. Fluorescence—Kelly et al. (1969)

Discussion. This tetracycline assay is representative of a series of similar methods for the tetracycline congeners in plasma. Most of the tetracyclines possess only weak native fluorescence; however, the addition of chelating cations to make certain bonds rigid enhances the fluorescent properties. In addition, Kelly et al. (1969) utilize acid dehydration of tetracycline to form 6-anhydrotetracycline with a still higher fluorescence yield in its aluminum chelate form. The long excitation wavelength of the aluminum-anhydrotetracycline complex (475 nm) provides a high degree of specificity, since few endogenous substrates produce fluorescence under these assay conditions. An exception is riboflavin with an activation maximum at 450 nm. The tetracycline assay presented is sensitive to 0.1 μg/ml plasma, and plasma level results obtained in dogs after intravenous doses of ^{3}H-tetracycline correlate well with ^{3}H determinations and with a microbiological assay (correlation coefficient-.980).

Detailed Method. Plasma (0.2 ml), 9 ml water, and 1 ml 30% trichloroacetic acid are thoroughly mixed and centrifuged. The supernatant (8 ml) is transferred to another test tube and heated with 1 ml 5 N HCl in a boiling water bath for 30 min. After cooling, 1 ml 6 N NaOH and 1 ml 1 M sodium citrate buffer (pH 4.5) are added to give a pH between 3.5 and 5.5, and the resultant mixture is extracted with 2 ml chloroform. After centrifuging, 1 ml of the chloroform layer is transferred to another tube, 1 ml 0.1% $AlCl_3 \cdot 6H_2O$ in absolute ethanol is added, and the solution is allowed to stand for at least 1 hr. Fluorescence is read on an Aminco Bowman spectrophotofluorometer with an excitation wavelength of 475 nm and an emission wavelength of 550 nm.

2. Other Fluorescence Assays—Day et al. (1978), Gavend et al. (1972), Hall (1976, 1977), Kohn (1961), Lever (1972), Murthy and Goswami (1973), Scales and Assinder (1973), Schatz (1973), Wilson et al. (1972)

Discussion. Kohn (1961) first observed the ability of tetracyclines to form organic-solvent-extractable, highly fluorescent complexes with Ca^{2+} and barbiturates. Most tetracyclines can be measured with a sensitivity of 0.1 $\mu g/ml$ plasma, although oxytetracycline gives only a poor fluorescence yield under the assay conditions. Among the potential causes of interference of such fluorescence assays are certain other polyvalent cations and phosphate which may precipitate the tetracyclines. Kohn (1961) includes a phosphate precipitation with Pb^{2+} to prevent tetracycline losses due to precipitation. Subsequent reports deal with slightly varying assay conditions to optimize the procedure for individual tetracyclines (doxycycline: Gavend et al., 1972; minocycline: Hall, 1977; oxytetracycline: Murthy and Goswami, 1973; Scales and Assinder, 1973; tetracycline: Schatz, 1973; several tetracyclines: Hall, 1976; Lever, 1972; Wilson, 1972). In addition to Ca^{2+} and Al^{3+}, magnesium chelates in the presence of sodium barbital have also been used for several tetracyclines (Lever, 1972). The magnesium chelate of minocycline gives high fluorescence yields only in the presence of base and citric acid instead of barbital. Hall (1977) employs an aluminum-minocycline chelate. The specific detection of several tetracyclines in the same plasma sample can be achieved without organic solvent extraction by measuring the fluorescence of aluminum chelates of the free tetracycline bases and of the acid-generated anhydro forms (Hall, 1976).

The mechanism of the fluorimetric analysis of tetracycline involving the formation of a terniary calcium complex with barbital sodium and tryptophane has been studied by Day et al. (1978). On the basis of the characteristics of the terniary complexes and of the tetracycline degradation products, these authors conclude that only the active form of tetracycline can be complexed and extracted for fluorescence analysis (e.g., by the method of Kohn, 1961).

The problem of photocatalyzed oxidative decomposition of tetracyclines in stored plasma samples can be overcome by adding mercaptopropionic acid as the antioxidant; ascorbic acid cannot be used because of interference with the fluorescence assays (Scales and Assinder, 1973).

3. High Performance Liquid Chromatography—Magic (1976), Nilsson-Ehle et al. (1976), Sharma and Bevill (1978), Sharma et al. (1977a, 1977b), White et al. (1975)

Discussion. Good correlation between a specific tetracycline HPLC assay and a microbiological assay of plasma samples indicates the absence of active metabolites (Nilsson-Ehle et al., 1976). Sharma et al. (1977a) measure chlortetracycline, oxytetracycline, and tetracycline by anion-exchange HPLC-UV assay in urine after organic solvent extraction as their calcium chelates. The sensitivity is only 4 to 12 μg/ml, which is sufficient for urine, but not plasma, samples. The same group (Sharma et al., 1977b) also proposes a similar assay employing octadecylsilane reverse phase HPLC with an improved sensitivity of 1 μg/ml urine and 1.5 μg/ml plasma for these tetracyclines. The addition of ethylenediamine tetraacetic acid (0.005 M) to the reverse phase HPLC eluent may prevent peak tailing caused by complexing of the tetracyclines with the metal tubing (White et al., 1975). Sharma and Bevill (1978) describe an improved procedure for extracting tetracyclines from urine and plasma, involving the formation of a phenylbutazone-tetracycline ion pair that is readily extractable into ethyl acetate. The increased extraction yield results in increased sensitivity of the HPLC assay.

The reverse phase HPLC assay of β-cetotetrine (Magic, 1976) is several orders of magnitude more sensitive because of thin-layer electrochemical detection. The minimum detectable amount injected onto the column is 100 pg, while the assay sensitivity in plasma is 25 ng/ml.

4. Microbiological Assays—Altmann et al. (1968), Braito and Bassetti (1968), Brice and Hammer (1969), Dormia Cristallo et al. (1972)

Discussion. These are selected reports on the microbiological assay of tetracyclines in plasma and tissues (doxycycline: Braito and Bassetti, 1968; Dormia Cristallo et al., 1972; oxytetracycline: Brice and Hammer, 1969; tetracycline: Altmann et al., 1968). Microbiological assays and chemical assays (fluorescence: Altmann et al., 1968) usually yield equivalent results, indicating the absence of active metabolites. The assay of Brice and Hammer (1969) has been used to demonstrate the bioinequivalence of several oxytetracycline pharmaceutical preparations.

5. **Comments.** Fluorescence assays appear to be sufficiently specific and sensitive for most biomedical applications. The excellent procedure of Magic (1976), employing HPLC with electrochemical detection of β-cetotetrine, may find more general applicability in pharmacological and toxicological research on the tetracyclines.

e. References

Altmann, A. E. et al., Serum Levels of Tetracycline after Different Administration Forms, *Clin. Chim. Acta*, **20**, 185–188 (1968).

Braito, A. and D. Bassetti, Determination of Doxycycline and Metacycline Blood Levels in Children Following a Single Oral Administration, *Minerva Pediatr.*, **20**, 1580–1584 (1968).

Brice, G. W. and H. F. Hammer, Therapeutic Nonequivalence of Oxytetracycline Capsules, *J.A.M.A.*, **208**, 1189–1190 (1969).

Day, S. T., W. G. Crouthamel, L. C. Martinelli, and J. K. H. Ma, Mechanism of Fluorometric Analysis of Tetracycline Involving Metal Complexation, *J. Pharm. Sci.*, **67**, 1518–1523 (1978).

Dormia Cristallo, L., F. Craveri, and V. De Pascale, Microbiological Determination of Hematic, Urinary, Biliary and Tissue Levels of Doxycycline in Pigs after Single-Dose and Prolonged Administration, *Boll. Chim. Farm.*, **111**, 777–784 (1972).

Gavend, M. et al., Spectrofluorimetric Study of Plasma Concentration and Urinary Excretion of Intravenously Administered Doxycycline, *Therapy*, **27**, 969–973 (1972).

Hall, D., Fluorimetric Assay of Tetracycline Mixtures, *J. Pharm. Pharmacol.*, **28**, 420–423 (1976).

Hall, D., Rapid Fluorimetric Assay of Minocycline in Plasma or Serum: Comparison with Microbiological Assay, *Br. J. Clin. Pharmacol.*, **4**, 57–60 (1977).

Kelly, R. G., L. M. Peets, and K. D. Hoyt, A Fluorometric Method of Analysis for Tetracycline, *Anal. Biochem.*, **28**, 222–229 (1969).

Kohn, K. W., Determination of Tetracyclines by Extraction of Fluorescent Complexes: Application to Biological Material, *Anal. Chem.*, **33**, 862–866 (1961).

Lever, M., Improved Fluorometric Determination of Tetracyclines, *Biochem. Med.*, **6**, 216–222 (1972).

Magic, S. E., Determination of Beta-Cetotetrine in Plasma and Urine Using High-Performance Liquid Chromatography with Electrochemical Detection, *J. Chromatogr.*, **129**, 73–80 (1976).

Murthy, V. V. and S. L. Goswami, A Modified Fluorimetric Procedure for the Rapid Estimation of Oxytetracycline in Blood, *J. Clin. Pathol.*, **26**, 548–550 (1973).

Neuvonen, P. J., O. Penttila, M. Roos, and J. Tirkkonen, Effect of Long-Term

Alcohol Consumption on the Half-Life of Tetracycline and Doxycycline in Man, *Int. J. Clin. Pharmacol. Biopharm.*, **14**, 303–307 (1976).

Nilsson-Ehle, I., T. T. Yoshikawa, M. C. Schotz, and L. B. Guze, Quantitation of Antibiotics Using High-Pressure Liquid Chromatography: Tetracycline, *Antimicrob. Agents Chemother.*, **9**, 754–760 (1976).

Raghuram, T. C. and K. Krishnaswamy, Influence of Nutritional Status on Plasma Levels and Relative Bioavailability of Tetracycline, *Eur. J. Clin. Pharmacol.*, **12**, 281–284 (1977).

Scales, B. and D. A. Assinder, Fluorometric Estimation of Oxytetracycline in Blood and Plasma, *J. Pharm. Sci.*, **62**, 913–917 (1973).

Schatz, F., Fluorometric Determination of Tetracycline in the Serum and Urine, *Arzneim.-Forsch.*, **23**, 426–428 (1973).

Sharma, J. P. and R. F. Bevill, Improved High-Performance Liquid Chromatographic Procedure for the Determination of Tetracyclines in Plasma, Urine and Tissues, *J. Chromatogr.*, **166**, 213–220 (1978).

Sharma, J. P., G. D. Koritz, E. G. Perkins, and R. F. Bevill, High-Pressure Liquid Chromatographic Determination of Tetracyclines in Urine, *J. Pharm. Sci.*, **66**, 1319–1322 (1977a).

Sharma, J. P., E. G. Perkins, and R. F. Bevill, Reversed-Phase High-Performance Liquid Chromatographic Determination of Tetracyclines in Urine and Plasma, *J. Chromatogr.*, **134**, 441–450 (1977b).

Simon, C. et al., The Pharmacokinetics of Doxycycline in Kidney Insufficiency in Geriatric Patients Compared to Younger Adults, *Schweiz. Med. Wochenschr.*, **105**, 1615–1620 (1975).

White, E. R., M. A. Carroll, J. E. Zarembo, and A. D. Bender, Reverse Phase High-Speed Liquid Chromatography of Antibiotics, *J. Antibiot.*, **28**, 205–214 (1975).

Wilson, D. M., M. Lever, E. A. Brosnan, and A. Stillwell, A Simplified Tetracycline Assay, *Clin. Chim. Acta*, **36**, 260–261 (1972).

TETRAHYDROCANNABINOL—by B. Silber

Tetrahydro-6,6,9-trimethyl-3-pentyl-6H-dibenzo(b,d)pyran-1-ol. THC (Δ^9-THC). Psychotropic agent.

$C_{21}H_{30}O_2$; mol. wt. 314.45; pK_a 10.6.

a. Pharmacologic Concentration Range. Pharmacological effects after smoking marijuana are associated with plasma levels in the range of 1 to 50 ng THC/ml plasma (Vinson et al., 1977).

b. Metabolism. Hydroxylation reactions occur at various positions of the THC moiety, including C-8 and C-11. The C-11-OH metabolite is further metabolized to the carboxylic acid (Lemberger et al., 1970). The plasma elimination of THC in human beings is biphasic; the alpha and beta half-lives are approximately 30 min and 50 to 60 hr, respectively (Lemberger et al., 1971). The prolonged terminal phase of THC elimination is caused by a slow release of the drug from fat tissue depots (Garrett and Hunt, 1977b). Little unchanged THC is excreted into the urine. The pharmacological contributions of metabolites to THC effects are poorly understood.

c. Analogous Compound. Cannabidiol.

d. Analytical Methods

1. High Performance Liquid Chromatography-Gas Chromatography—Garrett and Hunt (1977a)

Discussion. The procedure of Garrett and Hunt (1977a) utilizes silica gel HPLC for initial separation of THC from metabolites and other potentially interfering compounds. The THC eluent fraction is collected and derivatized with pentafluorobenzoate, and quantitation is accomplished by electron capture detection. Cannabidiol is selected as the internal standard because of its structural similarity to THC. Also, it is neither a metabolite nor a contaminant of THC.

Detailed Method. Two milliliters plasma is transferred to a glass tube (all glassware is silanized) and alkalinized to pH 9.5 to 11.0 by the addition of 0.2 to 0.5 ml 0.1 N Na$_2$CO$_3$. The volume is adjusted to 2.5 ml with water purified by HPLC, using a Bondapak C$_{18}$ Corasil column. After 15 ml heptane containing 1.5% isoamyl alcohol is added, the solution is vortexed for 15 min and centrifuged. An aliquot (5 to 14 ml) of the upper organic layer is transferred to another glass tube and dried under nitrogen. The sides of the tube are then rinsed with 100 μl ethanol and again dried under nitrogen. The residue is dissolved in 15 to 20 μl ethanol, and an aliquot (10 to 14 μl) is used for the separation of THC by normal phase HPLC under the following conditions.

A Waters Model ALC 202 liquid chromatograph is used, equipped with a 6000 psi flow pump and a 254 nm UV detector. The column is a 30 cm μ-Porasil with SiOSiOH surface functionality designed for normal phase chromatography. The isomers Δ^9-THC and Δ^8-THC can be separated using 5% tetrahydrofurane in hexane at a flow rate of 0.5 ml/min (A. C. Hunt, unpublished). The THC is eluted with 20% chloroform in heptane at a rate of 1.5 ml/min. The fraction eluting between 8 and 12 ml is collected in a glass tube and dried under nitrogen. The residue is dissolved in 0.5 ml ethanol and stored at $-20°C$ until further analysis. Samples are then dried under nitrogen in a 50°C water bath. The residue is dissolved in 0.1 ml dry benzene. A 100 x M excess of pyridine in benzene is added, and the mixture allowed to stand for 15 min; then to it is added 100 x M excess of pentafluorobenzoyl chloride in 5 to 20 μl benzene. After this solution is allowed to stand for 15 min, 100 μl 0.1 M Na_2CO_3 is added. After vortexing for 1 min and then centrifuging, the aqueous layer is removed, and 200 μl benzene is added, vortexed, and centrifuged. A 250 μl aliquot of the benzene layer is transferred to a 5 ml centrifuge tube and dried in a water bath. The residue is dissolved in toluene (0.2 to 2.0 ml) containing 50 pg cannabidiol pentafluorobenzoate (internal standard)/ml. Aliquots of 0.5 to 1.5 μl of this final solution are then analyzed by GC-EC under the following conditions.

A Varian Aerograph Model 1200 is used, fitted with a ^{63}Ni electron capture detector set at 270°C. The glass column, 30 cm x 2.0 mm (i.d.), is packed with 3% OV-17 on 100/120 mesh Gas Chrom Q. The glass column and injector port temperatures are set at 210 to 230 and 245°C, respectively. The nitrogen gas flow is set at 35 to 45 ml/min. Preparation of the cannabidiol internal standard is accomplished first by its purification, using normal phase HPLC (inject 1 mg). The cannabidiol peak is collected and dried, and then dissolved in 10 ml dry benzene. Cannabidiol pentafluorobenzoate is prepared by adding 2 μl pyridine and pentafluorobenzoate chloride to 1 ml of this solution. After 5 min, 1 ml of a 0.1 M Na_2CO_3 solution is added; the solution is shaken for 1 min and centrifuged. The organic layer is separated, transferred, and dried under nitrogen. This stable residue is dissolved in dry toluene to give concentrations between 50 and 500 pg/ml when needed for use as an internal standard. Standard curves with known amounts of THC added to control plasma are similarly processed. Peak area ratios for the standard curve are utilized to determine the concentration of THC in plasma. Preparation of the standard curve must account for dilutions, reconstitutions, and efficiencies of the extraction and HPLC elution procedures. The standard curve is linear in the range of 1 to 100 ng/ml.

2. Other GC Methods—Fenimore et al. (1973), McCallum (1973)

Discussion. Fenimore et al. (1973) describe a GC-EC assay for THC in plasma with a sensitivity of 100 pg/ml, using a 10 ml sample volume. After organic solvent extraction, THC and its internal standard, hexahydrocannabinol, are detected as the heptafluorobutyrates on a dual-column–dual-oven gas chromatograph, using a capillary column as the final resolving component. The method is presumably specific for THC, which was measured in rat plasma after injection of 0.1 mg/kg.

McCallum (1973) proposes a GC method with flame photometric detection of a C-1–OH diethyl phosphate derivative. The sensitivity of this technique (2 ng THC/ml plasma) is inferior to that of GC-EC detection; however, the author states that flame photometric detection offers good sensitivity with an exceptionally stable baseline, making it more amenable to routine analysis.

3. Gas Chromatography-Mass Spectrometry—Agurell et al. (1973), Rosenfeld et al. (1974), Rosenthal et al. (1978), Valentine et al. (1977)

Discussion. The method of Agurell et al. (1973) includes a light petroleum extraction from plasma and further purification of THC and dideutero-THC as the internal standard by column chromatography, using Sephadex LH-20. The GC-MS analysis is performed with the underivatized extract, using selected ion monitoring (m/e 299 and 301) under electron impact ionization. Rosenthal et al. (1978) present a similar GC-MS method with a sensitivity of 500 pg THC/ml plasma. These authors also describe chemical ionization of the pentafluoropropionate derivative of THC, which, however, offers no advantage over the electron impact method.

Purification of biological extracts by gradient elution HPLC prior to GC-MS (electron impact) is used by Valentine et al. (1977), who achieve a sensitivity of 2.5 ng THC/ml plasma. Rosenfeld et al. (1974) introduce a chemical derivatization step by methylation of the phenolic C-1–OH. The internal standard is C-1-O-CD$_3$-THC. This type of internal standard is readily synthesized not only from THC, but also from its many metabolites bearing the free phenolic OH function.

4. Thin-Layer Chromatography-Fluorescence—Vinson et al. (1977)

Discussion. Following a double back-extraction of THC from urine or serum, a derivative is formed at the phenolic C-1-OH with 2-*p*-chlorosulfophenyl-3-phenylindone. The sulfonyl derivative is chromatographed on thin-layer plates, which are subsequently treated with methoxide to produce a fluorescent spot. Detection limit is 0.2 ng THC/ml in a sample of 5 ml serum.

5. Radioimmunoassay—Gross et al. (1974), Teale et al. (1975)

Discussion. These assays utilize antibodies with varying degrees of specificity for THC. Antibodies were raised in sheep by immunizing with a conjugate of THC hemisuccinate and bovine serum albumin (Teale et al., 1975). Sensitivity is in the range of 50 pg THC per sample incubation, or 7.5 and 1.0 ng THC/ml plasma and urine, respectively (Teale et al., 1975). Without the use of concurrent chromatography, RIA methods are not specific for THC in the presence of its metabolites.

6. Homogeneous Enzyme Immunoassay—Rodgers et al. (1978)

Discussion. Malate dehydrogenase is labeled with a derivative of THC. The antibodies are raised in sheep after injection of a bovine γ-globulin conjugate of the same THC derivative. Binding of the enzyme-THC conjugate to the antibody partially inhibits the enzyme reaction, which can be monitored by UV absorbance at 340 nm of the cofactor NADH. Competition of the enzyme-THC conjugate and free THC for antibody binding sites gives a displacement curve that permits quantitation of amounts of THC as low as 15 ng/ml urine. Specificity is equivalent to that observed with RIA method, and the homogeneous enzyme immunoassay lends itself to rapid, automated analysis of THC in biological specimens.

7. Comments.

Major difficulties arise in the analysis of THC in biological samples because of its very high lipophilicity, the presence of closely related analogs and isomers (Δ^9- and Δ^8-THC), the formation of a large number of metabolites of similar structure, and finally its presence at very low active concentrations in the body. Highly specific detection systems like selected ion monitoring GC-MS, therefore, require further prepurification steps to assure specific detection of unchanged THC. The method of choice depends on the biomedical problem to be solved; several viable alternatives are presented here. Care has to be given to the proper storage of THC samples, which may rapidly sequester into rubber stoppers or similar lipophilic materials.

e. References

Agurell, S. et al., Quantitation of Δ^1-Tetrahydrocannabinol in Plasma from Cannabis Smokers, *J. Pharm. Pharmacol.*, **25**, 554–558 (1973).

Fenimore, D. C., R. R. Freeman, and P. R. Loy, Determination of Δ^9-Tetrahydrocannabinol in Blood by Electron Capture Gas Chromatography, *Anal. Chem.*, **45**, 2331–2335 (1973).

Garrett, E. R. and C. A. Hunt, Separation and Analysis of Δ^9-THC in Biological Fluids by High-Pressure Liquid Chromatography and GLC, *J. Pharm. Sci.*, **66**, 20–26 (1977a).

Garrett, E. R. and C. A. Hunt, Pharmacokinetics of Δ^9-THC in Dogs, *J. Pharm. Sci.*, **66**, 395–407 (1977b).

Gross, S. J., J. R. Soares, S-L. R. Wong, and R. E. Schuster, Marijuana Metabolites Measured by a Radioimmune Technique, *Nature*, **252**, 581–582 (1974).

Lemberger, L., S. D. Silberstein, J. Axelrod, and I. J. Kopin, Marijuana Studies on the Disposition and Metabolism of Delta-9-Tetrahydrocannabinol in Man, *Science*, **170**, 1320–1322 (1970).

Lemberger, L., J. Axelrod, and I. J. Kopin, Metabolism and Disposition of the Tetrahydrocannabinol in Naive Subjects and Chronic Marijuana Users, *N.Y. Acad. Sci. Ann.*, **119**, 142–152 (1971).

McCallum, N. K., The Measurement of Cannabinols in the Blood by Gas Chromatography, *J. Chromatogr. Sci.*, **11**, 509–511 (1973).

Rodgers, R. et al., Homogeneous Enzyme Immunoassay for Cannabinoids in Urine, *Clin. Chem.*, **24**, 95–100 (1978).

Rosenfeld, J. J., B. Bowins, J. Roberts, J. Perkins, and A. S. Macpherson, Mass Fragmentographic Assay for Δ^9-Tetrahydrocannabinol in Plasma, *Anal. Chem.*, **46**, 2232–2234 (1974).

Rosenthal, D., T. M. Harvey, J. T. Bursey, D. R. Brine, and M. E. Wall, Comparison of Gas Chromatographic Mass Spectrometry Methods for the Determination of Δ^9-Tetrahydrocannabinol in Plasma, *Biomed. Mass Spectrom.*, **5**, 312–316 (1978).

Teale, J. D., E. J. Forman, L. J. King, E. M. Piall, and V. Marks, The Development of a Radioimmunoassay for Cannabinoids in Blood and Urine, *J. Pharm. Pharmacol.* **27**, 465–472 (1975).

Valentine, J. L. et al., High-Pressure Liquid Chromatographic-Mass Spectrometric Determination of Δ^9-Tetrahydrocannabinol in Human Plasma Following Marijuana Smoking, *J. Pharm. Sci.*, **66**, 1263–1266 (1977).

Vinson, J. A., D. D. Patel, and A. H. Patel, Detection of Tetrahydrocannabinol in Blood and Serum Using a Fluorescent Derivative and Thin-Layer Chromatography, *Anal. Chem.*, **49**, 163–165 (1977).

THEOPHYLLINE

1,3-Dimethylxanthine.
Theal, Theolix, Theocin, and so on; Amesec (Aminophyllin). Bronchodilator.

$C_7H_8N_4O_2$; mol. wt. 180.17; pK_a 8.77 and 0.7.

a. Therapeutic Concentration Range. From 5 to 20 μg/ml (Mitenko and Ogilvie, 1973). Saliva concentrations of theophylline are about 50% of serum levels and are similar to the concentration of free (not protein-bound) drug in the serum (Koysooko et al., 1974). Toxic symptoms occur normally above 25 μg/ml and are usually absent below 15 μg/ml (Jacobs et al., 1976).

b. Metabolism. Major metabolic pathways include N-1,3 demethylation and C_8 oxidation to uric acid derivatives (Jenne et al., 1976). The predominant metabolites are 3-methylxanthine (36 ± 7% of the dose), 1-methyluric acid (16.5 ± 3.3%), and 1,3-dimethyluric acid (40 ± 5%) (Ogilvie, 1978). 3-Methylxanthine is pharmacologically active but between 1 and 5 times less potent than theophylline (Ogilvie, 1978). There is no predictable relationship between daily doses and theophylline serum concentrations among different patients (Jacobs et al., 1976). The plasma elimination half-life ranges from 2.9 to 20.7 hr (mean 5 to 6 hr), the total body clearance from 0.004 to 0.2l/kg hr, and the maintenance dose for the treatment of broncho-obstruction from 0.1 to 0.9 mg/kg·hr (Ogilvie, 1978). Theophylline serum level monitoring is considered to be a valuable aid in guiding the effective and safe usage of theophylline (Jacobs et al., 1976; Hendeles et al., 1978).

c. Analogous Compounds. Xanthines (e.g., caffeine, theobromine, monomethylxanthine, mono- and dihydroxypropyltheophylline,

7-theophyllineacetic acid, β-hydroxyethyltheophylline, 8-chlorot-heophylline); purines.

d. Analytical Methods

1. High Performance Liquid Chromatography—Adams et al. (1976), Dungan et al. (1978), Orcutt et al. (1977), Robinson et al. (1978)

Discussion. This assay (Orcutt et al., 1977) involves direct injection of serum samples on a reversed phase column following protein precipitation and centrifugation, and UV detection at 254 nm. Hydroxyethyltheophylline serves as the internal standard, although satisfactory results can also be achieved using the external standard method. The method is reproducible (C.V. 2.2%, $n = 10$), specific in the presence of other xanthines (with the exception of 1,7-dimethylxanthine), and sensitive to about 1 μg/ml, using 30 μl serum. Dungan et al. (1978) evaluated the specificity of this assay in 128 patient serum samples by using dual-wavelength measurements at 254 and 280 nm. Only one sample had a 254/280 ratio different from that of pure theophylline, indicating assay interference. Assay interferences can be caused by citrated plasma, ampicilline, and methicilline, as well as cefazoline and cephalothine (Kelly et al., 1978). Interferences by the last two agents may not be detectable by dual-wavelength measurements because their UV spectra are similar to the spectrum of theophylline. A preextraction into organic solvents can prevent these potential interferences (e.g., Adams et al., 1976); however, it has not yet been clarified whether this additional step is necessary for therapeutic drug level monitoring. Robinson et al. (1978) report no theophyllin interference by cephalosporins in their similar HPLC assay.

Detailed Method. Heparinized plasma (30 μl) and 30 μl of a solution of β-hydroxyethyltheophylline (36.7 μg/ml in 90 mM sodium acetate buffer (pH 4)-acetonitrile, 43:7) are mixed and centrifuged, and 5 to 9 μl of the supernatant is injected into the chromatograph. The HPLC analysis was performed on a liquid chromatograph (Waters Assoc., M-6000A) equipped with a 30 cm × 4 mm (i.d.) μ-Bondapak C_{18} reverse phase column (Waters Assoc.) and a 254 nm UV detector. A solution of 10 mM sodium acetate buffer (pH 4)-acetonitrile (93:7) serves as eluent at a flow rate of 2 ml/min and 1500 psi.

2. Other Reverse Phase HPLC Methods—Adams et al. (1976), Peng et al. (1978)

Discussion. The method of Adams et al. (1976) uses the same column as proposed by Orcutt et al. (1977); however, the internal standard is 8-chlorotheophylline, and the UV absorbance is measured at the UV maximum of theophylline (273 nm). An organic solvent extraction is included, providing for better specificity. No assay interferences were reported. Peng et al. (1978) also utilize detection at 275 nm, but no solvent extraction.

3. Additional HPLC Methods—Evenson and Warren (1976), Manion et al. (1974), Sitar et al. (1975), Weinberger and Chidsey (1975)

Discussion. These assays utilize silica gel adsorption columns (Sitar et al., 1975; Evenson and Warren, 1976), cation-exchange columns (Weinberger and Chidsey, 1975), and columns with oxypropionitrile bonded to Porasil C (Manion et al., 1974). They offer no advantage over reverse phase C_{18} bonded HPLC, and are more time consuming.

4. Gas Chromatography—Johnson et al. (1975), Kowblansky et al. (1973), Least et al. (1976), Perrier and Lear, (1976), Schwertner (1979), Shah and Riegelman (1974), Vinet and Zizian (1979)

Discussion. Gas chromatographic assays require organic solvent extraction and derivatization by alkylation such as propylation, butylation, and pentylation. Methylation converts mono- and dimethylxanthines into caffeine and therefore has to be avoided. With a nitrogen-sensitive FID, the sensitivity can be as good as 0.5 μg/ml with a serum sample of only 20 μl (Least et al., 1976). No assay interferences have been reported. Analysis time is longer than that of HPLC. However, a recent GC assay by Vinet and Zizian (1979) requires only 7 min total analysis time and 100 μl serum. The procedure involves no centrifugation or solvent evaporation after extraction. Theophylline and the internal standard, 3-isobutyl-1-methylxanthine, are injected onto the GC column, followed by an on-column butylating mixture containing 1-iodobutane. The compounds are measured with a nitrogen-phosphorus flame ionization detector. The method is precise, accurate, and free of interferences. It should be well suited for clinical applications. The rapid GC-FID analysis of theophylline on 2% SP 10-DA (Supelco) as the stationary phase can be performed on the underivatized drug with sensitivity equal to that of GC assays employing chemical derivatization (Schwertner, 1979). These studies indicate that GC assays of theophylline might be quite competitive with HPLC assays in the clinical laboratory.

5. Gas Chromatography-Mass Spectrometry—Horning et al. (1975), Sheehan et al. (1977)

Discussion. Because of the availability of sufficiently sensitive and specific alternative assays, GC-MS assays serve only as control procedures. Horning et al. (1975) use deuterated theophylline, and Sheehan et al. (1977) 3-isobutyl-1-methylxanthine, as the internal standard. Stable isotopic labels can also be useful to differentiate between exogenous (theophylline metabolites) and endogenous purines of identical chemical structure.

6. Homogeneous Enzyme Immunoassay (EMIT)—Castro et al. (1978), Henry et al. (1978), Stone and Gillilan (1978)

Discussion. The "enzyme multiplied immunoassay technique" (EMIT: Syva, Palo Alto, Calif.) for theophylline can be readily adapted to semiautomated procedures (Henry et al., 1978; Stone and Gillilan, 1978) and yields good precision and excellent correlation with standard methods using UV detection (Jatlow, 1975) and HPLC. The assay requires about 100 μl serum and can be completed in 20 min. The correlation coefficient of a fully automated system for the enzyme immunoassay of theophylline in serum, compared to a UV method, was .984 (Castro et al., 1978).

7. Nephelometric Immunoassay—Nishikawa et al. (1979)

Discussion. The principle of this assay is based on inhibition of the immunoprecipitation of a theophylline-labeled antigen-antitheophyllin antibody complex. The haptene theophylline competitively inhibits the immunoprecipitation reaction, an effect that can be measured by the decrease of scattered light (nephelometry). The assay is rapid (15 min incubation time), requires only 10 μl plasma, and can be performed without separation of antibody-bound from free antigen. The correlation coefficient with an HPLC method is .971.

8. Ultraviolet Absorption—Fellenberg and Pollard (1979), Jatlow (1975), Owen and Nakatsu (1978), Schack and Waxler (1949)

Discussion. The UV assays typically involve an organic solvent extraction and a back-extraction into NaOH. The UV absorbance is read at 275 nm. The classical method of Schack and Waxler (1949) is subject to serious assay interferences by other xanthines as well as barbiturates, particularly phenobarbital. Jatlow (1975) could eliminate phenobarbital interference by lowering the pH with NH_4Cl for UV absorbance measurements, which reverses the ionization of phenobarbital and causes a hypsochromic shift of its UV maximum. Owen and Nakatsu (1978) report on selective extractions of theophylline and theophylline-7-acetic acid in the same sample. Lack of specificity, however, remains the major problem when using these assays for therapeutic drug level monitoring. The extraction method of Fellenberg and Pollard (1979) removes most of the drug interferences, except for purines such as theobromine. The procedure employs chloroform extraction at pH 5, followed by back-extraction of an organic phase with 1 N HCl and UV absorbance measurement at 267 nm.

9. Comments. Direct injection reverse phase HPLC with dual-wavelength detection appears to be the method of choice for clinical applications at present. If further interferences are discovered, an organic solvent extraction may become necessary.

e. References

Adams, R. F., F. L. Vandemark, and G. J. Schmidt, More Sensitive High-Pressure Liquid-Chromatographic Determination of Theophylline in Serum, *Clin. Chem.*, **22**, 1903–1906 (1976).

Castro, A., J. Ibanez, W. Voight, T. Noto, and H. Malkus, A Totally Automated System for Enzyme Immunoassay of Theophylline in Serum, *Clin. Chem.*, **24**, 944–946 (1978).

Dungan, S. M., N. Powers, and D. K. Jansen, Quantitation of Theophylline in Serum by HPLC: An Evaluation of a Method Featuring Dual-Wavelength Detection, in *Biological/Biomedical Applications of Liquid Chromatography*, G. Hawk, Ed., Marcel Dekker, New York, in press (1978).

Evenson, M. A. and B. L. Warren, Serum Theophylline Analysis by High-Pressure Liquid Chromatography, *Clin. Chem.*, **22**, 851–855 (1976).

Fellenberg, A. J. and A. C. Pollard, A Rapid Ultraviolet Spectrophotometric Procedure for the Microdetermination of Theophylline (1,3-Dimethylxanthine) in Plasma or Serum, *Clin. Chim. Acta*, **92**, 267–272 (1979).

Hendeles, L., M. Weinberger, and G. Johnson, Monitoring Serum Theophyllin Levels *Clin. Pharmacokin.*, **3**, 294–312 (1978).

Henry, V., J. Deutsch, and G. Lum, Enzyme Immunoassay of Theophylline with a Centrifugal Analyzer, and Comparison with an Ultraviolet Method, *Clin. Chem.*, **24**, 514 (1978).

Horning, M. G. et al., Metabolic Switching of Drug Pathways as a Consequence of Deuterium Substitution, in *Proceedings of the 2nd International Conference on Stable Isotopes*, E. R. Klein and P. D. Klein, Eds., Oak Brook, Ill., 1975, pp. 41–54.

Jacobs, M. H., R. M. Senior, and G. Kessler, Clinical Experience with Theophyllin: Relationships Between Dosage, Serum Concentration, and Toxicity, *J.A.M.A.*, **235**, 1983–1986 (1976).

Jatlow, P., Ultraviolet Spectrophotometry of Theophylline in Plasma in the Presence of Barbiturates, *Clin. Chem.*, **21**, 1518–1520 (1975).

Jenne, J. W., H. T. Nagasawa, and R. D. Thompson, Relationship of Urinary Metabolites of Theophylline to Serum Theophylline Levels, *Clin. Pharmacol. Ther.*, **19**, 375–381 (1976).

Johnson, G. F., W. A. Dechtiaruk, and H. M. Solomon, Gas-Chromatographic Determination of Theophylline in Human Serum and Saliva, *Clin. Chem.*, **21**, 144–147 (1975).

Kelly, R. C., D. E. Prentice, and G. M. Hearne, Cephalosporin Antibiotics Interfere with the Analysis for Theophylline by High-Performance Liquid Chromatography, *Clin. Chem.*, **24**, 838 (1978).

Kowblansky, M., B. M. Scheintau, G. D. Cravello, and L. Chavetz, Specific Gas Chromatographic Determination of Xanthines and Barbiturates by Flash-Heater *N*-Butylation, *J. Chromatogr.*, **76**, 467–479 (1973).

Koysooko, R., E. F. Ellis, and G. Levy, Relationship Between Theophylline Concentration in Plasma and Saliva of Man, *Clin. Pharmacol. Ther.*, **15**, 454–460 (1974).

Least, C. J., G. F. Johnson, and H. M. Solomon, Gas-Chromatographic Micro-Scale Procedure for Theophylline, with Use of a Nitrogen-Sensitive Detector, *Clin. Chem.*, **22**, 765–768 (1976).

Manion, K. V., D. W. Shoeman, and D. L. Azarnoff, High-Pressure Liquid Chromatographic Assay of Theophylline in Biological Fluids, *J. Chromatogr.*, **101**, 169–174 (1974).

Mitenko, P. A. and R. I. Ogilvie, Rational Intravenous Doses of Theophylline, *N. Engl. J. Med.*, **289**, 600–603 (1973).

Nishikawa, T., H. Kubo, and M. Saito, Competitive Nephelometric Immunoassay of Theophylline in Plasma, *Clin. Chim. Acta*, **91**, 59–65 (1979).

Ogilvie, R. I., Clinical Pharmacokinetics of Theophylline, *Clin. Pharmacokin.*, **3**, 267–293 (1978).

Orcutt, J. J., P. P. Kozak, S. A. Gillman, and L. H. Cummins, Micro-Scale Method for Theophylline in Body Fluids by Reversed-Phase, High-Pressure Liquid Chromatography, *Clin. Chem.*, **23**, 599–601 (1977).

Owen, J. A. and K. Nakatsu, Spectrophotometry of Theophylline-7-acetic Acid and Theophylline, *Clin. Chem.*, **24**, 367–368 (1978).

Peng, G. W., M. A. F. Gadalla, and W. L. Chiou, High-Performance Liquid-Chromatographic Determination of Theophylline in Plasma, *Clin. Chem.*, **24**, 357–360 (1978).

Perrier, D. and E. Lear, Gas-Chromatographic Quantitation of Theophylline in Small Volumes of Plasma, *Clin. Chem.*, **22**, 898–900 (1976).

Robinson, C. A., Jr., B. Mitchell., J. Vasiliades, and A. L. Siegel, Cephalosporin Antibiotic Interference with Analysis for Theophylline by High-Performance Liquid Chromatography, *Clin. Chem.*, **24**, 1847 (1978).

Schack, J. A. and S. H. Waxler, An Ultraviolet Spectrophotometric Method for the Determination of Theophylline and Theobromine in Blood and Tissues, *J. Pharmacol. Exp. Ther.*, **97**, 283–291 (1949).

Schwertner, H. A., Analysis of Underivatized Theophylline by Gas-Chromatography on a Silicone Stationary Phase, SP-25 10-DA, *Clin. Chem.*, **25**, 212–214 (1979).

Shah, V. P. and S. Riegelman, GLC Determination of Theophylline in Biological Fluids, *J. Pharm. Sci.*, **63**, 1283–1285 (1974).

Sheehan, M., R. H. Hertel, and C. T. Kelly, Gas-Chromatographic/Mass-Spectrometric Determination of Theophylline in Whole Blood, *Clin. Chem.*, **23**, 64–68 (1977).

Sitar, D. S., K. M. Piafski, R. E. Rangno, and R. I. Ogilvie, Plasma Theophylline Concentrations Measured by High Pressure Liquid Chromatography, *Clin. Chem.*, **21**, 1774–1776 (1975).

Stone, H. and B. Gillilan, Enzyme Immunoassay of Theophylline with the ABA-100, *Clin. Chem.*, **24**, 520 (1978).

Vinet, B. and L. Zizian, Improved Determination of Serum Theophylline by Gas Chromatography with Use of a Nitrogen-Phosphorus Detector, *Clin. Chem.*, **25**, 156–158 (1979).

Weinberger, M. and C. Chidsey, Rapid Analysis of Theophylline in Serum by Use of High-Pressure Cation-Exchange Chromatography, *Clin. Chem.*, **21**, 834–837 (1975).

TOLBUTAMIDE—by M. Raeder-Schikorr

1-Butyl-3-(*p*-tolylsulfonyl)urea.
Diabetamid, Diabetoral, Orinase, Rastinon, Tolbutone. Hypoglycemic agent.

CH_3—⟨benzene⟩—SO_2NH—$\overset{O}{\overset{\|}{C}}$—$NH$—$CH_2CH_2CH_2CH_3$

$C_{12}H_{18}N_2O_3S$; mol. wt. 270.35; pK_a 5.43.

a. Therapeutic Concentration Range. From 50 to 100 μg/ml plasma (Nation et al., 1978; Pond et al., 1977).

b. Metabolism. Tolbutamide is mainly metabolized at the *p*-methyl group to hydroxytolbutamide as the rate limiting step in the elimination of the drug from the body. Further oxidation to carboxytolbutamide (1-butyl-3-(*p*-carboxyphensulfonyl)urea) and urinary excretion of this final metabolite are rapid and account for practically all of the administered dose of tolbutamide (Weinkam et al., 1977; Williams et al., 1977). The plasma elimination half-life in normal human adults is 4 to 6 hr, and the apparent volume of distribution 0.15 l/kg (Williams et al., 1977).

c. Analogous Compounds. Oral antidiabetics (e.g., acetohexamide, chlorpropamide, tolazamide); probenecid.

d. Analytical Methods

1. High Performance Liquid Chromatography—Sved et al. (1976)

Discussion. A specific and sensitive reverse phase HPLC determination of tolbutamide in plasma in the presence of its metabolites is described. The plasma sample is extracted with ether at acidic pH and analyzed on a reverse phase column with a UV absorbance detector at 254 nm. Tolbutamide metabolites are separated from the parent drug and do not interfere with the assay. Chlorpropamide serves as the internal standard. The procedure is also applicable to the analysis of chlorpropamide, and the two drugs serve as mutual internal standards. Sensitivity limits are 5 μg tolbutamide/ml and 0.5 μg chlorpropamide/ml plasma. This difference in sensitivity is due to a bathochromic shift and increase in intensity of the UV absorbance maximum of tolbutamide (230 nm) by Cl substitution of the *p*-methyl substituent in chlorpropamide ($\epsilon_{232.5 \text{ nm}}$ = 16,500 in 0.01 N HCl).

Detailed Method. One milliliter plasma is mixed with 0.25 ml internal standard solution (4 mg chlorpropamide in 1 ml 0.1 N NaOH diluted with water to a total volume of 50 ml) and acidified to pH 3 by the addition of 1 ml 0.15 M phosphoric acid. The drugs are extracted with 10 ml ether and centrifuged, and the ether layer is separated. The ether is evaporated, and the residue redissolved in 0.2 ml 0.1 M ammonium carbonate and 0.2 ml ethylene dichloride and mixed. Then 0.02 ml of the aqueous phase is chromatographed.

Reverse phase HPLC analysis is performed on silica gel-octadecylsilane support, using 17% acetonitrile in 0.05 M aqueous ammonium formate as the mobile phase, at a flow rate of 2.5 ml/min. The eluent is monitored with a fixed-wavelength (254 nm) UV absorbance flow cell detector. Standard curves are constructed by the peak height ratio method.

2. Other HPLC Methods—Nation et al. (1978), Weber (1976)

Discussion. The method of Weber (1976) is similar to that of Sved et al. (1976) with a reported sensitivity limit of 2 μg tolbutamide/ml plasma. The internal standard is N-(p-methoxybenzenesulfonyl)-N'-cyclohexylurea. Results obtained by HPLC and GC assays of tolbutamide in plasma show good agreement between these two procedures.

Nation et al. (1978) achieved a considerably increased UV detection sensitivity for tolbutamide using a variable-wavelength UV detector. Of the two UV absorbance maxima at 230 and 200 nm, the latter yields a 2.5-fold greater absorbance than the former. The plasma assay procedures involves acetonitrile protein precipitation, followed by direct HPLC analysis on a μ-Bondapak C_{18} reverse phase column and 200 nm UV absorbance detection. This technique permits the simultaneous detection of tolbutamide (retention time, 5 min) and carboxy-tolbutamide (retention time, 2.5 min) with a sensitivity of 0.5 μg/ml, using only 0.1 ml plasma. The specificity of this assay under clinical conditions remains to be evaluated, as the low UV detection wavelength increases the risk of interference by other chemicals.

3. Gas Chromatography—Prescott and Redmann (1972), Sabih and Sabih (1970), Simmons et al. (1972)

Discussion. All three reports describe the determination of tolbutamide in serum, blood, and urine after solvent extraction at low pH and N-methylation with dimethyl sulfate. The resulting N-methyltolbutamide can be analyzed by GC-FID with chlorpropamide as the internal standard. The sensitivity is as good as 0.1 μg tolbutamide/ml in plasma. Simmons et al. (1972) note that under the GC conditions tolbutamide partially pyrolyzes to N-methyl-p-toluenesulfonamide. Gas chromatographic-mass spectrometric analysis of N-methylated (diazomethane) tolbutamide confirmed the thermolytic cleavage to the free sulfonamide (Midha et al., 1976). This thermolytic reaction may make it difficult to reproduce previous GC assay results, although the internal standard, chlorpropamide, undergoes the same reaction. Therefore Simmons et al. (1972) propose high GC-inlet temperatures which afford quantitative pyrolysis of tolbutamide and chlorpropamide. Metabolic formation of N-methyl-p-toluenesulfonamide has also been reported and causes tolbutamide assay interference; however, this metabolite represents only a minor fraction of the total dose. Sabih and Sabih (1970) report that commonly used drugs (e.g., phenobarbital, pentobarbital, glutethimide, diphenylhydantoin) show no interfering peaks, and that the GC method can be used in clinical laboratories.

4. Mass Spectrometry—(Weinkam et al. (1977)

Discussion. Plasma and urine samples are acidified and extracted into ether after addition of dideuterated samples of each tolbutamide, hydroxytolbutamide, and carboxytolbutamide as the internal standards. Ether extracts are derivatized with diazomethane, in order to increase volatility and decrease thermolysis of underivatized tolbutamide, and are analyzed by direct insertion chemical ionization (isobutane) MS, which generates abundant MH^+ ions for all three compounds. The MH^+ m/e region is repetitively scanned, and the sulfonamides are quantitated by measuring MH^+ peak height ratios of the unlabeled to the labeled sulfonamide. Methylation of tolbutamide with diazomethane produces two isomers: the N-methyl and the O-methyl enol ether isomer (Midha et al., 1976), which are not differentiated by direct insertion chemical ionization MS because of formation of identical MH^+ ions. The assay has a sensitivity of at least 1 μg/ml and is specific and rapid.

5. Colorimetric Assay—Nelson and O'Reilly (1960)

Discussion. This method provides a quantitative determination of carboxytolbutamide in urine which may account for over 90% of the tolbutamide dose. Carboxytolbutamide is extracted from the acidic urine sample into amyl acetate. After adding dinitrofluorobenzene (0.1%) to the organic phase, the sample is heated for 5 min and measured at 380 nm. Standard curves constructed over the concentration range between 2 and 12 μg/ml are linear.

6. Comments.
Current HPLC assays are superior to GC analysis, since thermolytic reactions are avoided. However, lack of sufficient absorptivity above 254 nm dictates the use of a rather low UV wavelength (230 or 200 nm) for the sensitive detection of tolbutamide.

e. References

Midha, K. K., D. V. C. Awang, I. J. McGilveray, and J. Kleber, Gas Chromatographic and Mass Spectrometric Analysis of the Products of

Methylation of Sulfonylurea Drugs, *Biomed. Mass Spectrom.*, **3**, 100–109 (1976).

Nation, R. L., G. W. Peng, and W. L. Chiou, Simple, Rapid and Micro-High Pressure Liquid Chromatographic Method for the Simultaneous Determination of Tolbutamide and Carboxy Tolbutamide in Plasma, *J. Chromatogr.*, **146**, 121–131 (1978).

Nelson, E. and I. O'Reilly, Determination of Carboxytolbutamide in Urine, *Clin. Chim. Acta*, **5**, 774–776 (1960).

Pond, S. M., D. J. Birkett, and D. N. Wade, Mechanism of Inhibition of Tolbutamide Metabolism: Phenylbutazone, Oxyphenbutazone, Sulfaphenazole, *Clin. Pharmacol. Ther.*, **22**, 573–579 (1977).

Prescott, F. F. and D. R. Redmann, Gas Liquid Chromatographic Estimation of Tolbutamide and Chlorpropamide in Plasma, *J. Pharm. Pharmacol.*, **24**, 713–716 (1972).

Sabih, K. and K. Sabih, Gas Chromatographic Method for Determination of Tolbutamide and Chlorpropamide, *J. Pharm. Sci.*, **59**, 782–784 (1970).

Simmons, D. L., R. J. Ranz, and P. Picotte, Determination of Serum Tolbutamide by Gas Chromatography, *J. Chromatogr.*, **71**, 421–426 (1972).

Sved, S., I. J. McGilveray, and N. Beaudoin, Assay of Sulfonylureas in Human Plasma by High-Performance Liquid Chromatography, *J. Pharm. Sci.*, **65**, 1356–1359 (1976).

Weber, D. J., High-Pressure Liquid Chromatographic Analysis of Tolbutamide in Serum, *J. Pharm. Sci.*, **65**, 1502–1505 (1976).

Weinkam, R. J., M. Rowland, and P. J. Meffins, Determination of Phenylbutazone, Tolbutamide and Metabolites in Plasma and Urine Using Chemical Ionization Mass Spectrometry, *Biomed. Mass Spectrom.*, **4**, 42–47 (1977).

Williams, R. L., T. F. Blaschke, P. J. Meffin, K. L. Melmon, and M. Rowland, Influence of Acute Viral Hepatitis on Disposition and Plasma Binding of Tolbutamide, *Clin. Pharmacol. Ther.*, **21**, 301–309 (1977).

VALPROIC ACID—by G. M. Wientjes

2-Propylpentanoic acid. Sodium valproate used therapeutically. Depakene, Depakine, Epilim, Ergenyl. Antiepileptic drug.

$$CH_3CH_2CH_2-\underset{\underset{CH_3CH_2CH_2}{|}}{CH}-COOH$$

$C_8H_{15}O_2$; mol. wt. 144.21; pK_a 4.95.

a. Therapeutic Concentration Range. From 20 to 100 μg valproate/ml (Pinder et al., 1977; Meyer and Hessing-Brand, 1973; Vree and van der Kleijn, 1977).

b. Metabolism. Valproate is extensively metabolized according to general pathways of lower fatty acid metabolism. Major routes include ω-oxidation to 2-*n*-propylglutaric acid and glucuronidation, followed by renal excretion (review: Pinder et al., 1977; Vree and van der Kleijn, 1977). The mean plasma elimination half-life of valproate is 16 hr in normal human subjects (Gugler et al., 1977); capacity-limited elimination kinetics occur after administration of high doses (Vree and van der Kleijn, 1977). Klotz et al. (1977) reported plasma half-lives of 12 ± 4 hr, while interindividual plasma levels varied over a much wider range. Individual plasma levels should be titrated to about 50 to 100 μg valproate/ml by drug level monitoring. The drug level appears to remain rather constant over time with the same dosage regimen.

c. Analogous Compounds. Fatty acids.

d. Analytical Methods

1. Gas-Liquid Chromatography—Fellenberg and Pollard (1977)

Discussion. The assay procedure consists of extraction of acidified plasma with *n*-heptane (70% yield) and analysis of the extracted valproate, together with *n*-octanoic acid as the internal standard, by GC-FID. The detection limit is 1 μg valproate/ml plasma, while the intra-assay precision has a C.V. of 4.4% at about 50 μg valproate/ml plasma. The assay requires very small sample volumes (50 μl) with no significant interference from endogenous fatty acids.

Detailed Method. To a 300 μl vial are added 50 μl plasma, 10 μl 6 M H_2SO_4, and, after vortexing for 5 sec, 50 μl *n*-heptane containing 10 μg *n*-octanoic acid. After vortexing for 30 sec and centrifuging at 2000 g for 10 sec, the organic layer is transferred to another vial, and 5 μl is injected onto the GC column.

The GC analysis is performed on a Varian Aerograph Series 1400 fitted with a 1.5 m x 3 mm (i.d.) glass column packed with 10% diethylene glycol adipate and 2% phosphoric acid on Diatomite C 80/100 mesh. The GC temperatures are as follows: oven, 150°C; injection port, 190°C; and detector, 250°C. Flow rates are 30 (N_2), 50 (H_2), and 450 (air) ml/min. Retention times of valproate and *n*-octanoic acid are 4.5 and 7.5 min, respectively. Repetitive samples require only 8 min analysis time each, since no endogenous peaks appear after the internal standard peak. Peak area ratios are used for quantitation.

2. Other GC Methods—Meyer and Hessing-Brand (1973), Vree and van der Kleijn (1976, 1977)

Discussion. The procedure of Meyer and Hessing-Brand (1973) includes isothermal distillation of valproate and its internal standard, 3,3,5-tri-methylcaproic acid, from plasma at 90°C in a microdiffusion cell. The GC-FID analysis is performed on a glass column packed with 5% free fatty acid phase (FFAP) on 60/80 mesh Gas Chrom Q at 130°C.

Vree and van der Kleijn (1976) utilize a chloroform extraction of 200 μl plasma with cyclohexane carboxylic acid as the internal standard, and the same FFAP as suggested by Meyer and Hessing-Brand (1973).

Several additional variations of the valproate GC assay in plasma can be found in a monograph edited by Vree and van der Kleijn (1977).

3. Comments. The apparent lack of correlation between valproate doses and achieved plasma levels in epileptic patients, as well as nonlinear elimination kinetics, makes it desirable to monitor individual patients with routine valproate plasma level determinations. For the assay of valproate in plasma, GC-FID after organic solvent extraction is the method of choice.

e. References

Fellenberg, A. J. and A. C. Pollard, A Rapid and Sensitive Gas-Liquid Chromatographic Procedure for the Micro Determination of Sodium Valproate (Sodium Di-n-propylacetate) in Plasma or Serum, *Clin. Chim. Acta*, **81**, 203–208 (1977).

Gugler, R., A. Schell, M. Eichelbaum, W. Froscher, and H.-U. Schulz, Disposition of Valproic Acid in Man, *Eur. J. Clin. Pharmacol.*, **12**, 125–132 (1977).

Klotz, U., et al., Pharmacokinetics and Bioavailability of Sodium Valproate, *Clin. Pharmacol. Therap.*, **21**, 736–743 (1977).

Meyer, J. W. A. and L. Hessing-Brand, Determination of Lower Fatty Acids, Particularly the Anti-epileptic Dipropyl-Acetic Acid, In Biological Materials by Means of Micro Diffusion and Gas Chromatography, *Clin. Chim. Acta*, **43**, 215–222 (1973).

Pinder, R. M., R. N. Brogden, T. M. Speight, and G. S. Avery, Sodium Valproate: A Review of Its Pharmacological Properties and Therapeutic Efficacy in Epilepsy, *Drugs*, **13**, 81–123 (1977).

Vree, T. B. and E. van der Kleijn, Rapid Determination of 4-Hydroxybutyric Acid (Gamma OH) and 2-Propyl Pentanoate (Depakine) in Human Plasma by Means of Gas-Liquid Chromatography, *J. Chromatogr.*, **121**, 150–152 (1976).

Vree, T. B. and E. van der Kleijn, Eds., *Pharmacokinetics and Metabolism of the Antiepileptic Drug Sodium Valproate (Depakine, Epilim)*, Bohre, Scheltema and Holkema, Utrecht, 1977.

WARFARIN—by M. Cohen

3-(α-Acetonylbenzyl)-4-hydroxycoumarin.
Coumadin, Panwarfin (sodium salt). Anticoagulant.

$C_{19}H_{16}O_4$; mol. wt. 308.32; pK_a 4.78.

a. Therapeutic Concentration Range. From 1 to 10 μg/ml serum. However, plasma prothrombin activity, rather than warfarin plasma concentration, is used clinically to measure the anticoagulant effect. The drug is administered as the racemic mixture, although the S-enantiomer of warfarin is more active than the R-enantiomer (Breckenridge et al., 1974; Hewick and McEwen, 1973).

b. Metabolism. The keto function of warfarin is reduced to the corresponding four diastereoisomeric warfarin alcohols, 3-[α-(2-hydroxypropyl)benzyl]-4-hydroxycoumarin, which are less active than warfarin. Chan et al. (1972) indicate that the reduction is stereoselective in human beings. Hydroxylation at C_6 and C_7 yields inactive warfarin metabolites (Trager et al., 1970). The elimination half-life in human plasma is 45 hr for R-warfarin and 33 hr for S-warfarin (Breckenridge et al., 1974; Hewick and McEwen 1973). The potentially serious side effects of warfarin, which have been observed after adding additional drugs to the therapy, may be caused by an alteration of warfarin metabolism or by displacement of the tightly plasma protein-bound warfarin by other drugs (review: Kelly and O'Malley, 1979).

c. Analogous Compounds. Vitamin K inhibitors (e.g., coumarins: cyclocoumarol, dicumarol, phencoproumon).

d. Analytical Methods

1. High Performance Liquid Chromatography—Fasco et al. (1977)

Discussion. The reverse phase HPLC assay of Fasco et al. (1977) employs UV (313 nm) analysis of warfarin in plasma extracts. Plasma samples are deproteinized, washed with chloroform at alkaline pH, and directly injected onto the HPLC column. Dilute solutions up to a volume of 0.5 ml are concentrated on the column prior to elution to avoid peak spreading. The sensitivity for warfarin, its diastereomeric alcohols, and 4'-, 6-, 7-, 8-, and benzylic hydroxylated metabolites is approximately 0.02 μg/ml plasma.

Detailed Method. A 1.3 ml plasma sample is mixed with 0.2 ml saturated solution of sodium borate, vigorously shaken with 3 ml chloroform, and centrifuged at 20,000 g for 30 min. An appropriate volume of the aqueous phase is chromatographed.

The HPLC analysis is performed with a liquid chromatograph (Waters Assoc., Model 6000A), equipped with a three-channel solvent delivery system, a U6K injector, and a Model 440 absorbance detector equipped with a 313 nm filter. The UV absorption maximum of warfarin is at 308 nm in a pH 10 aqueous buffer. The column, 30 cm x 4 mm (i.d.), is packed with 10 μ chemically bonded silica (μ-Bondapak C_{18}, Waters Assoc.). Separation of warfarin and its metabolites is achieved with 1.5% acetic acid (adjusted to pH 4.7 with concentrated NH_3)-acetonitrile (69:31) as the eluent at a flow rate of 2.0 ml/min. Dilute solutions can be concentrated on-column before elution by a solvent of 1.5% acetic acid (adjusted to pH 4.7 with concentrated NH_3)-acetonitrile (9:1) at a flow rate of 2.0 ml/min. Concentrations of warfarin and its metabolites are assessed by comparing the resulting peak heights with those obtained from an external standard curve.

2. Other HPLC Methods—Bjornsson et al. (1977), Vessel (1974), Wong et al. (1977)

Discussion. The HPLC assays involve the extraction of warfarin from acidified plasma with organic solvents (e.g., ether, ethylene dichloride). The UV detection usually requires a variable-wavelength UV detector set at the warfarin absorbance maximum at 308 nm (Bjornsson et al., 1977). Internal standards include *p*-chlorowarfarin (Bjornsson et al.,

1977) and 4-hydroxymethylwarfarin (Wong et al., 1977). All of the stationary HPLC phases are reverse phase chemically bonded silica packings. Sensitivity for warfarin ranges from 0.1 to 0.5 μg/ml; warfarin metabolites are separated from the parent drug but not measured.

3. Gas Chromatography—Kaiser and Martin (1974), Midha et al. (1974)

Discussion. The assay of Kaiser and Martin (1974) includes extraction of warfarin into ethylene chloride, purification by TLC, derivatization with pentafluorobenzyl bromide, and GC with electron capture detection. With p-chlorowarfarin as the internal standard, the assay is specific for warfarin and sensitive to 0.02 μg warfarin/ml plasma.

Midha et al. (1974) use a double back-extraction followed by diazomethane derivatization and GC with flame ionization detection. With phenylbutazone as the internal standard, the technique is sensitive to 0.25 μg warfarin/ml plasma.

4. Fluorescence—Lewis et al. (1970)

Discussion. The method is based on warfarin extraction from plasma and urine into ethylene dichloride, TLC separation from its equally fluorescent metabolites, and the loss of warfarin fluorescence after acidification for the measurement of background fluorescence. Previous fluorescence assays without TLC separation are nonspecific in the presence of warfarin metabolites, a finding that is also supported by the results obtained with an HPLC method (Vessel, 1974).

5. Spectrophotometry—O'Reilly et al. (1962)

Discussion. Warfarin is extracted from plasma into ethylene dichloride, and the organic phase is washed with mild alkali and back-extracted with a strong alkali solution. The UV absorption is measured in a spectrophotometer; however, the method is nonspecific in the presence of warfarin metabolites.

6. Plasma Prothrombin Activity—Quick (1974)

Discussion. The "prothrombin time," developed by Quick in 1938, has provided the most useful method for monitoring anticoagulant therapy. The accurate clotting time of oxalated plasma is measured after addition of thromboplastin reagent and calcium ions. It measures the plasma prothrombin complex activity, which is reduced by treatment with vitamin K antagonists, for example, warfarin. Turnover rates of prothrombin and other factors determine the rate of onset of the anticoagulant effect of warfarin, which is therefore independent of warfarin pharmacokinetics after the initial doses. The "prothrombin time" is commonly used to titrate the patient to a satisfactory level of anticoagulant effect.

7. Comments. The HPLC-UV assay of Fasco et al. (1977) is the present state-of-the-art technique because of its sensitivity and its ability to separate and quantify warfarin and its metabolites. The plasma prothrombin activity assay is the clinical method of choice. However, because of a lag time, prothrombin activity response may not be suitable to detect immediate changes in warfarin concentrations caused by coadministration of other drugs.

e. References

Bjornsson, T. D., T. F. Blaschke, and P. J. Meffin, High-Pressure Liquid Chromatographic Analysis of Drugs in Biological Fluids. I: Warfarin, *J. Pharm. Sci.*, **66**, 142–144 (1977).

Breckenridge, A., M. Orme, H. Wesseling, R. J. Lewis, and R. Gibbons, Pharmacokinetics and Pharmacodynamics of the Enantiomers of Warfarin in Man, *Clin. Pharmacol. Ther.*, **15**, 424–430 (1974).

Chan, K. K., R. J. Lewis, and W. F. Trager, Absolute Configurations of the Four Warfarin Alcohols, *J. Med. Chem.*, **15**, 1264–1270 (1972).

Fasco, M. J., L. J. Piper, and L. S. Kaminski, Biochemical Applications of a Quantitative High-Pressure Liquid Chromatographic Assay of Warfarin and its Metabolites, *J. Chromatogr.*, **131**, 365–373 (1977).

Hewick, D. S. and L. McEwen, Plasma Half-Lives, Plasma Metabolites, and Anticoagulant Efficacies of the Enantiomers of Warfarin in Man, *J. Pharm. Pharmacol.*, **25**, 458–465 (1973).

Kaiser, D. G., and R. S. Martin, GLC Determination of Warfarin in Human Plasma, *J. Pharm. Sci.*, **63**, 1579–1581 (1974).

Kelly, J. G. and K. O'Malley, Clinical Pharmacokinetics of Oral Anticoagulants, *Clin. Pharmacokin.*, **4**, 1–15 (1979).

Lewis, R. J., L. P. Ilnicki, and M. Carlstrom, The Assay of Warfarin in Plasma or Stool, *Biochem. Med.*, **4**, 376–382 (1970).

Midha, K., I. J. McGilveray, and J. K. Cooper, GLC Determination of Plasma Levels of Warfarin, *J. Pharm. Sci.*, **63**, 1725–1729 (1974).

O'Reilly, R. A., P. M. Aggeler, M. S. Hoag, and L. Leong, Studies on the Coumarin Anticoagulant Drugs: The Assay of Warfarin and Its Biological Application, *Thromb. Diath. Haematol.*, **8**, 82–95 (1962).

Quick, A. J., *The Hemorrhagic Diseases and the Pathology of Hemostasis*, Charles C Thomas, Springfield, Ill., 1974, pp. 51–71.

Trager, W. F., R. J. Lewis, and W. A. Garland, Mass Spectral Analysis in the Identification of Human Metabolites of Warfarin, *J. Med. Chem.*, **13**, 1196–1204 (1970).

Vessel, E. S., Liquid Chromatographic Assay of Warfarin: Similarity of Warfarin Half-Lives in Human Subjects, *Science*, **184**, 466–468 (1974).

Wong, L. T., G. Solomonraj, and B. H. Thomas, Analysis of Warfarin in Plasma by High Pressure Liquid Chromatography, *J. Chromatogr.*, **135**, 149–154 (1977).

Table XXII. List of Journals Publishing Articles on Drug Level Monitoring That Are Cited in the Monographs of Chapter 5

Journal	Number of citations	Journal	Number of citations
Clin. Chem.	171	*Res. Commun. Chem. Pathol. Pharmacol.*	9
J. Chromatogr.	137		
J. Pharm. Sci.	134	*Biochem. Med.*	8
Clin. Pharmacol. Ther.	60	*Chemotherapy*	8
Clin. Chim. Acta	64	*Clin. Pharmacokin.*	8
J. Pharm. Exp. Ther.	43	*J. Antimicrob. Agents Chemother.*	8
Anal. Chem.	34		
Biomed. Mass Spectrom.	31	*Br. J. Clin. Pharmacol.*	7
J. Pharm. Pharmacol.	30	*Cancer Treat. Rep.*	7
Eur. J. Clin. Pharmacol.	20	*J. Antimicrob. Chemother.*	7
Cancer Res.	15	*Anal. Chim. Acta*	6
J. Clin. Pathol.	15	*J. Infect. Dis.*	6
Arzneim.-Forsch.	14	*Steroids*	6
Drug Metab. Dispos.	14	*Anal. Clin. Biochem.*	5
Anal. Biochem.	13	*Circulation*	5
Biochem. Pharmacol.	10	*Clin. Biochem.*	5
J.A.M.A.	10	*J. Biochem.*	5
Lancet	10	*J. Chromatogr. Sci.*	5
Science	10	*J. Pharmacokin. Biopharm.*	5
J. Med. Chem.	9	*Life Sci.*	5
N. Engl. J. Med.	9	*Nature*	5
		Remainder	391
		Total citations	1244

CITATION ANALYSIS

Table XXII contains a list of the major journals publishing articles in the field of drug level monitoring. The journals are ranked according to their frequency of citation in the monographs (Chapter 5) of this book. The total number of cited articles is 1244. The ranking order reflects the emphasis on the discussion of analytical techniques; therefore chemically oriented journals are cited more frequently than pharmacologic journals. The variety of cited scientific journals attests the interdisciplinary nature of drug level monitoring. The 40 journals with 5 or more citations account for 853 articles or 69% of all references.

ADDENDUM: LITERATURE UPDATE TO NOVEMBER 1979

The references listed here have been carefully selected to cover important new developments and continuing trends in drug level monitoring during 1979. Some of the papers deal with our improved understanding of the factors underlying intra- and intersubject variability and of the clinical interpretation of drug level data. An area of increasing concern is the variability of drug binding to plasma proteins, which may limit the use of total drug plasma concentrations as a guide in drug therapy (e.g., diazepam). Two papers on disopyramide are also included because of the recent clinical interest in therapeutic drug level monitoring of this antiarrhythmic agent.

Among the analytical techniques, development still focuses on HPLC and GC. The increasing use of suitable chemical derivatizations (e.g., hydrocortisone and valproic acid) or electrochemical detection (e.g., morphine) has been predicted in the preceeding chapters; moreover, rapid and sensitive GC assays continue to compete with HPLC procedures (e.g., barbiturates, anticonvulsants, chlorpromazine, nicotine, tetrahydrocannabinol). The barbiturate GC assay includes an automated plasma extraction with use of a mini-column cartridge, an area of rapid development and promise in drug level monitoring.

Various other drug assay techniques are growing, namely, fluorescence and enzyme immunoassays, bioluminescence (gentamicin), radioreceptor assays, and fluorescence. Of particular interest are the radioreceptor assays (diazepam, propranolol), which have evolved from recent discoveries of specific drug receptors. We can anticipate more radioreceptor assays of high sensitivity for a variety of drugs including their active metabolites. Last but not least, several new fluorescence procedures of good sensitivity and specificity (gentamicin, 6-mercapto-purine, methadone, theophylline) serve as a reminder that relatively simple techniques may be suitable for drug level monitoring.

DRUG METABOLISM

Key Words: Intraindividual variability, antipyrine, phenylbutazone, phenacetin, half-lives, "first-pass" absorption effect.

Alvares, A. P. et al., Commentary: Intraindividual Variation in Drug Disposition. *Clin. Pharmacol. Ther.*, **26,** 407–419 (1979).

Key Words: Liver disease, aminopyrine half-life.

Farrell, G. C., W. G. E. Cooksley, and L. W. Powell, Drug Metabolism in Liver Disease: Activity of Hepatic Microsomal Metabolizing Enzymes, *Clin. Pharmacol. Ther.*, **26,** 483–492 (1979).

Key Words: Antipyrine half-life, use as predictor of drug disposition.

Vesell, E. S., Commentary: The Antipyrine Test in Clinical Pharmacology: Conceptions and Misconceptions, *Clin. Pharmacol. Ther.*, **26,** 275–286 (1979).

PHARMACOKINETICS, CLINICAL PHARMACOKINETICS

Key Words: Drug level monitoring, clinical application, use of drug plasma concentrations in dosage calculations.

Sheiner, L. B., S. Beal, B. Rosenberg, and V. V. Marathe, Forcasting Individual Pharmacokinetics, *Clin. Pharmacol. Ther.*, **26,** 294–305 (1979).

ANALYTICAL TECHNIQUES

Key Words: Fluorescence immunoassay, antigen-fluorescein complex, antibody immobilized on polyacrylamide beads, automated.

Curry, R. E., H. Heitzman, D. H. Riege, R. V. Sweet, and M. G. Simonsen, A Systems Approach to Fluorescent Immunoassay: General Principles and Representative Applications, *Clin. Chem.*, **25,** 1591–1595 (1979).

Key Words: Fluorescence immunoassay, homogeneous assay, fluorescer labeled antigen immune complex, fluorescence quenching by fluorescein antibody.

Zuk, R. F., G. L. Rowley, and E. F. Ullman, Fluorescence Protection Immunoassay: A New Homogeneous Assay Technique, *Clin. Chem.*, **25**, 1554-1560 (1979).

Key Words: Bioluminescence.

Whitehead, T. P., L. J. Kricka, T. J. N. Carter, G. H. G. Thorpe, Analytical Luminescence: Its Potential in the Analytical Laboratory, *Clin. Chem.*, **25**, 1531-1546 (1979).

ANTICONVULSANTS

Key Words: GC analysis, isothermal, 5 min separation, no derivatization.

Bredesen, J. E., Rapid Isothermal Determination of Some Antiepileptic Drugs by Gas Liquid Chromatography, *Clin. Chem.*, **25**, 1669-1670 (1979).

Key Words: GC analysis, solid phase extraction cartridge, automated.

St.Onge, L. R., E. Dolar, M. A. Anglim, and C. J. Least, Jr., Improved Determination of Phenobarbital, Primidone and Phenytoin by Use of a Preparative Instrument for Extraction, Followed by Gas Chromatography, *Clin. Chem.*, **25**, 1373-1376 (1979).

AMINOPYRINE

Key Words: HPLC analysis, reverse phase, UV detection, aminopyrine internal standard.

Shargel, L., W.-M. Cheung, and A. B. C. Yu, High Pressure Liquid Chromatographic Analysis of Antipyrine in Small Plasma Samples, *J. Pharm. Sci.*, **68**, 1052-1054 (1979).

BCNU

Key Words: Nitrosoureas, alkylating activity.

Asami, M., K.-I Nakamura, K. Kawada, and M. Tanaka, Evaluation of the Alkylating Activity of Nitrosureas by Thin-Layer Densitometry, *J. Chromatogr.*, **174**, 216-220 (1979).

CHLORDIAZEPOXIDE

Key Words: HPLC analysis, metabolites (N-desmethylchlordiazepoxide, demoxepam, and N-desmethyldiazepam), sensitivity 50 ng/ml plasma.

Peat, M. A., B. S. Finkle, and M. E. Deyman, High Pressure Liquid Chromatographic Determination of Chlordiazepoxide and Its Major Metabolites in Biological Fluids, *J. Pharm. Sci.*, **68**, 1467–1468 (1979).

CHLORPROMAZINE

Key Words: GC-*N*-FID analysis, metabolites (mono- and di-*N*-desmethylchlorpromazine, chlorpromazine sulfoxide), sensitivity 5–20 ng/ml serum.

Bailey, D. N., and J. J. Guba, Gas Chromatographic Analysis for Chlorpromazine and Some of Its Metabolites in Human Serum, with Use of a Nitrogen Detector, *Clin. Chem.*, **25**, 1211–1215 (1979).

CYCLOPHOSPHAMIDE

Key Words: Review, pharmacokinetics.

Grochow, L. B., and M. Colvin, Clinical Pharmacokinetics of Cyclophosphamide, *Clin. Pharmacokin.* **4**, 380–394 (1979).

Key Words: HPLC analysis, reverse phase, UV detection at 200 nm.

Kensler, T. T., R. J. Behme, and D. Brooke, High Performance Liquid Chromatographic Analysis of Cyclophosphamide, *J. Pharm. Sci.*, **68**, 172–174 (1979).

DEXAMETHASONE

Key Words: HPLC analysis, pharmacokinetics.

Tsuei, S. E., R. G. Moore, J. J. Ashley, and W. G. McBride, Disposition of Synthetic Corticosteroids. I. Pharmacokinetics of Dexamethasone in Healthy Adults, *J. Pharmacokin. Biopharm.*, **7**, 249–264 (1979).

DIAZEPAM

Key Words: ^{3}H-Diazepam, specific binding to rat brain homogenates, nitrazepam, clobazepam, active metabolites, sensitivity 0.5 ng/ml serum.

Hunt, P., J.-M. Husson, and J.-P. Raynaud, A Radioreceptor Assay for Benzo-diazepines, *J. Pharm. Pharmacol.*, **31**, 448–451 (1979).

Key Words: Benzodiazepines, variability of plasma protein binding.

Johnson, R. F., S. Schenker, R. K. Roberts, P. V. Desmond, and G. R. Wilkinson, Plasma Protein Binding of Benzodiazepines in Humans, *J. Pharm. Sci.*, **68**, 1320–1322 (1979).

DIAZOXIDE

Key Words: HPLC analysis, UV detection at 270 nm, sensitivity 0.1 μg/ml plasma.

Vree, T. B. and B. Lenselink, Rapid Determination of Diazoxide in Plasma and Urine of Man by Means of High-Performance Liquid Chromatography, *J. Chromatogr.*, **164**, 228–234 (1979).

DISOPYRAMIDE

Key Words: HPLC analysis, metabolism, sensitivity 0.2 μg/ml plasma.

Nygard, G., W. H. Shelver, and S. K. W. Khalil, Sensitive High-Pressure Liquid Chromatographic Determination of Disopyramide and Mono-*N*-dealkyldisopyramide, *J. Pharm. Sci.*, **68**, 1318–1320 (1979).

Key Words: GC-*N*-FID analysis, HPLC, GC-MS, comparison of assays.

Vasiliades, J., C. Owens, and F. Ragusa, Disopyramide Determination by Gas-Chromatography, Liquid Chromatoggraphy, and Gas Chromatography-Mass Spectrometry, *Clin. Chem.*, **25**, 1900–1904 (1979).

DOXORUBICIN

Key Words: HPLC analysis, fluorescence detection, silica gel, doxorubicinol, aclacinomycin, sensitivity 2–4 ng/ml plasma.

Peters, J. H. and J. F. Murray, Jr., Determination of Adriamycin and Aclacino-mycin by High Pressure Liquid Chromatography and Spectrophoto-fluorometry, *J. Liquid Chromatogr.*, **2**, 45–52 (1979).

ETHOSUXIMIDE

Key Words: Review, pharmacokinetics, nonlinear kinetics.

Smith, G. A., L. McKauge, D. Dubetz, J. H. Tyrer, and M. J. Eadie, Factor Influencing Plasma Concentrations of Ethosuximide, *Clin. Pharmacokin.*, **4**, 38–52 (1979).

5-FLUOROURACIL

Key Words: HPLC analysis, 5-fluorodeoxyuridine, sensitivity $10^{-8} M$.

Jones, R. A. et al., Potential Clinical Applications of a New Method for Quantitation of Plasma Levels of 5-Fluorouracil and 5-Fluorodeoxyuridine, *Bull. Cancer*, **66**, 75–78 (1979).

GENTAMICIN

Key Words: Review, pharmacokinetics, aminoglycosides.

Pechere, J.-C. and R. Dugal, Clinical Pharmacokinetics of Aminoglycoside Antibiotics, *Clin. Pharmacokin.*, **4**, 170–199 (1979).

Key Words: Review of assays, aminoglycosides.

Maitra, S. K., T. T. Yoshikawa, L. B. Guze, and M. C. Schotz, Determination of Aminoglycoside Antibiotics in Biological Fluids: A Review, *Clin. Chem.*, **25**, 1361–1367 (1979).

Key Words: Fluorescence analysis, aminoglycosides, ion exchange column chromatography, dihydrolutidine derivatives.

Csiba, A., Spectrofluorimetric Method for Aminoglycoside Antibiotics, *J. Pharm. Pharmacol.*, **31**, 115–117 (1979).

Key Words: HPLC analysis, aminoglycosides, *o*-phthalaldehyde precolumn derivatization.

Bäck, S. E., I. Nilsson-Ehle, and P. Nilsson-Ehle, Chemical Assay, Involving Liquid Chromatoggraphy, for Aminoglycoside Antibiotics in Serum, *Clin. Chem.*, **25**, 1222–1225 (1979).

Key Words: Bioluminescence, aminoglycosides, based on ATP adenyltransferase and acetyltransferase.

Daigneault, R., A. Larouche, and G. Thibault, Aminoglycoside Antibiotic Measurement by Bioluminescence, with Use of Plasmid-Coded Enzymes, *Clin. Chem.*, **25**, 1639–1643 (1979).

HALOPERIDOL

Key Words: GC-MS analysis, EI and CI modes, high sensitivity.

Moulin, M. A. et al., Gas-Chromatography-Electron Impact and Chemical Ionization Mass Spectrometry of Haloperidol and Its Chlorinated Homologue, *J. Chromatogr.*, **178**, 324–329 (1979).

HYDRALAZINE

Key Words: Metabolism, hydrazine, GC-MS analysis.

Timbrell, J. A., and S. J. Harland, Identification of Hydrazine in the Urine of Patients Treated with Hydralazine, *Clin. Pharmacol. Ther.*, **26**, 81–88 (1979).

Key Words: HPLC analysis, derivatization with *p*-anisaldehyde, specific in the presence of hydrazone metabolites.

Ludden, T. M., L. K. Goggin, J. L. McNay, Jr., K. D. Haegele, and A. M. M. Shepherd, High-Pressure Liquid Chromatographic Assay of Hydralazine in Human Plasma, *J. Pharm. Sci.*, **68**, 1423–1425 (1979).

HYDROCORTISONE

Key Words: HPLC analysis, dansyl hydrazine derivatization, fluorescence detection, sensitivity 5 ng/ml plasma with a sample volume of 100 μl.

Goehl, T. J., G. M. Sundaresan, and V. K. Prasad, Fluorimetric High-Pressure Liquid Chromatographic Determination of Hydrocortisone in Human Plasma, *J. Pharm. Sci.*, **68**, 1374–1376 (1979).

Key Words: HPLC analysis, UV, silica gel, cortisone, prednisone, prednisolone, sensitivity 10 ng/ml plasma.

Frey, F. G., B. M. Grey, and L. Z. Benet, Liquid Chromatographic Measurement of Endogenous and Exogenous Glucocorticoids in Plasma, *Clin. Chem.*, **25**, 1944–1947 (1979).

IMIPRAMINE

Key Words: Review, clinical pharmacokinetics, tricyclic antidepressants.

Orsulak, P., and J. J. Schildkraut, Guidelines for Therapeutic Monitoring of Tricyclic Antidepressant Plasma Levels, *Therap. Drug Monitor.*, **1**, 199–208 (1979).

MELPHALAN

Key Words: Pharmacokinetics, HPLC analysis, ^{14}C label.

Alberts, D. S. et al., Kinetics of Intravenous Melphalan, *Clin. Pharmacol. Ther.*, **26**, 73–80 (1979).

6-MERCAPTOPURINE

Key Words: Fluorescence Assay, chemical reactions, azathioprine, sensitivity 10 ng/ml plasma.

Maddocks, J. L., Assay of Azathioprine, 6-Mercaptopurine and a Novel Thiopurine Metabolite in Human Plasma, *Br. J. Clin. Pharmacol.*, **8**, 273–278 (1979).

METHADONE

Key Words: Fluorescence assay, reaction in H_2SO_4-paraformaldehyde, specific in the presence of metabolites.

Chi, C. H. and B. N. Dixit, Sensitive Fluorescence Assay for *d,l*-Methadone, *J. Pharm. Sci.*, **68**, 1097–1099 (1979).

METHOTREXATE

Key Words: Pharmacokinetics, therapeutic drug level monitoring.

Stoller, R. G. et al., Use of Plasma Pharmacokinetics to Predict and Prevent Methotrexate Toxicity, *N. Engl. J. Med.*, **297**, 630–634 (1977).

Key Words: Metabolism, competitive protein binding assay and RIA interference.

Donehower, R. C., K. R. Hande, J. C. Drake, and B. A. Chabner, Presence of 2,4-Diamino-N^{10}-methylpteroic Acid After High-Dose Methotrexate, *Clin. Pharmacol. Ther.*, **26**, 63–72 (1979).

Key Words: Enzyme immunoassay, heterogeneous, β-D-galactosidase, sensitivity 1–10 ng/ml plasma.

Al-Bassam, M. N., M. J. O'Sullivan, J. W. Bridges, and V. Marks, Improved Double-Antibody Enzyme Immunoassay for Methotrexate, *Clin. Chem.*, **25**, 1448–1452 (1979).

MORPHINE

Key Words: HPLC assay, electrochemical detection, sensitivity 1 ng per injection.

White, M. W., Determination of Morphine and Its Major Metabolite, Morphine-3-glucuronide, in Blood by High Performance Liquid Chromatography with Electrochemical Detection, *J. Chromatogr.*, **178**, 229–240 (1979).

NICOTINE

Key Words: GC-N-FID assay, sensitivity 0.5 ng/ml, sample requirement 0.1 ml plasma.

Feyerabend, C., and M. A. H. Russell, Improved Gas-Chromatographic Method and Micro-Extraction Technique for the Measurement of Nicotine in Biological Fluids, *J. Pharm. Pharmacol.*, **31**, 73–76 (1979).

PHENYTOIN

Key Words: Review, pharmacokinetics.

Richens, A., Clinical Pharmacokinetics of Phenytoin, *Clin. Pharmacokin.*, **4**, 153–169 (1979).

PROPRANOLOL

Key Words: Review, pharmacokinetics.

Routledge, P. A., and D. G. Shand, Clinical Pharmacokinetics of Propranolol, *Clin. Pharmacokin.*, **4**, 73–90 (1979).

Key Words: Bioavailability, "first-pass" absorption effect, glucuronidation.

Walle, T. et al., Presystemic and Systemic Glucuronidation of Propranolol, *Clin. Pharmacol. Ther.*, **26,** 167–172 (1979).

Key Words: Competitive protein binding assay, turkey erythrocyte plasma membranes, ^{125}I-iodo-hydroxybenzylpindolol, sensitivity 0.5 ng/ml serum.

Bilezikian, J. P., D. E. Gammon, C. L. Rochester, and D. G. Shand, A Radioreceptor Assay for Propranolol, *Clin. Pharmacol. Ther.*, **26,** 173–180 (1979).

Key Words: HPLC assay, 4-hydroxypropranolol.

Wong, L., R. L. Nation, W. L. Chiou, and P. K. Mehta, Plasma Concentrations of Propranolol and 4-Hydroxypropranolol During Chronic Oral Propranolol Therapy, *Br. J. Clin. Pharmacol.*, **8,** 163–167 (1979).

QUINIDINE

Key Words: Review, assay specificity, pharmacokinetic results.

Guentert, T. W., R. A. Upton, N. H. G. Holford, and S. Riegelman, Divergence in Pharmacokinetic Parameters of Quinidine Obtained by Specific and Nonspecific Assay Methods, *J. Pharmacokin. Biopharm.*, **7,** 303–311 (1979).

SPIRONOLACTONE

Key Words: Metabolism, canrenone, pharmacological activity, HPLC analysis.

Dahlöf, C. G., P. Lungborg, B. A. Persson, and C. G. Regårdh, Reevaluation of the Antimineralocorticoid Effect of the Spironolactone Metabolite, Canrenone, from Plasma Concentrations Determined by a New High-Pressure Liquid Chromatographic Method, *Drug Metab. Dispos.*, **7,** 103–107 (1979).

Key Words: Canrenoate potassium, canrenone, metabolism.

Vose, C. W., D. R. Boreham, G. C. Ford, N. J. Haskins, and R. F. Palmer, Identification of Some Human Urinary Metabolites of Orally Administered Potassium Canrenoate by Stable Isotope-Labeling Techniques, *Drug Metab. Dispos.*, **7,** 226–232 (1979).

TETRACYCLINES

Key Words: HPLC-UV analysis, doxycycline, demeclocycline, sensitivity 50 ng/ml serum.

Leenheer, A. P., De, and H. J. C. F. Nelis, Doxycycline Determination in Human Serum and Urine by High-Performance Liquid Chromatography, *J. Pharm. Sci.*, **68**, 999–1002 (1979).

TETRAHYDROCANNABINOL

Key Words: Acidic metabolites.

Agurell, S. et al., Chemical Synthesis and Biological Occurrence of Carboxylic Acid Metabolites of $\Delta^{1(6)}$-Tetrahydrocannabinol, *Drug Metab. Dispos.*, **7**, 155–161 (1979).

Key Words: GC assay, electron capture detection, specific for Δ^9-THC.

Bachmann, E. B., A. F. Hofmann, and P. G. Waser, Identification of Δ^9-Tetrahydrocannabinol in Human Plasma by Gas Chromatography, *J. Chromatogr.*, **178**, 320–323 (1979).

THEOPHYLLINE

Key Words: HPLC analysis, 254 + 273 nm UV detection, high selectivity.

Butrimovitz, G. P., and V. A. Raisys, An Improved Micromethod for Theophylline Determination by Reverse-Phase Liquid Chromatography, *Clin. Chem.*, **25**, 1461–1464 (1979).

Key Words: Fluorescence assay, based on Ce(IV) reduction, good correlation with HPLC and EIA.

Meola, J. M., H. H. Brown, and T. Swift, Fluorimetric Measurement of Theophylline, *Clin. Chem.*, **25**, 1835–1837 (1979).

Key Words: Enzymatic assay, inhibition of alkaline phosphatase, good correlation with EIA.

Vinet, B., and L. Zizian, Enzymic Assay for Serum Theophylline, *Clin. Chem.*, **25**, 1370–1372 (1979).

VALPROIC ACID

Key Words: Pharmacokinetics, anticonvulsants, drug interactions.

Mihaly, G. W., F. J. Vajda, J. L. Miles, and W. J. Louis, Single and Chronic Dose Pharmacokinetic Studies of Sodium Valproate in Epileptic Patients, *Eur. J. Clin. Pharmacol.*, **16**, 23–29 (1979).

Key Words: GC analysis, clinical assay.

Hershey, A. E., J. R. Patton, and K. H. Dudley, Gas Chromatographic Method for the Determination of Valproic Acid in Human Plasma, *Therap. Drug Monitor.*, **1**, 217–241 (1979).

Key Words: HPLC analysis, phenacyl ester derivatives, UV detection.

Gupta, R. N., P. M. Keane, and M. L. Gupta, Valproic Acid in Plasma, as Determined by Liquid Chromatography, *Clin. Chem.*, **25**, 1984–1985 (1979).

WARFARIN

Key Words: GC-MS assay, chemical ionization, high sensitivity.

Duffield, P. H., D. J. Birkett, D. N. Wade, and A. M. Duffield, Quantitation of Plasma Warfarin Levels by Gas Chromatography Chemical Ionization Mass Spectrometry, *Biomed. Mass Spectrom.*, **6**, 101–104 (1979).

INDEX

Note: Page numbers in italic refer to the actual monographs on specific drugs.